Mammalian Teeth

Form and Function

edited by
Thomas Martin and
Wighart v. Koenigswald

Mammalian Teeth

Form and Function

edited by
Thomas Martin and
Wighart v. Koenigswald

Verlag Dr. Friedrich Pfeil · München 2020

Bibliographic information published by the Deutsche Nationalbibliothek

The Deutsche Nationalbibliothek lists this publication in the Deutsche Nationalbibliografie;
detailed bibliographic data are available in the Internet at
http://dnb.dnb.de.

Druckvorstufe: Verlag Dr. Friedrich Pfeil, München
Druck: PBtisk a.s., Příbram I – Balonka

Printed in the European Union

ISBN 978-3-89937-266-3

Verlag Dr. Friedrich Pfeil, Wolfratshauser Straße 27, 81379 München
phone: +49 89 5528600-0 – fax: +49 89 5528600-4 – e-mail: info@pfeil-verlag.de – www.pfeil-verlag.de

Contents

Preface 7

Authors 9

Occlusion and Wear

1 **A new wear facet terminology for mammalian dentitions**
Julia A. Schultz, Ulrike Anders, Caroline Braune, Janka J. Brinkkötter, Ivan Calandra, Sandra Engels, Eva Findeisen, Juan Pablo Gailer, Jürgen Hummel, Kai R. K. Jäger, Thomas M. Kaiser, Daniela C. Kalthoff, Wighart v. Koenigswald, Ottmar Kullmer, Christina Landwehr, Markus Mau, Ulrike Menz, Irina Ruf, Anne M. Schubert, Ellen Schulz-Kornas, Achim H. Schwermann, Leonie C. Schwermann, Mirella Skiba, Patrick Steuer, Karl-Heinz Südekum, Daniela E. Winkler, and Thomas Martin 11

2 **Occlusal Fingerprint Analysis (OFA) reveals dental occlusal behavior in primate molars**
Ottmar Kullmer, Ulrike Menz, and Luca Fiorenza 25

3 **A brief history of quantitative wear analyses with an appeal for a holistic view on dental wear processes**
Ellen Schulz-Kornas, Thomas M. Kaiser, Ivan Calandra, and Daniela E. Winkler 44

Mastication and Jaw Mechanics

4 **Mandibular biomechanics as a key factor to understand diet in mammals**
Jordi Marcé-Nogué 54

5 **Interaction between food, saliva, and tooth surface**
Marcus Mau, Achim Johann, Thomas M. Kaiser, and Karl-Heinz Südekum 81

6 **Aspects of food comminution in ungulates and their consequences for energy budget**
Jürgen Hummel, Marcus Clauss, and Karl-Heinz Südekum 87

Ontogeny and Heterochrony

7 **Case studies on functional aspects and constraints in early and late tooth ontogeny**
Irina Ruf, Anne M. Schubert, and Wighart v. Koenigswald 102

8 **Enamel ridge alignments in ungulates: a cut above**
Daniela E. Winkler, Jordi Marcé-Nogué, and Thomas M. Kaiser 125

9 **Construction and wear of mammalian teeth in terms of heterochrony**
Wighart v. Koenigswald 171

Evolutionary Changes

10 **Molar diversity and functional adaptations in Mesozoic mammals**
Thomas Martin, Kai R. K. Jäger, Thorsten Plogschties, Achim H. Schwermann, Janka J. Brinkkötter, and Julia A. Schultz 187

11 **Evolutionary trends in the mastication patterns in some perissodactyls, cetartiodactyls, and proboscideans**
Julia A. Schultz, Sandra Engels, Leonie C. Schwermann, and Wighart v. Koenigswald 215

12 **Short review of dental microstructure and dental microwear in xenarthran teeth**
Daniela Kalthoff 231

Index 243

T. Martin & W. v. Koenigswald: Mammalian Teeth – Form and Function. – München (Pfeil) 2020 – ISBN 978-3-89937-266-3

Preface

Teeth are the hardest and most durable parts of the mammalian body. Even after millions of years they provide a wealth of paleobiologic information. Teeth reflect the interaction of a mammal with the environment, as evident from their shape and traces of mastication on their surface. Teeth provide information on the diet, mode of food processing, biomechanics of mastication, and energy gain. Modern analytical techniques such as micro-computed tomography, high-resolution surface analysis, and 3D imaging have greatly boosted the research on dental function in recent years. They reveal sub-microscopic surface structures such as striations and micro-textures in great detail, providing insight into dietary adaptations of long-extinct mammalian clades. The shortcomings of 2D imaging that formerly have hampered the depiction of complex occlusal relationships have been overcome by 3D analysis and virtual modelling of the chewing cycle with the Occlusal Fingerprint Analyser (OFA) software.

The DFG Research Unit 771 "Function and performance enhancement in the mammalian dentition – phylogenetic and ontogenetic impact on the masticatory apparatus" has been active since 2008, with groups at the University of Bonn, the Senckenberg Research Institute and Natural History Museum Frankfurt, the Zoological Museum of the University of Hamburg and the Natural History Museum in Stockholm. In the framework of Research Unit 771 an interdisciplinary group of researchers from paleontology, zoology, anthropology, animal nutrition and biomechanics bundled their expertise and efforts for the study of the function of the mammalian dentition. Results of the Research Unit have been presented and discussed at various conferences and at the international workshop "Abrasion and Attrition: Traces of Mastication in Fossil Mammals", held in Mayschoß (Ahr Valley) in 2014.

The book "Mammalian Teeth – Form and Function" is a synopsis of the achievements of Research Unit 771 which have been published in more than 100 technical papers in many different journals. In order to make the general results more accessible for non-specialists, they have been cast in 12 chapters arranged according to four major research areas: Occlusion and Wear, Mastication and Jaw Mechanics, Ontogeny and Heterochrony, and Evolutionary Changes. These chapters are based on the published papers of the Research Unit, which are marked with an asterisk (*) in the respective reference lists.

The German Research Foundation (DFG) has generously funded Research Unit 771 and the publication of this book, which is gratefully acknowledged. We also would like to thank the following colleagues for agreeing to act as reviewers and for their constructive comments: Marcus Clauss, Zurich; Alistair Evans, Melbourne; Mikael Fortelius, Helsinki; Madeleine Geiger, Zurich; Helder Gomez-Rodriguez, Paris; Dave Grossnickle, Seattle; Luke Holbrook, Glassboro; Christine Janis, Bristol; Stephan Lautenschlager, Birmingham; Vincent Lazzari, Poitiers; Peter Lucas, Kuwait City; Zhe-Xi Luo, Chicago; Peder Nørgaard, Copenhagen; Florent Rivals, Tarragona; Gordon Sanson, Melbourne; Angela Schwarm, Ås; Gina Semprebon, Longmeadow; Carsten Staszyk, Gießen; Mark Teaford, Vallejo; Jessica Theodor, Calgary; Jack Tseng, Buffalo; and Indrė Žliobaitė, Helsinki. We are grateful to Friedrich Pfeil and Hubert Hilpert from Verlag Dr. Friedrich Pfeil (Munich) for the high standard get-up and excellent print quality of the book.

Bonn, November 2020

Thomas Martin
and Wighart von Koenigswald,
editors

T. Martin & W. v. Koenigswald: Mammalian Teeth – Form and Function. – München (Pfeil) 2020 – ISBN 978-3-89937-266-3

Authors

Ulrike Anders, Jüterbog
[formerly at Section Paleontology
Institute of Geosciences, Rheinische Friedrich-Wilhelms-Universität Bonn]

Caroline Braune, Berlin
[formerly at Department of Mammalogy/Paleoanthropology, Center of Natural History, Universität Hamburg]

Janka J. Brinkkötter
Koordinationsstelle wissenschaftliche Infrastruktur
University of Bonn Medical Center
Venusberg-Campus 1
53127 Bonn
[formerly at Section Paleontology, Institute of Geosciences, Rheinische Friedrich-Wilhelms-Universität Bonn]

Ivan Calandra
TraCEr, Laboratory for Traceology and Controlled Experiments at Monrepos Archaeological Research Centre and Museum for Human Behavioural Evolution
Schloss Monrepos
56567 Neuwied, Germany
calandra@rgzm.de
[formerly at Department of Mammalogy/Paleoanthropology, Center of Natural History, Universität Hamburg]

Marcus Clauss
Clinic for Zoo Animals, Exotic Pets and Wildlife
Vetsuisse Faculty
Universität Zürich
Winterthurerstr. 260
8057 Zurich, Switzerland
mclauss@vetclinics.uzh.ch

Sandra Engels, Seeheim-Jugenheim
sandy.engels@gmx.net
[formerly Section Paleontology, Institute of Geosciences, Rheinische Friedrich-Wilhelms-Universität Bonn]

Eva Findeisen, Grafschaft
[formerly at Institute of Animal Science, Rheinische Friedrich-Wilhelms-Universität Bonn]

Luca Fiorenza
Department of Anatomy and Developmental Biology
Monash University
Melbourne VIC 3800, Australia
and
School of Environmental and Rural Science
University of New England
Armidale NSW 2351, Australia
luca.fiorenza@monash.edu

Juan Pablo Gailer
Department of Mammalogy/Paleoanthropology
Center of Natural History
Universität Hamburg
Martin-Luther-King-Platz 3
20146 Hamburg, Germany
juan.pablo.gailer@uni-hamburg.de

Jürgen Hummel
Department of Animal Sciences
Georg-August-Universität Göttingen
Kellnerweg 6
37077 Göttingen, Germany
jhummel@gwdg.de
[formerly at Institute of Animal Science, Rheinische Friedrich-Wilhelms-Universität Bonn]

Kai R. K. Jäger
Section Paleontology, Institute of Geosciences
Rheinische Friedrich-Wilhelms-Universität Bonn
Nussallee 8
53115 Bonn, Germany
jaegerk@uni-bonn.de

Achim Johann
NaturZoo Rheine
Salinenstraße 150
48432 Rheine
info@naturzoo.de

Christina Landwehr, Hamburg
[formerly at Department of Mammalogy/Paleoanthropology, Center of Natural History, Universität Hamburg]

Thomas M. Kaiser
Department of Mammalogy/Paleoanthropology
Center of Natural History
Universität Hamburg
Martin-Luther-King-Platz 3
20146 Hamburg
thomas.kaiser@uni-hamburg.de

Daniela C. Kalthoff
Department of Zoology
Swedish Museum of Natural History
Stockholm, Sweden
Daniela.Kalthoff@nrm.se

Wighart v. Koenigswald
Section Paleontology, Institute of Geosciences
Rheinische Friedrich-Wilhelms-Universität Bonn
Nussallee 8
53115 Bonn, Germany
koenigswald@uni-bonn.de

Ottmar Kullmer
Division Paleoanthropology, Senckenberg Forschungsinstitut und Naturmuseum Frankfurt
Senckenberganlage 25
60325 Frankfurt a.M., Germany
and Department of Paleobiology and Environment
Institute of Ecology, Evolution, and Diversity
Goethe-Universität
Max-von-Laue-Straße 13
60438 Frankfurt a.M., Germany;
ottmar.kullmer@senckenberg.de

Jordi Marcé-Nogué
Universitat Rovira i Virgili, Department of Mechanical Engineering
Av. Països Catalans 26
43007 Tarragona, Spain
and
Institut Català de Paleontologia M. Crusafont,
Universitat Autònoma de Barcelona
Cerdanyola del Vallès
Barcelona, 08193, Spain
jordi.marce@urv.cat
[formerly at Department of Mammalogy/Palaeoanthropology, Center of Natural History, Universität Hamburg]

Thomas Martin
Section Paleontology, Institute of Geosciences
Rheinische Friedrich-Wilhelms-Universität Bonn
Nussallee 8
53115 Bonn, Germany
tmartin@uni-bonn.de

Marcus Mau, Potsdam
marcus.mau@gmx.de
[formerly at Institute of Animal Science, Rheinische Friedrich-Wilhelms-Universität Bonn]

Ulrike Menz, Geisenheim
[formerly at Division Paleoanthropology, Senckenberg Forschungsinstitut und Naturmuseum Frankfurt]

Thorsten Plogschties
Section Paleontology, Institute of Geosciences
Rheinische Friedrich-Wilhelms-Universität Bonn
Nussallee 8
53115 Bonn, Germany
t.plogschties@uni-bonn.de

Irina Ruf
Department Messel Research and Mammalogy
Senckenberg Forschungsinstitut und Naturmuseum Frankfurt
Senckenberganlage 25
60325 Frankfurt a.M., Germany
irina.ruf@senckenberg.de
[formerly at Section Paleontology, Institute of Geosciences, Rheinische Friedrich-Wilhelms-Universität Bonn]

Anne M. Schubert, Plochingen
[formerly Section Palaeontology, Institute of Geosciences, Rheinische Friedrich-Wilhelms-Universität Bonn]

Ellen Schulz-Kornas
Department of Cariology, Endodontology and Periodontology
Universität Leipzig
Liebigstraße 12
and
Department of Human Evolution
Max-Planck-Institut für Evolutionäre Anthropologie
Deutscher Platz 6
04103 Leipzig, Germany
ellen.schulz-kornas@medizin.uni-leipzig.de
[formerly at Department of Mammalogy/Paleoanthropology, Center of Natural History, Universität Hamburg]

Achim H. Schwermann
LWL-Museum für Naturkunde
Westfälisches Landesmuseum mit Planetarium
Sentruper Straße 285
48161 Münster
achim.schwermann@lwl.org
[formerly at Section Paleontology, Institute of Geosciences, Rheinische Friedrich-Wilhelms-Universität Bonn]

Leonie C. Schwermann
LWL-Museum für Naturkunde
Westfälisches Landesmuseum mit Planetarium
Sentruper Straße 285
48161 Münster
leonie.schwermann@lwl.org
[formerly at Section Paleontology, Institute of Geosciences, Rheinische Friedrich-Wilhelms-Universität Bonn]

Julia A. Schultz
Section Paleontology, Institute of Geosciences
Rheinische Friedrich-Wilhelms-Universität Bonn
Nussallee 8
53115 Bonn, Germany
jaschultz@uni-bonn.de

Mirella Skiba, Hamburg
[formerly at Department of Mammalogy/Paleoanthropology, Center of Natural History, Universität Hamburg]

Patrick Steuer
Senzyme GmbH
Gierlichsstraße 6
53840 Troisdorf
[formerly at Institute of Animal Science, Rheinische Friedrich-Wilhelms-Universität Bonn]

Karl-Heinz Südekum
Institute of Animal Science
Rheinische Friedrich-Wilhelms-Universität Bonn
Endenicher Allee 15
53115 Bonn, Germany
ksue@itw.uni-bonn.de

Daniela E. Winkler
Department of Natural Environmental Studies
Graduate School of Frontier Sciences
The University of Tokyo
5-1-5 Kashiwanoha
Kashiwa, Chiba 277-8563
Japan
daniela.eileen.winkler@k.-utokyp.ac.jp
and
Institute for Geosciences
Johannes Gutenberg-Universität Mainz
J.-J.-Becher-Weg 21
55128 Mainz, Germany
daniela.winkler@uni-mainz.de
[formerly at Department of Mammalogy/Paleoanthropology, Center of Natural History, Universität Hamburg]

A new wear facet terminology for mammalian dentitions

Julia A. Schultz, Ulrike Anders, Caroline Braune, Janka J. Brinkkötter, Ivan Calandra, Sandra Engels, Eva Findeisen, Juan-Pablo Gailer, Jürgen Hummel, Kai R. K. Jäger, Thomas M. Kaiser, Daniela C. Kalthoff, Wighart v. Koenigswald, Ottmar Kullmer, Christina Landwehr, Markus Mau, Ulrike Menz, Irina Ruf, Anne M. Schubert, Ellen Schulz-Kornas, Achim H. Schwermann, Leonie C. Schwermann, Mirella Skiba, Patrick Steuer, Karl-Heinz Südekum, Daniela E. Winkler, and Thomas Martin

Introduction

Teeth of vertebrates are under constant use and therefore show wear-induced traces on the surface. According to Williams (2005), the mechanism by which wear occurs is a mechanical and/or chemical process resulting in material loss. Wear facets, the smooth and polished contact areas between opposing teeth, form during force-fit chewing activity between antagonists due to attritional (tooth-tooth) and abrasional (tooth-substrate-tooth) contacts involving either some kind of resistant alimentary bolus or some other dental behavior with an empty mouth cavity (Maier & Schneck 1981). In mammals, the occlusion of dental surfaces produces characteristic facet patterns on the crown's surface. Orientation and inclination of a facet are determined by tooth morphology, while the texture of the surface (e.g., complexity and roughness) is the result of the food ingested (e.g., Schulz et al. 2010, Winkler et al. 2019, Schulz-Kornas et al. 2020, this volume). Complementary patterns of wear facets on the occlusal surface of opposing teeth change in shape and size during ontogeny and therefore can be used for age determination (e.g., Baumann 1949, Grau et al. 1970, Habermehl 1985, Anders et al. 2011a, b, Ruf et al. 2020, this volume).

Wear facet patterns begin to form as soon as the tooth crown erupts and antagonists occlude. Fortelius (1985) differentiated between primary and secondary occlusal surfaces. Facets on primary surfaces develop gradually on the enamel cover (i.e., enamel facets) and only later expose the dentin with increasing wear due to age. In those cases, the exposed dentin has no explicit function. Facets on a secondary surface are composed of enamel and dentin. Dentin exposure happens relatively fast and it forms depressions in close proximity to protruding enamel, and the combination of both has a specific function. Those facets remain constant for a relatively long period of time in the life of an animal (i.e., dentin facets interrupted and/or surrounded by enamel bands). Facets on secondary surfaces are specialized occlusal surfaces that are mostly developed in herbivorous taxa with a horizontal power stroke movement, regardless of the direction (Koenigswald 2020, this volume). In these specialized cases enamel and exposed dentin function as a unit, and neighboring facets merge and may cover large areas composed of both tissues (e.g., occlusal surface of elephant molars). In correspondence, some mammal groups developed specialized dentitions (e.g., hypsodonty, euhypsodonty, thickened enamel layer, enamel infoldings) to keep the function of dental structures including the facets stable over longer periods of time, thereby extending the lifespan of the individuals (heterochrony). However, the larger such composed areas become, the more difficult it is to identify individual facets. Extreme cases can be found in some rodents or modern equids where the flat occlusal surface forms basically one large rasp facet (Koenigswald 2018). In such cases it is very difficult to identify individual facets, making comparisons almost impossible. Here, the Modular Wear Facet Nomenclature may be applied to the functional parts of enamel bands only (see examples of *Apodemus* and *Bison* below).

In the past, the diversity of dental wear facet patterns of mammalian molars has led to the erection of diverse nomenclatory systems for comparative and descriptive purposes. The Modular Wear Facet Nomenclature for mammalian and mammal-like dentitions proposed by Schultz et al. (2018) is a general, descriptive and modular concept, offering the user flexibility, including the use of already established molar cusp terminologies. It primarily serves as a labelling system for wear facets, derived exclusively from the position in relation to major crown structures. It is therefore based on existing terms of primary tooth elements (e.g., cusps, crests, basins), because the conventionally accepted terms of cusps and crests are well established and widely used in the scientific literature and by many disciplines, like paleontology, biology, anthropology, archeology, human medicine and dentistry. The advantage of the modularity of the new nomenclature is that it is applicable to a wide variety of dentitions independent from specific taxonomic designations. It is an intuitive understanding that cusp and crest names determine the first element for naming the location of the facet on the tooth surface, and the second element depicts the facet's position in relation to the structure of the first element. This system is flexible; it therefore can be expanded with additional information if needed (e.g., phase of chewing cycle [1 for phase I; 2 for phase II] etc.). In this chapter we demonstrate the infinite applicability of the Modular Wear Facet Nomenclature to various mammalian cheek teeth and show the many options for comparative purposes (for illustration purposes 3D surface models of different mammalian dentitions were used and colored in 2D using Photoshop CS6).

T. Martin & W. v. Koenigswald (eds.): Mammalian Teeth – Form and Function. Pp. 11–24, 7 figs.

DOI: http://doi.org/10.23788/mammteeth.01 Published 22 December 2020

Institutional abbreviations

Gui Mam Guimarota collection currently housed in the Institute of Geosciences, Section Paleontology, Rheinische Friedrich-Wilhelms-Universität Bonn, Germany
IGPB Mammal collection of the Institute of Geosciences, Section Paleontology, Rheinische Friedrich-Wilhelms-Universität Bonn, Germany
NHM Natural History Museum, London, UK
NMW Mammal collection of the Naturhistorisches Museum, Vienna, Austria
NRM Mammal collection of the Swedish Museum of Natural History, Stockholm, Sweden
SMF Mammal collection of the Senckenberg Forschungsinstitut und Naturmuseum Frankfurt, Germany
UMZC University Museum of Zoology, Cambridge, UK
ZFMK Mammal collection of the Zoologisches Forschungsinstitut Alexander Koenig, Bonn, Germany
ZMB Mammal collection of the Museum für Naturkunde Berlin, Germany

Anatomical abbreviations

AC/acd anticone, anticonid
AL/ald anteroloph, anterolophid
AS anterostyle
CCA centrocrista
CNG/cng cingulum, cingulid
ecsd ectostylid
en entoconid
ENS entostyle
ERS enterostyle
hfd hypoflexid
hld hypoconulid
HY/hy hypocone, hypoconid
L lower
M/m upper molar, lower molar
ME/me metacone, metaconid
MSL/msld mesoloph, mesolophid
MTCL metaconule
MTL/mtld metaloph, metalophid
P/p upper premolar, lower premolar
PA/pa paracone, paraconid
PAC/pacd paracrista, paracristid
PCL paraconule
POS posterostyle
PPAC postparacrista
PR/pr protocone, protoconid
prcd protocristid
PRL/prld protoloph, protolophid
PSL/psld posteroloph, posterolophid
TA talon
U upper

History of facet description and analysis

At the end of the 19th century, Cope (1883) and Osborn (1888), among others, developed the dental cusp nomenclature to describe features of teeth and to have a common language for discussion (for a detailed review see Ungar 2010). Butler (1952) and Mills (1955) were the first to document corresponding facet patterns on opposing upper and lower molars. Mills (1955) described these areas as follows: "These facets appear, macroscopically, as flat, highly polished areas on the cusps of the molar teeth. They can be seen more readily under low-power magnification, if the specimen is slowly turned about until the light strikes it at a favorable angle." Butler (1973) restricted the term 'facet' to areas on which wear produces traces when teeth are nearly or actually in contact with their antagonists. He described facets as flat, light-reflecting areas with striations on the surface and he also proposed the functional interpretation that both facets and striations are indicative of relative movements, though never tested in in-vivo or in-vitro experiments. Fortelius (1985) followed Butler's (1973) functional interpretation supporting the fact that facets are wear-dependent and have an orientation, which is dictated by the interactive wear process occurring between one or several facet pairs. During the last century, various attempts to refine the dental-cusp nomenclature were taken in order to better reflect mammalian dental evolution (e.g., Vandebroek 1967, Crompton & Jenkins 1968, Szalay 1969, Hershkovitz 1971, Butler 1978, Maier & Schneck 1981, Van Valen 1982). However, describing the occlusal surface in detail is a challenging task, because various factors influence the formation and a wide variation exists among different taxa. It was frequently discussed that ontogeny (Winkler & Kaiser 2011), ingested diet (Fortelius 1987), habitat (Kaiser & Schulz 2006, Schulz & Kaiser 2013), and even tooth position (Kullmer et al. 2009, Taylor et al. 2013, Gailer & Kaiser 2014) have an influence on the formation of the occlusal surface and its characteristic wear features.

Historically, Butler (1952) was the first to number corresponding wear facet pairs of upper and lower molars (of perissodactyls). When Mills (1955) mapped wear facets of primate molars he did not use a numbering system at all, but instead he attributed each facet to the phase of mastication in which it was used. Crompton (1971) identified homologous facets in his comprehensive study on the evolution of the mammalian dentition and numbered them sequentially to their appearance during evolution. Gingerich (1974) followed this system and described how wear facets form on molars of the fossil primate *Plesiadapis*, but added a buccal phase (B) and a lingual phase (L) to each facet number specifying the phase of mastication. He described two orientations of parallel striations formed during the two power stroke phases (buccal and lingual phase) of the chewing cycle. The two movements were also recognized by Kay & Hiiemäe (1974) as "phase I" (i.e., buccal phase) and "phase II" (i.e., lingual phase) of the power stroke. Crompton (1971) defined in total six wear facets in the hyaeonodontid *Didelphodus*, and Kay & Hiiemäe (1974) – working on primate molars – identified facets 7 to 10. In both studies the letters "a" and "b" were added to the numbers of the wear facets if two facets belong to one wear area, and were formed by the same antagonistic structure during the same movement. Kay (1977) identified 11 wear facets for cercopithecid molars. In the same publication, he described facet 7n ("n" for "new"); the homologous wear facet on molars of the propliopithecid primate *Aegyptopithecus* had been numbered 7 earlier by Kay & Hiiemäe (1974). More confusing, Kay (1977) also mentioned facet 10n, and an additional wear surface 9 with an anterior extension X. Several following studies added new abbreviations and letters. Maier (1980) basically identified the same wear facets but did not add a and b. Adding to the confusion, facet X identified by Kay (1977) corresponds to facet 11 of Maier (1980) and facet 10n of Kay (1977) corresponds to facet 12 of Maier (1980). Maier & Schneck (1981) also identified an additional facet 13, only occurring in hominid molars (see Kullmer et al. 2020, this volume: 16, Tab. 2.1; Schultz et al. 2018: 35, tab. 1). Over the last 50 years numerous studies proposed tooth facet nomenclatures for various taxa including early mammaliaforms and stem mammals. However, in these taxa tooth structures are difficult to be put in homologous relation to that of crown mammals (e.g., tritylodonts [Crompton 1972], "symmetrodontans" [Crompton & Jenkins 1967], haramiyidans [Butler & MacIntyre 1994], non-mammalian mammaliaforms [Kermack et al. 1965, Crompton & Jenkins 1968, Gingerich 1973], "pseudotribosphenidans" [Chow & Rich 1982, Wang et al. 1998], etc.). In most of these studies the authors simply numbered the identified wear facets, suggesting a homology of structures which does not necessarily exist.

In summary, the past studies often proposed contradicting labeling systems for wear facets and used different numbers for the same facet (see Schultz et al. 2018: 35, tab. 1), which complicates comparisons and discussions of functional questions and bears the risk to lead to misunderstanding and misinterpretation. In some cases, different systems were used for the description of tooth morphologies and dental wear facets of mammals, depending on the specific morphology of the teeth, historical establishment or individual author affinities. Pinto-Llona (2013) for example reconstructed the paleodiet of Pleistocene cave bears from Spain using 2D macro- and microwear analyses. In her study she used the classical numbering system of Butler (1952) followed by the system of Kay (1977) in parenthesis in order to avoid inconsistencies. For studies like the one of Pinto-Llona (2013), the Modular Wear Facet Nomenclature may serve as a system for comparative and descriptive purposes, which aims to improve and foster a more transparent scientific communication.

How to apply the modular system

The Modular Wear Facet Nomenclature for mammalian and mammal-like dentitions incorporates conventionally accepted tooth terminology in connection with a modular organization. Cusps and crests are the initial elements for naming the location of the facet on the tooth. Since this system is flexible and works as a modular system it can be expanded with additional information if needed (e.g., phase of chewing cycle [1 for phase I, 2 for phase II], enamel [e], dentin [de]). In our approach, the first part of the facet name is an acronym of the cusp or crest the facet is associated with. In case of the protocone, this would be 'PR' (see Schultz et al. 2018: 36, tab. 2 for acronyms and page 12 this volume) following the conventional usage of capital letters for the upper dentition. Accordingly, for the protoconid, 'pr' would be used following the convention of using lowercase letters for the lower dentition. We distinguish five main directions depending on the location of the facet on the cusps or crests, that is 'm' for mesial, 'd' for distal, 'b' for buccal, 'l' for lingual, and 'h' for horizontal in case of flat apical wear. Those directions can be combined in order to precisely specify the location of the facet. For example 'mb' describes a facet that is mesiobuccally located on a cusp, if the facet is located more in buccal direction than in mesial, the letters can be exchanged to "bm" in order to clarify, i.e. buccomesially. In combination, for the protocone this results in 'PR-mb' as facet term, for the protoconid this results in 'pr-mb' as facet term. If a user wishes to describe the facet location in relation to a crest because the facet occupies a large area below a crest, the first module can be the acronym of the crest's name. This is also expandable to larger elements on the occlusal surface like lophs/lophids, fossettes/fossetids, styles/stylids or cingula/cingulids depending on the user's intentions. It further allows addressing dental wear surfaces on both enamel and dentin by adding a final module with '-e' or '-de'.

Formula: [structure on tooth surface]–[location of facet in relation to tooth structure]–[tissue (if important)]–[chewing phase (if important)]–etc.

Example: Facet 1 of Crompton (1971) of a moderately worn lower tribosphenic molar occupying nearly the entire area of the distal flank of the trigonid below the protocristid visible by the striated polished enamel surface, and being formed during phase I of the chewing cycle translates to: prcd-d-e-1

As an addition, here we propose a color code which is based on the standard 12-color circle, but for simplification and easy applicability we reduced it to eight colors (Fig. 1.1). In this circle for the upper dentition yellow is always assigned to mesial, purple to distal, red to buccal, and turquoise to lingual. Colors between the four main directions (clockwise: orange, pink, blue, green) indicate the directions lying between the main directions. For example, mesiolingual is green and distobuccal is pink. In order to color match opposing wear surfaces the color circle needs to be swapped for the lower dentition with mesial being purple and distal yellow (Fig. 1.1B). Accordingly, buccal is depicted by turquoise and lingual is red. Subsequently, mesiolingual is pink and distobuccal green etc. In our approach, horizontal facets are kept in white, to clearly distinguish them from inclined facets. Often horizontal facets are produced during abrasional processes (not necessarily in direct contact with an antagonist) resulting in blunted cusps. The same as for apical wear applies in the extreme cases of horizontal secondary occlusal surfaces with confluent facet areas composed of enamel and dentin areas mostly showing flat enamel cross-sections (i.e., enamel bands, for example Fig. 1.1C). If enamel bands appear inclined around cusps (Figs. 1.4A, 1.5B), parts of the enamel bands should be colored according to where the parts are located, thus treated as individual facets.

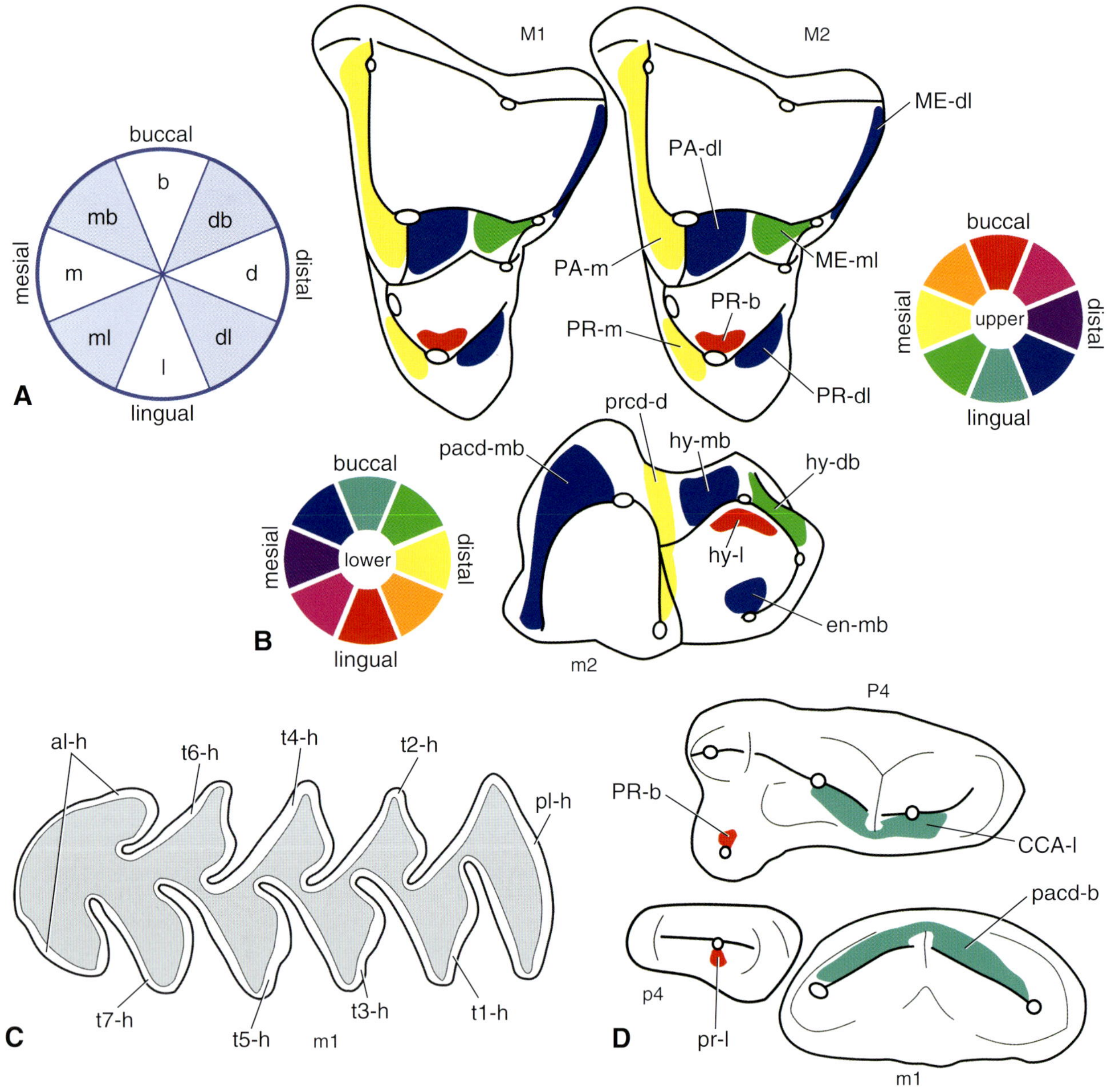

Fig. 1.1. Schematic instructions how to use the Modular Wear Facet Nomenclature. **A,** Main directions that indicate the position of the facet in relation to the structure on the tooth surface. The center of the circle is anchored to the structure of interest (e.g., cusp, basin, or crest). **B,** Example of facet labelling for two upper (M) and one lower molar (m) of the marsupial *Didelphis virginiana* (based on Crompton 1971). Proposed color-coding based on the standard twelve-color circle (here reduced to eight colors) for upper and lower teeth. The colors are swapped for the lower dentition to color match opposing wear surfaces of the upper. **C,** Lower molar of the cricetid rodent *Arvicola*, example for horizontal occlusal surface, with h being the acronym for 'horizontal'. **D,** Facet labelling on the carnassials (P4/m1) of the feliform carnivoran *Panthera leo*. Not to scale.

Examples

Non-mammalian Mammaliaformes

Morganucodon watsoni

UMZC.Eo.CR.1, Early Jurassic (Fig. 1.2A)

Cusp B of the upper first molar has one wear surface on the cingulum: distolingual (BCNG-dl). Cusp A also has one wear surface on the cingulum: mesiolingual (ACNG-ml). The crest connecting cusp C and cusp D has one large wear surface: distolingual (CD-dl). Cusp B of the second molar has one wear surface: distolingual (B-dl). The cingulum bordering cusp B on the lingual side has one wear surface: distolingual (BCNG-dl). Cusp A has two wear surfaces on the cingulum: mesiolingual (ACNG-ml) and distolingual (ACNG-dl). The cingulum bordering cusp C on the lingual side has one wear surface: mesiolingual (CCNG-ml). Cusp C has one wear surface: mesiolingual (C-ml). The crest connecting cusp C and cusp D has one large wear surface: distolingual (CD-dl)

Cusp b of the second lower molar has one wear surface: mesiobuccal (b-mb). Cusp a has two wear surfaces:

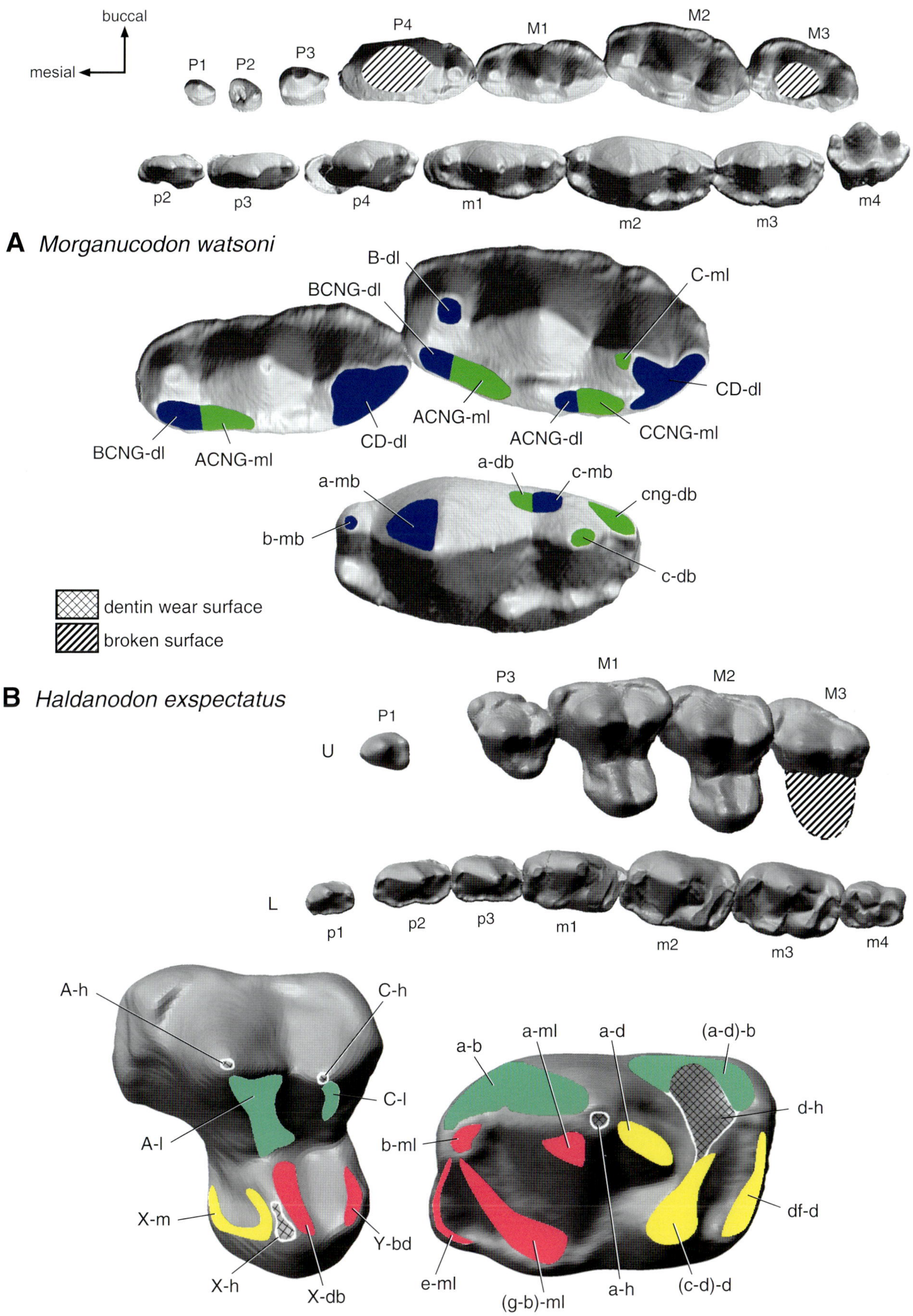

Fig. 1.2. Example for the application of the Modular Wear Facet Nomenclature in two mammaliaforms. **A,** premolars (P/p) and triconodont molars (M/m) of the morganucodontan *Morganucodon watsoni* (facet location and cusp terminology after Jäger et al. 2019, for chapter purposes slightly modified) and **B,** premolars and complex "pseudotribosphenic" molars of the docodontan *Haldanodon exspectatus* (terminology after Luo & Martin 2007). For close up views M2/m2 were chosen for colored facet locations, here and in the following figures if not labelled otherwise. Not to scale.

mesiobuccal (a-mb) and distobuccal (a-db). Cusp c has one wear surface: distobuccal (c-db). The cingulum bordering cusp c on the buccal side has one wear surface: distobuccal (cng-db)

Haldanodon exspectatus

Gui Mam 30/79 and Gui Mam 6/82,
Late Jurassic (Fig. 1.2B)

Cusp A of the M2 has two wear surfaces: lingual (A-l) and one flat apical dentin field (A-h). Cusp C has two wear surfaces: lingual (C-l), one flat apical dentin field (C-h). Cusp X has three wear surfaces: mesial (X-m), distobuccal (X-db) and one apical dentin field (X-h). Cusp Y has one wear surface: buccodistal (Y-bd).

Cusp b of the m2 has one wear surface: mesiolingual (b-ml). Cusp a has four wear surfaces: buccal (a-b), distal (a-d), mesiolingual (a-ml), and one flat apical dentin field (a-h). Cusp d has one wear surface: one large apical dentin field (d-h). Crest a-d has one wear surface: buccal ((a-d)-b). Cusp df has one wear surface: distal (df-d). Crest c-d has one wear surface: distal ((c-d)-d). Crest g-b has one wear surface: mesiolingual ((g-b)-ml). Cusp e has one wear surface: mesiolingual (e-ml). Terminology after Luo & Martin (2007).

Marsupialia

Virginia opossum (*Didelphis virginiana*)

Theoretical example in Fig. 1.1B, Recent

The paracone of the M2 has two wear surfaces: mesial (PA-m), distolingual (PA-dl). The metacone has two wear surfaces: mesiolingual (ME-ml), distolingual (ME-dl). The protocone has three wear surfaces: buccal (PR-b), distolingual (PR-dl), and mesial (PR-m).

The paracristid has one wear surface: mesial (pacd-m). The protocristid has one large wear surface: distal (prcd-d). The hypoconid has three wear surfaces: mesiobuccal (hy-mb), distobuccal (hy-db) and lingual (hy-l). The entoconid has two wear surfaces: mesiobuccal (en-mb) and mesial (en-m).

Afrosoricida

Greater hedgehog tenrec (*Setifer setosus*)

SMF 55333, Recent (Fig. 1.3A)

The paracone of the M2 has two wear surfaces: mesial (PA-m), distolingual (PA-dl). Alternatively, the paracrista has one wear surface: mesial (PAC-m). The postparacrista has one wear surface: distal (PPAC-d).

The paraconid m2 has one wear surface: mesiobuccal (pa-mb). The protocristid has one wear surface: distal (prcd-d). The entoconid (or hypoflexid) has one wear surface: an apical dentin field (en(or hfd)-h).

Sirenia

West Indian Manatee (*Trichechus manatus*)

ZFMK MAM 73223, Recent (Fig. 1.3B)

The paracone of the upper molar has one wear surface: distolingual (PA-dl). The paraconule has two wear surfaces: mesial (PCL-m) and one distal (PCL-d). The protocone has two wear surfaces: mesial (PR-m) and one distal (PR-d). The crest metaconule has one wear surface: mesial (MTCL-m). The hypocone has one wear surface: a flattened apical enamel field (HY-h).

The protoconid of the lower molar has one wear surface: distal (pr-d). The metaconid has one wear surface: distal (me-d). The hypoconid has two wear surfaces: mesial (hy-m) and distal (hy-d). The entoconid has two wear surfaces: mesial (en-m) and distal (en-d). The hypoconulid has one wear surface: a flattened apical enamel field (hld-h).

Scandentia

Common tree shrew (*Tupaia glis*)

IGPB M70, Recent (Fig. 1.4A)

The paracone of the M2 has three wear surfaces: mesial (PA-m), distolingual (PA-dl) and an apical dentin field (PA-h). The metacone has three wear surfaces: mesial (ME-m), distolingual (ME-dl) and an apical dentin field (ME-h). The protocone has one wear surface: buccal (PR-b).

The paracristid of the m2 has one wear surface: a mesiobuccal enamel facet (pacd-mb). The protocristid has one large wear surface: mesial (prcd-m). The protoconid has one apical wear surface: an apical dentin field (pr-h). The hypoconid has four wear surfaces: mesiobuccal (hy-mb), distal (hy-d), lingual (hy-l) and an apical dentin field (hy-h). The entoconid has one wear surface: an apical dentin field (en-h).

Rodentia

Yellow-necked fieldmouse (*Apodemus flavicollis*)

IGPB 1077, Recent (Fig. 1.4B)

The protocone of the M2 has two connected wear surfaces forming an enamel band: lingual (PR-l) and buccal (PR-b). Buccally and lingually the combined wear surface is connected with the paracone and the enterostyle. The anticone has two wear surfaces forming an enamel band: distolingual (AC-dl) and lingual (AC-l). Paracone and metacone have one wear surface forming an enamel band: lingual (PA-l+ME-l). The hypocone has two connected wear surfaces forming an enamel band: a large buccal one (HY-b) and a smaller lingual one (HY-l). The posterostyle has one wear surface: buccal (POS-b); it is connected with the lingual wear surface of the hypocone. The enterostyle has one wear surface: buccal (ERS-b); it is connected to the lingual wear surface of the protocone. The anterostyle has two wear surfaces: lingual (AS-l) and buccal (AS-b).

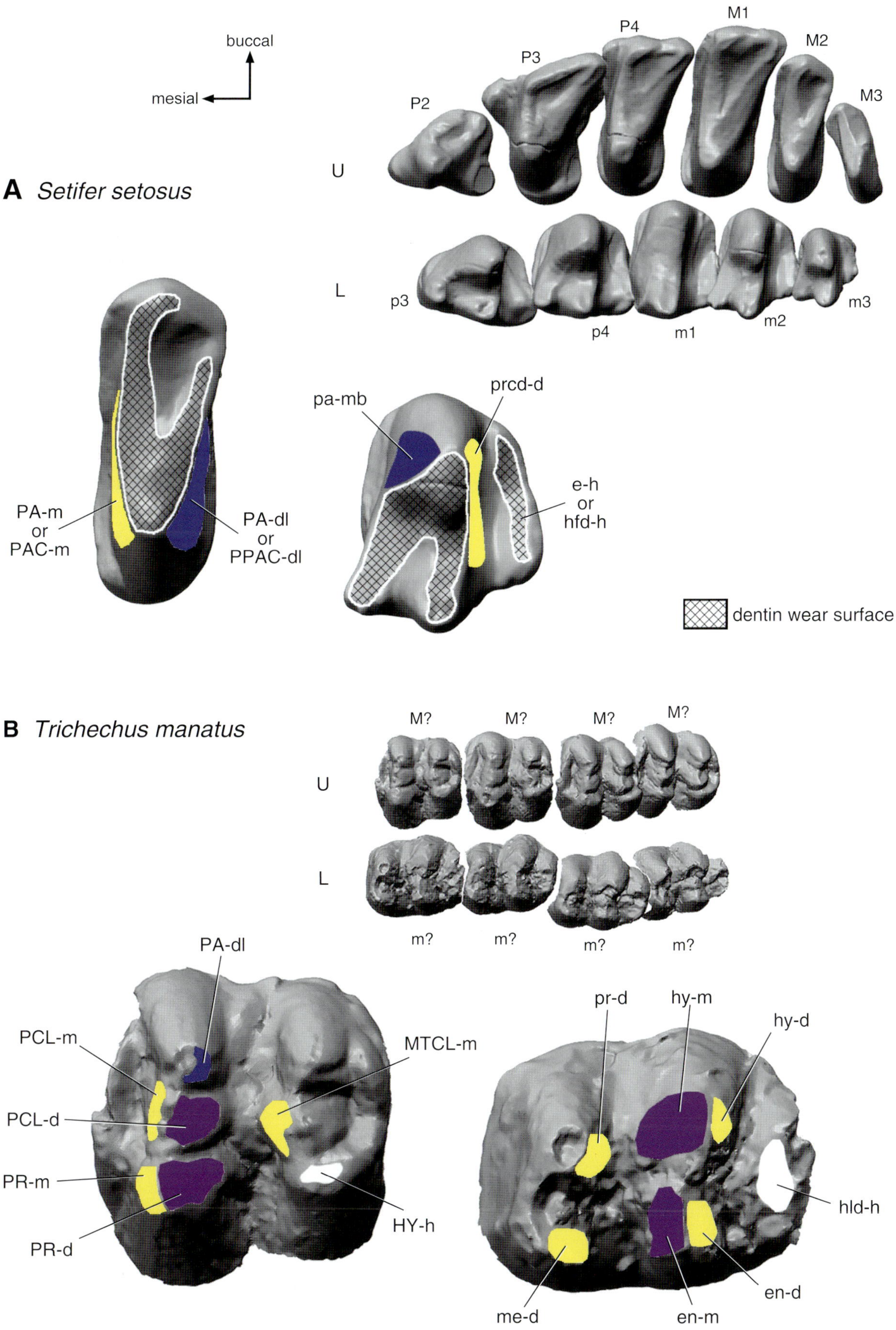

Fig. 1.3. Example for the application of the Modular Wear Facet Nomenclature in two afrotherians. **A,** zalambdodont premolars (P/p) and molars (M/m) of the afrosoricid *Setifer setosus* and **B,** bunodont molars of the manatee *Trichechus manatus*. Because of the horizontal tooth replacement mode in Sirenia it is not possible to determine the exact tooth positions, here indicated by a question mark. Not to scale. Terminology after Thenius (1989).

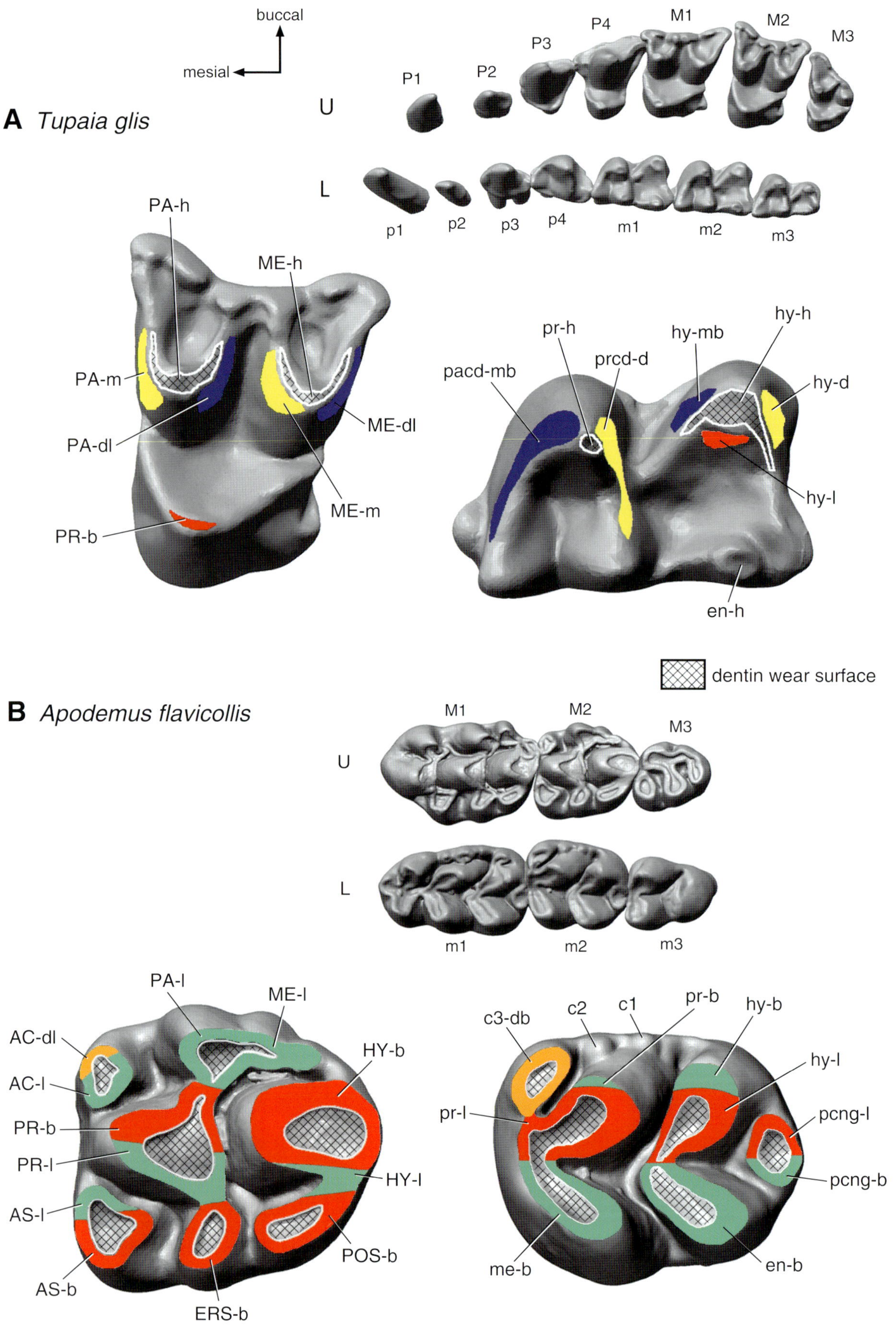

Fig. 1.4. Example for the application of the Modular Wear Facet Nomenclature in *Tupaia* and *Apodemus*. **A,** premolars (P/p) and molars (M/m) of the common tree shrew *Tupaia glis* and **B,** brachyodont molars of the muroid rodent *Apodemus flavicollis*. Not to scale.

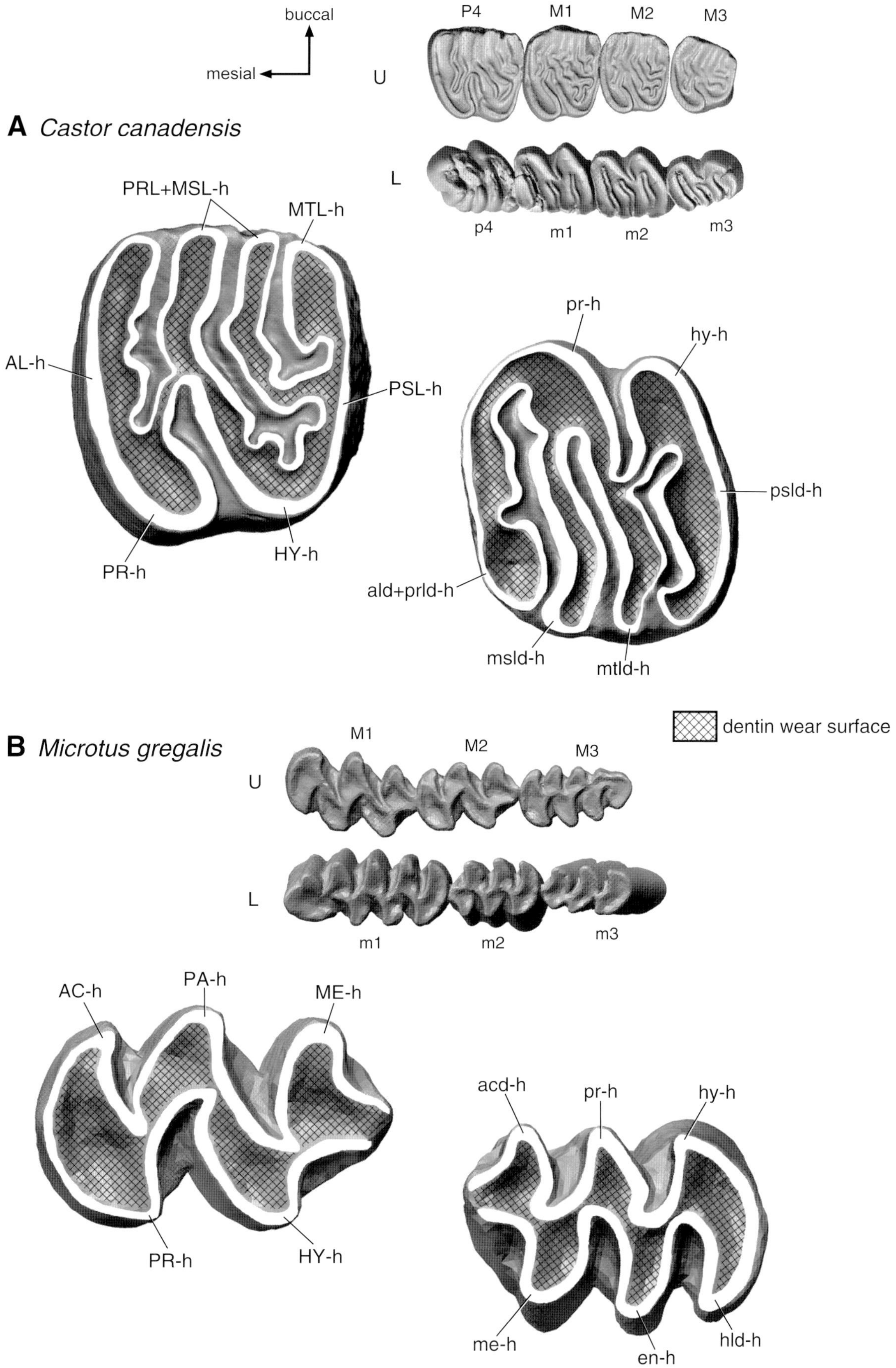

Fig. 1.5. Example for the application of the Modular Wear Facet Nomenclature in two rodents with high crowned teeth. **A,** hypsodont premolars (P/p) and molars (M/m) of the North American beaver *Castor canadensis* and **B,** euhypsodont molars of the vole *Microtus gregalis*. Not to scale.

The cusp c3 of the m2 has one wear surface: distobuccal (c3-db). The protoconid has two connected wear surfaces forming an enamel band: buccal (pr-b) and lingual (pr-l). The metaconid has one large wear surface: buccal (me-b). The wear surfaces of cusp c3, protoconid, and metaconid are connected. The hypoconid has two connected wear surfaces forming an enamel band: buccal (hy-b), lingual (hy-l). The entoconid has one wear surface: buccal (en-b). The wear surfaces of hypoconid and entoconid are connected. The posterior cingulum has two connected wear surfaces forming an enamel band: lingual (pcng-l), buccal (pcng-b).

North American beaver (*Castor canadensis*)

Upper jaw specimen IGPB MaÜ2, lower jaw specimen IGPB M3346, Recent (Fig. 1.5A)

The occlusal surface of the upper cheek teeth is flat and shows only faint relief, forming one confluent horizontal wear surface. There are no isolated facets developed, but the whole enamel part on the occlusal surface forming lophs is equally worn. Therefore the wear surface can be described as following: AL-h, PRL+MSL-h, MTL-h, PSL-h, PR-h, HY-h.

The lower teeth show the same pattern like the upper cheek dentition, a flat confluent wear surface with all components worn. Similar to the upper dentition the occlusal surface of the m2 can be described with ald+prld-h, msld-h, mtld-h, psld-h, pr-h, hy-h.

Narrow-headed vole (*Microtus gregalis*)

IGPB 1004, Recent (Fig. 1.5B)

The occlusal surface of the upper cheek teeth consists of a confluent, flat plane. Due to the flat occlusal surface, typical isolated facets are not developed. The enamel band, forming triangles, surrounds one large dentin field. The wear surface can be described as: AC-h, PR-h, PA-h, HY-h, ME-h.

Similar to the upper dentition the lower dentition as well shows a flat occlusal surface. Here are also no isolated facets developed, and the whole surface is one confluent wear surface. Therefore the facets can be described as followed: acd-h, me-h, pr-h, en-h, hy-h, hld-h.

Soricomorpha

European mole (*Talpa europaea*)

ZFMK 93.76, Recent (Fig. 1.6A)

The paracone of the M2 has three wear surfaces: mesial (PA-m), distolingual (PA-dl), and an apical dentin field (PA-h). The metacone has three wear surfaces: mesial (ME-m), distolingual (ME-dl), and an apical dentin field (ME-h). The protocone has three wear surfaces: buccal (PR-b), lingual (PR-l), and an apical dentin field (PR-h).

The paraconid of the m2 has one wear surface: an apical dentin field (pa-h). The protoconid has two wear surfaces: mesiobuccal (pr-mb), and an apical dentin field (pr-h). The protocristid has one large wear surface: distal (prcd-d). The hypoconid has four wear surfaces: mesiobuccal (hy-mb), distal (hy-d), lingual (hy-l), and an apical dentin field (hy-h). The entoconid has two wear surfaces: buccal (en-b) and an apical dentin field (en-h). The metaconid has one wear surface: an apical dentin field (me-h).

Carnivora

Wolf (*Canis lupus*)

NRM 2011 5027, Recent (Fig. 1.6B)

The paracone of the M2 has two wear surfaces: mesial (PA-m) and distolingual (PA-dl). The protocone has one wear surface: distobuccal (PR-db). The cingulum has one wear surface: mesiobuccal (CNG-mb).

The paracristid of the m1 (large cutting edge) has one wear surface: mesiobuccal (pacd-mb). The metaconid has one wear surface: distal (me-d). The entoconid has one wear surface: mesiobuccal (en-mb). The hypoconid has two wear surfaces: mesiolingual (hy-ml) and distolingual (hy-dl).

Lion (*Panthera leo*)

NRM A595061, Recent (Fig. 1.1D)

The protocone of the P4 has one wear surface: buccal (PR-b). The crest connecting paracone and metacone (i.e., centrocrista) has one large connected wear surface: lingual (CCA-l).

The crest connecting the protoconid and paraconid (i.e., paracristid) of the m1 has one large wear surface: buccal (pacd-b).

Polar bear (*Ursus maritimus*)

NMW 7140, Recent (Fig. 1.7A)

The paracone of the M2 has one wear surface: lingual (PA-l). The protocone has two wear surfaces: mesiobuccal (PR-mb) and distobuccal (PR-db). The hypocone has two wear surfaces: mesial (HY-b) and distal (HY-d). The metacone has two wear surfaces: mesial (ME-m) and mesiolingual (ME-ml). The talon has one wear surface: buccal (TA-b).

The protoconid of the m2 has three wear surfaces: mesiolingual (pr-ml), distolingual (pr-dl), and buccal (pr-b). The metaconid has two wear surfaces: mesiobuccal (me-mb) and distal (me-d). The hypoconid has three wear surfaces: mesial (hy-m), distobuccal (hy-db), and lingual (hy-l). The entoconid has one wear surface: mesiobuccal (en-mb).

Artiodactyla

European bison (*Bison bonasus*)

NHM-92-3-15-1, Recent (Fig. 1.7B)

The paracone of the M2 has four wear surfaces, all connected forming enamel bands: mesiobuccal (PA-mb), mesiolingual (PA-ml), distobuccal (PA-db), and distolingual (PA-dl). The protocone also has four wear surfaces form-

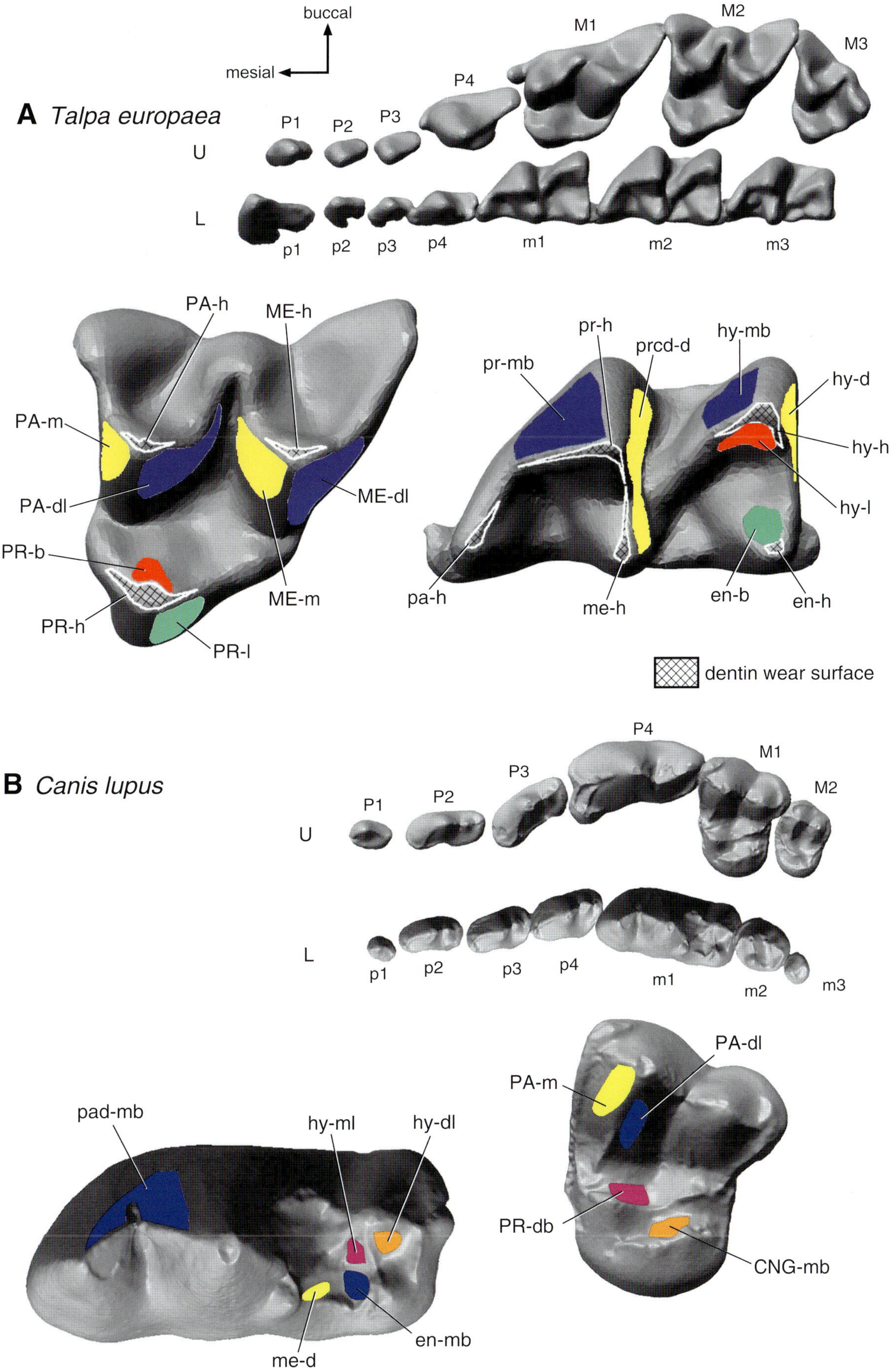

Fig. 1.6. Example for the application of the Modular Wear Facet Nomenclature in *Talpa* and *Canis*. **A,** premolars (P/p) and molars (M/m) of the soricomorph lipotyphlan *Talpa europaea* and **B,** of the carnivoran *Canis lupus*. Close up pair M1/m1 for the wolf example. Not to scale.

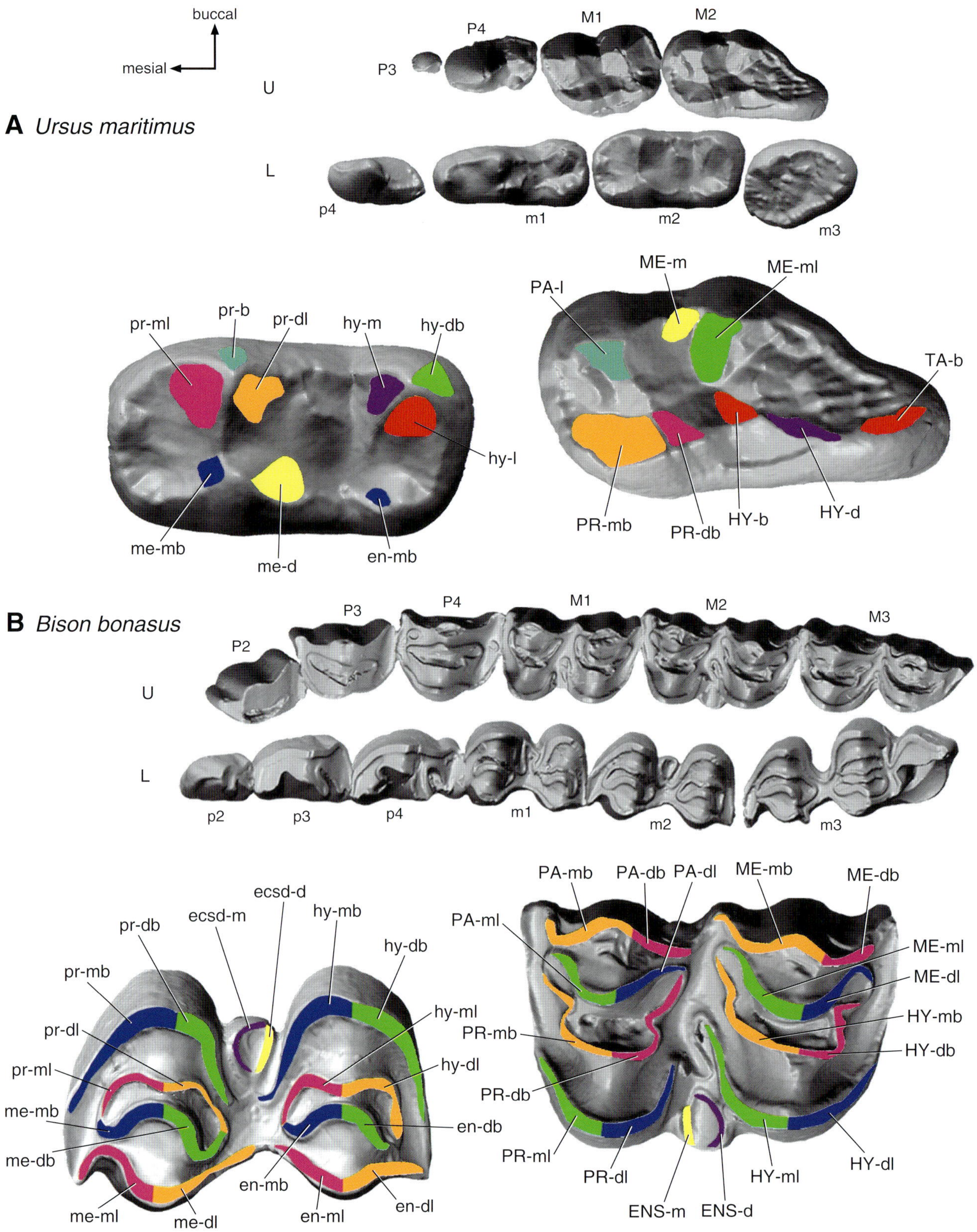

Fig. 1.7. Example for the application of the Modular Wear Facet Nomenclature in *Ursus* and *Bison*. **A,** premolars (P/p) and molars (M/m) of the arctoid carnivoran *Ursus maritimus* and **B,** of the bovid artiodactyl *Bison bonasus*. Not to scale.

ing enamel bands: mesiobuccal (PR-mb), mesiolingual (PR-ml), distobuccal (PR-db), and distolingual (PR-dl). The hypocone also has four wear surfaces forming enamel bands: mesiobuccal (HY-mb), mesiolingual (HY-ml), distobuccal (HY-db), and distolingual (HY-dl). The metacone also has four surfaces forming enamel bands: mesiobuccal (ME-mb), mesiolingual (ME-ml), distobuccal (ME-db), and distolingual (ME-dl). The entostyle has two wear surfaces: mesial (ENS-m) and distal (ENS-d).

The protoconid of the m2 has four wear surfaces, all connected forming continuous enamel bands: mesiobuccal (pr-mb), mesiolingual (pr-ml), distobuccal (pr-db), and distolingual (pr-dl). The metaconid has also four surfaces forming a continuous enamel band: mesiobuccal (me-mb), mesiolingual (me-ml), distobuccal (me-db), and distolingual (me-dl). The hypoconid also has four wear surfaces forming an enamel band: mesiobuccal (hy-mb), mesiolingual (hy-ml), distobuccal (hy-db), and distolingual (hy-dl). The entoconid also has four wear surfaces forming an enamel band: mesiobuccal (en-mb), mesiolingual (en-ml), distobuccal (en-db), and distolingual (en-dl). The ectostylid has two wear surfaces: mesial (ecsd-m) and distal (ecsd-d).

Conclusion

With the examples shown above, we demonstrate that the Modular Wear Facet Nomenclature is broadly applicable to a wide variety of mammalian and mammaliaform cheek teeth. It can easily be expanded to the facets on incisors and canines and serves primarily as a description to convey topographic location of facets in relation to primary structures on the occlusal surface with previously established homology, such as cusps and crests. The homology of wear facets is inevitably determined by the structures they occur on (e.g. cusps, basin surfaces, crests, lophs etc.) and thus the system (and the implicit homology of wear facets) can be further changed and re-adapted if the homology hypotheses of the primary structures should change. As shown above, the modular system allows high flexibility and can be expanded with further information if necessary (e.g., phase of chewing cycle I [1] or II [2], or tissue type like enamel ['e'] or dentin, ['de']). If the homology is uncertain, taxon related terms can be used (e.g., cusp numbering system in rodents) for the first module depicting the structure a facet is located on. The simple modular structure makes this nomenclature applicable to a wide array of occlusal surfaces. A common nomenclature is the basis for communication, allowing comparisons between studies and across taxa, and thereby permitting functional inferences. Even more so, a common nomenclature facilitates the interpretation of evolutionary changes in tooth crown morphology.

Acknowledgments

We are deeply thankful to the responsible curators of the museum collections that we visited to study the various dentitions of fossil and extant mammals. Reviewers Mikael Fortelius and Luke Holbrook provided constructive comments that greatly improved the manuscript.

References

* indicates publications that originated from the DFG Research Unit 771

*Anders, U., Koenigswald, W. v., Ruf, I. & Smith, B. H. (2011a): Generalized individual dental age stages for fossil and extant placental mammals. Paläontologische Zeitschrift 85: 321–339.

*Anders, U., Koenigswald, W. v., Ruf, I. & Smith, B. H. (2011b): Erratum. Paläontologische Zeitschrift 85: 341.

Baumann, D. F. (1949): Die freilebenden Säugetiere der Schweiz. Verlag Hans Huber, Bern.

Butler, P. M. (1952): The milk-molars of Perissodactyla with remarks on molar occlusion. Proceedings of the Zoological Society London B 121: 777–817.

Butler, P. M. (1973): Molar wear facets of early Tertiary North American primates. In: Zingeser, M. R. (ed.). Craniofacial Biology of Primates. Symposia of the Fourth International Congress of Primatology, Portland, Oregon, August 15–18, 1972, vol. 3. Karger, Basel: 1–27.

Butler, P. M. (1978): Molar cusp nomenclature and homology. In: Butler, P. M. & Joysey, K. A. (eds.). Studies in the development, function and evolution of teeth. Academic Press, London: 441–453.

Butler P. M. & Macintyre, G. T. (1994): Review of the British Haramiyidae (?Mammalia, Allotheria), their molar occlusion and relationships. Philosophical Transactions of the Royal Society London B 345: 433–458.

Chow, M. & Rich, T. H. (1982): *Shuotherium dongi*, n. gen. and sp., a therian with pseudo-tribosphenic molars from the Jurassic of Sichuan, China. Australian Mammalogy 5: 127–142.

Cope, E. D. (1883): On the trituberculate type of molar tooth in the Mammalia. Proceedings of the American Philosophical Society 21: 324–326.

Crompton, A. W. (1971): The origin of the tribosphenic molar. In: Kermack, D. M. & Kermack, K. A. (eds.). Early mammals. Zoological Journal of the Linnean Society, Academic Press 50, supplement 1: 65–87.

Crompton A. W. (1972): Postcanine occlusion in cynodonts and tritylodontids. Bulletin of the British Museum of Natural History, Geology 21: 29–71.

Crompton A. W. & Jenkins, F. A. (1967): American Jurassic symmetrodonts and Rhaetic "pantotheres". Science 155: 1006–1009.

Crompton, A. W. & Jenkins, F. A. (1968): Molar occlusion in Late Triassic mammals. Biological Reviews 43: 427–458.

Fortelius, M. (1985): Ungulate cheek teeth: development, functional, and evolutionary interrelations. Acta Zoologica Fennica 180: 1–76.

Fortelius, M. (1987): A note on the scaling of dental wear. Evolutionary Theory 8: 73–75.

*Gailer, J. P., & Kaiser, T. M. (2014): Common solutions to resolve different dietary challenges in the ruminant dentition: the functionality of bovid postcanine teeth as a masticatory unit. Journal of Morphology 275: 328–341.

Gingerich P. D. (1973): Molar occlusion and function in the Jurassic mammal *Docodon*. Journal of Mammalogy 54: 1008–1013.

Gingerich, P. D. (1974): Dental function in the Palaeocene primate *Plesiadapis*. In: Martin, R. D., Doyle, G. A. & Walker, A. C. (eds.). Prosimian Biology. Duckworth, London: 531–541.

Grau, G. A., Sanderson, G. C. & Rogers, J. P. (1970): Age determination of raccoons. Journal of Wildlife Management 34: 364–372.

Habermehl, D. K.-H. (1985): Altersbestimmung bei Wild- und Pelztieren. Paul Parey, Berlin.

Hershkovitz, P. (1971): Basic crown patterns and cusp homologies of mammalian teeth. In: Dahlberg, A. A. (ed.). Dental morphology and evolution. University of Chicago Press, Chicago: 95–150.

Jäger, K. R. K., Gill, P. G., Corfe, I. & Martin, T. (2019): Occlusion and dental function of *Morganucodon* and *Megazostrodon*. Journal of Vertebrate Paleontology 39: e1635135.

Kaiser, T. M. & Schulz, E. (2006): Tooth wear gradients in zebras as an environmental proxy – a pilot study. Mitteilungen aus dem Hamburgischen Zoologischen Museum und Institut 103: 187-210.

Kay, R. F. (1977): The evolution of molar occlusion in the Cercopithecidae and early Catarrhines. American Journal of Physical Anthropology 46: 327-352.

Kay, R. F. & Hiiemäe, K. M. (1974): Jaw movement and tooth use in Recent and fossil primates. American Journal of Physical Anthropology 40: 227-256.

Kermack, K. A., Lees, P. M. & Mussett, F. (1965): *Aegialodon dawsoni*, a new trituberculosectorial tooth from the Lower Wealden. Proceedings of the Royal Society B 162: 535-554.

Kullmer, O., Benazzi, S., Fiorenza, L., Schulz, D., Bacso, S. & Winzen, O. (2009): Technical note: occlusal fingerprint analysis: quantification of tooth wear pattern. American Journal of Physical Anthropology 139: 600-605.

*Kullmer, O., Menz, U. & Fiorenza, L. (2020): Occlusal Fingerprint Analysis (OFA) reveals dental occlusal behavior in primate molars. In: Martin, T. & Koenigswald, W. v. (eds.). Mammalian Teeth – Form and Function. Pfeil, Munich: 25-43.

Koenigswald, W. v. (2018): Specialized wear facets and late ontogeny in mammalian dentitions. Historical Biology 30: 7-29.

*Koenigswald, W. v. (2020): Construction and wear of mammalian teeth in terms of heterochrony. In: Martin, T. & Koenigswald, W. v. (eds.). Mammalian Teeth – Form and Function. Pfeil, Munich: 171-186.

Luo, Z. X. & Martin, T. (2007): Analysis of molar structure and phylogeny of docodont genera. Bulletin of Carnegie Museum of Natural History 39: 27-47.

Maier, W. (1980): Konstruktionsmorphologische Untersuchungen am Gebiß der rezenten Prosimiae (Primates) [Constructional and morphological investigations of the dentition of extant Prosimiae (Primates)]. Abhandlungen der senckenbergischen naturforschenden Gesellschaft 538: 1-158.

Maier, W. & Schneck, G. (1981): Konstruktionsmorphologische Untersuchungen am Gebiß der hominoiden Primaten. [Constructional and morphological investigations of the hominoid dentition]. Zeitschrift für Morphologische Anthropologie 72: 127-169.

Mills, J. R. E. (1955): Ideal dental occlusion in the primates. Dental Practitioner 6: 47-61.

Osborn, H. F. (1888): The evolution of mammalian molars to and from the tritubercular type. American Naturalist 22: 1067-1079.

Pinto-Llona, A. C. (2013): Macrowear and occlusal microwear on teeth of cave bears *Ursus spelaeus* and brown bears *Ursus arctos*: inferences concerning diet. Palaeogeography, Palaeoclimatology, Palaeoecology 370: 41-50.

*Ruf, I., Schubert, A. M. & Koenigswald, W. v. (2020): Case studies on functional aspects and constraints in early and late tooth ontogeny. In: Martin, T. & Koenigswald, W. v. (eds.). Mammalian Teeth – Form and Function. Pfeil, Munich: 102-124.

*Schultz, J. A., Menz, U., Winkler, D. E., Schulz-Kornas, E., Engels, S., Kalthoff, D. C., Koenigswald, W. v., Ruf, I., Kaiser, T. M., Kullmer, O., Südekum, K.-H. & Martin, T. (2018): Modular wear facet nomenclature for mammalian post-canine dentitions. Historical Biology 30: 30-41.

*Schulz, E., Calandra, I., & Kaiser, T. M. (2010): Applying tribology to teeth of hoofed mammals. Scanning 32: 162-182.

*Schulz, E. & Kaiser, T. M. (2013): Historical distribution, habitat requirements and feeding ecology of the genus *Equus* (Perissodactyla). Mammalian Review 43: 111-123.

*Schulz-Kornas, E., Kaiser, T. M., Calandra, I. & Winkler, D. E. (2020): A brief history of quantitative wear analyses with an appeal for a holistic view on dental wear processes. In: Martin, T. & Koenigswald, W. v. (eds.). Mammalian Teeth – Form and Function. Pfeil, Munich: 44-53.

*Taylor, L. A., Kaiser, T. M., Schwitzer, C., Müller, D. W., Codron, D., Clauss, M. & Schulz, E. (2013): Detecting inter-cusp and inter-tooth wear patterns in rhinocerotids. PLoS One 8: e80921.

Szalay, F. S. (1969): Origin and evolution of function of the mesonychid condylarth feeding mechanism. Evolution 23: 703-720.

Thenius, E. (1989): Zähne und Gebiß der Säugetiere. De Gruyter, Berlin.

Ungar, P. (2010): Mammal teeth. The Johns Hopkins University Press, Baltimore, MD.

Vandebroek, G. (1967): Origin of the cusps and crests of the tribosphenic molar. Journal of Dental Research 46: 796-804.

Van Valen, L. M. (1982): Homology and causes. Journal of Morphology 173: 305-312.

Wang, Y.-Q., Clemens, W. A., Hu, Y.-M. & Li, C.-K. (1998): A probable pseudotribosphenic upper molar from the Late Jurassic of China and the early radiation of the Holotheria. Journal of Vertebrate Paleontology 18: 777-787.

Williams, J. A. (2005): Wear and wear particles – some fundamentals. Tribology International 38: 863-870.

Winkler, D. E. & Kaiser, T. M. (2011): A case study of seasonal, sexual and ontogenetic divergence in the feeding behavior of the moose (*Alces alces* Linné, 1758). Verhandlungen des naturwissenschaftlichen Vereins Hamburg 46: 331-348.

Winkler, D. E., Schulz-Kornas, E., Kaiser, T. M., De Cuyper, A., Clauss, M., & Tütken, T. (2019): Forage silica and water content control dental surface texture in guinea pigs and provide implications for dietary reconstruction. Proceedings of the National Academy of Sciences 116: 1325-1330.

Occlusal Fingerprint Analysis (OFA) reveals dental occlusal behavior in primate molars

Ottmar Kullmer, Ulrike Menz, and Luca Fiorenza

Introduction

A comprehensive understanding of the occlusal relationship between upper and lower tooth crowns in mammals is essential to interpret the evolution of dental function and adaptations. In fossil species, we can only hypothesize occlusal dynamics based on the analysis of tooth structures, the interpretation of dental macro- and microwear patterns, and the reconstruction of jaw mechanics. We are not able to test and comparably render in vivo details of the chewing physiology and other dental behaviour.

Dental wear patterns are the result of the biomechanical interplay between dental hard tissue, ingested food, incorporated environmental grit, and chemical processes induced by saliva and food itself (Lussi 2002). In this respect it is well known that wear patterns can inform about properties of ingested food materials, and aspects of general environmental setup, such as dust present in the air and adhered to food (Baker et al. 1959, Ungar et al. 1995, Sanson et al. 2007, Lucas et al. 2013). A detailed analysis of dental wear patterns provides evidence about masticatory and para-masticatory occlusal dynamics, based on the upper and lower antagonistic tooth relation during occlusal movements (Butler 1952, Mills 1955). This is of particular interest for the extraction of dietary and behavioural signals encoded in the macro-relief morphology of worn dental surfaces. Since the early 1970s detailed conclusions were drawn from analyses of food composition in the context of macro- and microwear on surfaces of dental crowns (e.g., Crompton & Hiiemäe 1970, Crompton 1971, Butler 1973, Kay 1973, 1975, 1981, Kay & Hiiemäe 1974a, b, Janis 1979, 1984, 1990, Smith 1984). Numerous studies confirmed a distinct correspondence between dental biomechanics, occlusal movements and crown relief structures (M'Kirera & Ungar 2003, Maier 1977a, b, 1978, 1980, Maier & Schneck 1981, 1982, Strait 1993a, b, Teaford 1983, Ungar & Williamson 2000, Ungar & M'Kirera 2003, Ulhaas et al. 2004, 2007).

Biologically, the dentition represents an organismic functional unit for the ingestion and disintegration of food, biomechanically adapted to various food materials (Lucas 2004). Mechanical processes during chewing and other behavioural activities on the tooth surface lead to crown abrasion and attrition (Kay & Hiiemäe 1974a). Consequently, pits and scratches (microwear) emerge on the crown surface, evidencing the loss of hard tissue (enamel/dentin) due to mechanical stress. The morphogenetic crown relief (primary relief) in a freshly erupted occluding crown is permanently altered (secondary relief) throughout the lifetime of an individual. Physical strain combined with saliva and food-induced chemical erosion results in gradual loss of dental structures on upper and lower crowns in respect to the pattern of sequential antagonistic approaches for the duration of power stroke movements (Fig. 2.1). Tooth-to-tooth contacts (dental two-body contacts) and tooth-food/tooth-grit contacts (three-body contacts) (d'Incau et al. 2012) regulate overall mechanical dental wear activities, while the morphology of the antagonistic occlusal surface permanently adapts to individual occlusal dynamics. The contacts occur as integrative effects during repeated dental antagonistic intercuspation, resulting in a recurring precise

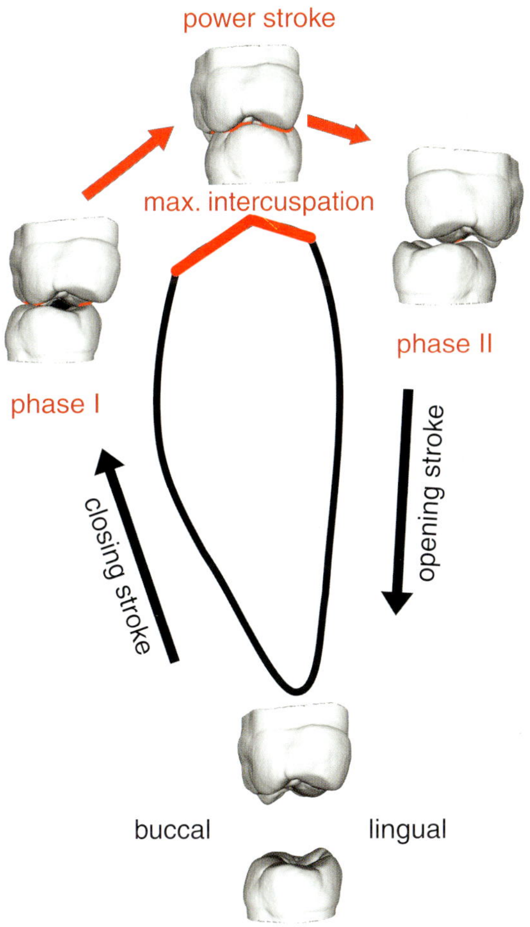

Fig. 2.1. The chewing strokes in primate molars. During the power stroke antagonistic tooth contacts may occur. The power stroke is subdivided into an incursive phase (phase I closing stroke) from the first contact until maximum intercuspation, and an excursive phase (phase II opening stroke) until the last contact occurs. Complementary wear facet pairs on the occlusal surfaces of the molars are typically attributed to phase I or phase II movements.

T. Martin & W. v. Koenigswald (eds.): Mammalian Teeth – Form and Function. Pp. 25–43, 7 figs., 1 tab.
 – ISBN 978-3-89937-266-3
DOI: http://doi.org/10.23788/mammteeth.02 Published 22 December 2020

occlusal fitting of complementary force-fit on the worn surfaces (wear facets = flat and polished enamel areas with well-defined margins created by loss of material) during occlusion (Maier 1978).

The mammalian dentition evolved as a variable and adaptive system, permanently changing functional abilities due to fluctuating environmental and behavioural selective stress for more than 200 million years. Morphological structures of crowns such as cusps, lophs, basins, notches, and fissures determine the major pathways of relief-guided occlusal movements.

Chewing in mammals consists of sequential relatively stereotypic cycles of motion (Maier 1978) operated by a neuromuscular system which is controlled in the brain stem (Luschei & Goodwin 1974, Lund 1976, Luschei & Goldberg 1981, Agrawal et al. 1998). Each chewing cycle (Fig. 2.1) is divided into a sequence of three strokes (Hiiemäe 1978). The opening stroke unlocks upper and lower teeth, and increases the gape of the mouth cavity for food ingestion preparing for the closing stroke, which describes the initial chewing cycle when the lower jaw moves upward until first occlusal contact occurs. That is the moment when the power stroke starts, often divided in two phases (phase I and phase II). In phase I (mandibular incursion) the cusps of the lower and upper molars slide past one another ending up in force-fit individual maximum intercuspation (max IC ≙ centric occlusion). Phase II of the power stroke follows when the lower jaw moves out of max IC (mandibular excursion) until the final contact occurs. The molar wear pattern of complementary facet pairs reflects the power stroke only. In phase I of bunodont primate molars the palatinal surfaces of the upper molar cusps contact with the buccal surfaces of lower molar's lingual cusps (Hillson 1996). In phase II the lingual surfaces of buccal lower molar cusps move along the buccal slopes of the lingual upper molar cusps until the antagonists separate (Hillson 1996). Finally, a new chewing cycle starts with the next opening stroke. Kay & Hiiemäe (1974a) described how the system of occlusal wear facets in extant and extinct primate molars develops during repeated power strokes in phase I and phase II applying shearing, crushing and grinding actions to the trapped food particles, while adding saliva to produce a more or less homogeneous and mushy bolus before swallowing.

A number of comparative studies of phase I and phase II macrowear facets have led to numerous labelling systems (Tab. 2.1) for complementary wear facet pairs in upper and lower molars (Butler 1952, Mills 1955, Crompton & Hiiemäe 1970, Crompton 1971, Gingerich 1974, Kay 1977, Maier 1980, Maier & Schneck 1981, Kullmer et al. 2009, Schultz et al. 2018). Typically, the spatial position of wear facets on a molar crown reveals the phase of the power stroke in which they developed (Mills 1955, Butler 1956, 1967, 1973, Kay & Hiiemäe 1974a, b, Kay 1975, 1977, 1978, Grine 1981, Kullmer et al. 2009). Two distinguished sets of wear facets corresponding to phase I and phase II movements are produced in primate molars (Fig. 2.2). Since chewing consists of a number of cycles (Maier 1978) it is favourable to start in max IC as a feasible and definable moment for modelling the pathways of occlusal movements and compare the kinematics of antagonistic tooth contacts (Kullmer et al. 2009, 2012). Max IC has been used as initial position for the concept of the occlusal compass of *Homo sapiens* molars, sketching latero-, pro-, medio- and retrusive fields of movements for the description of occlusal motion in a horizontal (occlusal) plane (Kordaß & Velden 1996, Douglass & DeVreugd 1997). In addition, surtrusion and detrusion upward and downward occlusal movements (Kordaß & Velden 1996, Douglass & DeVreugd 1997) complement each horizontal movement of the occlusal compass in three dimensions. The dental occlusal compass enables researchers to relate each wear facet pair to an occlusal direction by moving upper or lower crowns out of max IC, while controlling antagonistic contacts (Schulz 2003, 2008, Schulz & Winzen 2004, Kullmer et al. 2009, 2012, 2013, Fiorenza 2009, Fiorenza et al. 2010, 2011 a, b, c). The occlusal compass helps to equate possible directions and tolerance space during an individual power stroke, and for its resulting facet pattern. Even though potential positions of wear facets on the molar crowns are known for most mammalian tooth types, the final detailed appearance of the wear facets, including their size and shape is highly variable and multifactorially influenced. Besides the individual age, environmental and behavioural factors involved, the expression of wear facets depends also on the morphology of the primary relief structures and the exact occlusal relationship of the antagonists (Fiorenza et al. 2010, 2011c, Fiorenza & Kullmer 2013, 2015). Consequently, the individual macrowear pattern,

Table 2.1. List of published comparative occlusal wear facet labelling systems.

Butler (1952), Mills (1955)	Crompton & Hiiemäe (1970)	Crompton (1971)	Gingerich (1974)	Kay (1977)	Maier (1980)	Maier & Schneck (1981)	Kullmer et al. (2009)	Schultz et al. 2018
1	1	2a, 2b	B2	2a, 2b	2	2	F2	ME-d/pr-mb
2	2	1a, 1b	B1	1a, 1b	1	1	F1	PA-m/pr-db
3	4	5	B5	5	5	5	F5	PR-ml/md-db
4, 4‘				7	7	7	F7	HY-dl/md-m
5			L2	10	10	10	F10	HY-db/pr-ml
				10n	12	12	F12	HY-mb/hyl-l/hy-l
6	3	3a, 3b	B3	3a, 3b	3	3	F3	PA-dl/hy-mb
7	6	4a, 4b	B4	4a, 4b	4	4	F4	ME-ml/hy-d
8, 8‘			B7	8, 8n	8	8	F8	HY-l/ed-d
9	5	6	B6	6	6	6	F6	PR-dl/ed-b
10			L1	9	9	9	F9	PR-b/hy-l
				x	11	11	F11	PR-mb/pr-dl
						13	F13	PR-db/hyl-ml
						4‘	F2'	ME-ml/hyl-b

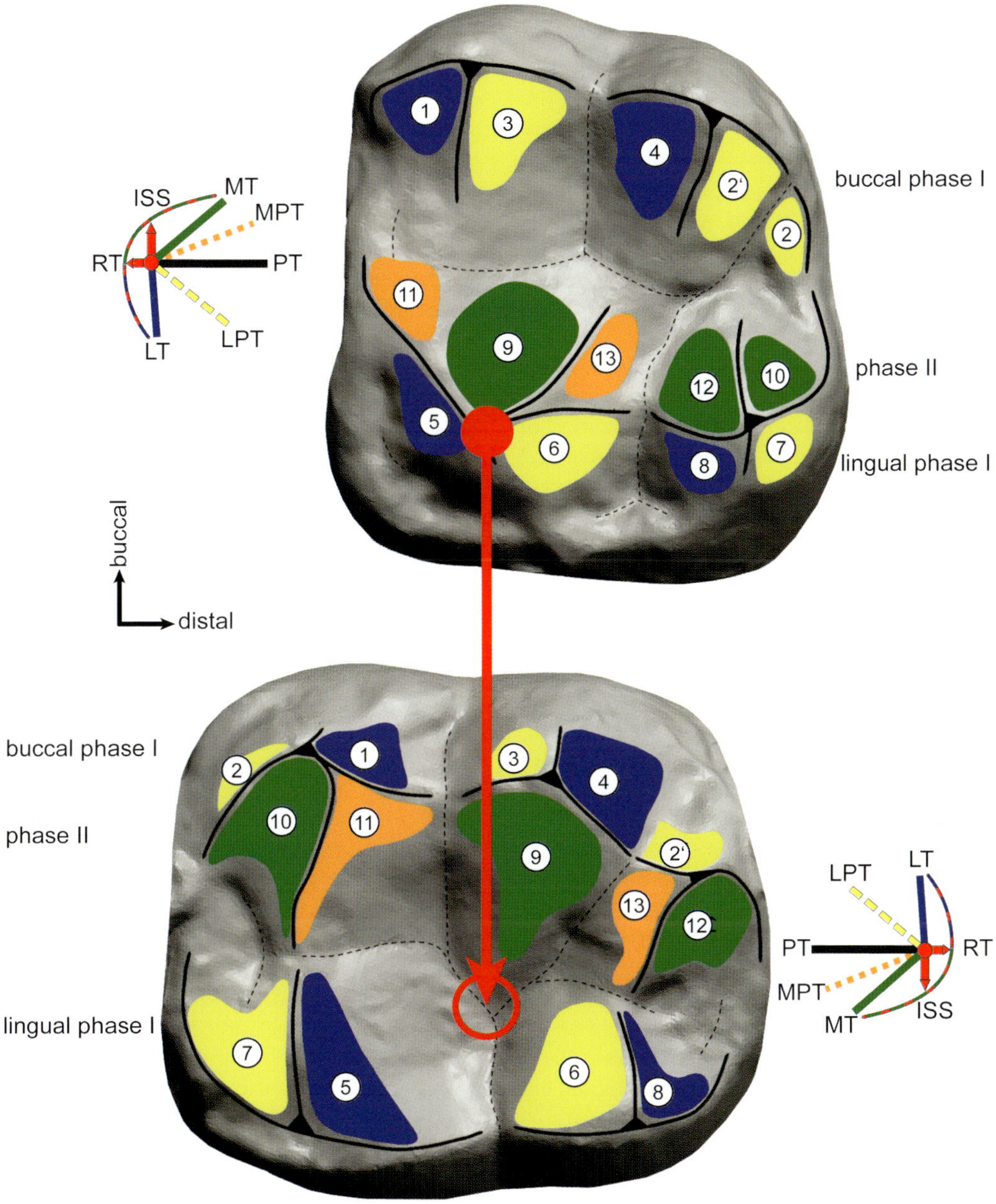

Fig. 2.2. Map of buccal phase I, lingual phase I, and phase II wear facets in virtual crown models of *Proconsul africanus* KNM-RU 7290. Upper and lower right M2 in occlusal view aligned in occlusal relationship. The general occlusal compass is oriented for each molar. Major directions of horizontal occlusal movements starting in maximum intercuspation (maximum intercuspation position is indicated by red arrow) are shown by color-coded lines following Kullmer et al. (2009); lateroretrusion (blue), lateroprotrusion (yellow), protrusion (black), medioprotrusion (orange), mediotrusion (green), immediate side shift and retrusion (red). The green-red dashed line in the compass describes the retrusive space. Wear facets are color-coded corresponding to the major directions in the occlusal compass. Each wear facet possesses a complementary facet in the antagonistic tooth. All facets are labelled for comparison. Facets 1 to 4 belong to buccal phase I, 5 to 8 to lingual phase I, and 9 to 13 are identified as phase II facets (Kullmer et al. 2009). No scale.

including the wear facets in their morphometric details, is unique. Therefore, it is described as the occlusal fingerprint of an individual (Kullmer et al. 2009).

The alteration of the occlusal fingerprint of each crown throughout an individual's life represents changing occlusal dynamics from moderate to advanced wear, from a pattern of isolated wear facets towards a single enlarged worn area as a result of merged wear facets in late adult individuals (Smith 1984). With progressive tooth wear dentin is exposed, and dentin basins start to grow until in heavily worn crowns occasionally only a single large dentin basin, surrounded by a marginal rim of enamel, completely obscures the original wear facet pattern. In most heavily worn crowns the occlusal primary relief has changed into a secondary relief where the remaining enamel surface occurs flat or even concave with a reduced antagonistic indentation, while in crowns with moderate wear the dental articulation is often well-guided by interlocking relief structures along complementary wear facet pairs in upper and lower teeth.

Gradual relief alteration through dental wear indicates changes also in occlusal motion with a gradual increase in the spatial degree of freedom with progressing crown wear. Size, inclination, and inclination direction angles of wear facets inform us about individual three-dimensional occlusal kinematics (Kullmer et al. 2012). A comparison of quantitative wear facet data from large samples in differing and similar wear stages enables us to discuss the variability of the macrowear pattern and dental behaviour for the interpretation of chewing biomechanics, processed food, para-masticatory behaviour, and evolutionary dento-functional adaptations (e.g. Smith 1984, Kaifu et al. 2003, Fiorenza 2009, Fiorenza et al. 2010, 2011a,b,c, 2015a, 2019a,b, 2020, Fiorenza & Kullmer 2013, Kullmer 2015, Oxilia et al. 2015, Fung et al. 2020).

Those facts inspired us to develop the Occlusal Fingerprint Analysis (OFA) for the quantification and visualization of antagonistic occlusal relationships derived from the analysis of virtual crown surface models. OFA provides a spatial map of the macrowear pattern on tooth crowns, but it also facilitates the size and the three-dimensional orientation of the wear facets, which allows the reconstruction of occlusal kinematics. It records complementary wear facet pairs supporting interpretations of antagonistic approaches during phase I and phase II occlusal movements of the power stroke.

In this chapter, we provide the functional background of quantitative wear facet analyses and the workflow protocol of two principal OFA steps, i.e. the determination and interpretation of the static wear facet pattern in molars, and the virtual simulation and quantification of the sequential occlusal contacts applying the Occlusal Fingerprint Analyser software. Further, we discuss the potential of the OFA approach in respect of specific primate molar adaptations.

Reconstruction of occlusal movements

A satisfactory quantification of macrowear patterns including quantitative assessment of occlusal kinematics from wear facet patterns was for a long time out of reach, although quantitative methods in microwear analyses on wear facets have developed rapidly due to advances in digital microscopy and confocal profilometry (e.g., Kay 1987, Pedreschi et al. 2000, Ungar et al. 2003, Scott et al. 2005, 2006, Schulz et al. 2013, Calandra & Merceron 2016). A few studies focused on the interpretation of microwear pattern for the reconstruction of occlusal movements (e.g. Mills 1955, Ryan 1979, Gordon 1984, Krueger et al. 2008). Many earlier approaches, using cinematography, cineradiography and electromyography (Crompton & Hiiemäe 1970, Hiiemäe & Crompton 1971, Luschei & Goodwin 1974, Herring 1976, Gorniak & Gans 1980) had investigated the general pattern of chewing cycles, but those techniques were not able to visualize occluding dental contacts and wear facets in function with their details at a reasonable resolution. In addition, imaging techniques in those days were mostly limited to two-dimensional visualizations only. The visualization of occlusal kinematics is challenging because occlusion usually occurs in a concealed place within the mouth cavity: cheeks, lips, tongue, food material and saliva surround the teeth, and occlusal contacts are not appropriately assessable (Evans & Fortelius 2008). To overcome this problem Evans & Fortelius (2008) reconstructed jaw movements and occlusion of Recent carnivore species, simulating chewing cycles with virtual dental 3D models. They were able to visualize for example details of the occluding pathway of carnassial tooth pairs in a number of carnivore specimens. They could demonstrate that the carnassials in their sample do not, as expected, describe a linear pathway during vertical occlusal movements. The occlusal trajectories gradually change their angular direction until the crowns reach the max IC, often following concave or convex occlusal pathways. This and other observations would not have been possible without a detailed view on the articulating crown surfaces. They also could show that during the occluding pathway of the carnassial articulation, contacts usually do not occur over a complete facet area in the same moment, but sequentially travel over a facet surface covering just a relatively small facet area at the time. Evans & Fortelius (2008) had to cope with some accuracy limits in their study, including the fact that encountered tooth positions and dental relationship derived from digital scans of dry collection specimens are generally questionable, often requesting precise repositioning of crowns before applying a digital approach. Additionally, at the time of their study the Virtual Reality Modeling Language (VRML) simulation techniques could only handle relatively low-resolution models, which limited the resolution and accuracy on occlusal details.

In the meantime, advances in surface scanning technologies, digital data processing and three-dimensional reconstruction favored the development of new virtual methods for the quantification of occlusion and dental wear facet patterns. Today, the Occlusal Fingerprint Analysis (OFA) determines the size, position and orientation of wear facets in space, and supports the interpretation of occlusal developmental kinematics (Kullmer et al. 2009). Fiorenza (2009) applied the OFA method to a large sample of hominin upper molars, collecting OFA data to determine variability in molar macrowear patterns of Neanderthals and various modern and fossil *Homo sapiens* hunter-gatherer groups who inhabited different geographic regions. The OFA results showed that the wear facet data contain eco-geographic signals, reflecting dietary preferences and habitat settings (Fiorenza et al. 2011a, 2015a).

A demanding validation of reconstructed trajectories of occlusal movements from OFA data (Kullmer et al. 2009) inspired the programming of a specific OFA software application, the Occlusal Fingerprint Analyser, for reconstruction, visualization and quantitative assessment of sequential antagonistic occlusal approaches, showing that antagonistic contacts on complementary wear facet pairs exist (Benazzi et al. 2011f). Previous occlusal reconstructions of dental movements with no option for a kinematic validation, derived mostly from static considerations of wear facet patterns (Butler 1952, 1973, Mills 1955, Crompton 1971, Kay & Hiiemäe 1974a, Rensberger 1973, Kay 1975, Fortelius 1985), or from simulations with no real relief guidance (Evans & Fortelius 2008). Now, the OFA software combines 3D simulation by relief-guidance and a contemporaneous recording of area size, position, and spatial orientation of occlusal contacts for functional comparison. The occlusal simulation in the OFA software integrates the relief-guided motion of crown models derived from collision detection, deviation, and break free algorithms, generating an individual occlusal trajectory, while collecting sequential data from user-defined timesteps of antagonistic colliding dental contacts.

The Occlusal Fingerprint Analysis (OFA)

The recent OFA workflow protocol can be divided in three parts, a qualitative description, the quantification of wear facets, and the reconstruction of the occlusal kinematics.

An OFA analysis usually starts with an investigation and selection of reasonable dental specimens. In this respect it is favourable to work either directly with original specimens, or with high resolution dental casts (Fiorenza et al. 2009). Depending on the specific research question, scholars decide to choose samples in specimens with moderate wear representing fully developed wear facet patterns for most accurate reconstruction of the occlusal pathways, or to use a sample of crowns in various wear stages for studies of occlusal alteration through ontogeny and wear pattern gradients (Grant 1982, Smith 1984, Brown & Chapman 1990, Kullmer et al. 1999, Anders et al. 2011). Ideal for an OFA are usually molars in late wear stage 2 to early wear stage 4 following the categories of Smith (1984). Comparative samples should represent a similar range of wear stages.

An OFA always starts with a macroscopic inspection of the worn crown areas including segmentation of wear facets (Kullmer et al. 2009). An accurate mapping of wear facets and a determination of the number of wear facets is an essential procedure for the study of macrowear pattern and a reconstruction of occlusal kinematic relationship (e.g., Schultz et al. 2018). Initially, the macrowear mapping can be performed in 2D by simply drawing a sketch map of identified wear facets (e.g., Kullmer et al. 2013). Using a binocular microscope usually one can directly identify and draw complementary wear facet pairs (Fig. 2.2), if we work with upper and lower crowns of the same individual by tracing the slightly edged margins of wear facets (e.g., Maier 1977, 1978, Maier & Schneck 1981, Kullmer et al. 2009, 2013). The identification of relative antagonistic occlusal contacts sometimes allows a direct attribution of major directions of occlusal movements for a complementary wear facet pair, if a functional occlusal 3D compass for occlusal articulation, such as it is established for bunodont hominid molars, is known (Kullmer et al. 2009, 2012, 2013). In a wear facet map the individual occlusal fingerprint for a molar or antagonistic tooth pair is visualized (Fig. 2.2). A specific wear facet labelling system combined with a color-code for major directions of occlusal movements helps to standardize the visualisation for qualitative comparisons. The occlusal compass directions starting from the point of max IC in primate molars are divided in protrusion (black), a parallel forward movement of both lower jaw condyles, retrusion (red), a backward motion, laterotrusion (blue), buccal movements, and mediotrusion (green), lingual occlusal motions (Kullmer et al. 2009). In most primate molars we will find combinations of the major directions of movements, such as lateroprotrusion (yellow), laterotrusion (blue), and medioprotrusion (orange). An occlusal compass can be adapted to any crown morphology.

The collection of presence/absence data, size and spatial orientation of wear facets in a sample of crowns reveals the variability of occlusal patterns. Besides a determination of the occlusal relationship in upper and lower crowns and a reconstruction of dental motion, the occlusal fingerprint can help to enlighten the expression of wear stages, frequency and appearance of malocclusion, as well as para-masticatory use of teeth (Fiorenza 2009, Fiorenza et al., 2011c, 2019a, b, Krueger-Janson et al. 2013, Fiorenza & Kullmer 2013, 2015, 2016). Two-dimensional facet maps derived from investigation of physical tooth surfaces support the determination of wear facets in virtual 3D models (Kullmer et al. 2013).

3D software packages such as Polyworks™ (Innovmetric Inc.) or Geomagic™ (Geomagic Inc.) enable a segmentation of wear facets on virtual 3D surface models (Kullmer et al. 2009, Fiorenza et al. 2011a). Virtual triangulated surface models consist of a varying number of connected triangles. Despite challenges in identifying wear facets for segmentation, the model quality of reconstructed relevant functional structures on a crown surface depends on the resolution and accuracy of the raw 3D data (Kullmer 2008). The higher the resolution of a virtual model is, the shorter is the triangle edge length between single data points. A model resolution of $\leq$0.065 mm is good enough for morpho-functional analysis of dental crowns the size of human molars (Kullmer et al. 2002, 2009, 2012, Ulhaas et al. 2004, 2007, Fiorenza 2010, 2011a, b, c).

A morphometric challenge for functional comparison of dental relief structures and wear facet patterns is the use of a standardized crown orientation, and the comparison of gradually changing wear stages. Classical landmark methods for spatial comparison of shape such as mathematical-statistical approaches of modern geometric morphometrics (e.g., GMM after Bookstein 1991) including "Procrustes Superimposition" and "Thin Plate Spline Warping" applied for biological shape and size analysis in taxonomical studies of unworn crowns (Weber & Bookstein 2011), or shape comparison of cervical lines of single crowns and jaw bones (Benazzi et al. 2011a, b, c, d, e), fail in the analyses of functional wear facets. 3D GMM methods are based on spatial comparison of landmarks, semi-landmarks and curves which are likely homologous in sequence and number in a sample. Landmarks can be difficult to define on worn tooth surfaces. Wood & Abbott (1983), Wood (1993) and Ungar et al. (2002) defined on molars a reference plane through the cervix for the morphometry of crown height and crown relief in hominoid molars. The cervical plane is in most wear stages independent of dental tissue loss

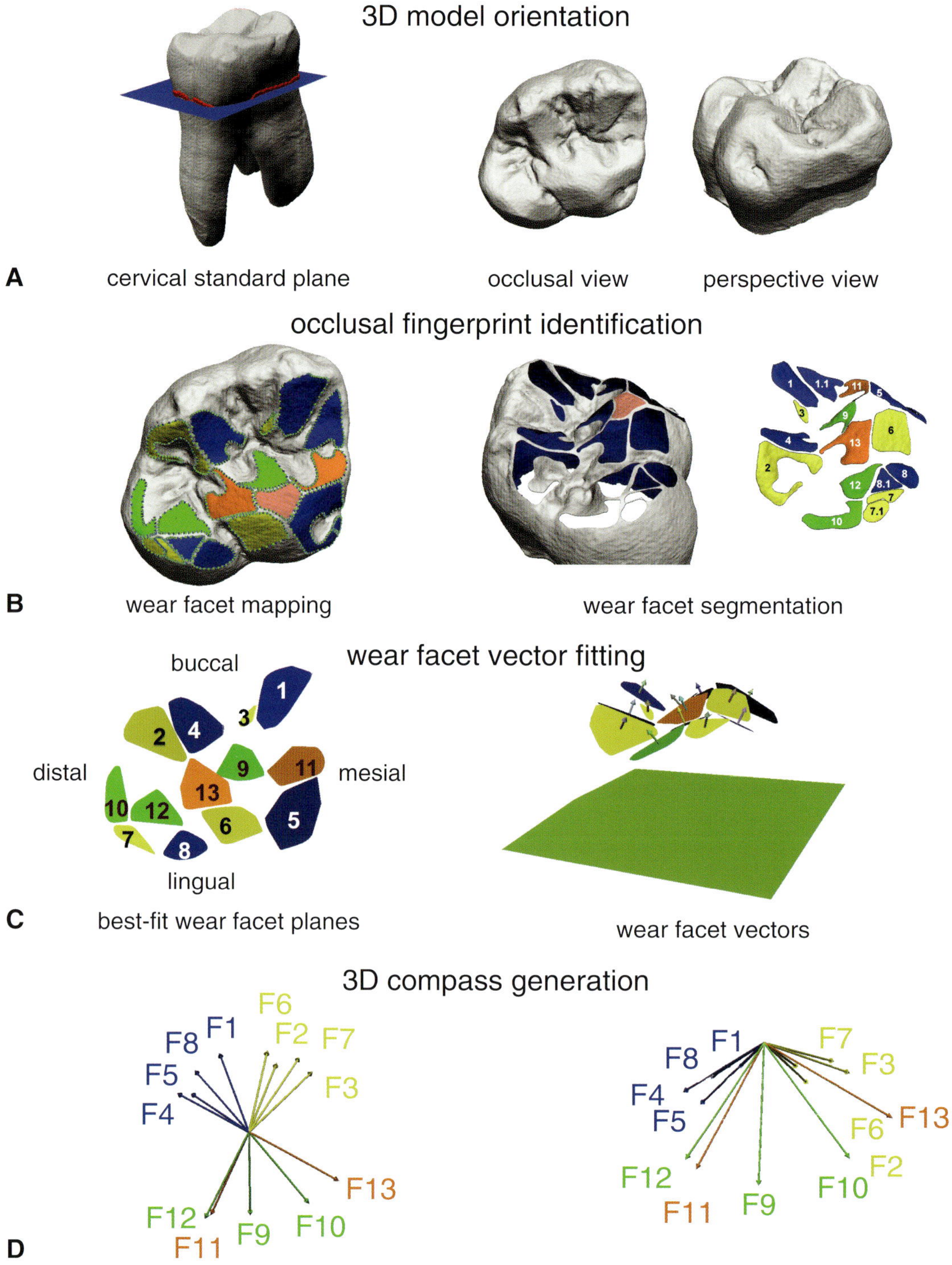

Fig. 2.3. Generation of an OFA 3D compass of the wear facet pattern in four steps. **A,** Orientation of virtual 3D model of right M2 (*H. neanderthalensis*, Krapina 137) with cervical standard plane (left), occlusal view (middle), perspective view (right); **B,** (left) 3D model (Krapina 137) in occlusal view with mapped and color-coded wear facet pattern, lateroretrusion wear facets (blue), lateroprotrusion (yellow), mediotrusion (green), medioprotrusion (orange) and tip crushing area (pink), (right) triangulated and segmented occlusal fingerprint, and cut wear facets labelled; **C,** (left) computed best-fit planes of wear facets (Krapina 137), (right) perspective view on best-fit planes of wear facets with normal vectors and cervical reference plane in green; **D,** (left) wear facet vectors (Krapina 137) with normalized length in occlusal view (3D occlusal compass), the shorter the vector appears, the steeper the wear facet, (right) 3D occlusal wear facet compass in perspective view. All images are extracted from Polyworks™ Modeller and Inspector (Innovmetric Inc.). Not to scale.

through wear (Benazzi et al. 2011c) and therefore it can be used also in most isolated primate molars as standard plane for comparison of functional relief structures (Spears 1994, Spears & Crompton 1996). Kullmer et al. (2002) described the use of a cervical plane and a mesial-distal axis as reference for the spatial orientation of bunodont molar crowns. The cervical plane is calculated in virtual models from a best-fit plane fitted with the least-squares method through a point-cloud of surface data points acquired in a defined distance from a cervical line placed

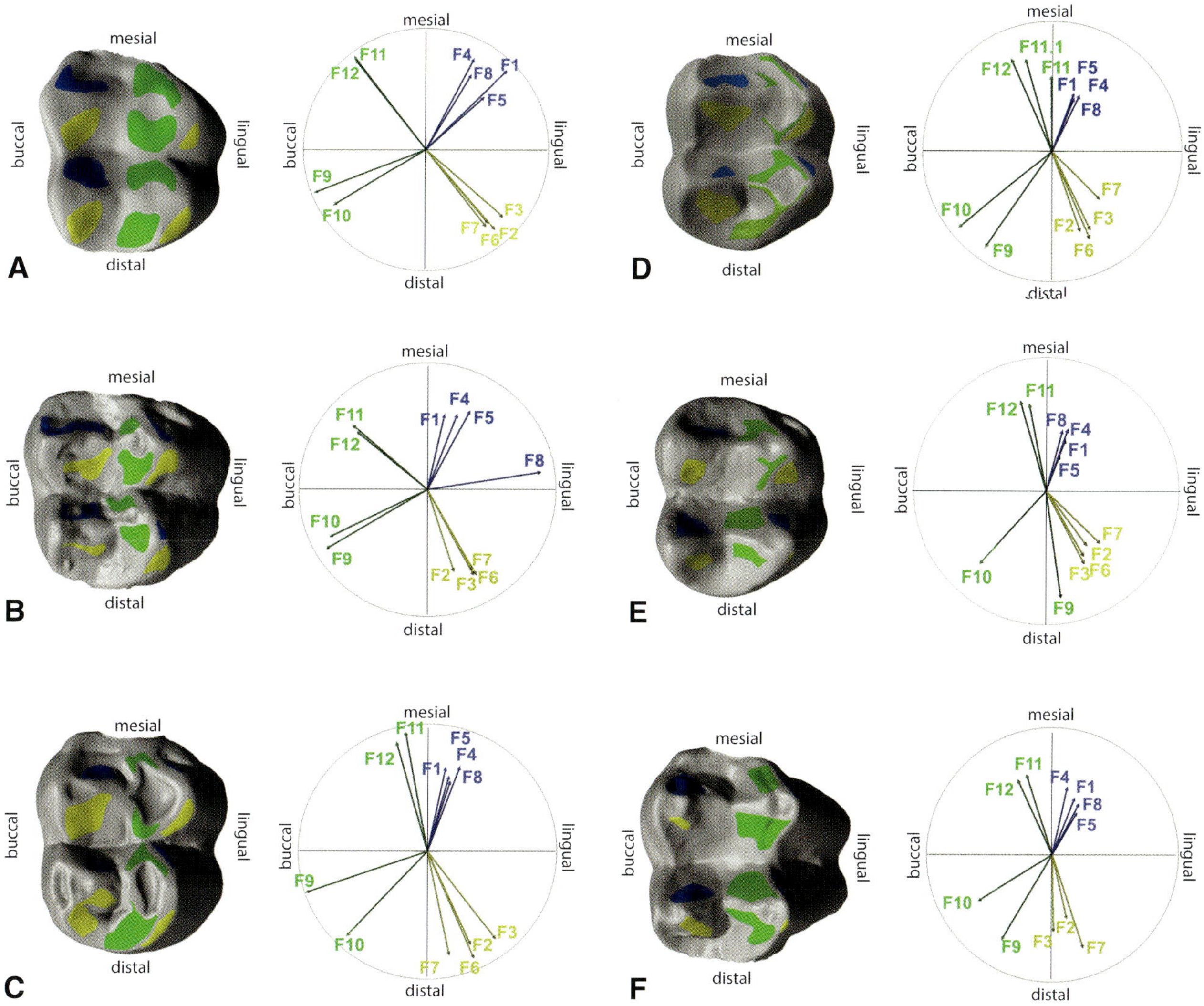

Fig. 2.4. Wear facet maps and 3D occlusal compasses of six upper right M1 from cercopithecoid species (not to scale). **A,** *Chlorocebus aethiops* (left M1 mirrored); **B,** *Macaca maurus*; **C,** *Papio ursinus*; **D,** *Colobus guereza*; **E,** *Nasalis larvatus*; **F,** *Presbytis cristatus* (left M1 mirrored). All bilophodont molars of Cercopithecidae monkeys show a similar occlusal fingerprint pattern in number of wear facets and the distribution of wear facet directions. But, cercopithecines (A, B, C) with a high fruit content in their diet and more leaf browsing colobines (D, E, F) differ by distinctly steeper wear facets in the latter. No to scale.

manually on the model surface (Fig. 2.3). Afterwards the crown model is virtually rotated in the coordinate system, so that the cervical plane matches the x-y-plane of the coordinate system. In the next step the model is rotated around the z-axis until the mesio-distal axis of the crown is oriented parallel to the x-axis of the coordinate system (Ulhaas et al. 2004, 2007). In this standardized orientation the functional areas such as wear facets in isolated crowns are compared (Kullmer et al. 2009, 2012). This orientation method is also practical for OFA. For functional repositioning of crowns in dental arches scholars choose a standardized occlusal plane for the dentition. Therefore, a plane is fitted through three defined prominent points on the relief of molars and incisors in the dental rows, for example the metaconid cusp tips of the right and left lower second molar, and the incisal point between the edges of the first incisors (Schulz 2003, 2008, Kullmer et al. 2012, 2013). Such an occlusal reference plane is valid also for the reconstruction of the upper dental arch, if the dental relationship in maximum intercuspation with the antagonists is reproducible (Kullmer et al. 2013).

Since the quantification of area size, phase I and phase II area ratio and spatial position of the wear facet patterns (Kullmer et al. 2002, 2009, Fiorenza et al. 2011a) derives from the segmentation of the wear facet on the virtual 3D models (Kullmer 2002, 2009, Ulhaas et al. 2004, 2007), it is important to define the boundary of a wear facet along their slightly edged margins, representing an almost planar, sometimes slightly concave or convex worn enamel surface region. Real edges along wear facet boundaries may develop through a relatively high rate of wear through relief-guidance in occlusal motion with complementary regions in the antagonists. In crown areas with flat slopes such as in basin structures, the facet boundary is frequently obscured, possibly through abrasive polishing of the worn edges (Fig. 2.3).

Preservation of facets does not only depend on the morphology of the relief and the wear stage of a tooth, but also on other factors such as enamel erosion processes, for example due to a high acidity in food and saliva. In those cases, qualitative pre-inspection of the original surfaces, and drawing of a wear facet map, can assist the wear facet segmentation in virtual models. Applying various software tools for visualization such as surface rendering and light source positioning may support wear facet boundary segmentation. Placement of polylines along facet boundaries is the first step of segmentation. Afterwards, re-triangulation of the model with an integration of the polylines into the triangulated grid of the surface model allows a precise selection of all data points within a wear facet for area size determination. The average inclination and direction of inclination of a wear facet can be calculated by using the best-fit plane through all data points of a wear facet (Fig. 2.3). The presence-absence, spatial position, and size of all wear facets describe the occlusal fingerprint of a molar (Kullmer et al. 2009). The occlusal fingerprint can be visualized in classical stereoplot diagrams (Kullmer et al. 2002, Ulhaas et al. 2004, 2007), such as those frequently used in geology and mineralogy for a visual documentation of sedimentary layers or mineral crystal orientation. The stereoplot presentation of wear facets was later slightly refined for easier readability (Kullmer et al. 2009, 2012). Meanwhile normalized vectors for each wear facet are used to represent the spatial facet position in a stereoplot (Figs. 2.3, 2.4). The origin of each facet vector in the stereoplot is placed in the centre of the half sphere and the length of the vector is normalized to the radius length of the sphere. The software module Inspector of Polyworks™ (Innovmetric Inc.) enables a direct generation of wear facet vectors and drawing of the individual 3D occlusal compass (Kullmer et al. 2009). In the stereoplot circle 0° is orientated at the top, indicating the mesial direction of the crown in occlusal view. The steeper the wear facet, the shorter is the vector in occlusal view in a stereoplot diagram of a 3D occlusal compass. It is also possible to turn and orient on screen a 3D compass in perspective view, for example, to explore average inclination and directions of wear facet orientation from buccally, or any other point of view (Kullmer et al. 2009). Facet directionality is readable in relation to the crown orientation and the directions of wear facets can be determined by applying the color code of the general occlusal compass (Schulz, 2003, 2008, Kullmer et al. 2009). The 3D occlusal compass of individual molars represents advancement of the original occlusal compass concept of movements in two dimensions. It enables a comparative visual presentation of individual wear facet orientation (Kullmer 2009, Fiorenza et al. 2010, 2011a,b).

The extraction of wear facet area size attributable to phase I and phase II of the power stroke provides additional functional information. Bunodont primate molars develop two rows of phase I and one row of phase II wear facets in a normal occlusal relationship. Phase I wear facets are divided in buccal phase I wear facets and lingual phase I wear facets. The percentage of size distribution of buccal phase I, lingual phase I, and phase II wear facets allows an interpretation of power stroke portions in phase I and phase II of the occlusal motion. Usually, in molars with a dominance of incursive phase I movements, the wear facets of phase I show an increased surface area compared to the excursive phase II wear facets. The three power stroke proportions are comparably visualized in ternary plots (Fiorenza et al. 2010, 2011a,b).

Occlusal kinematics derived from Occlusal Fingerprint Analysis

Detailed studies of occlusal motion for a more comprehensive understanding of dental articulation and relief guided movements were based for a long time mostly on visual observations from in vivo manual manipulation in patients and volunteers, and handling of physical models in the laboratory. Significant progress was made in modern dentistry, when virtual models were used for the visualization of occlusal distances between upper and lower crowns. Thanks to modern virtual reality applications it is possible to move virtual dental models in the computer along predefined occlusal pathways and record the moment of model surface intrusion (Kordaß & Gärtner 2001, Kordaß et al. 2002, Bisler et al 2002). Meanwhile, digital records of patients' jaw motion are used to simulate natural occlusal jaw movements and dental contacts for dental crown restorations with modern CAD/CAM systems (Solaberrieta 2009).

In the Deutsche Forschungsgemeinschaft (DFG) Research Unit 771 we developed the virtual Occlusal Fingerprint Analyser, a software to digitally establish occlusal contacts and complementary kinematic relationship between wear facet pairs in dental antagonists (Benazzi et al. 2011f, 2012, 2014, Koenigswald et al. 2013, Kullmer et al. 2013, 2015). In the OFA software, two or more 3D models are positioned in a scene window on the computer screen in any arbitrary view. One dental surface model is then moved towards the other along a user-defined rough occlusal pathway, consisting of a number of set path points. As soon as a collision of the models is detected, the moving model is deflected along the surface relief of the other, attempting to reach the next path point. An individual trajectory of relief-guided deflections is generated by calculating the new position of the model per user-defined time steps (Kullmer et al. 2013). The OFA software allows a virtual calculation and simulation of movements with diverse modes for a reconstruction of the potential individual occlusal kinematics. It augments the understanding how wear facet positions and morphology relate to occlusal antagonistic approaches during power stroke motion. The realistic situation of any optional occlusal moment is testable as long as a pair of opposing virtual crowns is aligned in one timestep in a known occlusal position such as the moment of max IC.

There are few requirements of importance to note before working with the OFA software. Two key characteristics of digital surface models are essential to know for using the software application to its full potential. In this respect basic background knowledge about the occurrence of collisions on triangulated surface models, and how collisions can be manipulated, is helpful to evaluate OFA software results. The accuracy of the collision calculation depends on the model resolution and surface roughness in correlation to the software set up of the collision detection distance.

A collision is easily detectable as long as the surface model resolution is higher than the defined minimum collision distance. The collision distance needs to be set up by the user in the OFA software, depending on the quality of the virtual models. Almost any triangulated polygon model in Standard Triangle Language (STL) and Polygon File Format (PLY) format can be collided in the OFA software.

At the occurrence of a collision all affected triangles in both surface models are highlighted and recorded. The raw collision data need to be transformed into practical information for occlusal contact comparison. Therefore, the OFA software merges neighbouring colliding polygon triangles into collision groups and quantifies the number of engaged model triangles, area size, and angles of direction and inclination, if a reference plane is setup in the OFA software. The collision groups of triangles are saved for each model together with the resultant trajectory coordinates for each model position per time step of the occlusal sequence. Additional to parallel model approaches, various path point criteria can be set up by the user, including a simulation of rotational lower jaw motion by defining a virtual condyle axis.

In the latest OFA software version (OFA 2.0), more than two models can collide in one OFA project scene. Single tooth models from one dentition can be combined into a trajectory group following the same path, while e.g. recording the dental contacts in complete dental arches in single OFA software projects (Fig. 2.5). In this manner it is also possible to simulate the periodontal motion of each tooth in a dental row in its alveolar socket. When a collision is detected in a trajectory group an evasive movement is calculated depending on the parameter setups of each model in the scene. This allows for a more realistic reconstruction of potential occlusal contacts between opposing tooth crowns. The collision computation can be repeated arbitrarily until model repositioning and the path point adjustment after each calculation lead to a satisfying result in occlusal contacts. Usually this is reached at the moment when the sequential occlusal contacts cover most of the functional wear facet areas during a sequential occlusion.

Trajectory results contain the direction angles of phase I and phase II occlusal movements. This enables a quantitative comparison of occlusal trajectories and the generation of the mastication compass (Koenigswald et al. 2013) based on measured data from occlusal kinematics. The collision data allow a direct comparison of changes in contact area size and position during an occlusal sequence. One can directly follow the contact activity of facet positions during phase I and phase II movements for the discussion of functional aspects and differences in respect to the data of occlusal direction and inclination. The OFA software is freely available on the website of the DFG Research Unit FOR 771 (https://www.for771.uni-bonn.de/for771-en/ofa). In the latest version of the OFA software scholars can display the individual 3D occlusal compass of the dental contact areas of each time step for a direct comparison functional changes during a power stroke sequence.

Perspectives of Occlusal Fingerprint Analysis

Meanwhile, quantitative macrowear analyses using OFA or related methods were applied by scholars initially studying variability and function of wear facet patterns. OFA was used in various modern and fossil mammals in an inter- and intraspecific context in order to discuss wear facet patterns in molars with respect to evolutionary functional adaptation and dietary ecology (Engel 2005, Huck 2006, Ulhaas 2006, Fiorenza 2009, 2010, Fiorenza et al. 2011a, Schwermann 2015, Menz 2016, Engels & Schultz 2018). Originally, general wear facet analysis was developed to review occlusal concepts established for primate dentitions (Mills 1955, 1963, 1967, 1973, Hiiemäe & Kay 1972, 1973, Kay 1975, 1977, 1978, Kay & Hiiemäe 1974a, Stones 1948, Maier 1977a, b, Maier & Schneck, 1981, Douglass & DeVreugd 1997, Schulz 2003, 2008).

Recent OFA results from a cross-sectional study of primate upper molars in comparable wear stages from more than 40 species of strepsirrhines and haplorrhines show that primate molars vary highly in their phase I buccal, phase I lingual, and phase II proportional distribution of wear facet size (Fig. 2.6).

Some prosimian upper molars without a hypocone, such as molars from Lemuridae or Tarsiidae species, surprisingly possess a relatively high percentage of phase II wear facets compared to species with a hypocone cusp (Menz 2016). The buccolingual slope of the protocone in strepsirrhine species without hypocone is steep and enlarged, usually developing a strong elongated facet 9 with progressing tooth wear. This is reflected not only in the OFA wear facet data, but also in dental contact data from kinematic OFA in upper and lower molar pairs (Menz 2016). A close inspection on the OFA data of occlusal kinematics reveals that only a very short large antagonistic contact occurs at the phase II facet 9 position towards the end of phase I at maximum intercuspation (Fig. 2.7). In this case, a large facet 9 develops rather from vertical induced crushing activity along steep slopes of the cusps than from a grinding motion as normally expected during phase II through a horizontally dominated occlusal shift. A comparison with the general dietary spectrum of the species (fruits, leaves, insects), demonstrated that strepsirrhini with a high proportion of insects in their diet (e.g., *Galago senegalensis*, *Galago demidoff*, *Loris tardigradus*) possess less phase II wear facet area. This is sometimes the case even if a hypocone is present in their molars, and one would expect an increase in phase II wear. In those cases the facet number can be up to three facets higher than in those species without a hypocone, and the phase II portion is still low. The occlusal sequences and the percentage distribution of phase I and phase II facet areas show a stronger phase I component in insectivorous species than in those with a more variable diet, including a large amount of plant materials. Some insectivorous species (>70 % insects) represent a phase II wear facet proportion of less than 2 %. In those species which are characterized by less insectivorous diet (up to

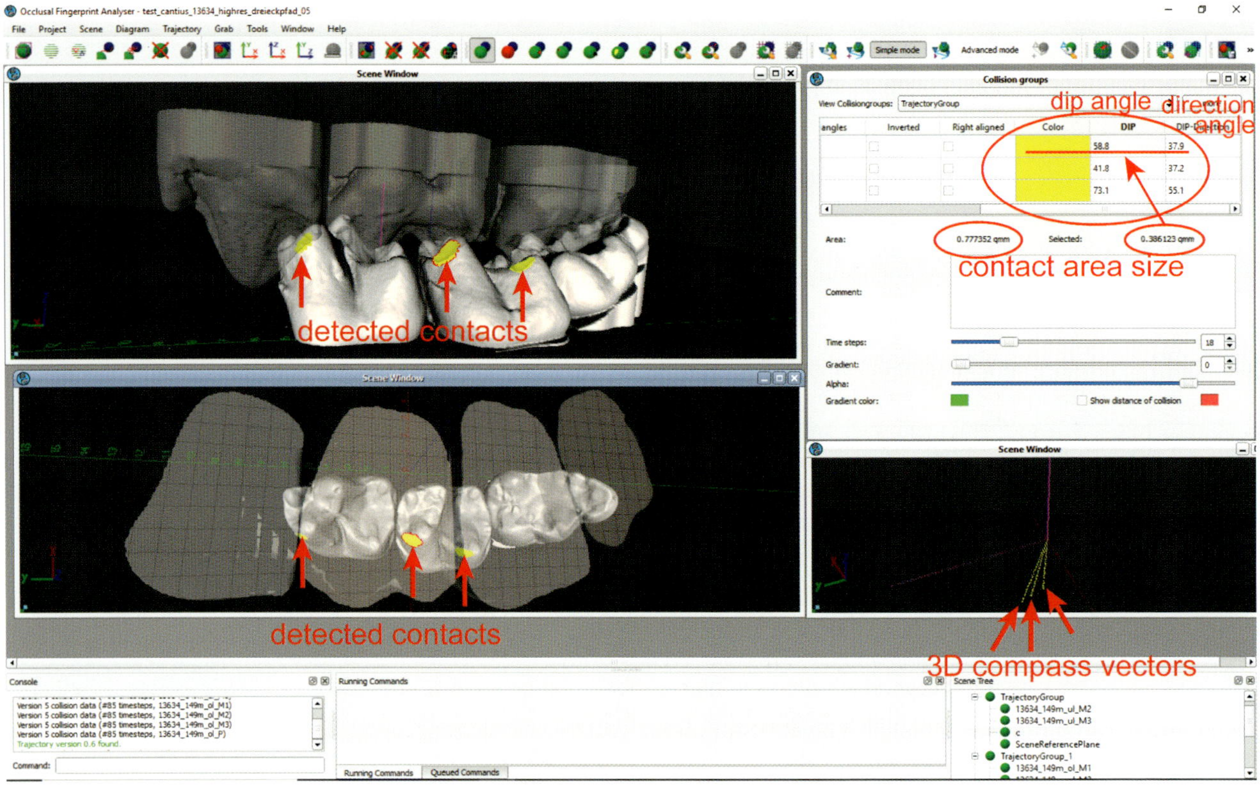

Fig. 2.5. Screenshot OFA software user-interface. Three open scene windows (dark grey), and collision group window (upper right). OFA project with lower left M1-M3 and P4-M3 single crown 3D models of an Eocene primate (*Cantius ralstoni*, Big Horn Basin, Wyoming). Three antagonistic contacts detected in time-step 18 of the occlusal sequence are visible (yellow), indicated by red arrows in two scene windows in mesiolateral and occlusal view. Scene windows at the lower right shows the normalized vectors of the three detected contact areas in the lower molar row in perspective view (3D occlusal compass of time step). In the collision group window, dip angle and dip direction are displayed for each of the three detected contact areas. Further, the contact area size is provided. All data exportable to comma separated values (CSV) or Excel spreadsheet formats.

35 % insects), the phase II proportion is slightly increased up to almost 9 %. In some species with less insectivorous diet the percentage of the phase II proportion still stays low, even the number of phase II wear facets is increased. This also indicates that if a hypocone is present in some insectivorous strepsirrhines, its phase II wear facets properly do not necessarily enlarge with respect to stronger crushing or grinding activity. This matches the idea that pointed and steep sloped cusps producing steeply inclined facets in insectivorous primates may serve as effective perforation tools to open and cut insect cuticules along small and elongated phase facets (Kay 1975, Maier 1980). It also stimulates a discussion that the invention of a hypocone in molars was maybe not solely coupled with the development of grinding.

Upper molars of *Cantius* species from different stratigraphic levels of the Bighorn basin in Wyoming demonstrate an excellent example in early Eocene euprimates for dental adaptation, and the integration of a hypocone into occlusion (Jones et al. 2014). There is good evidence for parallel environmental and dento-morphological changes during approximately three million years of *Cantius* evolution, but no consensus exists about the dietary preferences of *Cantius* spp. (Gregory 1920, Covert 1986, 1995, Jones et al. 2014). The wear facet pattern of their upper and lower molars and the comparison with those from modern prosimians feeding mostly on insects demonstrates that early *Cantius* spp. also retained a low phase II with dominant and steeply inclined buccal phase I facets (Menz 2016).

Even early *Cantius* spp. do not possess a hypocone, such as it is present in modern insectivorous Strepsirrhini; they possess a very similar percentage in the distribution of power stroke phases. In later *Cantius* spp. the hypocone is less pointed and more occlusally integrated. The proportion of phase II facets increases through the development of an additional relatively flat wear facet 10 on the hypocone. This indicates a functional shift in the hypocone which leads together with an increase in tooth size to the assumption that later *Cantius* spp. were more frugivorous. It reflects a timely shift in *Cantius* ecology contemporary to a rise of the average annual temperature in the upper stratigraphic levels of the *Cantius* fossil-bearing deposits (Chew 2009).

There is no doubt that differences in primary morphology between species are expressed in the general appearance of the occlusal fingerprint, since the articulation and indentation of crowns differ (e.g. in depth and position of the antagonistic cusps and basins).

Bilophodont molars of Cercopithecoidea with a well-developed hypocone show wear facet areas with a distinct higher percentage of phase II areas than in prosimian monkeys (Fig. 2.6). An increased phase II proportion with two additional wear facets (facet 10 and 11) (Maier 1977b),

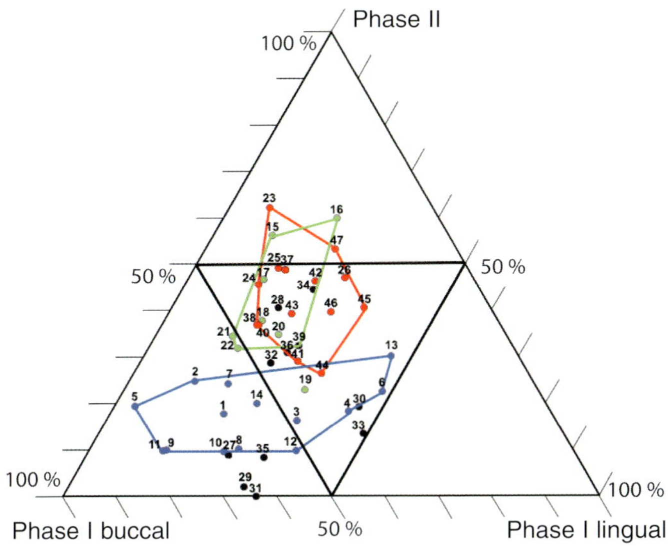

1	*Tupaja glis* M1
2	*Cheirogaleus major* M1
3	*Galago demidovii* M1
4	*Galago senegalensis* M1
5	*Hapalemur* sp. M1
6	*Lemur catta* M1
7	*Lemur mongoz* M1
8	*Lemur variegatus* M1
9	*Lepilemur mustelinus* M1
10	*Loris tardigradus* M1
11	*Microcebus murinus* M1
12	*Otolemur crassicaudatus* M1
13	*Perodicticus potto* M1
14	*Tarsius syrichta* M1
15	*Presbytis cristatus* M1
16	*Cercocebus torquatus* M1
17	*Chlorocebus aethiops* M1
18	*Colobus guereza* M1
19	*Indri indri* M1
20	*Macaca maurus* M1
21	*Nasalis larvatus* M1
22	*Papio ursinus* M1
23	*Pongo* sp. M1
24	*Gorilla gorilla* M1
25	*Hylobates lar* M1
26	*Pan troglodytes* M1
27	*Absarokius abbotti* M1
28	*Aegyptopithecus zeuxis* M1
29	*Afradapis longicristatus* M1
30	*Apidium* sp. M1
31	*Cantius* sp. M1
32	*Cantius* sp. M1
33	*Catopithecus browni* M1
34	*Dendropithecus* sp. M1
35	*Palenochtha minor* M1
36	*Paraptihecus grangeri* M1
37	*Proconsul africanus* M1
38	*Sivapithecus* sp. M1
39	*Victoriapithecus* M1
40	*Victoriapithecus* M1
41	*Proconsul* sp. M1
42	*Proconsul* sp. M2
43	*Proconsul* sp. M2
44	*Proconsul* sp. M2
45	*Proconsul* sp. M2
46	*Proconsul* sp. M2
47	*Proconsul* sp. M2

Fig. 2.6. Ternary plot of power stroke phase area proportions in primate upper molars. Modern prosimian species in blue, modern cercopithecines with bilophodont molars in green, modern great apes, proconsulids, and *Sivapithecus* in red, and other fossil primates in black.

and the larger size of Cercopithecoidea molars is not only expressed in the larger number of wear facets but also in the wear facet values (Menz 2016). Differences in primary relief height and cusp slope steepness in bilophodont molars of frugivorous cercopithecines versus folivorous colobines (Maier 1977b) are also reflected in the expression of their occlusal fingerprint (Fig. 2.4). Leaf eating colobine monkeys show steeper wear facets at similar wear facet positions than most cercopithecines (Ulhaas et al. 2004).

A very similar development was observed in modern great ape molars with crowns showing the typical Y-pattern fissure pattern of hominids (Engel 2005, Menz 2016, Knight-Sadler & Fiorenza 2017). *Gorilla* molars with the highest relief produce steeper wear facets than less steep sloped *Pongo* or *Pan* molars. Although all of them have a similar background facet pattern consisting of maximum 13 wear facets (Maier & Schneck 1981), the three genera differ in their food spectra. OFA results confirm that dietary ecology of ape and cercopithecine species is recorded in their occlusal fingerprint pattern of the molars. The wear facet inclination in teeth with similar basic morphologies can verify food categories, such as folivory versus frugivory in cercopithecines and great apes. A comparison of the occlusal fingerprint patterns of several species from both groups illustrated some common evolutionary adaptive characteristics in general principles of functional morphology of the wear facet pattern (Fig. 2.6), although basic morphology of the cusp patterns distinctly differs in bilophodont and dryopithecine molar types. *Gorilla* and colobine monkeys feeding mainly on leaves show very similar functional molar relief with steeply inclined and elongated phase I wear facets, whereas *Pongo* with its relatively low inclined and enlarged wear facets demonstrates similarities in its OFA results closer to fruit loving cercopithecines (Menz 2016).

Hominoidea molars differ from the bilophodont molars of Cercopithecoidea having a separated hypocone, and usually have a more rhomboid cusp pattern in contrast to a rectangular arrangement of cusp positions in Cercopithecoidea molars. In the upper molars of Hominoidea a new wear facet 13 is formed along the distolingual slope of the crista obliqua (Maier & Schneck 1981), a crest connecting the protocone and the metacone cusps. Functional OFA studies provided new evidence that wear facet 13 develops in the direct neighbourhood of wear facet 12, enhancing the phase II occlusal contacts (Fig. 2.2). This facet was found in only one specimen of the early Old World monkey *Aegyptopithecus* from the Oligocene, but not in early Cercopithecoidea such as *Victoriapithecus* from the Middle Miocene (Menz 2016). The OFA data demonstrate an evolutionary shift to a higher variability and an increased phase II compression and grinding capability with the presence of a wear facet 13 in molars of early Hominoidea such as Proconsulidae from the Early and Middle Miocene. In comparison to modern apes, Proconsulidae molars may represent a transitional stage towards a higher compression and grinding capability in later Hominoidea. In modern great apes, *Pongo* spp. have the lowest relief with relatively flat wear facets and the strongest expression of phase II (Fiorenza et al. 2015b), but we also can find single *Pan* and *Gorilla* individuals with a very similar phase II proportion in their occlusal fingerprint (Zanolli et al. 2019). This is also evidenced through a relatively long sequence of contacts during phase II analyzed in OFA data from kinematic analyses (Menz 2016). Upper molars with a dryopithecine cusp pattern display substantial functional differences compared to prosimian monkey teeth, but strongly overlap with monkeys having bilophodont molars (Fig. 2.6).

The comparison of the functional macrowear patterns shows that from the point of split and dichotomous contemporary evolution of the dryopithecine and bilophodont cusp pattern both functional tooth types followed a very similar pattern of functional adaptation. Probably since the end of the Oligocene through the Miocene phase II of the power stroke enhanced. Later bilophodont teeth developed a functional variability mainly in the relief height to cope dominantly with leaves, fruits, and other above-ground growing plant materials. The occlusal fingerprints of great ape molars show that further differentiation with even higher functional variability occurred, probably for dealing with new food resources, i.e. roots and tubers. External food preparation techniques, including cooking, probably has increased the functional variability in hominin molars. The differentiations in wear facet patterns mirror differences in tooth crown architecture, enamel thickness and occlusal dynamics (Schwartz 2000). Further investigations have to clarify, if the hominoid tooth type maybe provides a larger adaptive versatility for variable occlusal pathways than in the strongly interlocking bilophodont teeth of cercopithecines. Studies including OFA data from hominin molars confirm a large variability in the facet pattern (Engel 2005, Huck 2006, Ulhaas et al. 2007, Fiorenza 2009), highlighting also basic differences in the wear facet pattern between great apes and hominins in particular in the distribution of area size distribution in context of the power stroke phases (Zanolli et al. 2019). This still needs to be investigated in more detail.

The comparison of molar wear facet patterns in Late Pleistocene *Homo sapiens* and *Homo neanderthalensis* underlines the observations of a high diversity within the genus *Homo* (Fiorenza 2009, Fiorenza et al. 2011a). *H. sapiens* hunter-gatherer and early agriculturalist groups exhibit a high variability in inclination and direction of inclination in wear facets (Fiorenza et al. 2009, Fiorenza et al. 2018). OFA data on wear facet size proportions and inclination also illustrate that cultural evolution with the invention of and increasing versatility in preparation of foods reflect a higher food choice mirrored in the functional adaption of the macrowear pattern in *Homo* (Fiorenza et al. 2011c). Wear facet patterns of molars from *H. sapiens* of hunter-gatherer groups with known food spectra match with *H. neanderthalensis* samples from similar eco-geographic regions and time intervals (Fiorenza et al. 2011c). Molars from groups that lived in temperate and cooler conifer-forest environments in Central Europe are distinguishable in their occlusal fingerprint from groups that thrived in evergreen deciduous forests in the Mediterranean areas (Fiorenza et al. 2011c). Relative steep buccal phase I wear facets occur in fossil *H. sapiens*, Neanderthals, and modern hunter-gatherers, such as Inuit and Vancouver Islanders feeding generally on a diet rich in meat with a low plant contribution. In contrast, a high portion of plant foods obviously promotes the preparation of larger lingual phase I wear facets such as it was reported in Australian aborigines and African bushmen as examples of mixed-feeders. Some simple functional considerations may support these observations. More rough, fibrous and tough meat probably is easier to chew with a strong, vertical and relief guided bite producing dominant edge contacts along steep buccal phase I shearing facets. Harder and brittle plant food is better crushed and ground with powerful high pressure in vertical and lateral occlusal movement. This is expressed in flat buccal phase I wear facets and the integration of increased lingual phase I and enlarged phase II wear facet areas (Fiorenza et al. 2011c). Neanderthals from steppe and coniferous forests of Central and Eastern Europe show the lowest variability in their wear facet patterns, with large and relatively steep buccal phase I facets, very similar to meat-eating Inuit and Vancouver Islanders. Neanderthal groups from warmer deciduous forest and evergreen habitats are similar in their phase I wear facet area distribution to the mix-feeding attributed to bushmen and Australian aborigines. In both fossil groups, Neanderthals and Pleistocene *H. sapiens*, the significant differences can be assigned to differences in ecogeographic habitats and nutrition strategies.

Previous OFA results of hominoid molars emphasize that ecological factors (for example climate and vegetation) which are to be interpreted in connection with the existing food resources are manifested in the expression of specific occlusal tooth contact patterns. These can be understood as variants of the general wear facet pattern as described by Maier & Schneck (1981). Considering the evolutionary development of the general pattern of tooth contact surfaces in the bunodont hominid molars as a function morphological adaptation (Maier & Schneck 1981), the specific expression of the contact facets can be regarded as a functional reaction of the tooth construction. Teaford & Ungar (2000) and Ungar et al. (2006) summarized the ecological adaptation of hominid tooth construction in tooth size, crown shape, enamel structure, enamel thickness, and general biomechanics as an example of a flexible adaptation to the environment and cultural development. The extent to which the ontogenetic variability and the different variants of contact facet pattern reflect dietary flexibility of a primary construction can only be understood, if the function of wear facets is apprehended in the context of biomechanics of the entire mastication system. The comparison of the absolute and relative wear facet areas with the cusp area and cusp height of the most discriminating cusp (hypocone) in *H. sapiens* and Neanderthal molars shows that the primary morphology of the cusps has little effect on wear facet size (Fiorenza et al. 2011a), and the appearance of large wear facets and their leveling depends essentially on environmental factors (habitat, climate, food resources), finally triggering chewing and cultural behavior. However, it is confirmed that the development of the general wear facet pattern of hominoid molars is closely related to the primary morphology of tooth relief and occlusal jaw movement (Kullmer et al. 2012). The experimental simulation of the occlusal movement confirms a direct correspondence of the movement space described in the dental occlusal compass with the emergence of the wear facet patterns (Kullmer et al. 2012). Deviations in the space of movement, e.g. because of chewing different diets, leads to changes in the shape and absolute position of the wear facets (Fiorenza et al. 2011a). Variations in the inclination direction of the facets are therefore attributable to differences in the movement pattern (Kullmer et al. 2012). The presence of a complete hominoid wear facet pattern (facets 1–13) on a molar depends not only on the presence of all cusps and the occlusal relationship, but also on the wear stage of the tooth, and on the molar position (Fiorenza et al. 2010, 2011a). Only when the tooth is erupted and in occlusion, the contact facet pattern develops. In the case of the first and second permanent molars (M1 and M2), a complete pattern with 13 facet positions is established

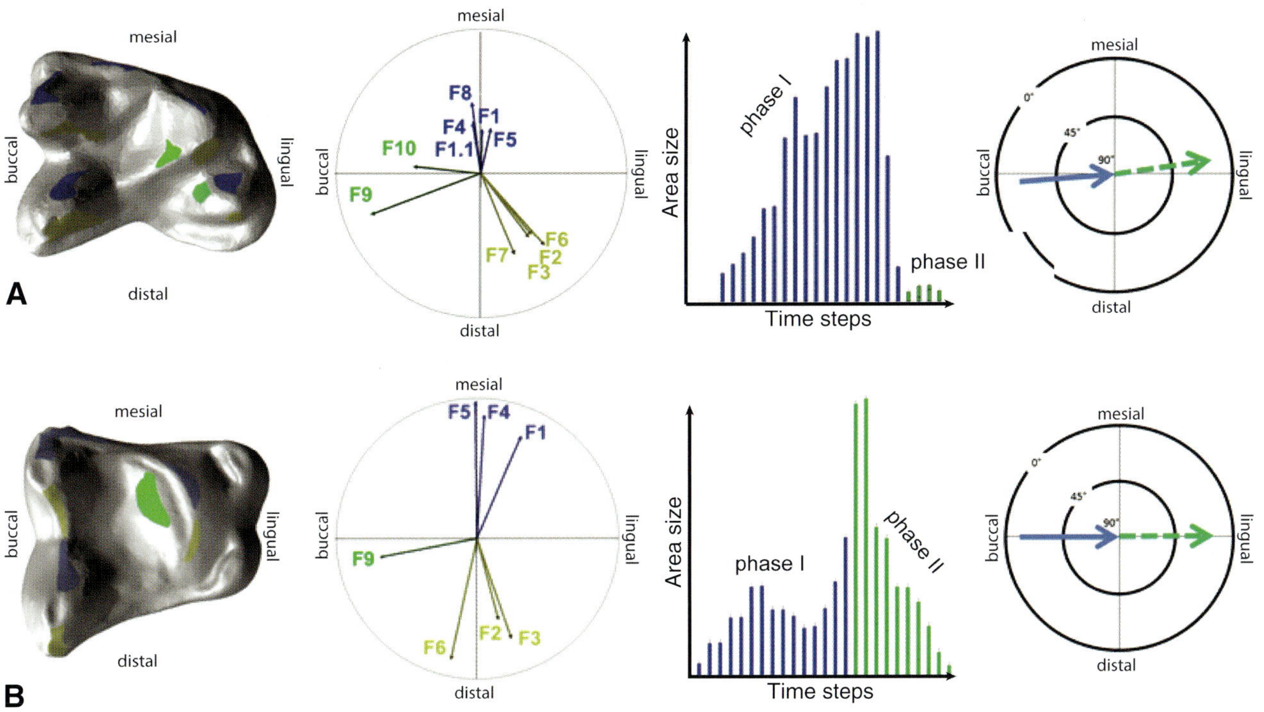

Fig. 2.7. Comparative OFA visualization of an insectivore (*Loris tardigradus*) and a fruit and leaf eating (*Lemur mongoz*) strepsirhine primate right M1. From left, 3D model with wear facet map in occlusal view, 3D occlusal compass, sequential power stroke contact area size kinematics and mastication compass with phase I (blue) and phase II (green). Not to scale.

during normal occlusal relationships (Ulhaas et al. 2007, Kullmer et al 2009, 2012).

In OFA of upper third molars (M3) it was found (Fiorenza et al. 2010) that facets 7 and 10 are absent in the fully formed pattern in *H. sapiens* and Neanderthals, since the upper M3 distally contacts with the lower m3 only. Therefore, there are no antagonistic crown surfaces present in the mandibular dentition for positions 7 and 10 at the distal slope of the hypocone (Fiorenza et al. 2010). The OFA results and wear facet map may thus be helpful in the determination of isolated third molars. This was the case with the Neanderthal tooth D58 from Krapina in Croatia (Fiorenza et al. 2010), where the OFA results could identify the crown as a third molar underlined by its reduced hypocone and a missing distal interproximal contact facet. Distally, the molar has distinct facets 7 and 10 whose inclination direction and angle of inclination indicate that, in addition to its direct antagonist in the lower jaw, there were also two contacts to a more distally positioned opposing tooth, and therefore it should be rather identified as an upper M2 than an M3 (Fiorenza et al. 2010). Taking into account the principles of intercuspation of the molars, it is possible to determine the exact positional relationship of the antagonists via the occlusal fingerprint pattern (Kullmer et al. 2013).

If the antagonists are available, the determination of the angles of the facets (inclination direction and inclination angle) in the 3D occlusal compass for both teeth can be established and the correspondence of the complementary patterns can be quantitatively demonstrated (Kullmer et al. 2012). Thus, the tooth position of the premolars as well as M1 and M2 can be verified. The assessment of facet correspondence not only gives an objective evaluation of the association of isolated teeth (Fiorenza et al. 2010, Kullmer et al. 2012), but it can also be used for evidence of paramasticatory activity, such as the use of the teeth as a third hand for holding and manipulating objects (Fiorenza et al. 2011b). In addition, the identification of facet patterns and OFA data is important for the most realistic functional repositioning of dental crowns in the dental arch (Kullmer et al. 2013, Benazzi et al. 2013a). This is of particular relevance for the reconstruction of the occlusal mechanics and the comparison of the chewing motion of rows of teeth and complete dental arches. Despite the numerous studies on the occlusal relationship of tooth contacts in primates (Mills 1955, 1963, Hiiemäe & Kay 1973, Kay 1975, 1977, 1978, 1981, Kay & Hiiemäe 1974, Stones 1948, Maier 1977a,b, Maier & Schneck 1981), the quantitative OFA data and the occlusal 3D compass (Kullmer et al. 2009) offer for the first time the possibility to directly compare and visualize the multiple facet patterns of teeth in a standardized diagram (Fiorenza et al. 2010, 2011a,b, Kullmer et al. 2012).

A comparative diagram for occlusal trajectories has also been missing. Inspired by the 3D occlusal compass, Koenigswald et al. (2013) proposed a new terminology and visualization of the occlusal pathways in mammalian molars based on the spatial orientation of wear facets and of "microwear" striations on the tooth surfaces (Fig. 2.7). The mastication compass combines the number of movements during the occlusal phases, the direction and the inclination of each individual phase into a three-dimensional overall movement pattern of the "power stroke" in a single diagram (Koenigswald et al. 2013). An overview comparison of mastication compasses from different mammalian molars in a wear stage with full wear facet development shows the enormous variability in power stroke movements as a function of the different crown shapes. In most bilophodont and secodont molars, the incursive phase I is very dominant. Examples of such dentitions are found in arvicolids, elephantids, and modern Equidae. In lophodont, selenodont, and secodont teeth with a high relief, the movement is clearly guided in comparison to relatively flat bunodont teeth. It remains unclear whether mammalian mastication is actually controlled by the composition and quality of food or by phylogenetic preconditioning (Koenigswald et al. 2013). Initial observations in different mammals suggest that high quality (high energy) food correlates with a prominent phase I during the power stroke seen in many insectivorous and carnivorous mammals, and the combination of phase I and phase II is apparently appropriate, if a larger amount of slightly less energetic food is on the menu, such as in Proboscidea, Ruminantia, Equidae and some Rodentia (Koenigswald et al. 2013). The connection of two phases can lead to synchronization in the direction of movements when comminuting very low energy food. In the biphasic bunodont teeth of hominoids, a large variability in occlusal motions is demonstrated by a large variability in the OFA patterns (e.g., Fiorenza et al. 2011c). Here it turns out that particularly bunodont dentitions of hominoids can react within their construction and show the possibility of adaptations of motion in their wear facet patterns to variable foods.

In order to be able to show the correlation between a flexible chewing behavior and the adaptation in the primate masticatory apparatus, it is important to reconstruct the occlusal trajectories in complete rows of teeth. A precise 3D mapping of the wear facet positions and the measurement of the spatial facet directions (OFA), in the sense of the dental 3D compass, provides the methodological prerequisite for quantitatively detecting the correspondence in the occlusal movement pattern and tooth wear in primate teeth and complete tooth rows. Beyond doubt, the positions of the antagonistic wear facets of hominoid molars show occlusal contacts during the occlusal power stroke and in maximum intercuspation in a natural dentition on both sides of the dental arch. In dentistry, one speaks of a balanced occlusion (Douglass & DeVreugd 1997, Schulz 2008). In many anisognatic mammalian tooth systems (e.g., Bovidae, Cervidae, Equidae, many Rodentia), there is no or only a partially balanced occlusion, as we can see it in most prosimians with an unfused symphysis.

This does not necessarily mean that bilateral chewing predominates in hominoid dentitions, but it is rather possible due to the fact that it does not produce contact interferences during chewing on the respective working side. This is observable in a perfect functional interplay of the phase I wear facets on one side and the phase II wear facets on the other side of the jaw. The documentation of the occlusal fingerprint in complete crown rows of hominoid individuals (Kullmer et al. 2013) and physical testing with dental models in natural rows of teeth, show that the spatial position of phase I wear facets on one side coincides with the alignment of the wear facets of phase II on the opposite side in the occlusal motion. This can be explained also by the translation and rotation in the two condyles of the joints during the "power stroke" (Kullmer et al. 2012, 2013). However, it is still debated at what exact time of the occlusion the phase II facets actually arise (Wall et al. 2006). The two-sided sliding over the tooth surfaces during the occlusion supports as precisely as possible a functional reconstruction of the occlusal movement of complete dental arches of hominoid individuals. Thus, through the functional understanding of individual wear facet relations, derived from OFA, the complete dentition of a specimen of the Miocene ape *Rudapithecus hungaricus* from Rudabánya could be restored. The occlusion has been functionally tested in a physical model in a dental articulator, and in parallel with virtual 3D models in the OFA software (Kullmer et al. 2013). The occlusal fingerprint of the individual shows a good fit of the occlusal contacts and leaves almost no room for variation in the positioning of the molars. The static and kinematic occlusion was first qualitatively assessed with the physical models in a dental articulator, a device commonly used in dental laboratories. A quantitative evaluation of the reconstruction was made by virtual measurement of the tooth contacts after repositioning. The comparison of the wear facet pattern with the contacts in the static occlusion was carried out by a virtual surface distance measurement of the maxillary and mandibular arch in the maximum intercuspation. The pattern of the minimal distances (<0.1 mm) of both models in comparison with the facet pattern reflects the precision of the interlocking in the dentition of the young adult *Rudapithecus* individual (Kullmer et al. 2013). An examination of the kinematic occlusion was carried out with the OFA software by means of relief-guided collision detection. Thus, it could be shown with the virtual motion simulation, and the detection of the tooth contacts, that during the sequential occlusion of the reconstructed mandibular arch there is no interference in the occlusal jaw movement. Such a functional dental arch reconstruction was first performed with *Rudapithecus*, and it provided a functional basis with a well articulating dentition for a full skull reconstruction of the fossil specimen (Gunz et al. 2020), such as it was put into practice with the *Australopithecus africanus* specimen Sts 52 (Benazzi et al. 2013a). A review of the trajectory of the occlusal sequence from the virtual OFA simulation in *Rudapithecus* indicates that the bite of the individual at this stage of tooth wear was characterized by a prominent incursion from buccally with a strong vertical movement. After minimal lingual translation of the lower jaw during phase II the crowns separate rapidly, quite similar as it can be observed in young gorilla individuals, feeding on young juicy soft plant shoots and fruits (Kullmer et al. 2013).

Computer simulation and virtual motion analysis in combination with OFA data is still in its infancy, but it can already be predicted that virtual 3D crown morphometry in combination with virtual motion concepts fundamentally expands the functional analysis of dentitions. The OFA analysis with its quantitative possibilities of wear facet patterns with the reconstruction of the occlusal trajectories allows the study of entire occlusal kinematic patterns of the mastication system in its complex mechanical context of food comminution.

In the latest version of the OFA software (OFA 2.0) scholars can consider the temporomandibular joint geometry and also the movement of single teeth in their alveolar socket for a more realistic simulation of real occlusal. Future software versions should incorporate simulations of food particles over the occlusal surfaces, to quantitatively study crushing and shearing performance.

In the meantime, results from kinematics analysis of the occlusion are incorporated in simulations of the occlusion dynamics, including dental loading scenarios. The aim is to test effects of occlusal motion patterns and tooth wear pattern on force reaction patterns and adaptations in the 3D structures of the masticatory system. First results from combination of OFA and finite element analyses (FEA) were drawn from *H. sapiens* molar pairs (Benazzi et al. 2011f, 2012, 2016). Based on individual occlusal movement contacts on lower first molars of two individuals in two different stages of wear, the distribution of the loads in the tooth and jaw bone at different moments of occlusion was compared for the first time. From microcomputed tomography data of teeth and jawbones, different tissues in the tooth crown structure were segmented for a numerical computation of loading scenarios. Finite element models for enamel, dentin, cementum, pulp, periodontium, and compact and spongy jawbone were calculated and brought together in an overall model for the application of occlusal force (Benazzi et al. 2011f). After the physical properties (data from literature) were assigned to each tooth material, the strain vectors for the tooth contact areas, from data of time steps of the motion simulation with the OFA software, were calculated as resulting vectors. The occlusal load was calculated during three occlusal contact moments: (1) initial occlusion (phase I), (2) maximal intercuspation, and (3) in the excursive phase (phase II). The pattern of stress distribution in the dental bone system showed significant differences in tensile stress distribution as a function of the occlusal position, especially in the cervical region of the crown, at the enamel-dentin junction, in the fissure pattern, and in the roots (Benazzi et al. 2011f, 2012).

This was the first FEA applying an individual loading simulation derived from OFA kinematic results in a complex FEA tooth model. Principal stresses were observed along the fissures of the tooth surface and in the crown walls (Benazzi et al. 2011f). Contrary to previous studies in unworn *H. sapiens* molars, the highest tension in worn tooth crowns with wear facets occurred cervically at the mesial and distal walls rather than buccally or lingually as previously assumed (Benazzi et al. 2011f). This suggests that facet patterns can contribute to a more favourable distribution of occlusal forces due to their enlargement and flattening of the contact areas, whereby pits and fissures are critical localities, since a concentration of tensile stress takes place there, as recently underlined through a study with premolars (Benazzi et al. 2013b). This also suggests that the fissures with their profiles and varying enamel thickness contribute to the general stiffening of the crown. Similarly, it could also be related to the functionality of the mid-trigonid crest, an edge structure connecting metaconid and protoconid (Benazzi et al. 2013b). This structure is also found in most Neanderthal molars while it lacks, or is only very weakly developed, in *H. sapiens*.

Conspicuous tensile stress occurred in the bifurcation of the roots at maximum intercuspation. This is maybe an indication for adaptive occurrences of a strong expression of taurodontism with a shift of the bifurcation in Neanderthals, who were interpreted to have produced higher loads in their pointed face (Trinkaus 1987, Benazzi et al. 2015). Probably, an apical displacement of the root separation and also root fusion can contribute to a reduction of tensile stress in the roots at maximum intercuspation (Benazzi et al. 2013b). An increase in facet size and a flatter inclination in the advanced worn tooth crowns results in a more tooth-axis-oriented distribution of forces so that the buccal and mesial cervical margins are less conspicuously stressed. This load image provides e.g., a relatively simple explanation for the frequent occurrence of cervical enamel chipping at the buccal enamel margins of human teeth in modern industrialized societies. We have critically reduced natural tooth wear as a result of eco- and socio-esthetical behaviours. The abridged masticatory activity due to externally processed and softened food, as well as a prolongation of the individual longevity may contribute to common fatigue in enamel and dentin observed in modern *H. sapiens*, resulting in an overload reaction in the dental crowns in the cervix area towards tooth and root fractures (Benazzi et al. 2013b). Stress distribution patterns from premolars with crowns in various wear stages provide a simple biomechanical explanation for the frequent occurrence of cervical enamel chipping (non-carious cervical lesions) at the buccal enamel margins in humans from industrialized societies.

The results from previous studies of virtual OFA describe an innovative approach to capture relevant functional structures for the reconstruction and visualization of occlusion and structural adaptation in teeth. The acquisition and data processing of three-dimensional structural data still require a relatively large amount of time, making it difficult to evaluate larger samples. It will be a major endeavour for the future to shorten time for OFA data acquisition by an intensive work on the automation of OFA processes. A larger throughput of sample quantities will significantly enhance OFA applications. Nevertheless, OFA methods are useful to simulate and compare evolutionary developments in tooth structures and to evaluate their adaptive effect. A future challenge in functional analysis of dental structures will be to experimentally validate results from virtual biomechanical simulations such as FEA.

OFA offers a range of innovative methods that contribute to a comprehensive paleobiological understanding of functional relationship between tooth structures and occlusal biomechanics, and evolutionary dental adaptations and behaviour. OFA studies on hominid molars including *H. sapiens* are closely related to societal applications in modern dentistry (e.g., prosthodontics, orthodontics, implantology) but also nutritional sciences, and they are maybe also supportive for other disciplines in human medicine. From an evolutionary perspective, OFA can provide important quantitative data to understand the increase in modern occlusal dysfunctions, temporomandibular joint problems, and possibly even digestive disorders, occurring frequently in modern industrialized societies. Some of them may be closely related to dietary behaviour and the consequent reduction of use in our masticatory system in its original biological sense.

Acknowledgments

We want to thank Christine Hemm for 3D surface scanning of specimens and Stefan Bacso, Stefano Benazzi, Dieter Schulz, Olaf Winzen, and all members of the DFG FOR 771 for discussions and ideas concerning the OFA protocol and software concept. We express our gratitude also to Stephan Lorsbach and Heiko Zimmermann from ZiLoX-IT for programming the OFA software and excellent cooperation.

We are very thankful to Peter Lucas and Alistair Evans for reviewing a previous version of the manuscript, and appreciate their helpful comments. We are very grateful to Dick Byer for English revision.

This research was supported by the Senckenberg Research Institute and Natural History Museum Frankfurt and the EU Marie Curie Training Network MRTN-CT-2005-019564 EVAN. The German Research Foundation (DFG) financed OFA research on primate teeth and the OFA software programming in the frame of the DFG Research Unit 771.

References

* indicates publications that originated from the DFG Research Unit 771

Agrawal, K. R., Lucas, P. W., Bruce, I. C., & Prinz, J. F. (1998): Food properties that influence neuromuscular activity during human mastication. Journal of Dental Research 77: 1931–1938.

*Anders, U., Koenigswald, W. v., Ruf, I. & Smith, B. H. (2011): Generalized individual dental age stages for fossil and extant placental mammals. Paläontologische Zeitschrift 85: 321–339.

Baker, G., Jones, L. H. P. & Wardrop, I. D. (1959): Cause of wear in sheeps' teeth. Nature 184: 1583–1584.

*Benazzi, S., Douka, K., Fornai, C., Bauer, C.C., Kullmer, O., Svoboda, J., Pap, I., Mallegni, F., Bayle, P., Coquerelle, M., Condemi, S., Ronchitelli, A., Harvati, K. & Weber, G. W. (2011a). Early dispersal of modern humans in Europe and implications for Neanderthal behavior. Nature 479: 525–528.

*Benazzi, S., Fiorenza, L., Katina, S., Bruner, E. & Kullmer, O. (2011d): Quantitative assessment of interproximal wear facet outlines for the association of isolated molars. American Journal of Physical Anthropology 144: 309–316.

*Benazzi, S., Fiorenza, L., Kozakowski, S. & Kullmer, O. (2011e): Comparing 3D Virtual Methods for Hemimandibular Body Reconstruction. The Anatomical Record 294: 1116–1125.

*Benazzi, S., Fornai, C., Bayle, P., Coquerelle, M., Kullmer, O., Mallegni, F. & Weber, G. (2011c): Comparison of dental measurement systems for taxonomic assignment of Neanderthal and modern human lower second deciduous molars. Journal of Human Evolution 61: 320–326.

*Benazzi, S., Grosse, I. R., Gruppioni, G., Weber, G. W. & Kullmer, O. (2014): Comparison of occlusal loading conditions in a lower second premolar using three-dimensional finite element analysis. Clinical Oral Investigations 18: 369–375.

*Benazzi, S., Kullmer, O., Grosse, I. R. & Weber, G. W. (2011f): Using occlusal wear information and finite element analysis to investigate stress distributions in human molars. Journal of Anatomy 219: 259–272.

*Benazzi, S., Kullmer, O., Grosse, I. R. & Weber, G. W. (2012): Brief Communication: Comparing Loading Scenarios in Lower First Molar Supporting Bone Structure Using 3D Finite Element Analysis. American Journal of Physical Anthropology 147: 128–134.

*Benazzi, S., Kullmer, O., Schulz, D. & Weber, G. W. (2013a): Individual tooth macrowear pattern guides the reconstruction of Sts 52 (*A. africanus*) dental arches. American Journal of Physical Anthropology 150: 324–329.

*Benazzi, S., Nguyen, H. N., Kullmer, O. & Hublin, J.-J. (2013b): Unravelling the functional biomechanics of dental features and tooth wear. PLoS One 8 (7): e69990. https://doi.10.1371/journal.pone.0069990

*Benazzi, S., Nguyen, H. N., Kullmer, O. & Hublin, J.-J. (2015): Exploring the biomechanics of taurodontism. Journal of Anatomy 226:180–188.

*Benazzi, S., Nguyen H. N., Kullmer, O. & Kupczik, K. (2016): Dynamic modelling of tooth deformation using occlusal kinematics and finite element analysis. PLoS One 11 (3): e0152663. https://doi.10.1371/journal.pone.0152663

*Benazzi, S., Nguyen, N. H., Schulz, D., Grosse, I. R. Gruppioni, G., Hublin, J.-J. & Kullmer, O. (2013b): The evolutionary paradox of tooth wear: simply destruction or inevitable adaptation? PLoS One 8(4): e62263. https://doi.10.1371/journal.pone.0062263.

*Benazzi, S., Viola, B., Kullmer, O., Fiorenza, L., Harvati, K., Paul, T., Gruppioni, G., Weber, G. W. & Mallegni, F. (2011b): A reassessment of the Neanderthal teeth from Taddeo cave (southern Italy). Journal of Human Evolution 61: 377–387.

Bisler, A., Bockholt, U., Kordass, B., Suchan, M. & Voss, G. (2002): The virtual articulator. International Journal of Computerized Dentistry 5: 101–106.

Bookstein, F. L. (1991): Morphometric Tools for Landmark Data. Geometry and Biology. Cambridge University Press, New York.

Brown, W. A. B. & Chapman, N. G. (1990): The dentition of fallow deer (*Dama dama*): a scoring scheme to assess age from wear of the permanent molariform teeth. Journal of Zoology 221: 659–682.

Butler, P. M. (1952): The milk-molars of Perissodactyla, with remarks on molar occlusion. Proceedings of the Zoological Society of London 121: 777–817.

Butler, P. M. (1956): The ontogeny of molar pattern. Biology Reviews 31: 30–70.

Butler, P. M. (1967): Dental merism and tooth development. Journal of Dental Research 46: 845–850.

Butler, P. M. (1973): Molar wear facets of early tertiary North American primates. In: Zingeser, M. R. (ed.). Craniofacial Biology of Primates. Symposia of the Fouth International Congress of Primatology, Portland, Oregon, August 15–18, 1972, vol. 3. Karger, Basel: 1–27.

Calandra, I. & Merceron, G. (2016): Dental microwear texture analysis in mammalian ecology. Mammal Review 46: 215–228.

Chew, A. E. (2009). Paleoecology of the early Eocene Willwood mammal fauna from the central Bighorn Basin, Wyoming. Paleobiology 35: 13–31.

Covert, H. H. (1986): Biology of the early Cenozoic primates. In: Swindler, D. W. & Erwin, J. (eds.). Comparative Primate Biology, Vol. 1: Systematics, Evolution, and Anatomy. Liss, New York: 335–359.

Covert, H. H. (1995): Locomotor adaptations of Eocene primates: adaptive diversity among the earliest prosimians. In: Alterman, L., Doyle, G. A., Izard, M. K. (eds.). Creatures of the Dark: The Nocturnal Prosimians. Plenum, New York: 495–509.

Crompton, A. W. (1971): The origin of the tribosphenic molar. In: Kermack, D. M. & Kermack, K. A. (eds.). Early Mammals. Zoological Journal of the Linnean Society 50, supplement 1: 65–87.

Crompton, A. W. & Hiiemäe, K. M. (1970): Functional occlusion and mandibular movements during occlusion in the American opossum, *Didelphis marsupialis* L. Zoological Journal of the Linnean Society 49: 21–47.

Douglass, G. D. & DeVreugd, R. T. (1997): The dynamics of occlusal relationships. In: McNeill, C. (ed.). Science and Practice of Occlusion, Quintessence Publishing, Chicago: 69–78.

Engel, K. (2005): Vergleichende 3-D-Morphometrie, Konstruktion und Funktion der Molaren von *Gorilla*, *Pan*, *Pongo* und *Hylobates* (Primates: Hominoidea). Doctoral dissertation, Johann Wolfgang Goethe-Universität Frankfurt a.M.

*Engels, S. & Schultz, J. A. (2018): Evolution of the power stroke in early Equoidea (Perissodactyla, Mammalia). Palaeobiodiversity and Palaeoenvironments: 1-21.

Evans, A. R. & Fortelius, M. (2008). Three-dimensional reconstruction of tooth relationships during carnivoran chewing. Palaeontologia Electronica 11: 1-11.

*Fiorenza, L. (2009): Occlusal wear pattern analysis of functional morphology in Neanderthals and early *Homo sapiens* dentition. Doctoral dissertation, Johann Wolfgang Goethe-Universität, Frankfurt a.M.

*Fiorenza, L. & Kullmer, O. (2013): Dental wear and cultural behavior in Middle Paleolithic humans from the Near East. American Journal of Physical Anthropology 152: 107-117.

*Fiorenza, L. & Kullmer, O. (2015): Dental wear patterns in early modern humans from Skhul and Qafzeh: A response to Sarig and Tillier. Homo – Journal of Comparative Biology 66: 414-419.

*Fiorenza L. & Kullmer, O. (2016): Occlusion in an adult male gorilla with a supernumerary maxillary premolar. International Journal of Primatology 37: 762-777.

Fiorenza, L., Benazzi, S., Estalrrich, A. & Kullmer, O. (2019b): Diet and cultural diversity in Neanderthals and modern humans from dental macrowear analyses. In: Schmidt, C. & Watson, J. T. (eds.). Dental Wear in Evolutionary and Biocultural Contexts. Academic Press, London: 39-72.

*Fiorenza, L., Benazzi, S., Henry, A. G., Salazar-García, D. C., Blasco, R., Picin, A., Wroe, S. & Kullmer, O. (2015a): To meat or not to meat? New perspectives on Neanderthal ecology. American Journal of Physical Anthropology156: 43-71.

*Fiorenza, L., Benazzi, S. & Kullmer, O. (2009): Morphology, wear and 3D digital surface models: materials and techniques to create high-resolution replicas of teeth. Journal of Anthropological Sciences 87: 211-218.

*Fiorenza, L., Benazzi, S. & Kullmer, O. (2011c): Para-masticatory wear facets and their functional significance in hunter-gatherer maxillary molars. Journal of Archaeological Sciences 38: 2182-2189.

*Fiorenza, L., Benazzi, S., Kullmer, O., Mazurier, A. & Macchiarelli, R. (2019a): Dental macrowear analysis of the Neanderthal mandible from Regourdou (Dordogne, Southwestern France). Journal of Human Evolution 132: 174-188.

*Fiorenza, L., Benazzi, S., Oxilia, G. & Kullmer, O. (2018): Functional relationship between dental macrowear and diet in Late Pleistocene and Recent modern human populations. International Journal of Osteoarchaeology 28: 153-161.

*Fiorenza, L., Benazzi, S., Tausch, J., Kullmer, O., Bromage, T. G. & Schrenk, F. (2011a): Molar macrowear reveals Neanderthal eco-geographic dietary variation. PLoS One 6: e14769.

*Fiorenza, L., Benazzi, S., Tausch, J., Kullmer, O. & Schrenk, F. (2010): Identification reassessment of the isolated tooth Krapina D58 through occlusal fingerprint analysis. American Journal of Physical Anthropology 14: 306-312.

*Fiorenza, L., Benazzi, S., Viola, B., Kullmer, O. & Schrenk, F. (2011b): Relationship between cusp size and occlusal wear pattern in Neanderthal and *Homo sapiens* first maxillary molars. The Anatomical Record 294: 453-461.

Fiorenza, L., Nguyen, H. N. & Benazzi, S. (2015b): Stress distribution and molar macrowear in *Pongo pygmaeus*: a new approach through finite element and occlusal fingerprint analyses. Human Evolution 30: 215-226.

*Fiorenza, L., Menter, C. G., Fung, S., Lee, J., Kaidonis, J., Moggi-Cecchi, J., Townsend, G. & Kullmer, O. (2020): The functional role of the Carabelli trait in early and late hominins. Journal of Human Evolution 145: 102816.

Fortelius, M. (1985): Ungulate cheek teeth: developmental, functional, and evolutionary interrelations. Acta Zoologica Fennica 180: 1-76.

Fung, S., Lee, J., Yong, R., Ranjitkar, S., Kaidonis, J., Pilbrow, V., Panagiotopoulou, & O. Fiorenza, L. (2020): Brief communications: a functional analysis of the Carabelli trait in Australian aboriginal dentition. American Journal of Physical Anthropology: 24120.

Gingerich, P. D. (1974): Dental function in the Palaeocene primate *Plesiadapis*. In: Martin, R. D., Doyle, G. A. & Walker, A. C. (eds.). Prosimian Biology. Duckworth, London: 531-541.

Gordon, K. D. (1984): The assessment of jaw movement direction from dental microwear. American Journal of Physical Anthropology 63: 77-84.

Gorniak, G. C., & Gans, C. (1980): Quantitative assay of electromyograms during mastication in domestic cats (*Felis catus*). Journal of Morphology 163: 253-281.

Grant, A. (1982): The use of tooth wear as a guide to the age of domestic animals. In: Wilson, B., Grigson, C., Payne, S. (eds.). Ageing and Sexing Animal Bones from Archaeological Sites. British Archaeological Reports Series 109, Oxford: 91-108.

Gregory, W. K. (1920): On the Structure and Relations of *Notharctus*, an American Eocene Primate. Memoirs of the American Museum of Natural History, New Series, Vol. III, Part II.

Grine, F. E. (1981): Trophic differences between 'gracile' and 'robust' australopithecines: A scanning electron microscope analysis of occlusal events. South African Journal of Science 77: 203-230.

*Gunz, P., Kozakowski, S., Neubauer, S., Le Cabec, A., Kullmer, O., Benazzi, S., Hublin, J.-J. & Begun, D. R. (2020): Skull reconstruction of the late Miocene ape *Rudapithecus hungaricus* from Rudabánya, Hungary. Journal of Human Evolution 138: 102687.

Herring, S. W. (1976): The dynamics of mastication in pigs. Archives of Oral Biology 21: 473-480.

Hiiemäe, K. M. (1978): Mammalian mastication: a review of the activity of the jaw muscles and movements they produce in chewing. In: Butler, P. M. & Joysey, K. A. (eds.). Development, Function, and Evolution of Teeth. Academic Press, London: 359-398.

Hiiemäe, K. M. & Crompton, A. W. (1971): A cinefluorographic study of feeding in the American opossum, *Didelphis marsupialis*. In: Dahlberg, A. A. (ed.). Dental Morphology and Evolution. University of Chicago Press, Chicago: 299-334.

Hiiemäe, K. M. & Kay, R. F. (1972): Trends in the evolution of primate mastication. Nature 240: 486-487.

Hiiemäe, K. M. & Kay, R. F. (1973). Evolutionary trends in the dynamics of primate mastication. Craniofacial Biology of Primates 3: 28-64.

Hillson, S. (1996): Dental Anthropology. Cambridge University Press, Cambridge.

Huck, M. (2006): Kaumechanik und Funktion der Molaren-Okklusion früher Hominiden. Doctoral dissertation, Johann Wolfgang Goethe-Universität Frankfurt a.M.

d'Incau, E., Couture, C. & Maureille, B. (2012): Human tooth wear in the past and the present: tribological mechanisms, scoring systems, dental and skeletal compensations. Archives of Oral Biology 57: 214-229.

Janis, C. M. (1979): Mastication in the hyrax and its relevance to ungulate dental evolution. Paleobiology 5: 50-59.

Janis, C. M. (1984): Predictions of primate diets from molar wear pattern. In: Chivers, D. J., Wood, B. A. & Bilsborough, A. (eds.). Food Acquisition and Processing in Primates. Plenum Publishing Corporation, New York: 331-340.

Janis, C. M. (1990): The correlation between diet and dental wear in herbivorous mammals, and its relationship to the determination of diets of extinct species. In: Boucot, A. J. (ed.). Evolutionary Paleobiology of Behaviour and Coevolution. Elsevier, Amsterdam: 241-259.

Jones, K. E., Rose, K. D. & Perry, J. M. (2014): Body size and premolar evolution in the early-middle Eocene euprimates of Wyoming. American Journal of Physical Anthropology 153: 15-28.

Kaifu, Y., Kasai, K., Townsend, G. C. & Richards, L. C. (2003): Tooth wear and the "design" of the human dentition: a perspective from evolutionary medicine. American Journal of Physical Anthropology 122: 47-61.

Kay, R. F. (1973): Mastication, molar tooth structure and diet in primates. Ph.D. dissertation, Yale University.

Kay, R. F. (1975): The functional adaptations of primate molar teeth. American Journal of Physical Anthropology 43: 195-216.

Kay, R. F. (1977). The evolution of molar occlusion in the Cercopithecoidea and early catarrhines. American Journal of Physical Anthropology 46: 327-352.

Kay, R. F. (1978): Molar structure and diet in extant Cercopithecidae. In: Butler, P. M & Joysey K. A (eds.). Development, Function and Evolution of Teeth. Academic Press, London: 309-339.

Kay, R. F. (1981): The nut-crackers – a new theory of the adaptations of the Ramapithecinae. American Journal of Physical Anthropology 55: 141–151.

Kay, R. F. (1987): Analysis of primate dental microwear using image processing techniques. Scanning Microscopy 1: 657–662.

Kay, R. F. & Hiiemäe, K. M. (1974a). Jaw movement and tooth use in Recent and fossil primates. American Journal of Physical Anthropology 40: 227–256.

Kay, R. F. & Hiiemäe, K. M. (1974b): Mastication in *Galago crassicaudatus*, a cinefluorographic occlusal study. In: Martin, R. D., Doyle, G. A. & Walker, A. (eds.). Prosimian Biology. Duckworth, London: 501–503.

Knight-Sadler, J. & Fiorenza, L. (2017): Tooth wear inclination in great ape molars. Folia Primatologica 88: 223–236.

*Koenigswald, W. v., Anders, U., Engels, S., Schultz, J. A. & Kullmer, O. (2013): Jaw movement in fossil mammals: analysis, description and visualization. Paläontologische Zeitschrift 87: 141–159.

Kordaß, B. & Velden, P. (1996): Der individuelle okklusale Kompaß. Computergestützte Aufzeichnungen der okklusalen Bewegungsfunktion für die Aufwachstechnik nach Polz und Schulz. Dental Lab 44: 1595–1601.

Kordaß, B. & Gärtner, C. (2001): The virtual articulator – concept and development of VR-tools to analyse the dysfunction of dental occlusion. International Congress Series Volume 1230, Computer Assisted Radiology and Surgery: 689–694.

Kordaß, B., Gärtner, C., Söhnel, A., Bisler, A., Voss, G., Bockholt, U. & Seipel, S. (2002): The virtual articulator in dentistry: concept and development. Dental Clinics of North America 46: 493–506.

Krueger, K. L., Scott, J. R., Kay, R. F. & Ungar, P. S. (2008): Dental microwear textures of "Phase I" and "Phase II" facets. American Journal of Physical Anthropology 137: 485–490.

*Krueger-Janson, U., Schulz, D. & Kullmer, O. (2013): Anteriore Ästhetik durch posteriore Funktion?! Quintessenz Zahntechnik 39: 1024–1064.

*Kullmer, O. (2008): Benefits and risks in virtual anthropology. Journal of Anthropological Sciences 86: 205–207.

*Kullmer, O. (2015): Der verlorene Biss: Biologische und kulturelle Evolution der Zähne. In: Böhme, H., Kordaß, B. & Slominski, B. (eds.). Das Dentale - Faszination des oralen Systems in Wissenschaft und Kultur. Quintessenz, Berlin: 65–86.

*Kullmer, O., Benazzi, S., Fiorenza, L., Schulz, D., Bacso, S. & Winzen, O. (2009): Technical note: Occlusal Fingerprint Analysis (OFA): quantification of tooth wear pattern. American Journal of Physical Anthropology 139: 600–605.

*Kullmer, O., Benazzi, S., Schulz, D., Gunz, P., Kordos, L. & Begun, D. R. (2013): Dental arch restoration using tooth macrowear patterns with application to *Rudapithecus hungaricus*, from the late Miocene of Rudabánya, Hungary. Journal of Human Evolution 64: 151–160.

*Kullmer, O., Huck, M., Engel, K., Schrenk, F. & Bromage, T. (2002): Hominid tooth pattern database (HOTPAD) derived from optical 3D topometry. In: Mafart, B. & Delingette, H. (eds.). Three-Dimensional Imaging in Paleoanthropology and Prehistoric Archaeology. Liege, Acts of the XIVth UISPP Congress, BAR International Series.1049: 71–82.

*Kullmer, O., Schulz, D. & Benazzi, S. (2012): An experimental approach to evaluate the correspondence between wear facet position and occlusal movements. The Anatomical Record 295: 846–852.

Lucas, P. W. (2004): Dental Functional Morphology: How Teeth Work. Cambridge University Press, Cambridge.

Lucas, P. W., Omar, R., Al-Fadhalah, K., Almusallam, A. S., Henry, A. G., Michael, S., Thai, L. A., Watzke, J., Strait, D. S. & Atkins, A. G. (2013): Mechanisms and causes of wear in tooth enamel: implications for hominin diets. Journal of the Royal Society Interface 10. https://doi.org/10.1098/rsif.2012.0923.

Lund, J. P. (1976): Evidence for a central neural pattern generator regulating the chewing cycle. In: Anderson, D. J. & Matthews, B. (eds.) Mastication. Wright, Bristol: 204–212.

Luschei, E. S. & Goldberg, L. J. (1981): Neural mechanisms of mandibular control: mastication and voluntary biting. In: Brookhart, J. M., Mountcastle, V. B., Brooks, V. B, & Geiger, S. R. (eds.) Handbook of Physiology – Section I: The Nervous System. Vol. II. Motor control, Part II. Bethesda, American Physiological Society: 1237–1273.

Luschei, E. S. & Goodwin, G. M. (1974): Patterns of mandibular movement and jaw muscle activity during mastication in the monkey. Journal of Neurophysiology 37: 954–966.

Lussi, A. (2006): Dental Erosion – From Diagnosis to Therapy. Monographs in Oral Science, vol. 20. Karger, Basel.

Maier, W. (1977a): Die bilophodonten Molaren der Indriidae (Primates) – ein evolutionsmorphologischer Modellfall. Zeitschrift für Morphologie und Anthropologie 68: 307–344.

Maier, W. (1977b): Die Evolution der bilophodonten Molaren der Cercopithecoidea. Eine funktionsmorphologische Untersuchung. Zeitschrift für Morphologie und Anthropologie 68: 26–56.

Maier, W. (1978): Zur Evolution des Säugetiergebisses – Typologische und konstruktionsmorphologische Erklärungen. Natur und Museum 108: 288–300.

Maier, W. (1980): Konstruktionsmorphologische Untersuchungen am Gebiß der rezenten Prosimiae (Primates). Abhandlungen der senckenbergischen naturforschenden Gesellschaft 538: 1–158.

Maier, W. & Schneck, G. (1981): Konstruktionsmorphologische Untersuchungen am Gebiß der hominoiden Primaten. Zeitschrift für Morphologie und Anthropologie 72: 127–169.

Maier, W. & Schneck, G. (1982): Functional morphology of hominoid dentitions. Journal of Human Evolution 11: 693–696.

*Menz, U. (2016): Evolution und Funktionsvariabilität von bunodonten Molaren bei Primaten. Doctoral dissertation, Johann Wolfgang Goethe-Universität Frankfurt a. M.

Mills, J. R. E. (1955): Ideal dental occlusion in primates. Dental Practitioner 6: 47–61.

Mills, J. R. E. (1963): Occlusion and malocclusion in the teeth of primates. In: Brothwell, D. R. (ed.). Dental Anthropology. Oxford University Press, Oxford: 29–51.

Mills, J. R. E. (1967): A comparison of lateral jaw movements in some mammals from wear facets on the teeth. Archives of Oral Biology 12: 645–661.

Mills, J. R. E. (1973): Evolution of mastication in primates. In: Zingeser, M. R. (ed.). Craniofacial Biology of Primates. Symposia of the Fouth International Congress of Primatology, Portland, Oregon, August 15–18, 1972, vol. 3. Karger, Basel: 65–81.

M'Kirera, F. & Ungar, P. S. (2003): Occlusal relief change with molar wear in *Pan troglodytes troglodytes* and *Gorilla gorilla gorilla*. American Journal of Primatology 60: 31–41.

*Oxilia, G., Peresani, M., Romandini, M., Matteucci, C., Spiteri, C. D., Henry, A. G., Schulz, D., Archer, W., Crezzini, J., Boschin, F., Boscato, P., Jaouen, K., Dogandzic, T., Broglio, A., Moggi-Cecchi, J., Fiorenza, L., Hublin, J.-J., Kullmer, O. & Benazzi, S. (2015): Earliest evidence of dental caries manipulation in the Late Upper Palaeolithic. Scientific Reports 5. https://doi.org/10.1038/srep12150.

Pedreschi, F., Aguilera, J. M., Brown, C. A. (2000): Quantitative characterization of food surfaces using scale-sensitive fractal analysis. Journal of Food Process Engineering 23: 127–143.

Rensberger, J. M. (1973): An occlusion model for mastication and dental wear in herbivorous mammals. Journal of Paleontology 47: 515–528.

Ryan, A. S. (1979): A preliminary scanning electron microscope examination of wear striation direction on primate teeth. Journal of Dental Research 58: 525–530.

Sanson, G. D., Kerr, S. A. & Gross, K. A. (2007): Do silica phytoliths really wear mammalian teeth? Journal of Archaeological Science 34: 526–531.

Schulz, D. (2003): NAT – Die naturgemäße Aufwachstechnik. Teil 1: Der anteriore Bereich. Teamwork Media GmbH, Fuchstal.

Schulz, D. (2008): NAT – Die naturgemäße Aufwachstechnik. Teil 2: Der posteriore Bereich. Teamwork Media GmbH, Fuchstal.

Schulz, D. & Winzen, O. (2004): Basiswissen zur Datenübertragung. Teamwork Media GmbH, Fuchstal.

*Schulz, E., Piotrowski, V., Clauss, M., Mau, M., Merceron, G. & Kaiser, T. M. (2013): Dietary abrasiveness is associated with variability of microwear and dental surface texture in rabbits. PLoS One 8: e56167.

*Schultz, J. A., Menz, U., Winkler, D. E., Schulz-Kornas, E., Engels, S., Kalthoff, D. C., Koenigswald, W. v., Ruf, I., Kaiser, T. M., Kullmer, O., Südekum, K.-H. & Martin, T. (2018): Modular wear facet nomenclature for mammalian post-canine dentitions. Historical Biology 30: 30–41.

Schwartz, G. T. (2000): Taxonomic and functional aspects of the patterning of enamel thickness distribution in extant large-bodied hominoids. American Journal of Physícal Anthropology 111: 221–244.

*Schwermann, A. H. (2015): Über die Funktionsweise prätribosphenischer und tribosphenischer Gebisse. Doctoral dissertation, Rheinische Friedrich-Wilhelms-Universität Bonn.

Scott, R. S., Ungar, P. S., Bergstrom, T. S., Brown, C. A., Childs, B. E., Teaford, M. F. & Walker, A. (2006): Dental microwear texture analysis: technical considerations. Journal of Human Evolution 51: 339–349.

Scott, R. S., Ungar, P. S., Bergstrom, T. S., Brown, C. A., Grine, F. E., Teaford, M. F. & Walker, A. (2005): Dental microwear texture analysis reflects diets of living primates and fossil hominins. Nature 436: 693–695.

Smith, B. H. (1984): Patterns of molar wear in hunter–gatherers and agriculturalists. American Journal of Physical Anthropology 63: 39–56.

Solaberrieta, E., Etxaniz, O., Minguez, R., Muniozguren, J. & Arias, A. (2009): Design of a virtual articulator for the simulation and analysis of mandibular movements in dental CAD/CAM. In Proceedings of the 19th CIRP Design Conference – Competitive Design. Cranfield University Press, Bedford: 323–331.

Spears, I. R. (1994): Functional adaptations of hominoid molars: an engineering approach. PhD. Dissertation, University of Liverpool.

Spears, I. R. & Crompton, R.H. (1996): The mechanical significance of the occlusal geometry of great ape molars in food breakdown. Journal of Human Evolution 31 517–535.

Strait, S. G. (1993a): Differences in occlusal morphology and molar size in frugivores and faunivores. Journal of Human Evolution 25: 471–482.

Strait, S. G. (1993b): Molar morphology and food texture among small-bodied insectivorous mammals. Journal of Mammalogy 74: 391–402.

Stones, H. H. (1948): Oral & Dental Diseases. Livingston, Edinburgh.

Teaford, M. F. (1983): The morphology and wear of the lingual notch in macaques and langurs. American Journal of Physical Anthropology 60: 7–14.

Teaford, M. F. & Ungar, P. S. (2000): Diet and the evolution of the earliest human ancestors. Proceedings of the National Academy of Sciences of the United States of America 97: 13506–13511.

Trinkaus, E. (1987): The Neandertal face: evolutionary and functional perspectives on a recent hominid face. Journal of Human Evolution 16: 429–443.

Ulhaas, L. (2006): Vergleichende computergestützte funktionsmorphologische Analyse an Molaren cercopithecoider Primaten. Doctoral dissertation, Johannes Gutenberg Universität Mainz.

Ulhaas, L., Kullmer, O. & Schrenk, F. (2007): Tooth wear diversity in early hominid molars – a case study. In: Bailey, S. & Hublin, J. J. (eds.). Dental Perspectives on Human Evolution: State of the Art Research in Dental Anthropology. Springer, Dordrecht: 369–390.

Ulhaas, L., Kullmer, O., Schrenk, F. & Henke, W. (2004): A new 3-d approach to determine functional morphology of cercopithecoid molars. Annals of Anatomy 186: 487–493.

Ungar, P. S. & M'Kirera, F. (2003): A solution to the worn tooth conundrum in primate functional anatomy. Proceedings of the National Academy of Sciences of the United States of America 100: 3874–3877.

Ungar, P. S. & Williamson, M. (2000): Exploring the effects of tooth wear on functional morphology: A preliminary study using dental topographic analysis. Palaeontologia Electronica 3: 17.

Ungar, P. S., Brown, C. A., Bergstrom, T. S. & Walkers, A. (2003): Quantification of dental microwear by tandem scanning confocal microscopy and scalesensitive fractal analyses. Scanning 25: 185–193.

Ungar, P. S., Dennis, J. C., Wilson, J. & Grine, F. (2002): Quantification of tooth crown shape by dental topographic analysis. American Journal of Physical Anthropology, Supplement 34: 158–159.

Ungar, P. S., Grine F. E. & Teaford, M. F. (2006): Diet in early Homo: a review of the evidence and a new model of adaptive versatility. Annual Review of Anthropology 35: 209–228.

Ungar, P. S., Teaford, M. F., Glander, K. E., Pastor, R. F. (1995): Dust accumulation in the canopy: a potential cause of dental microwear in primates. American Journal of Physical Anthropology 97: 93–99.

Wall, C. E., Vinyard, C. J., Johnson, K. R., Williams, S. H. & Hylander, W. L. (2006): Phase II jaw movements and masseter muscle activity during chewing in *Papio anubis*. American Journal of Physical Anthropology 129: 215–224.

Weber, G. W., & Bookstein, F. L. (2011): Virtual Anthropology: A Guide to a New Interdisciplinary Field. Springer, Vienna.

Wood, B. A. (1993): Early *Homo*: How many species? In: Kimbel, W. H. & Martin, L. B. (eds.) Species, Species Concepts, and Primate Evolution. Plenum, New York: 485–522.

Wood, B. A., Abbott, S. A. (1983): Analysis of the dental morphology of Plio-Pleistocene hominids. I. Mandibular molars: crown area measurements and morphological traits. Journal of Anatomy 136: 197–219.

*Zanolli, C., Kullmer, O., Kelley, J., Bacon, A.-M., Demeter, F., Dumoncel, J., Fiorenza, L., Grine, F. E., Hublin, J.-J., Nguyen, A. T., Nguyen, T. M. H., Lei, P., Schillinger, B., Schrenk, F., Skinner, M. M., Xueping, J. & Macchiarelli, R. (2019): Evidence for increased hominid diversity in the Early to Middle Pleistocene of Indonesia. Nature Ecology & Evolution 3: 755–764.

A brief history of quantitative wear analyses with an appeal for a holistic view on dental wear processes

Ellen Schulz-Kornas, Thomas M. Kaiser, Ivan Calandra, and Daniela E. Winkler

Introduction

This chapter is a comprehensive overview about the development in the field of quantitative wear analysis during the last 20 years. It focuses on specific interdisciplinary approaches that either introduced new viewpoints or solved technical problems, and in particular guided our projects within the DFG research unit 771 at the University of Hamburg. The scope of our group's interest facilitates the historical shift from qualitative to quantitative wear analysis. Our research was driven by the urge to gain a deeper understanding of wear, which is much more than just traces, and develop a holistic view of the wear process on diverse materials. We start with a condensed historical review of selected developments, focus on the major debates that have influenced our understanding of tooth wear as a part of oral food processing, and set the context of our work within the larger theoretical framework. We show examples that the field of dental wear analyses is evolving on all scales, and that we are only beginning to comprehend the complex dental wear process as one factor amongst many driving evolution through dental adaptation. In particular, the combined studies based on museum material, in vivo, and in vitro experiments have shed new light on the importance of abrasives in the wear process, but also raised new questions for future research.

From traces to textures

Dental wear has been widely investigated in paleobiology and paleoanthropology to infer diet, habitat and climatic conditions. For decades, the approaches mostly revolved around qualitative description of wear in various research disciplines. In paleontology, dental wear analyses have a long history and the works of Simpson (1926, 1933), in which he relates striations on dental wear facets to chewing direction, might be considered a starting point. Since then, the literature on and interest in wear processes have rapidly expanded. Russian scientists used the term trace to describe use-wear marks on edges and surfaces of prehistoric stone and bone tools (Semenov 1957, Korobkova 1970). They employed traceology as a term to refer to the study of any trace (whether residues or surface alteration), usually in the context of tool use (Korobkova 1977, 1978, 1979, 1981). Later in archeology it has been used as a synonym for microwear to describe scarring, striations, edge rounding, smoothing, polishing and beveling (Hayden 1979, Keeley 1980, Kamminga 1982, Mecks et al. 1982). However, the fascination of use-wear analysis goes even further back in time, e.g. to the mid-19th century when the Swedish zoologist Sven Nilsson mentioned use-wear on flint-knapped artifacts by early inhabitants in Scandinavia (Nilsson, 1834–1843). Already during these early days of traceology Sergej A. Semenov and Galina L. Korobkova conducted extensive systematic experimental research using stereomicroscopy to describe the "microscopically small" wear traces they found (for a review, see Anderson et al. 2005). During the 1950s and 1960s wear traces in mammalian teeth were the object of considerable interest in the fields of vertebrate paleontology, paleoanthropology and also dentistry. Teeth were considered guiding structures of mastication and their wear reflected particular diets. In particular, Butler and Mills (Butler 1952, Mills 1955, Butler and Mills 1959, Mills 1963, 1966, 1967) as well as Kay and Hiiemäe (1974) suggested that striations as evidence of occlusal movement represent areas of contact or close approximations of antagonistic teeth. These studies focused preferentially on the mechanics of chewing per se. Later, Dahlberg and Kinzey (1962) studied human teeth using light microscopy and related the scar patterns on tooth enamel to the properties of the ingesta. Additionally, they differentiated between attritional (tooth-tooth) and abrasional (tooth-diet) contacts. Since then, dental microwear research has been widely conducted in paleobiology and paleoanthropology. For a review of the developments from 1950 to 1998, and with special reference given to application in hominins, see Rose and Ungar (1998). More details and particular aspects of facet striation and terminology are discussed in Schultz et al. (2018, 2020, this volume).

Here, we do not detail every single development made or every animal group microwear has been applied to. Instead, we focus on specific interdisciplinary approaches that either introduced new viewpoints or solved technical problems, and in particular influenced our approach and in turn, the design and execution of our projects within the DFG research unit 771 at the University of Hamburg, metaphorically speaking, "from traces to textures".

In his pioneering light microscopy work, Philip L. Walker (1976) discussed orientation and density of striations in non-human primates in relation to feeding substrate and

T. Martin & W. v. Koenigswald (eds.): Mammalian Teeth – Form and Function. Pp. 44–53, 3 figs.
 – ISBN 978-3-89937-266-3
DOI: http://doi.org/10.23788/mammteeth.03 Published 22 December 2020

mechanical properties of ingesta. Later studies unfortunately did not explore this aspect further. Instead, they focused on the technical improvement (higher resolution using scanning electron microscopy, SEM) and on the taxon-specific dietary signatures dependent on the enamel structure (Walker 1977, Rensberger 1978, Walker et al. 1978, Rensberger 1982). Based on their studies on hyraxes, Alan C. Walker et al. (1978) found that seasonal changes in diet are reflected in microwear signatures. Consequently, microwear has been suggested to mainly originate from opal phytoliths intrinsic to ingesta of certain kinds; this interpretation stimulated new research avenues in paleobiology and especially towards dietary reconstruction and ecological niche assignment in extant and extinct mammals.

The limits of microwear were discussed from the early days of microwear research and, more than once, the optimistic but simplistic notion that microwear traces reflected the immediate past diet or type of ingesta was challenged. Peters' (1982) in vitro experiments rejected the idea that some primate foods such as dicotyledonous seeds do not produce microwear by themselves. Recently, this criticism has been renewed by van Casteren et al. (2020) based on their results of in vitro nanoscale experiments and mechanical models. However, we should not forget that inference based on in vitro studies is limited to a specific species, the type of abrasive particles used in the experiments (hardness, shape, size), and the force setting and scale; the latter being difficult to control, and generalization should be cautious.

One of the earliest attempts to demonstrate the limitations of microwear were the three-month lab-based feeding experiments with American opossum, *Didelphis marsupialis*, by Covert and Kay (1981). Their experiments indicated that similar microwear patterns resulted from the three investigated diets (cat food alone and cat food mixed with soybean hulls or insect chitin). However, Gordon and Walker (1983) questioned these results, arguing that they reflected methodological limitations rather than fundamental limits of what microwear can tell us. In turn, Kay and Covert (1983) rebutted the arguments that grit and plant opalines should leave similar microwear patterns. This led to a series of experimental studies in primates that elucidated the process and duration of microwear formation (Teaford and Oyen 1989c) and helped to determine the causes of wear based on field studies in wild animals (e.g., Teaford & Glander 1991, Teaford & Runestad 1992, Strait & Overdorff 1994, Ungar et al. 1995). Additionally, the microwear analysis was constantly modified and transferred to vertebrate paleontology, the comparative dataset was expanded and the methods were adjusted for low-magnification application (Solounias & Hayek 1993, Solounias & Semprebon 2002).

With the increasing amount of data and the diversification of application modes (low-magnification versus high magnification, different field of views), the need arose for more quantified and automated methods to increase the comparability across studies. SEM microwear images were considered inadequate for pattern recognition at that time (e.g. see Kay 1987). However, it was suggested that edge detection or filtering could be useful for dealing with the complex background shading (Boyde & Fortelius 1991). A first attempt into this direction was the semi-automated microwear approach (Ungar et al. 1991). More than a decade later, new approaches using confocal microscopy and interferometry were used to three-dimensionally capture the information encapsulated in a surface area (the texture) rather than just scoring pre-defined scar morphs (e.g. scratch versus pit; Ungar et al. 2003, Scott et al. 2005, Scott et al. 2006).

Supported by technical advancements in engineering and surface profilometry, Kaiser & Brinkmann (2006) introduced a tactile measurement approach termed dental enamel micro-texture analysis. This approach was the first to apply standardized measuring routines based on 2D micro-profiles as described in ISO 4287 (1984, 1997). The advantage of the application of standardized measuring routines was two-fold: (1) they had the potential to enhance reproducibility and (2) the parameters were defined based on a functional understanding of the wear process with its technical application in mind. The challenge of the Hamburg-based projects within the DFG research unit 771 was to transfer the advancements made in 2D surface profilometry into 3D dental surface texture analysis (3DST) and apply it to a wide range of vertebrates. This approach was meant to complement scale-sensitive fractal analysis (SSFA) (Ungar et al. 2003, Scott et al. 2006) which had originally been intended for dietary reconstruction in primates. We aimed at testing dietary as well as functional interpretations of the 3D surface texture parameters and thus described dental surface textures in various extinct and extant taxa. From the beginning, our approach considered very closely the ISO engineering standards that were further improved during the years 2007–2012. The lengthy ratification of ISO 25178 was a stepwise bottom-up process, and when the first draft became available in 2007 (ISO/DIS 25178-2), this was the starting point that led to Schulz et al. (2010) – a study which applied 3D surface texture parameters for the first time to enamel wear surfaces of mammalian teeth.

New approaches and terminology

We use the umbrella term dental microwear texture analysis (DMTA) for all types of 3D dental microwear analysis. Where necessary, we introduce further terms and acronyms to refer to specific methods, e.g., scale-sensitive fractal analysis (SSFA) for the approach according to Ungar et al. (2003) and 3D surface texture analysis (3DST) for the approach according to Schulz et al. (2010). From the beginning, we have chosen a terminology that follows the descriptions in the ISO standards. We decided for the term 3D surface texture analysis (3DST) to emphasize that this method does not just serve to classify particular features such as scratches or pits (like in microwear *sensu lato*), but to quantify the whole surface, made possible by full automation. During the last decade 3DST was applied to natural (i.e. teeth and bone, see below) as well as additive manufactured materials (for a review see Townsend et al.

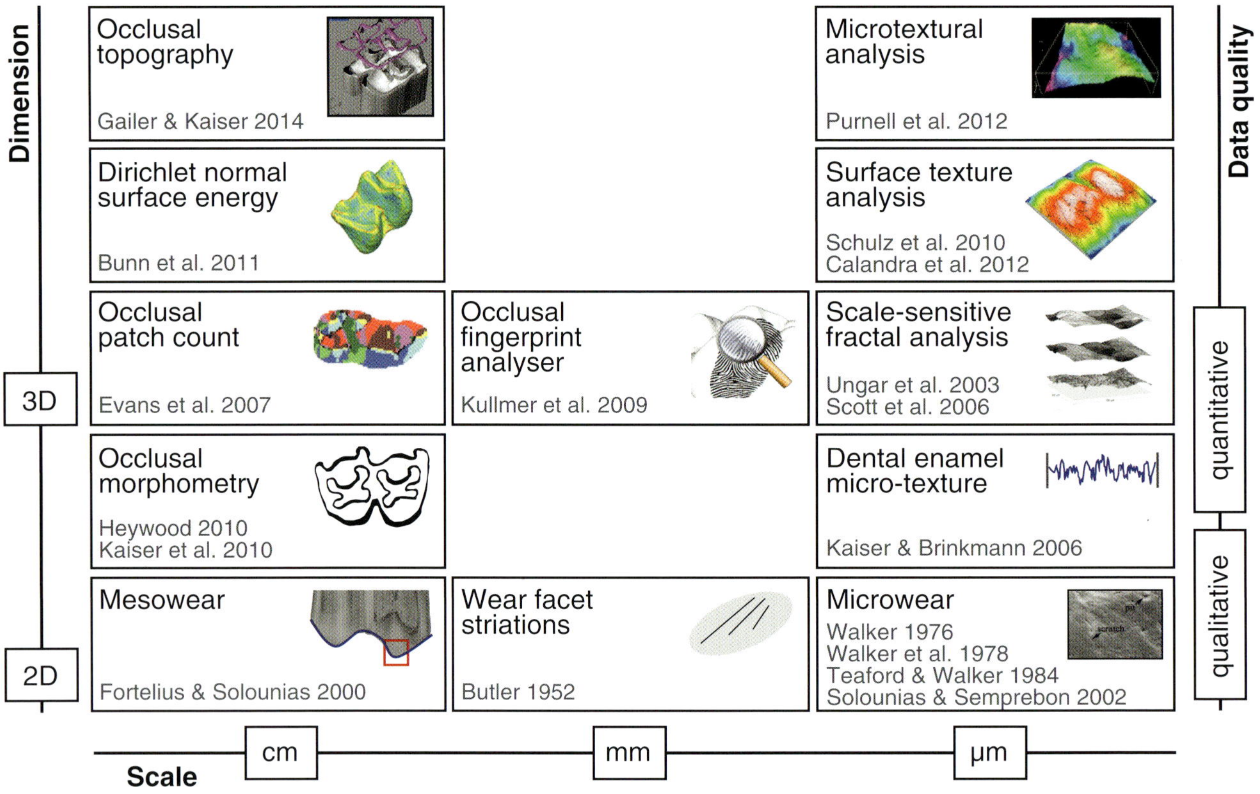

Fig. 3.1. Overview of the most recent tooth wear approaches. Images are sorted according to the following three categories scale (µm to cm), main character dimension (2D, 3D) and applied data (quantitative, qualitative).

2016). Hence, we simplified the term from 3D areal surface texture standards (Schulz et al. 2010) to 3D dental area surface texture analysis (DASTA, Calandra et al. 2012) and finally to 3D surface texture (3DST, Winkler et al. 2016).

One has to keep in mind that SSFA describes the change of surface features across scales, while 3DST is applied to one specific scale that is set. This results in relative advantages of the 3DST method compared to the SSFA method, i.e. as discussed for application in primates by Calandra et al. (2012). In general, in 3DST the advantages are (1) the huge variety of at least 46 standardized parameters from two international standards (ISO 25178, ISO 12781) and three additional industrial driven analyses (i.e. motif, furrow and texture direction); some even call it the parameter rush (Stout 2000); (2) the parameters quantify in exceptional details the geometry and topography of the whole surface texture; (3) the parameters were developed and linked with a functional interpretation without a certain taxonomic or dietary affiliation of the sampled individual in mind. The disadvantage of 3DST is that choosing and interpreting a parameter requires expert knowledge in tribology on pre-processing and wavelength filtering, which is challenging to achieve, while for SSFA the pre-processing is minor and SSFA parameters are directly related to food properties.

In the past each DMTA had its own software platform: for SSFA the affordable Sfrax and Toothfrax software (by Surfract, Worcester, USA) were used, while for 3DST the more expensive platform of MountainsMap (by digitalsurf, Besançon, France) was adapted by the instrument suppliers (e.g. µsoft analysis in case of Nanofocus, LeicaMap in case of Leica, SensoMap in case of Sensofar). All software solutions were very time-consuming in the past and only a few studies used both approaches (Schulz et al. 2010, Calandra et al. 2012, Winkler et al. 2013a). For researchers just embarking on this kind of work, it is important to know that a slightly modified SSFA was incorporated as an additional module to MountainsMap (version 7.4.8676) only recently in 2018, although not all SSFA parameters are available yet (*Tfv* is absent). Nevertheless, it is a huge technical improvement and for the first time one can conduct both SSFA and 3DST in a much more sophisticated way.

Another main advantage of using 3DST (Schulz et al., 2010) is the possibility to measure texture direction. To do so, we developed a measuring protocol that takes the facet orientation into account. We did so to follow the approach of Walker (1976). This allowed us to produce quantitative data in order to finally test for correlation between chewing direction and texture direction. We tested for measuring position (facet, tooth and jaw position) and cross-validated our approach with the SSFA approach of Ungar et al. (2003). In a next step, we transferred the approach to primates including the adaptation and application of new statistical tools for small samples (Calandra et al. 2012), tested for the influence of filtering routines to improve the data processing and selection of the scale to be analyzed (Schulz et al. 2013a), and applied the approach to data from a controlled feeding experiment (Schulz et al. 2013c).

We showed that this approach can provide insights into function, tribology and wear processes, i.e. aspects that are difficult to assess with SSFA. We also found the 3DST approach promising for dietary reconstruction in extant and extinct ungulates (Winkler et al. 2013a, b, Gailer et al. 2016) and rodents (Calandra et al. 2016, Winkler et al. 2016).

During the same period, several colleagues explored the new technological avenues of 3D high-resolution surface measurement (Fig. 3.1), e.g. using the focus variation method (Purnell et al. 2012) or a laser-scanning confocal microscope (Kubo et al. 2017) for data acquisition. This resulted in methodological diversification and an issue of multiple levels of data quality and comparability. The challenge for the future might possibly be related to the analysis of large datasets (i.e. "big data", for example via data mining and artificial intelligence). We need to find solutions for shared databases (e.g. open data initiative) and to combine datasets generated by various instruments (e.g. Arman et al., 2016).

During many years, we were involved in studies combining microwear (µm scale) and mesowear (mm/cm scale) signatures providing both short- and long-term dietary signals (Merceron et al. 2007, Schulz et al. 2007, Rivals et al. 2009a, b). We were fascinated by the easy applicability and the simple nature of mesowear method (no measurements) but we were also aware of the limitations of this approach (i.e. focus on the ectoloph, need for calibration when comparing higher taxa of herbivores, uncontrolled distortion imposed by the ruler approach). We tried to improve the mesowear method within the DFG research unit 771 by testing for relationships to habitat factors (Schulz & Kaiser 2013), by expanding and modifying the scoring system (Winkler & Kaiser 2011, Taylor et al. 2013, 2014) and by applying the method to field studies (Schulz et al. 2013b, Wronski & Schulz-Kornas 2015). We even adjusted it for the application to lagomorphs and rodents (Ulbricht et al. 2015). However, chewing is a three-dimensional process, and therefore, in addition to the 3D surface texture approach we put some effort into developing two 3D occlusal topometry approaches to investigate the extent to which adaptational value of dental morphology reflects tooth function (e.g., occlusal topography according to Winkler et al. 2013b, Gailer & Kaiser 2014; SAGA-GIS according to Nieberg et al. 2009, Bethune et al. 2019). This was a very active field of research at that time and many other researchers were engaged in the development of new 3D approaches to quantify the occlusal surface (Evans et al. 2007, Kullmer et al. 2009, Heywood 2010, Bunn et al. 2011, Hernesniemi et al. 2011, Saarinen et al. 2015). For an overview see Figure 3.1 with selected approaches sorted by scale (µm to cm), dimension (2D, 3D) and structure of data (quantitative, qualitative).

Causes of tooth wear – major debates and the need for a deeper understanding of the etiology of wear

As mentioned above, from the early days of microwear research, an active debate about the causes of tooth wear shook the community: Was it phytoliths, or grit, sand and dust? This debate was re-ignited at least three times during the last 50 years based on various pieces and lines of evidence. At first the focus was on the results of feeding experiments (Covert & Kay 1981, Kay & Covert 1983, Teaford & Walker 1983, Teaford & Oyen 1989a, b, Teaford & Oyen 1989c). The emphasis then shifted to the physical properties of abrasives with particular attention paid to silica phytoliths (Baker et al. 1959, Sanson et al. 2007, Lucas et al. 2013), which seemed to indicate that phytoliths are too soft to scratch enamel, while dental microwear has displayed a positive correlation between the consumption of phytolith-rich graminoids and abundance of scratches (i.e. see Merceron et al. 2007). More recently arguments have emerged based on experimental results from an in-vitro micro-loading and scratching study of phytoliths and metallic spheres on enamel with Xia et al. (2015, 2017, 2018) on the one hand and van Casteren et al. (2018, 2020) on the other.

Recently, we put together a set of hypotheses on the tribology of mammalian herbivore teeth based on the data we had generated during the period of the DFG research unit 771 (Kaiser et al. 2016). From these tribological hypotheses we suggest that there are parallels between artificial recurrent neural networks (e.g., Poznyak et al. 2019) and oral food processing. We started to develop a new theoretical framework and proposed that oral food processing has similarities with a multi-functional network that includes known and unknown variables (factors) which influence each other (Fig. 3.2). In Figure 3.2 variables are represented by dots and influences by lines, and all possible lines without any assessment are given. Based on large controlled experimental datasets it is possible to test and verify all possible connections and influences using mathematical algorithms. These algorithms are based on machine learning methods and can work in forward as well as backward propagation to tune the network with self-correcting methods. Five of these predictive classification methods have yielded promising results for the automatic image recognition of experimental cut- and bite marks (i.e. neural network, support vector machines, k-nearest neighbor, random forest, decision trees; Dominguez-Rodrigo & Baquedano 2018). However, in paleobiology, paleoanthropology, and archeology the large experimental datasets necessary to employ machine learning methods are not available yet. In addition, data exchange between the various DMTA methods based on interferometry, confocal and focus variation is not possible and not advisable (yet) due to technical limits and methodological differences. But it is most probably only a matter of time until large datasets and technical solutions are available.

The structure of artificial networks is referred to as architecture that is by definition organized in layers (i.e. input and output layers). Translated to oral food processing, this means that taxa, vegetation, morphology, abrasives and climate represent input layers (see layer A to D in Fig. 3.2). The resulting tooth wear is considered to be the result of the factors defined for each layer (output layer E in Fig. 3.2). We included all factors we are currently aware of into the

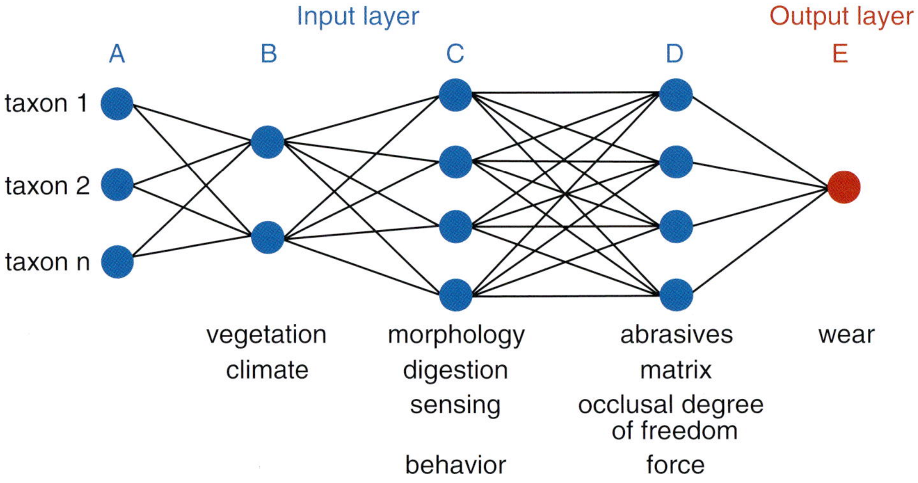

Fig. 3.2. Network model of oral food processing and wear. The model is based on the tribological hypotheses (Kaiser et al. 2016) and illustrates some of the known and unknown factors that influence each other (connecting lines). We choose three taxa, but the network could be extended to include more taxa (n) and more factors; a dot indicates the impact a certain factor might have.

network. Nonetheless the amount of factors is extendable, e.g., in case that new factors will be identified in the future. In many cases of oral food processing, we have only limited knowledge of how strong the variables and their influences are and how to consider their variation; taxa, for example, can adjust their behaviors (e.g., via sensing) and adapt to their environment. The self-correcting adaptive methods of machine learning approaches could be very helpful to run through various sets of influences and factors helping to identify the most promising combinations and seasonal scenarios.

Most dietary reconstructions published so far rely on fossil teeth or bones, which means that reconstructing input layers from measuring the output layer is the actualistic paleontology approach (as proposed by Richter 1928). Hopefully in the future, using the network structure concept of habitat and vegetation can be conducted in a more holistic way by following the evidence of connections between the identified factors within the layers. Feeding observations, in vitro and in vivo experiments have already helped to characterize the factors in more detail and to test the connection between these factors and the interplay between the layers. However, there is currently no consensus in the dental wear community on the importance of specific layers and interaction between the factors, especially when it comes to the contribution of ingested abrasives from different sources. It is well known that factors act on different temporal scales; e.g. microwear, and mesowear do not record the same timeframe in the life history of an individual. In addition, it could be concluded that our knowledge on abrasives, no matter if internal (phytoliths) or external (dust, grit, soil or ashes), as well as the chewing kinematics and the sensing of abrasives is still very limited. This lack of knowledge has led to conflicting results in studies employing different proxies for dietary reconstruction (e.g., tooth wear vs. stable isotopes). This is mostly because determining the local in vivo mechanical properties of biogenic materials, such as opaline phytoliths on the ingesta side of the antagonistic system, and dental hard tissues (enamel and dentin) plus protein pellicle on the tooth side, is a challenging task with many unknowns. To complicate matters further, material properties of materials that are wet in their native state (phytoliths and teeth) change during chemical or heat-mediated extraction processes necessary for material testing (Cabanes & Shahack-Gross, 2015). To overcome these problems, we established a new extraction protocol for opal phytoliths (Braune et al. 2012) and conducted nano-indentation hardness tests with phytoliths and tooth material under wet and dry conditions (Schulz-Kornas et al. 2017, Kaiser et al., 2018). Our data suggest that phytoliths, as well as enamel and dentin, are less hard when measured dry. Native phytoliths can easily indent native dentin and certainly contribute to the scouring of dentin. In nature the effects of internal abrasives almost always act together with the effect of external abrasives. Like Xia et al. (2018), we raised the question if phytoliths from living cells could act as inefficient wear agents (Schulz-Kornas et al. 2017, Kaiser et al. 2018), while grit had higher potential as an agent of tooth wear. We further proposed the idea that phytoliths impose a higher selective pressure on dentin as abrasives. In turn, this would affect the enamel ridges of herbivorous mammals because the dentin basins otherwise become too deep to maintain the structural integrity of the occlusal surface (Kaiser et al. 2018). Hypotheses relating the evolution of hypsodonty to increased roughage feeding, as frequently assumed (Strömberg 2006, Damuth & Janis 2011), are challenged by our findings. Also, we support the claim that phytoliths from living (wet) biomass may in fact play a subordinate role in tooth enamel wear. Animals at the end of droughts or prolonged dry seasons are forced to eat dry food. Dry phytoliths being harder, dry food leads to different tooth wear effects and signals (Winkler et al. 2019b). This hypothesis needs to be tested for dead (dry) biomass in field experiments. In addition, the impact of grit and dust (including its morphology and abundance) is still enigmatic (Winkler et al. 2020). Hence, the debate remains open.

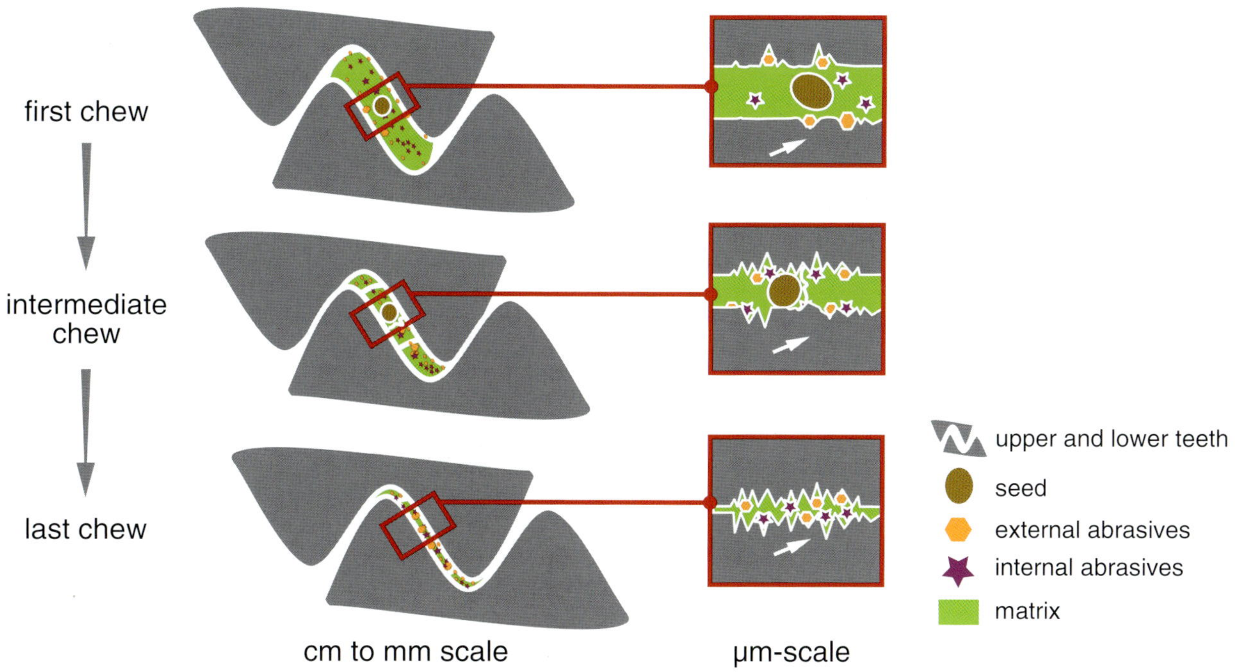

Fig. 3.3. Simplified model of occluding primate molars. The model illustrates that ingesta comminution and the resulting tooth wear are a multi-scale and multi-factorial process resulting in a cumulative wear signature.

New developments, future perspectives and broader implications

The DFG research unit 771 provided a great opportunity to exchange ideas with many researchers from various fields. But it also set the foundation for new developments and achievements made during the years after until today. Some of these will be highlighted in the following paragraphs.

As already pointed out, in vitro material testing as well as feeding experiments opened up new avenues for testing tooth wear-related hypotheses. New experimental designs, however, may involve methodological pitfalls and thus are debated. In vivo studies provide much-needed reference data, e.g. regarding the time involved in overwriting textures and variations in texture patterns (Ramdarshan et al. 2016, Schulz et al. 2013c). In vivo studies indicated that variable water content from consumed plant matter is a source of texture variation (Winkler et al. 2019a). But still, especially in the in vivo experiments, we found evidence for a dietary signal in tooth wear. That led us to expand DMTA to groups that have never been investigated with DMTA, i.e. fossil tritylodontids (Kalthoff et al. 2019) and extant lepidosaurs (Winkler et al. 2019b).

After recent experimental studies successfully measured tooth volume loss (Müller et al. 2014, 2015), as well as mesowear variation (Ackermans et al. 2018), it appears highly promising to combine both, realizing that wear processes act on different scales, which can lead to discrepancies observed in microwear and DMTA. These combined studies have facilitated our understanding and highlight that formation of microwear (texture) does not necessarily lead to substantial tooth volume loss (Martin et al. 2020, Schulz-Kornas et al. 2020). Additionally, field studies combining surface texture, feeding observations and chewing efficiency in chimpanzees pointed out that periodical dust loads (external abrasives) are detectable in the surface texture and lead to less efficient comminution (Schulz-Kornas et al. 2019, Stuhlträger et al. 2019). Further, Schulz-Kornas et al. (2019) showed that seeds in a frugivorous diet like that of chimpanzees could be very variable in size and less important for the chewing efficiency. They also showed that not every seed can be considered a hard particle, particularly if other tools are used to crack them as some animals have the option to process their foods with tools other than their teeth.

The hope is that our research will open up new research avenues to finally obtain a better understanding of the biomechanics and scaling of the complex and multi-factor interactions taking place during comminution. However, it remains a challenge for future research to characterize the repetitive kinematics of the chewing process, and at the same time the ingesta particles comminuted as well as the resulting complex changes of the occlusal surface. We are just at the beginning of understanding that tooth wear is a multi-scale and multi-factorial process that results in a cumulative wear signature (see Fig. 3.3).

The methods we developed for dental microwear texture analysis have also impacted other disciplines. Archeologists have been interested in wear patterns on the surface of archeological artifacts made from e.g. bones or stones (lithics). Traceology has a long history within archeology (see above "From traces to textures"), and while there were early attempts at automatic quantification (Dumont 1982, Beyries et al. 1988), has still not fully recognized the potential of surface texture analysis (3DST) and scale-sensitive fractal analysis (SSFA). Traceology could therefore learn a lot from dental microwear texture analysis (DMTA) (Calandra et al. 2019a), and we have already contributed to this exchange and collaboration between these fields (Pedergnana et al.

2020). Another promising way of experimental testing is the combination of mathematical modelling with actualistic experiments as conducted in archeology by Martisius et al. (2018, 2020). Here, we transferred 3DST on bone surfaces and employed multilevel multi-variate Bayesian models to explain surface texture variation under experimental conditions and showed that duration of use strongly affected the transformation (overwriting) of the surface texture. In general, 3DST is applicable to bone surfaces as well (Turcotte et al. 2019) to quantify the morphology of extrinsic fiber insertions sites of the biceps brachii of mice in relation to activity level.

The wider application of 3DST in various research fields shows us that there is a need for standardization, repeatability and reproducibility to guarantee high data quality and foster future exchange of data. This is apparent in traceology, as it still is in DMTA. Indeed, we showed that changing the numerical aperture of the objective used, one of the most basic properties of an objective, significantly modified the way surfaces are acquired with a laser-scanning confocal microscope (Calandra et al. 2019b). This study is therefore advocating for more transparency on the reporting of methods and results. We also developed a cheap and quick method to relocate the same area of a sample again and again (Calandra et al. 2019c) which is necessary when comparing the same spot before and after an experiment as well as for reproducibility. Although difficult to apply to live animals, this relocation method could prove very useful to acquire the same area with different instruments (in order to better compare the data), or in experiments with artificial chewing machines (Gügel et al. 2001, Hua et al. 2015, Karme et al. 2016).

Conclusion

Our projects within the DFG research unit 771 originally had a strong methodological focus. Since 2008, we developed a new method of dental microwear texture analysis and applied it to a variety of vertebrate groups. Nevertheless, this method was just a tool to answer an array of scientific questions, ranging from wear mechanics and etiology at the microscopic and short-time scales to dental evolution at the macroscopic and evolutionary scales. The field of dental wear analyses is evolving on all scales and our understanding of the complex wear process is only beginning to grow. The combined effort of studies based on museum material, in vivo, and in vitro experiments has shed light on the importance of abrasives in the wear process, but also raised new questions. With interdisciplinary collaborations involving animal nutritionists, veterinarians, dentists, and computer scientists, many more publications on dental wear and dietary adaptations are to be written.

Acknowledgments

We highly acknowledge the vivid discussions we had with all colleagues of the DFG research unit 771, especially during the yearly meetings in Visselhövede and Mayschoß. In particular, we thank Wighart von Koenigswald (University of Bonn) for initiating the DFG research unit 771, for keeping up his curiosity and enthusiasm and sharing it with us. We are greatly thankful to Marcus Clauss (University of Zurich) who set the highest standards in conducting feeding experiments and encouraged us to question and test many of the 'given facts' in the field and especially our own convictions.

Further, we would like to thank our laboratory team for their continuous support (in alphabetical order): Janina Bethge, Elehna Bethune, Caroline Braune, Alexander Daasch, Fenia De Bruin, Sven Fraas, Juan Pablo Gailer, Volker Halley, Lisa Krause, Susanne Krohn, Christina Landwehr, Julia Lingner, Nelson Mascarenhas, Mona Pinnow, Vanessa Piotrowski, Susanne Schwitzer, Mirella Skiba, Lucy A. Taylor, Nicole Töpfer, Jessica Westphalen, and Anna Zickert. We highly appreciate the constructive comments and helpful suggestions made by the two reviewers Florent Rivals and Gordon Sanson.

References

* indicates publications that originated from the DFG Research Unit 771.

Ackermans, N. L., Winkler, D. E., Schulz-Kornas, E., Kaiser, T. M., Müller, D. W. H., Kircher, P. R., Hummel, J., Clauss, M. & Hatt, J.-M. (2018): Controlled feeding experiments with diets of different abrasiveness reveal slow development of mesowear signal in goats (*Capra aegagrus hircus*). Journal of Experimental Biology 221: jeb186411.

Anderson, P. C., Korobkova, G. F., Longo, L., Plisson, H., Skakun, N. (2005): Various viewpoints on the work of S. A. Semenov. In: Longo, L., Skakun, N. (eds.). The roots of use-wear analysis: selected papers of S. A. Semenov. Memorie del Museo Civico di Storia Naturale di Verona (2. serie) 2005. Sezione Scienze Dell' Uomo: 11–19.

Arman, S. D., Ungar, P. S., Brown, C. A., DeSantis, L. R. G., Schmidt, C. & Prideaux, G. J. (2016): Minimizing inter-microscope variability in dental microwear texture analysis. Surface Topography: Metrology and Properties 4: 024007.

Baker, G., Jones, L. H. P. & Wardrop, I. D. (1959): Cause of wear in sheep's teeth. Nature 184: 1583–1584.

*Bethune, E., Kaiser, T. M., Schulz-Kornas, E. & Winkler, D. E. (2019): Multi proxy dietary trait reconstruction in Pleistocene Hippopotamidae from the Mediterranean Islands. Palaeogeography, Palaeoclimatology, Palaeoecology 553: 109210.

Beyries, S., Delamare, F. & Quantin, J.-C. (1988): Tracéologie et rugosimétrie tridimensionnelle. In: Beyries, S. (ed.). Industries Lithiques: Tracéologie et Technologie. BAR International Series, Oxford: 115–132.

Boyde, A. & Fortelius, M. (1991): New confocal LM method for studying local relative microrelief with special reference to wear studies. Zoological Journal of the Linnean Society 87: 182–214.

*Braune, C., Lieberei, R., Steinmacher, D. & Kaiser, T. M. (2012): A simple microwave extraction method for the isolation and identification of plant opal phytoliths. Biologia 67: 927–930.

Bunn, J. M., Boyer, D. M., Lipman, Y., St. Clair, E. M., Jernvall, J. & Daubechies, I. (2011): Comparing Dirichlet normal surface energy of tooth crowns, a new technique of molar shape quantification for dietary inference, with previous methods in isolation and in combination. American Journal of Physical Anthropology 145: 247–261.

Butler, P. M. (1952): The milk-molars of Perissodactyla with remarks on molar occlusion. Proceedings of the Zoological Society of London 121: 777–817.

Butler, P. M. & Mills, J. R. E. (1959): A contribution to the odontology of *Oreopithecus*. Bulletin of the British Museum of Natural History, Geology 4: 1–26.

Cabanes, D. & Shahack-Gross, R. (2015): Understanding fossil phytolith preservation: the role of partial dissolution in paleoecology and archaeology. PLoS One 10: e0125532.

*Calandra, I., Schulz, E., Pinnow, M., Krohn, S. & Kaiser, T. M. (2012): Teasing apart the contributions of hard dietary items on 3D dental microtextures in primates. Journal of Human Evolution 63: 85–98.

*Calandra, I., Labonne, G., Schulz-Kornas, E. & Kaiser, T. M. (2016): Tooth wear to quantify intra-specific variations in diet and chewing mechanics. Scientific Reports 6: 34037.

Calandra, I., Pedergnana, A., Gneisinger, W. & Marreiros, J. (2019a): Why should traceology learn from dental microwear, and vice-versa? Journal of Archeological Science 110: 105012.

Calandra, I., Schunk, L., Bob, K., Gneisinger, W., Pedergnana, A., Paixao, E., Hildebrandt, A. & Marreiros, J. (2019b): The effect of numerical aperture on quantitative use-wear studies and its implication on reproducibility. Scientific Reports 9: 6313.

Calandra, I., Schunk, L., Rodriguez, A., Gneisinger, W., Pedergnana, A., Paixao, E., Pereira, T., Iovita, R. & Marreiros, J. (2019c): Back to the edge: relative coordinate system for use-wear analysis. Archaeological and Anthropological Sciences 11: 5937–5948.

Covert, H. H. & Kay, R. F. (1981): Dental microwear and diet: Implications for determining the feeding behaviour of extinct primates with a comment on the dietary pattern of *Sivapithecus*. American Journal of Physical Anthropology 55: 331–336.

Dahlberg, A. A. & Kinzey, W. (1962): Etude microscopique de l'abrasion et de l'attrition sur la surface des dents. Bulletin du Groupement international pour la Recherche scientifique en Stomatologie 5: 242–251.

Damuth, J. & Janis, C. M. (2011): On the relationship between hypsodonty and feeding ecology in ungulate mammals, and its utility in palaeoecology. Biological Reviews of the Cambridge Philosophical Society 86: 733–758.

Dominguez-Rodrigo, M. & Baquedano, E. (2018) Distinguishing butchery cut marks from crocodile bite marks through machine learning methods. Scientific Reports 8: 5786.

Dumont, J. (1982): The quantification of microwear traces: a new use for interferometry. World Archaeology 14: 206–217.

Evans, A. R., Wilson, G. P., Fortelius, M. & Jernvall, J. (2007): High-level similarity of dentitions in carnivorans and rodents. Nature 445: 78–81.

*Gailer, J.-P. & Kaiser, T. M. (2014): Common solutions to resolve different dietary challenges in the ruminant dentition: The functionality of bovid postcanine teeth as a masticatory unit. Journal of Morphology 275: 328–341.

*Gailer, J.-P., Calandra, I., Schulz-Kornas, E. & Kaiser, T. M. (2016): Morphology is not destiny: discrepancy between form, function and dietary adaptation in bovid cheek teeth. Journal of Mammalian Evolution 23: 369–383.

Gordon, K. D. & Walker, A. C. (1983): Playing 'possum: a microwear experiment. American Journal of Physical Anthropology 60: 109–112.

Gügel, I. L., Grupe, G. & Kunzelmann, K.-H. (2001): Simulation of dental microwear: characteristic traces by opal phytoliths give clues to ancient human dietary behaviour. American Journal of Physical Anthropology 114: 124–138.

Hayden, B. (1979): Lithic Use-Wear Analysis. Academic Press, New York.

Hernesniemi, E., Blomstedt, K. & Fortelius, M. (2011): Multi-view stereo 3D reconstruction of lower molars of Recent and Pleistocene rhinoceroses for mesowear analysis. Palaeontologia Electronica 14: 1–15.

Heywood, J. J. N. (2010): Functional anatomy of bovid upper molar occlusal surfaces with respect to diet. Journal of Zoology 281: 1–11.

Hua, L. C., Brandt, E. T., Meullenet, J. F., Zhou, Z. R. & Ungar, P. S. (2015): Technical note: An in vitro study of dental microwear formation using the BITE Master II chewing machine. American Journal of Physical Anthropology 158: 769–775.

ISO 4287 (1984): ISO 4287-1:1984. Surface roughness – terminology: Part 1. Surface and its parameters. Replaced by ISO 4287:1997 by 1997-04-01. Beuth, Berlin.

ISO 4287 (1997): ISO 4287:1997. Geometrical Product Specifications (GPS) – Surface texture: Profile method – terms, definitions and surface texture parameters. Replaces ISO 4287-1:1984. Beuth, Berlin.

ISO/DIS 25178-2 (2007): Draft International Standard ISO/DIS 25178-2. Geometrical product specifications (GPS) – Surface texture: Areal – Part 2: Terms, definitions and surface texture parameters. International Organization for Standardization.

Kaiser, T. M .& Brinkmann, G. (2006): Measuring dental wear equilibriums – the use of industrial surface texture parameters to infer the diets of fossil mammals. Palaeogeography, Palaeoclimatology, Palaeoecology 239: 221–240.

*Kaiser, T. M., Clauss, M. & Schulz-Kornas, E. (2016): A set of hypotheses on tribology of mammalian herbivore teeth. Surface Topography Metrology Properties 4: 014003.

*Kaiser, T. M., Braune, C., Kalinka, G. & Schulz-Kornas, E. (2018): Nano-indentation of native phytoliths and dental tissues: implications for herbivore-plant combat and dental wear proxies. Evolutionary Systematics 2: 55–63.

*Kalthoff, D. C., Schulz-Kornas, E., Corfe, I., Martin, T., McLoughlin, S. & Schultz, J. A. (2019): Complementary approaches to tooth wear analysis in Tritylodontidae (Synapsida, Cynodontia) reveal a generalist diet. PLoS One 14: e0220188.

Kamminga, J. (1982): Over the edge: functional analysis of Australian stone tools. Occasional Papers in Anthropology 12. University of Queensland, St. Lucia.

Karme, A. J., Rannikko, J. C., Kallonen, A. P., Clauss, M. & Fortelius, M. (2016): Mechanical modelling of tooth wear. Journal of the Royal Society Interface 13: 120.

Kay, R. F. (1987): Analysis of primate dental microwear using image processing techniques. Scanning Microscopy 1: 657–662.

Kay, R. F. & Covert, H. H. (1983): True grit: A microwear experiment. American Journal Physical Anthropology 61: 33–38.

Kay, R. F. & Hiiemäe, K. M. (1974): Jaw movement and tooth use in recent and fossil primates. American Journal of Physical Anthropology 40: 227–256.

Keeley, L. H. (1980): Experimental Determination of Stone Tool Uses: a Microwear Analysis. Chicago University Press, Chicago.

Korobkova, G. F. (1970): Raboty eksperimental'noj gruppy Moldavskoj arkheologicheskoj ekspeditsii. In: Rybakov, B. A. (ed.). Arkheologicheskiie Otkrytia 1969 Goda. Nauka, Moscow: 455–456.

Korobkova, G. F. (1977): Eksperimental'no-trasologicheskoe izuchenie drevnikh serpov i mezoliticheskikh orudij. In: Rybakov, B .A. (ed.). Arkheologicheskiie Othrytia 1976 Goda. Nauka, Moscow: 455–456.

Korobkova, G. F. (1978): Drevnejshie zhatvennye orudiia i ikh proizvoditel'nost' (v svete eksperimental'no-trasologicheskogo izucheniia). Sovetskaia Arkheologiia 4: 36–52.

Korobkova, G. F. (1979): Eksperimentarno-trasologicheskoe izuchenie mezoliticheskikh i neoliticheskikh orudij. In: Rybakov, B. A. (ed.). Arkheologicheskiie Otkrytia I 1978 Goda. Nauka, Moscow: 347.

Korobkova, G. F. (1981): Ancient reaping tools and their productivity in the light of experimental tracewear analysis. In: Kohl, P. L. (ed.). The Bronze Age Civilization of Central Asia. Sharpe, New York.

Kubo, M. O., Yamada, E., Kubo, T. & Kohono, N. (2017): Dental microwear texture of extant sika deer with considerations on inter-microscope variability and surface preparation protocols. Biosurface and Biotribology 3: 155–165.

Kullmer, O., Benazzi, S., Fiorenza, L., Schulz, D., Bacso, S. & Winzen, O. (2009): Technical note: Occlusal fingerprint analysis: Quantification of tooth wear pattern. American Journal of Physical Anthropology 139: 600–605.

Lucas, P. W., Omar, R., Al-Fadhalah, K., Almusallam, A. S., Henry, A. G., Michael, S., Thai, L. A., Watzke, J., Strait, D. S. & Atkins, A. G. (2013): Mechanisms and causes of wear in tooth enamel:

implications for hominin diets. Journal of The Royal Society Interface 10: 20120923.
Martin, L. F., Krause L., Ulbricht, A., Winkler, D. E., Codron, D., Kaiser, T. M., Müller, J., Hummel, J., Clauss, M., Hatt, J.-M. & Schulz-Kornas, E. (2020) Dental wear at macro- and microscopic scale in rabbits fed diets of different abrasiveness: a pilot investigation. Palaeogeography, Palaeclimatology, Palaeoecology 556: 109886.
Martisius, N. L., Sidéra, I., Grote, M., Steele, T. E., McPherron, S. P. & Schulz-Kornas, E. (2018): Time wears on: understanding how bone wears using 3D surface texture analysis. PLoS One 13: e0206078.
Martisius, N. L., McPherron, S. P., Schulz-Kornas, E., Soressi, M. & Steele, T. E. (2020): A method for the taphonomic assessment of bone tools using 3D surface texture analysis of bone microtopography. Archaeological and Anthropological Sciences 12: 251.
Mecks, N. D., Sieveking, G. D. G., Tite, M. S. & Cook, J. (1982): Glass and usewear traces in flint sickles and similar phenomena. Journal of Archaeological Science 9: 317–340.
Merceron, G., Schulz, E., Kordos, L. & Kaiser, T. M. (2007): Paleoenvironment of *Dryopithecus brancoi* at Rudabánya, Hungary: evidence from dental meso- and micro-wear analyses of large vegetarian mammals. Journal of Human Evolution 53: 331–349.
Mills, J. R. E. (1955): Ideal dental occlusion in the primates. Dental Practitioner 6: 47–61.
Mills, J. R. E. (1963): Occlusion and Malocclusion of Teeth of Primates. Pergamon Press, Oxford.
Mills, J. R. E. (1966): The functional occlusion of the teeth of Insectivora. Zoological Journal of the Linnean Society 47: 1–25.
Mills, J. R. E. (1967): A comparison of lateral jaw movements in some mammals from wear facets on the teeth. Archives of Oral Biology 12: 645–661.
Müller, J., Clauss, M., Codron, D., Schulz, E., Hummel, J., Fortelius, M., Kircher, P. & Hatt, J.-M. (2014): Growth and wear of incisor and cheek teeth in domestic rabbits (*Oryctolagus cuniculus*) fed diets of different abrasiveness. Journal of Experimental Zoology. Part A, Ecological Genetics and Physiology 321: 283–298.
Müller, J., Clauss, M., Codron, D., Schulz-Kornas, E., Hummel, J., Fortelius, M., Kircher, P. & Hatt, J.-M. (2015): Tooth length and incisal growth and wear in guinea pigs (*Cavia porcellus*) fed diets of different abrasiveness. Animal Physilogy and Animal Nutrition 99: 591–604.
*Nieberg, C., Gailer, J.-P. & Kaiser, T. M. (2009): Quantifying functional traits in the ungulate dentition. Terra Nostra 3: 83.
Nilsson, S. (1834-43): Skandinaviska nordens ur-invanare, ett försök i komparativa ethnografien och ett bidrag till menniskoslägtets utvecklings-historia. Berlingska Boktryckerikt, Lund.
Pedergnana, A., Calandra, I., Bob, K., Gneisinger, W., Paixao, E., Schunk. L., Hildebrandt, A. & Marreiros, J. (in press): Evaluating the microscopic effect of brushing stone tools as a cleaning procedure. Quaternary International DOI: 10.1016/j.quaint.2020.06.031.
Peters, C. R. (1982): Electron-optical microscopic study of incipient dental microdamage from experimental seed and bone crushing. American Journal of Physical Anthropology 57: 283–301.
Poznyak, A., Chairez, I. & Poznyak, T. (2019): A survey on artificial neural networks application for identification and control in environmental engineering: Biological and chemical systems with uncertain models. Annual Reviews in Control 48: 250–272.
Purnell, M., Seehausen, O. & Galis, F. (2012): Quantitative three-dimensional microtextural analyses of tooth wear as a tool for dietary discrimination in fishes. Journal of The Royal Society Interface 9: 2225–2233.
Ramdarshan, A., Blondel, C., Gautier, D., Surault, J. & Merceron, G. (2016): Seeds, browse, and tooth wear: a sheep perspective. Ecology and Evolution 6: 5559–5569.
Rensberger, J. M. (1978) Scanning electron microscopy of wear and occlusal events in some small herbivores. In: Butler, P. M. & Josey, K. A. (eds.). Development, Function and Evolution of Teeth. Academic Press, London: 415–438.
Rensberger, J. M. (1982): Patterns of dental change in two locally persistent successions of fossil aplodontid rodents. In: Kurten, B. (ed.). Teeth: Form, Function, and Evolution. Columbia University Press, New York: 323–349.
Richter, R. (1928) Aktuopaläontologie und Paläobiologie, eine Abgrenzung. Senckenbergiana 10: 285–292.
Rivals, F., Schulz, E. & Kaiser, T. M. (2009a): Late and middle Pleistocene ungulates dietary diversity in Western Europe indicate variations of Neanderthal paleoenvironments through time and space. Quaternary Science Reviews 28: 3388–3400.
Rivals, F., Schulz, E. & Kaiser, T. M. (2009b): A new application for dental wear analyses: estimation of duration of hominid occupations in archaeological localities. Journal of Human Evolution 56: 329–339.
Rose, J. C. & Ungar, P. S. (1998): Gross dental wear and dental microwear in historical perspective. In: Alt, K. W., Rösing, F. W. & Teschler-Nicola, M. (eds.). Dental Anthropology: Fundamentals, Limits, and Prospects. Springer, New York: 349–384.
Sanson, G. D., Kerr, S. A. & Gross, K. A. (2007): Do silica phytoliths really wear mammalian teeth? Journal of Archaeological Science 34: 526–531.
Saarinen, J., Karme, A., Cerling, T., Uno, K., Säilä, L., Kasiki, S., Ngene, S., Obari, T., Mbua, E., Manthi, F. K. & Fortelius, M. (2015): A new tooth wear–based dietary analysis method for Proboscidea (Mammalia). Journal of Vertebrate Paleontology 35: e918546.
Schulz, E. & Kaiser, T. M. (2013): Historical distribution, habitat requirements and feeding ecology of the genus *Equus* (Perissodactyla). Mammal Review 43: 111–123.
*Schulz, E., Calandra, I. & Kaiser, T. M. (2010): Applying tribology to teeth of hoofed mammals. Scanning 32: 162–182.
*Schulz, E., Calandra, I. & Kaiser, T. M. (2013a): Feeding ecology and chewing mechanics in hoofed mammals: 3D tribology of enamel wear. Wear 300: 169–179.
Schulz, E., Fahlke, J. M., Merceron, G. & Kaiser, T. M. (2007): Feeding ecology of the Chalicotheriidae (Mammalia, Perissodactyla, Ancylopoda). Results from dental micro- and mesowear analyses A. Introduction. Verhandlungen des Naturwissenschaftlichen Vereins in Hamburg 43: 5–32.
*Schulz, E., Fraas, S., Kaiser, T. M., Cunningham, P. L., Ismail, K. & Wronski, T. (2013b): Food preferences and tooth wear in the sand gazelle (*Gazella marica*). Mammalian Biology 78: 55–62.
*Schulz, E., Piotrowski, V., Clauss, M., Mau, M., Merceron, G. & Kaiser, T. M. (2013c): Dietary abrasiveness is associated with variability of microwear and dental surface texture in rabbits. PLoS One 8: e56167.
*Schulz-Kornas, E., Braune, C., Winkler, D. E. & Kaiser, T. M. (2017): Does silica concentration and phytolith ultrastructure relate to phytolith hardness? Biosurface and Biotribology 3: 135–143.
Schulz-Kornas, E., Winkler, D. E., Clauss, M., Carlsson, J., Ackermans, N. L., Martin, L. F., Hummel, J., Müller, D. E. W., Hatt, J.-M. & Kaiser, T. M. (2020) Everything matters: molar microwear texture in goats (*Capra aegagrus hircus*) fed diets of different abrasiveness. Palaeogeography, Palaeclimatology, Palaeoecology 552: 109783.
Schulz-Kornas, E., Stuhlträger, J., Clauss, M., Wittig, R. M. & Kupczik, K. (2019): Dust affects chewing efficiency and tooth wear in forest dwelling Western chimpanzee (*Pan troglodytes verus*). American Journal of Physical Anthropology 169: 66–77.
*Schultz, J. A., Anders, U., Braune, C., Brinkkötter, J. J., Calandra, I., Engels, S., Findeisen, E., Gailer, J. P., Hummel, J., Jäger, K. R. K., Kaiser, T. M., Kalthoff, D. C., Koenigswald, W. v., Kullmer, O., Landwehr, C. Mau, M., Menz, U., Ruf, I., Schubert, A. M., Schulz-Kornas, E., Schwermann, A. H., Schwermann, L. C., Skiba, M., Steuer, P., Südekum, K.-H., Winkler D. E. & Martin, T. (2020): A new wear facet terminology for mammalian dentitions. In: Martin, T. & Koenigswald, W. v. (eds.). Mammalian Teeth – Form and Function. Pfeil, Munich: 11–24.
*Schultz, J. A., Menz, U., Winkler, D. E. Schulz-Kornas, E., Engels, S., Kalthoff, D. C., Koenigswald, W. v., Ruf, I., Kaiser, T., Kullmer, O., Südekum K.-H. & Martin, T. (2018): Modular Wear Facet Nomenclature for mammalian post-canine dentitions. Historical Biology 30: 30–41.

Scott, R. S., Ungar, P. S., Bergstrom, T. S., Brown, C. A., Grine, F. E., Teaford, M. F. & Walker, A. (2005): Dental microwear texture analysis shows within-species diet variability in fossil hominins. Nature 436: 693-695.

Scott, R. S., Ungar, P. S., Bergstrom, T. S., Brown, C. A., Childs, B. E., Teaford, M. F. & Walker, A. C. (2006): Dental microwear texture analysis: technical considerations. Journal of Human Evolution 51: 339-349.

Semenov, S. A. (1957): Pervobytnaia Tekhnika, Moscow.

Simpson, G. G. (1926) Mesozoic Mammalia, IV; the multituberculates as living animals. American Journal of Science 11: 228-250.

Simpson, G. G. (1933): Paleobiology of Jurassic mammals. Paleobiologica 5: 127-158.

Solounias, K. & Hayek, L.-A. C. (1993): New methods of tooth microwear analysis and application to dietary determination of two extinct antelopes. Journal of Zoology 229: 421-445.

Solounias, N. & Semprebon, G. (2002): Advances in the reconstruction of ungulate ecomorphology and application to early fossil equids. American Museum Novitates 3366: 1-49.

Stout, K. J. (2000): Development and methods for the characterization of roughness in three dimensions. Penton, London.

Strait, S. G. & Overdorff, D. J. (1994): A preliminary examination of molar microwear in strepsirhine primates. American Journal of Physical Anthropology 92: 63-79.

Strömberg, C. A. E. (2006): Evolution of hypsodonty in equids: testing a hypothesis of adaptation. Paleobiology 32: 236-258.

Stuhlträger, J., Schulz-Kornas, E., Wittig, R. M. & Kupczik, K. (2019): Ontogenetic dietary shifts and microscopic tooth wear in western chimpanzees. Frontiers in Ecology and Evolution 7: 298.

Taylor, L. A., Kaiser, T. M., Schwitzer, C., Müller, D. W. H., Codron, D., Clauss, M. & Schulz, E. (2013): Detecting inter-cusp and inter-tooth wear patterns in rhinocerotids. PLoS One 8: e80921.

*Taylor, L. A., Müller, D. W. H., Schwitzer, C., Kaiser, T. M., Codron, D., Schulz, E. & Clauss, M. (2014): Tooth wear in captive rhinoceroses (*Diceros*, *Rhinoceros*, *Ceratotherium*: Perissodactyla) differs from that of free-ranging conspecifics. Contributions to Zoology 83: 109-119.

Teaford, M. F. & Glander, K. E. (1991): Dental microwear in live, wild-trapped *Alouatta palliata* from Costa Rica. American Journal of Physical Anthropology 85: 313-319.

Teaford, M. F. & Oyen, O. J. (1989a): Differences in the rate of molar wear between monkeys raised on different diets. Journal of Dental Research 68: 1513-1518.

Teaford, M. F. & Oyen, O. J. (1989b): Live primates and dental replication: New problems and new techniques. American Journal of Physical Anthropology 80: 73-81.

Teaford, M. F. & Runestad, J. A. (1992): Dental microwear and diet in Venezuelan primates. American Journal of Physical Anthropology 88: 347-364.

Teaford, M. F. & Oyen, O. J. (1989c): In vivo and in vitro turnover in dental microwear. American Journal of Physical Anthropology 80: 447-460.

Teaford, M. F. & Walker, A. (1983): Dental microwear in adult and still-born guinea pigs (*Cavia porcellus*). Archives of Oral Biology 28: 1077-1081.

Townsend, A., Senin, N., Blunt, L., Leach, R. K. & Taylor, J. S. (2016) Surface texture metrology for metal additive manufacturing: a review. Precision Engineering 46: 34-47.

Turcotte, C. M., Green, D. J., Kupczik, K., McFarlin, S. C. & Schulz-Kornas, E. (2019): Elevated activity levels do not influence extrinsic fiber attachment morphology on the surface of muscle-attachment sites. Journal of Anatomy 236: 827-839.

Ulbricht, A., Maul, L. C. & Schulz, E. (2015): Can mesowear analysis be applied to small mammals? A pilot-study on leporids and murines. Mammalian Biology 80: 14-20.

Ungar, P. S., Brown, C. A., Bergstrom, T. S. & Walker, A. C. (2003): Quantification of dental microwear by tandem scanning confocal microscopy and scale-sensitive fractal analysis. Scanning 25: 185-193.

Ungar, P. S., Simon, J.-C. & Cooper, J. W. (1991): A semiautomated image analysis procedure for the quantification of dental microwear. Scanning 13: 31-36.

Ungar, P. S., Teaford, M. F., Glander, K. E. & Pastor, R. F. (1995): Dust accumulation in the canopy: a potential cause of dental microwear in primates. American Journal of Physical Anthropology 97: 93-99.

van Casteren, A., Lucas, P. W., Strait, D. S., Michael, S., Bierwisch, N., Schwarzer, N., Al-Fadhalah, K., Almusallam, A. S., Saji, S., Shekeban, A. & Swain, M. V., 2018. Evidence that metallic proxies are unsuitable for assessing the mechanics of microwear formation and a new theory of the meaning of microwear. Royal Society Open Science 5: 171699.

van Casteren, A., Strait, D. S., Swain, M. V., Michael, S., Thai, L. A., Philip, S. M., Saji, S., Al-Fadhalah, K., Almusallam, A. S., Shekeban, A., McGraw, W. S., Kane, E. E., Wright, B. W. & Lucas, P. W. (2020): Hard plant tissue do not contribute meaningfully to dental microwear: evolutionary implications. Scientific Reports 10: 582.

Walker, A. C. (1977): Scanning electron microscopy of early hominoid teeth. Proceedings of the forty-sixth annual meeting of the American Association of Physical Anthropologists, Washington, D.C., April 14-16, 1977. American Journal of Physical Anthropology 47: 165.

Walker, A. C., Hoeck, H. N. & Perez, L. (1978): Microwear of mammalian teeth as an indicator of diet. Science 201: 908-910.

Walker, P. L. (1976): Wear striations on the incisors of ceropithecid monkeys as an index of diet and habitat preferences. American Journal of Physical Anthropology 45: 299-308.

Winkler, D. E. & Kaiser, T. M. (2011): A case study of seasonal, sexual and ontogenetic divergence in the feeding behaviour of the moose (*Alces alces* Linné, 1758). Verhandlungen des Naturwissenschaftlichen Vereins in Hamburg 46: 331-348.

Winkler, D. E., Andrianasolo, T., Andriamandimbiarisoa, L., Ganzhorn, J., Rakotondranary, J., Kaiser, T. M. & Schulz-Kornas, E. (2016): Tooth wear patterns in black rats (*Rattus rattus*) of Madagascar differ more in relation to human impact than to differences in natural habitats. Mammalian Evolution 23: 369-383.

*Winkler, D. E., Schulz, E., Calandra, I., Gailer, J.-P., Landwehr, C. & Kaiser, T. M. (2013a): Indications for a dietary change in the extinct bovid genus *Myotragus* (Plio-Holocene, Mallorca, Spain). Geobios 46: 143-150.

Winkler, D. E., Schulz-Kornas, E., Kaiser, T. M., De Cuyper, A., Clauss , M. & Tütken, T. (2019a): Forage silica and water content control dental surface texture in guinea pigs: implications for diet reconstruction. Proceedings of the National Academy of Sciences of the Unites States of America 116: 1325-1330.

Winkler, D. E., Schulz-Kornas, E., Kaiser, T. M. & Tütken, T. (2019b): Tooth wear in extant Lepidosauria reflects dietary traits despite limited mastication behavior. Proceedings of the Royal Society B 286: 20190544.

*Winkler, D. E., van der Hoek Ostende, L. W., Schulz, E., Calandra, I., Gailer, J.P., Landwehr, C. & Kaiser, T. M. (2013b): Dietary divergence in space and time – Lessons from the dwarf-goat *Myotragus balearicus* (Plio-Holocene, Mallorca, Spain). Mammalian Biology 78: 430-437.

Winkler, D. E., Tütken, T., Schulz-Kornas, E., Kaiser, T. M., Müller, J., Leichliter, J., Weber, K., Hatt, J.-M. & Clauss, M. (2020): Shape, size, and quantity of external abrasives influence dental microwear texture formation in guinea pigs. Proceedings of the National Academy of Sciences of the United States of America 117: 22264-22273.

*Wronski, T. & Schulz-Kornas, E. (2015): The Farasan gazelle – a frugivorous browser in an arid environment? Mammalian Biology 80: 87-95.

Xia, J., Tian, Z. R., Hua, L. C., Chen, L., Zhou, Z. R., Qian, L. M. & Ungar, P. S. (2017): Enamel crystallite strength and wear: nanoscale responses of teeth to chewing loads. Journal of The Royal Society Interface 14: 20170456.

Xia, J., Zheng, J., Huang, D. D., Tian, Z. R., Chen, L., Zhou, Z. R., Ungar, P. S. & Qian, L. M. (2015): New model to explain tooth wear with implications for microwear formation and diet reconstruction. Proceedings of the National Academy of Sciences of the Unites States of America 112: 10669-10672.

Xia, J., Zhou, Z., Qian, L. & Ungar, P. S., 2018. Comment on van Casteren et al. (2018): Softer metallic spheres do abrade harder enamel. Royal Society Open Science 5: 181376.

Mandibular biomechanics as a key factor to understand diet in mammals

Jordi Marcé-Nogué

Introduction

The mammalian mandible plays a fundamental role in food processing, as it transmits the forces generated by masticatory muscles to the food through the teeth (Maynard-Smith & Savage 1959). Generally speaking, the greater the forces required to fracture ingesta, and the more often such forces are needed, the stronger the mandible has to be to maintain its structural integrity (Hogue 2008). In that line, it has been suggested that among primates, those that include hard-food in their diets have stiffer mandibles when compared to soft-food eaters (Marcé-Nogué et al. 2017b). However, other attempts to establish relationships between mandibular morphology and dietary categories have not been particularly successful (Ross et al. 2012). If the latter is true and biomechanics cannot be used to establish relationships between morphology and function, it has been argued that it is probably because these biomechanical analyses are flawed or incomplete, or that non-mechanical factors such as phylogeny are more important determinants of morphology (Ross & Iriarte-Diaz 2014). Nevertheless, there is a general consensus that trying to understand the interaction between the mammalian feeding mechanism and the processed ingesta represents a unique opportunity for the study of ecomorphological adaptations in extant species, as well as having the potential to acquire valuable tools for the reconstruction of oral behaviors in extinct taxa. Hence the importance of this particular topic.

The association between dietary preferences and biomechanical performance has traditionally been studied as the relationship between dental and mandibular morphology (Fig. 4.1). For example, carnivorous mammals have carnassials that act as blades allowing enlarged and often self-sharpening edges to pass by each other in a shearing manner. Herbivorous ungulates have developed reduced canine teeth and specialized molars to grind fibrous food by lateral movements. Among other examples, primates have teeth that exhibit an elaborated cusp pattern of the molars that crush the food (Kardong 2014).The way teeth work (e.g. blading, grinding, or crushing) is adapted to the diet as well as the morphology of the mandibles, thus enabling the generation of different models of distribution and position of the muscular forces and bite positions based on different geometries (Vizcaíno et al. 2016). The influence of muscular forces on the design of the shape of the mandible (e.g., Sella-Tunis et al. 2018 in humans), as well as the influence of the position of the condyle allowing different mandible movements (Maynard-Smith & Savage 1959) are evidence that reinforces the idea that food is processed differently depending on mandibular morphology.

For these reasons, feeding biomechanics represents the set of analyses performed to "[...] relate variation in feeding system morphology (size and shape of muscles, bones, teeth, tongues or joints) to variation in how animals feed (feeding behavior) and what they feed on (diet)." (Ross & Iriarte-Diaz 2014: 105). This relationship between muscular and bite forces, mandibular size and shape, and the consumed ingesta have previously been studied in the mandibles of different mammalian clades such as terrestrial Cetartiodactyla (Christiansen & Wroe 2007, Hogue & Ravosa 2001, Spencer 1995, Varela & Fariña 2015), Chiroptera (Freeman 1979, 1981, 1984, 1988, 2000), Primates (Bouvier 1986a, b, Taylor 2006) and Carnivora (Biknevicius & Ruff 1992, Radinsky 1981a, b). These works used classical biomechanics based on the laws of static mechanics and beam theory, hence not requiring the application of computer simulation techniques to model complex biological geometries, thus avoiding a high computational cost.

The eruption of computational methods opened a whole array of new possibilities to investigate feeding biomechanics. Specifically, computational biomechanics represents the application of engineering computational tools, such as the finite element analysis (FEA) (Zienkiewicz & Taylor 1981) to study the mechanics of biological systems. Computational models and simulations are particularly useful to predict the relationship between parameters that are otherwise challenging to test using experimental approaches and are also quite helpful when designing more relevant experiments, hence reducing the time and costs of such tests. The underlying premise of the method is that a complex geometry can be subdivided into a mesh consisting of a number of finite elements in which the respective mechanical equations are approximately solved. Although FEA (see Rayfield 2007 for a review) or multibody dynamic analysis (see Curtis 2011 for a review of MDA) have been common approaches in engineering and biomedicine for more than thirty years, they have also been applied to biological research in order to address questions related to biomechanics and the evolution of living and extinct vertebrates (using both 2D or high-resolution 3D models). Vertebrate paleontologists have also applied complementary non-invasive techniques, such as computed tomography (CT), which is a useful tool to generate accurate 3D images of living structures in a reverse engineering process, thus facilitating the application of computational biomechanics. The reliability and bio-fidelity of the biomechanical models has increasingly improved by capturing the real geometries of the specimens under analysis. Previous approaches

T. Martin & W. v. Koenigswald (eds.): Mammalian Teeth – Form and Function. Pp. 54–80, 13 figs., 8 tabs.
 – ISBN 978-3-89937-266-3
DOI: http://doi.org/10.23788/mammteeth.04 Published 22 December 2020

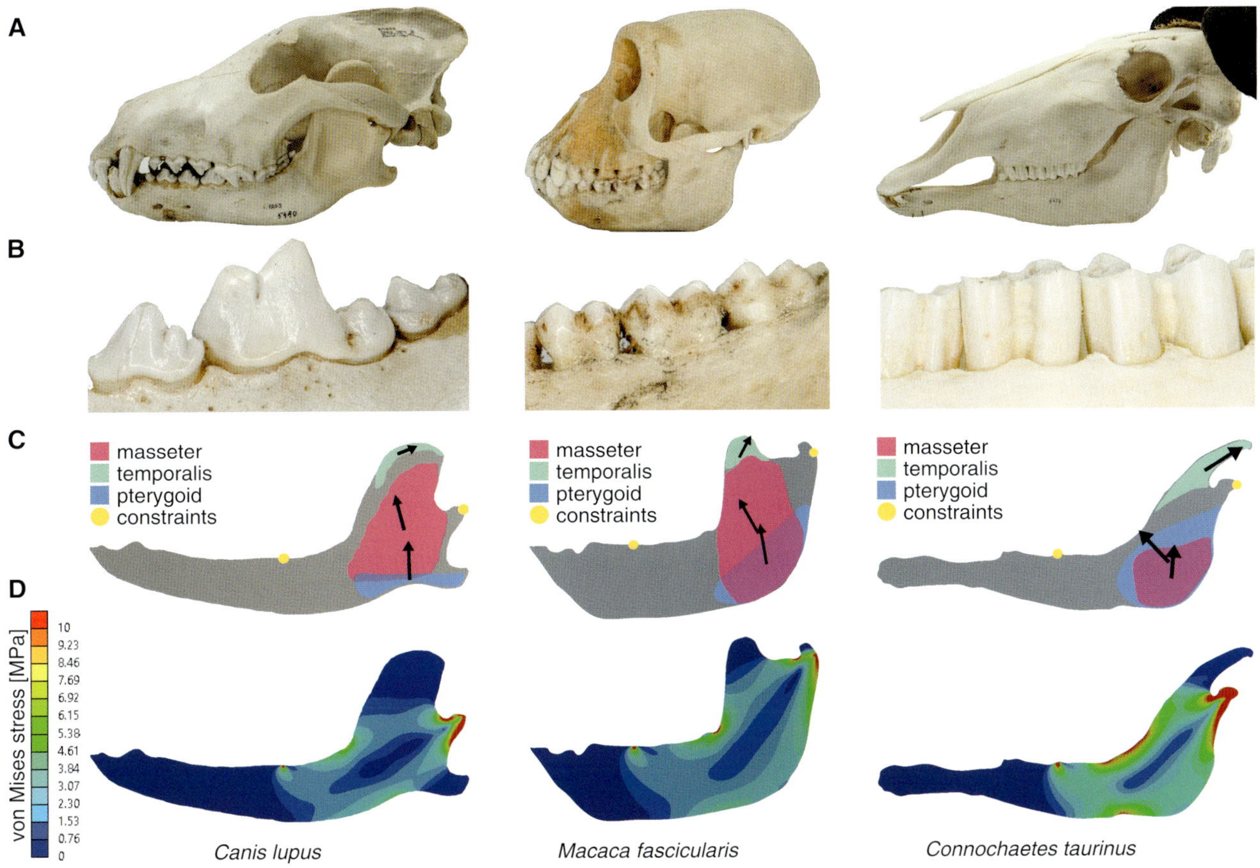

Fig. 4.1. Morphology and bite forces in mammalian crania and mandibles. **A,** cranium and mandible; **B,** details showing different tooth morphologies; **C,** free-body diagram of the biomechanical problem of the mandible during chewing; **D,** von Mises stress results after a finite element analysis of three different taxa: *Canis lupus* (carnivore), *Macaca fascicularis* (primate) and *Connochaetes taurinus* (ungulate). Not to scale.

relied on simplified lever or beam models, which although still useful, possess many known limitations that can be overcome by modern computational biomechanics.

To date, FEA has been conducted in a wide spectrum of vertebrates providing new insights into biomechanical function, particularly regarding the constraints and adaptive values of certain morphologies and bone structures. These bony structures are generally considered to be a force transmission system that can be studied by mean of mechanics laws. For instance, the biomechanics of the mandible in different mammalian families has been studied using FEA to obtain the state of stresses during mastication in 3D models (Gill et al. 2014, Gröning et al. 2011b, Lautenschlager et al. 2018, Püschel et al. 2018, Tseng & Binder 2010, Zhou et al. 2019), as well as using simplified planar models (Fletcher et al. 2010, Marcé-Nogué et al. 2017b, Neenan et al. 2014, Piras et al. 2013, Serrano-Fochs et al. 2015). Other studies have approached this problem including the study of both the skull and the mandible (Dumont et al. 2011, Strait et al. 2013, Tseng 2013, Veitschegger et al. 2018). Nonetheless, it is important to keep in mind that if a study aims to detect dietary preferences, it would be preferred to analyze mandibles as part of the antagonistic system, since the lower jaw may be more susceptible to failure than the cranium (Wroe et al. 2010), and also because cranium morphology is influenced by several other non-feeding variables (Tseng & Flynn 2018). In fact, it has been noted that cranial shape reflects a compromise between several diverse functions (e.g., phonation, cognition, respiration), which could mask a strong dietary signal (Püschel et al. 2018). By contrast, the mandible is primarily involved in food acquisition and consumption, and consequently it is to be expected that its morphology better reflects ingesta-related adaptations (Gröning et al. 2011b).

It is logical to wonder how paleontologists and paleoanthropologists have applied these biomechanical tools in fossil taxa, taking into account that many variables are unknown and impossible to obtain from skeletal morphology alone. The answer is relatively simple. If it is possible to study extant species that are phylogenetically related and/or share the same or similar diet to the fossil taxa of interest, and if it is possible to find correlations, associations or differences to their dietary preferences or their ingesta characteristics, then it would be possible to study the biomechanics of the fossil taxa and find the appropriate relationships with the biomechanical data. Nonetheless, there are two important questions that need to be addressed before carrying out analyses on fossil specimens. These questions are related to the intrinsic nature of biological data and make the study of biological structures completely different to the more common mechanical problems solved by engineers in other fields. The first one is variation, which focuses on how different models are properly compared when considering both intra and inter species variability. This means that it is necessary to create an appropriate comparative framework where a large number of different models and data can be analyzed and compared. The second one is how to post-process the extremely large amount of data derived from FEA results or classical biomechanics in a quantitative way that allows the obtained data to be used and combined with statistics, geometric morphometrics, phylogenetic models or even machine-learning algorithms to try to find affinities between the feeding behavior and the diet of the taxa.

Consequently, the aim of this chapter is to discuss classical and computational biomechanics of the mandible in the study of dietary adaptations in mammals. Firstly, a brief theoretical background in mechanics will be provided to facilitate understanding of the concepts used in this chapter. Afterwards, these concepts will be used in examples of biomechanics of the mandible. Secondly, the use of FEA in mammalian mandibles will be discussed, along with the concept of a comparative framework that enables the comparison of different models, as well as reviewing the latest methods to quantify and compare a large number of FEA models. Finally, an example using different mandibles of armadillos will be provided and discussed to further develop the understanding of the topics under discussion. This will provide insights about the connection between mandibular biomechanical performance and dietary preferences.

Brief introduction to biomechanics

Theoretical background

Biomechanics is the study of the structure and function of the mechanical aspects of biological systems, at any biologically relevant organization level, from whole organisms to small cells, using the methods of mechanics (Alexander 2005). By contrast, mechanics is a more general field of science concerned with the behavior of any physical body subjected to forces or displacements.

Biomechanics comprises several sub-areas, such as for example aerodynamics to study the flight of birds and insects, hydrodynamics to study how fish swim, locomotor mechanics to study how animals move, as well as the study of chewing mechanisms using static theory, among other sub-fields. Statics is the branch of mechanics that is concerned with the analysis of loads acting on physical systems that are in static equilibrium with their environment. It means that a body is in static equilibrium when it fulfils the first condition for equilibrium, which requires that the net force applied to the system must equal zero, as well as the second condition for equilibrium, meaning that the net torque applied to the system must also be zero. This is the application of Newton's second law when the body has a null acceleration.

Newton's three laws of motion are physical laws that, taken together, constitute the basis of classical mechanics.

- First law: An object in uniform motion remains in that state of motion unless an external force is applied to it.
- Second law: The vector sum of the external forces F on an object is equal to the mass of that object multiplied by the acceleration vector of the object.
- Third law: When one body exerts a force on a second body, the second body simultaneously exerts a force equal in magnitude and opposite in direction against the first body. They are called action force and reaction force.

Newton's laws describe the relationship between a body and the forces acting upon it. Therefore, they are the base of any biomechanical model, both classical as well as the computational ones.

Free-body diagram. A free-body diagram (FBD) is a graphical illustration which is used to visualize the applied forces, movements, and resulting reactions on a body in a given condition or mechanical scenario. They depict a body or connected bodies with all the applied forces and moments, as well as reactions that act on that/those bodies. An FBD is an important step in understanding statics, dynamics and other forms of classical mechanics, because is not meant to be a scaled drawing. It is a diagram that is modified as the problem is solved.

A free body diagram explicitly excludes some things: (1) Bodies other than the free body (2) Constraints; the body is not free from constraints; constraints have just been replaced by the forces and moments that they exert on the body (Newton's third law) (3) Forces exerted by the free body (supported by Newton's third law) (4) Internal forces and (5) in the case to solve a dynamic problem, the velocity or acceleration vectors.

Lever mechanics. Lever mechanics are based on Newton's laws and the condition for static equilibrium of statics. A lever is a body connected to a fixed point by a hinge or pivot called fulcrum and obeys the following physical law: the ratio of output to input force is given by the ratio of the distances from the fulcrum to the points of application of these forces.

Lever mechanics is useful in the context of chewing biomechanics, when studying the effect of muscles of the mandible on the bite force. There is a parameter called mechanical advantage (MA, equation 1) that is a measure of the force amplification achieved by using a mechanical system and is defined as the ratio of the force produced by the mechanical system (F_{output}) to the force applied to it (F_{input})

$$MA = \frac{F_{output}}{F_{input}} \quad \text{(equation 1)}$$

This parameter is useful because it is a dimensionless proportion (or ratio) of forces and, consequently, is not related to the size of the system. It allows to compare models without the influence of size on the results.

Elasticity: Stress and strain. Up to this point we have presented the concepts of biomechanics without considering what is happening inside the analyzed bodies, and without taking into consideration the effect of the forces on the deformation of a body. All the bodies in our surroundings experience deformations. For instance, there could be small deformations that we cannot see in concrete buildings, or there might be larger deformations in rubbers, such as in the case of a chewing gum.

In physics, elasticity is the ability of a body to resist reversible deformations when it is under the action of external forces, and to return to its original size and shape when that force is removed. Although solid objects will deform when adequate forces are applied to them, it is important to bear in mind that perfect elasticity is only an approximation of the real world because few materials remain purely elastic, even after very small deformations.

When we want to study what is happening inside a body, and how this body deforms under the application of a load, we need to study the laws of elasticity using the methods of continuum mechanics (Mase & Mase 1999). Continuum mechanics is a branch of mechanics that studies the mechanical behavior of materials when they are modelled not as discrete particles but as a continuous mass.

In biomechanics, bony structures can be considered bodies modelled as a continuous mass with small deformations, which means that one can apply to them all the equations derived from continuum mechanics and the linear theory of elasticity. Other biological structures such as soft tissues, are beyond the scope of this chapter and should be analyzed using non-linear theories.

When studying continuum mechanics, two concepts of great interest need to be mentioned, namely stress and strain. Stress is a quantity that corresponds to the internal forces that adjacent particles of a continuous material apply on each other. Stress is computed in a specific plane of the body and it is equivalent to the force divided by the surface of this plane. Strain is the ratio of the deformation of the material. It is computed as the proportion of displacement when a body is deformed with the original dimension of the body. Both are commonly used in biomechanics to study what is happening inside a body subjected to external loads.

$$\sigma = \frac{F_x}{S}\left[\frac{N}{mm^2}\right] \quad \text{(equation 2)}$$

$$\varepsilon = \frac{x}{l}\left[-\right] \quad \text{(equation 3)}$$

Where σ is the stress, F is the force, S the surface, ε the strain, x the distance of the elongation and l the original length of the bar (Fig. 4.2A). Compression or traction are obtained depending on how the balanced forces are applied (i.e. inwardly or outwardly) (Fig. 4.2B).

In general, stress and strain are computed in a specific point of the body in relation to a perpendicular plane and may be regarded as the sum of two components: the normal stress (compression or tension) perpendicular to this plane (Fig. 4.3A), and the shear stress that is parallel to this plane surface (Fig. 4.3B). Note that, for one point it is possible to define infinite normal and shear stress as function of the perpendicular plane in which they are computed (see tensorial notation below for further explanation).

$$\sigma = \frac{F_x}{S}\left[\frac{N}{mm^2}\right] \quad \text{(equation 4)}$$

$$\tau = \frac{F_y}{S}\left[\frac{N}{mm^2}\right] \quad \text{(equation 5)}$$

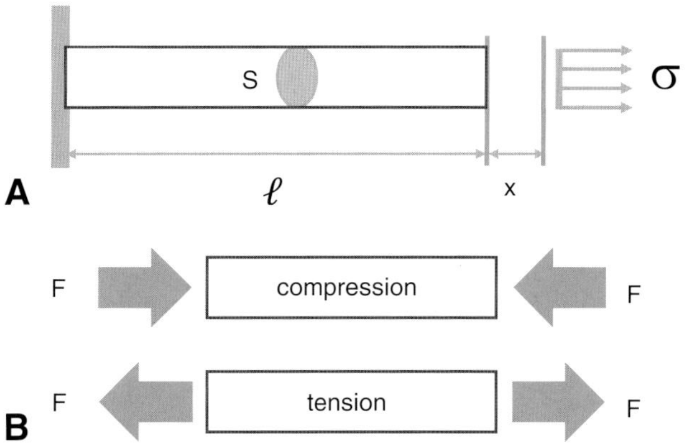

Fig. 4.2. Stress and strain. **A,** parameters of the definition of stress and strain and **B,** definition of tension and compression.

In linear elasticity, stress and strain are related via the Hooke's law that links both variables by means of the Young's modulus. E is the Young's modulus, a mechanical property of linear elastic solid materials. Constitutive equations are used to relate two physical quantities that are specific to a material and that describe the response of that material to external inputs such as applied fields or forces. They are combined with other equations governing physical laws to solve physical problems; the relationship between applied stresses to strains is a linear constitutive equation because the relationship between them is linear and proportional to the Young's modulus. Other materials such as soft tissues use other constitutive equations because they do not follow linear elasticity. For bones, the use of Hooke's law is quite common (Doblaré et al. 2004).

$$\sigma = E \cdot \varepsilon \quad \text{(equation 6)}$$

The Poisson coefficient is the ratio of transverse contraction strain to longitudinal extension strain in the direction of stretching force. It measures the Poisson effect, which is the phenomenon that happens when a material tends

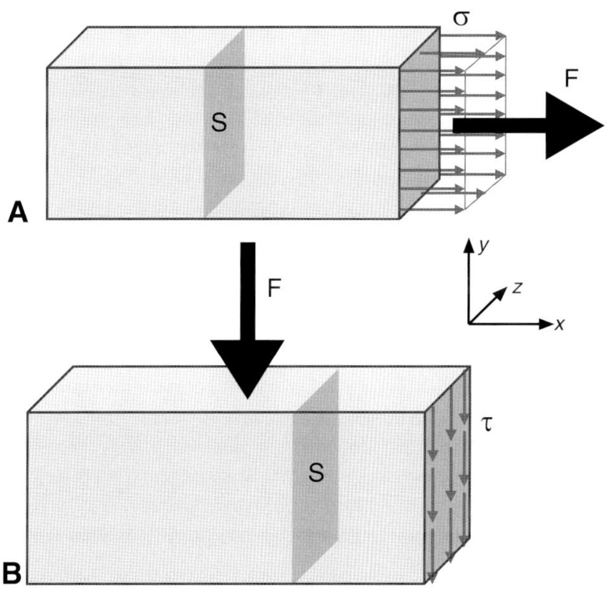

Fig. 4.3. Normal and shear stress. Definition of **A,** normal stress and **B,** shear stress, where F are the forces in the xyz coordinate axis and S is the surface.

to expand in directions perpendicular to the direction of compression or conversely, tends to contract if the material is stretched. This value is also necessary when defining the constitutive equation of the Hooke's law in two or three dimensions.

Beam mechanics. Classical beam theory is a simplification of the linear theory of elasticity that provides a way of calculating the loads and deflection of beams when they are subjected to outer forces. The main difference between lever mechanics and beam theory is that levers are not deformable whilst beam theory assumes the deformation of the body under study. Another important difference is that in lever mechanics only Newton's laws are used to solve the input and output forces whereas in beam theory, the use of the Newton's laws is needed to get the input and output forces and then the stresses and strains inside the beam. For this reason, a beam is a mechanical prism: a line with an associated cross-section where we calculate the stresses. It covers the case of small deflections of a beam that are subjected to lateral loads only. Moreover, Hooke's law is achieved between stresses and strains and plane sections keep on the plane after deformation. This means each cross-section of the beam is always at 90 degrees to the line that describes the beam (Timoshenko 1955).

Once we know the value of the input (actions) and output forces (reactions) solved by Newton's laws, the force that each cross-section of the beam is dealing with, is computed via force and moment laws keeping the equilibrium of forces in all the cross-sections. The equations of the forces and moment laws depend on the position of the section along the beam. They can be graphically displayed with force and moment diagrams to facilitate the interpretation. For example, a force applied in a specific point can vary its action in the cross-section due to the distance from this point creating a bending moment. The forces applied in the beam can create compression/tension, bending, shear and torsion.

The stresses (both normal and shear) in the cross-section of the beam depend on whether they are from compression/tension, bending, shear, and torsion or if they are linearly related with the forces applied via the Navier equations for bending, the Collignon theorem for shear, or the torsion equations (Timoshenko 1955). The diagrams are a good option to understand in which points the effects of the forces are higher in order to correlate them with anatomical or functional aspects (see the example below). In this case, the most important thing to consider is that the normal stress produced by the bending moment is parallel to the direction of the beam. If we decide that the axis of the beam is the x axis, then the normal stress is σ_x (see tensorial results below). And the shear or torsion stress are also lying on the plane of the cross-section.

Another important issue is that the transformation of shearing and normal forces, as well as bending and torsional moments to the stress values implies the geometrical definition of the beam's section. In structural mechanics, beams are easy to define because they have simple geometries, but when used in biological systems some simplifications need to be assumed. Mandibles can be modelled as hollow tubes (Porro et al. 2011) or the real cross-sectional properties at some locations of the model can be calculated following the method described by Cuff & Rayfield (2013).

In the above-mentioned examples, stress was measured for bending moments. Torsion, compression, or shear can also be studied using beam models. In this case, other diagrams must be obtained instead of the bending diagrams, and other equations (instead of Navier's) are needed to relate the diagram to stress values. All these equations to transform the results from beam theory to stress require the definition of the geometric properties of the section of the beam. In this case the beam is the mandible. A good example to understand the implications of these simplifications in the mammal jaw can be found in van Eijden (2000).

Classical biomechanical methods applied to mammalian mandibles

Three examples are presented here in which the authors have used classical biomechanics. The two first examples by Maynard-Smith & Savage (1959) and Crompton & Parkyn (2009) are based on lever mechanics to study mechanical advantage, whereas Thomason (1991) used beam theory.

Firstly, Maynard-Smith & Savage (1959) studied the biomechanics of the mandible of one carnivore (*Martes)* and one artiodactyl (*Strepsiceros*). They analyzed the efficiency during chewing via the mechanical advantage of these two taxa which feed in different ways and have different mandibular shapes. The models included the temporalis and the masseter muscular forces, as well as the bite force of the canines. A reaction force appeared on the articulation point in the glenoid fossa. They found big differences between the mechanical advantage of the temporal and masseter muscles in the carnivore and the herbivore. These differences were associated with the different geometries of the lower mandible to perform a high MA in relation to the jaw closing muscles. In the case of the carnivore, the temporalis muscle presents the higher MA because the lever arm (the distance to the articulation point) is higher. This corresponds to the muscle that allows the mandible to open in the sagittal plane creating a big and powerful bite without dislocating the mandible. In contrast, in *Strepsiceros* the masseter presents a higher MA because the lever arm is higher for the masseter. This is the muscle that allows the mandible to move laterally and antero-posteriorly making the grinding of herbivorous food possible. Therefore, the different geometries of the two mandibles and the position of the articulation point play an important role in the value of the MA, as well as for the efficiency of chewing.

By using lever models, Crompton & Parkyn (2009) studied the lower mandible of Triassic mammals in which both the reptilian and mammalian mandible articulations were defined. The models included the resultant forces of jaw closing muscles (temporalis, masseter, and pterygoid) and the bite force and the reaction force in the articulation thrust. They demonstrated how the component parts of the mandible musculature gradually changed their orientation (due to the changes in their geometry), in such a way that the reaction force to which the jaw joint was subjected in the articular thrust (the fulcrum of the lever) was progressively decreased whereas there was an increase in the bite force across the post-canine teeth.

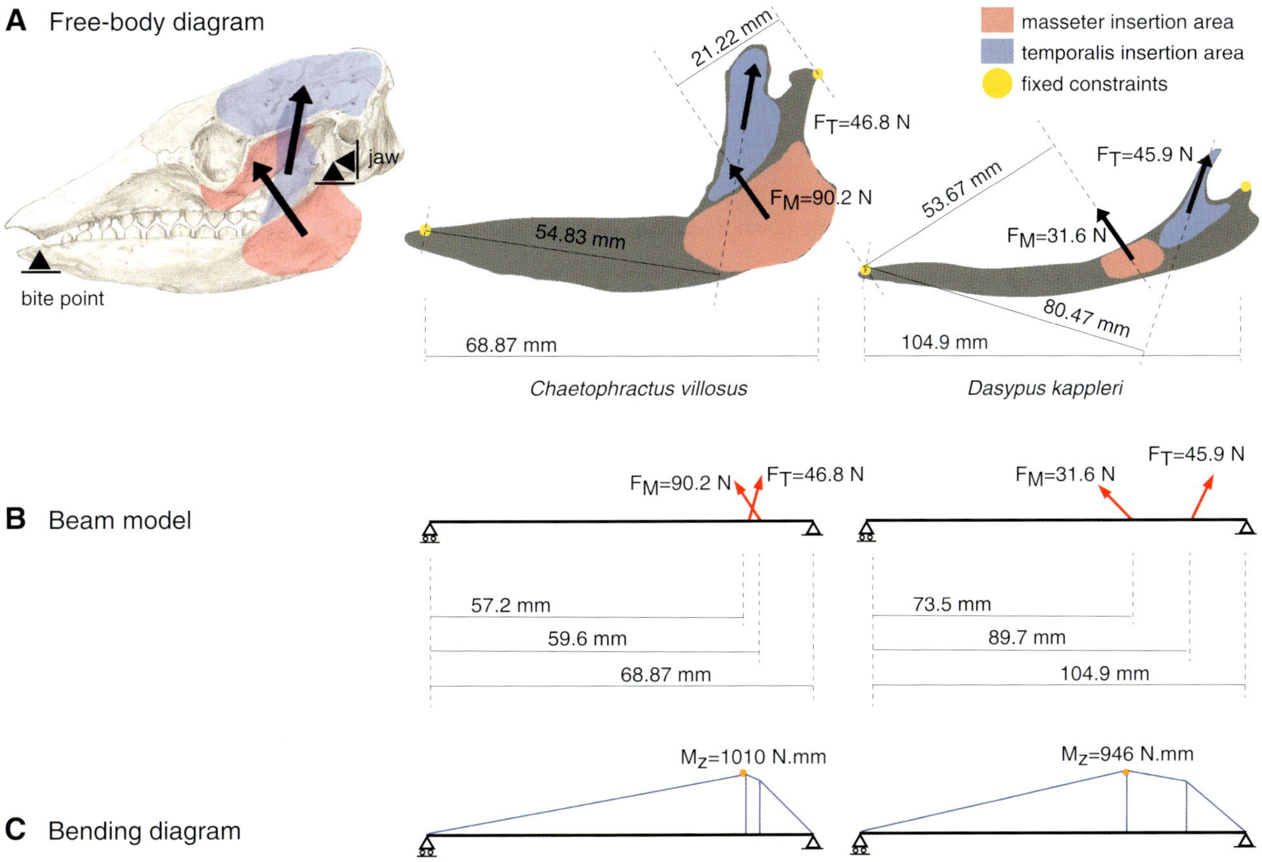

Fig. 4.4. Bite force diagrams of armadillo mandibles. **A,** free-body diagram of a bilateral bite using a two-dimensional approach; **B,** beam model; **C,** bending diagram of *Chaetophractus villosus* and *Dasypus kappleri* from Serrano-Fochs et al. (2015).

Finally, Thomason (1991) modeled the skull of Canidae and Felidae as tubular beams formed by the facial and neurocranial regions, with the attached 'jug handles' of the zygomatic arches. In that case, the model included the reaction force at the articulation thrust and the muscular forces from the masseter and the temporalis. The bite force was studied in two cases: on the canine and on the molar positions. The authors wanted to test the hypothesis that the bending strength of the skull in some mammals correlates with the maximal loads imposed through the masticatory apparatus. The application of the beam theory allowed to assess the cranial strength during bending and, for instance, the Navier-Bernouilli equation was used to relate muscle and bite forces with the stresses inside the skull. In that case, the author used between 20 and 30 transverse CT scan sections through each skull to calculate the cross-sectional properties for the equation. This is one of the first works to show that bending stresses might be of importance in the cranial design of mammals and also has set the stage for more detailed future modelling.

Practical example A. Two different armadillos are studied: *Dasypus kappleri* and *Chaetophractus villosus*, each one presenting different dietary preferences. *D. kappleri* is a generalist insectivore, whereas *C. villosus* has an omnivore diet. The free-body diagram of the mandible, including masseter and temporalis muscle forces, is shown in Figure 4.4. Fixed positions are located on the bite position and on the jaw joint, thus constraining both directions on the jaw-joint and the vertical displacement on the bite point.

Muscle forces are calculated using muscle areas, and a value of 0.3 MPa (force per unit area) is assumed as muscular contraction. This value, estimated by Alexander (1992), has been considered in several works as a valid estimate for the isometric contraction of each adductor muscle (Thomason 1991). In this example, we assumed that the insertion area of the muscle is a good proxy of the 3D muscle cross-section, thus enabling the computation of muscular forces.

Equilibrium of forces in the free-body diagram: According to the equations of equilibrium of forces and the condition for static equilibrium of the statics, the bite force – applying the equilibrium of moments in the jaw joint – can be solved using equation 7. Results can be found in Table 4.1. F_M and F_T are the forces of the masseter and temporalis, l_M and l_T the lever arms of the masseter and the temporalis and l_{Total} the length between the bite point and the jaw joint.

$$BF = \frac{F_M \cdot l_M + F_T \cdot l_T}{l_{Total}} = \frac{0{,}3A_M \cdot l_M + 0{,}3A_T \cdot l_T}{l_{Total}}$$

(equation 7)

The results from classical biomechanics using the free-body diagram show that *Chaetophractus villosus* generates a higher bite force and possesses a more efficient mechanical advantage. This higher biomechanical performance allows *Chaetophractus villosus* to eat a more demanding omnivore diet, whereas *Dasypus kappleri* probably does not need to perform a high bite force in order to crush insects before processing them.

Beam theory in a linear model: The use of the beam theory enables to find the maximum bending moment in the mandible by assuming that mandibular morphology can be represented as a straight line (Fig. 4.4). Curiously, both *Dasypus kappleri* and *Chaetophractus villosus* present similar values of maximum bending moment (1010 and 946 N · mm respectively) but this does not mean that there is a lack in functional differences, because the capacity to withstand this bending moment exclusively depends on the design of the mandible. According to Serrano-Fochs et al. (2015), the mandible of *Chaetophractus villosus* is thicker (5 mm vs. 3.5 mm) and the body of the mandible is broader than that of *Dasypus kappleri*. The corpus of the mandible of *Dasypus kappleri* is narrow in height and limiting its capacity to sustain the stresses derived from the bending moment. These results from beam models clearly suggest that *Chaetophractus villosus* is capable to withstand higher bending moments than *Dasypus kappleri* when chewing, therefore also is capable to process more demanding diets such as an omnivore one.

There is a difference between using the free-body diagram and the beam-model which explains the different bite force values (see Tab. 4.1 and 4.2 for the two approaches). The geometry assumed in both cases is different. In the free-body-diagram, the distances used from the muscles to the bite point and the condyle are based on the actual geometry, whereas in a beam model the position of the forces is in the line defined by the beam. This changes the values of the distances, and consequently, the value of the bite force. The free-body diagram is better suited to study muscle and bite forces, whereas the beam model allows the study of stresses in the cross-sections.

From the values of the maximum bending moment and from geometrical properties of the section of the mandible, normal stress values in the mandible can be found using the Navier equation (equation 8). In this example, we will assume a hollow ellipse as a constant cross-section of the mandible, and the second moment of inertia is obtained via equation 9.

$$\sigma_{xx} = \frac{M_Z}{I_Z} \cdot y = \frac{M_Z}{\frac{\pi(ab^3 - a_1 b_1^3)}{64}} \cdot \frac{b}{2} \qquad \text{(equation 8)}$$

$$I_Z = \frac{\pi(ab^3 - a_1 b_1^3)}{64} \qquad \text{(equation 9)}$$

Where M_z is the bending moment computed in the beam, I_z the second moment of inertia in the axis of the bending and y the distance from the centroid of the section to the point of the section under study.

High stresses are in the farthest point from the center of the cross-section during bending. Geometrical parameters of the section are described in Table 4.3 assuming, for this example, a constant cortical bone thickness of 1 mm in both models. The geometric parameters a, a_1, b, and b_1 are defined in Table 4.3 and are real measurements from the models (Serrano-Fochs et al. 2015). The normal stress obtained using equations 8 and 9 for both mandibles assuming a beam model (Tab. 4.3) also supports that the mandible of *Chaetophractus villosus* is stronger than the one of *Dasypus kappleri* during chewing.

Table 4.1. Results from equilibrium of forces.

Taxon	Masseter area (mm^2)	Temporalis area (mm^2)	Masseter force (N)	Temporalis force (N)	Bite force (N)	Mechanical advantage (MA)
Dasypus kappleri	105.37	153.18	31.6	45.9	15.66	0.2
Chaetophractus villosus	300.58	156.08	90.2	46.8	34.97	0.25

Table 4.2. Results from beam theory.

Taxon	Masseter area (mm^2)	Temporalis area (mm^2)	Masseter force (N)	Temporalis force (N)	Bite force (N)	Mechanical advantage (MA)	Mz max (N · mm)
Dasypus kappleri	105.37	153.18	31.6	45.9	12.87	0.16	-1010
Chaetophractus villosus	300.58	156.08	90.2	46.8	17.63	0.13	946

Table 4.3. Reometrical parameters for the mandibles of *Dasypus kappleri* and *Chaetophractus villosus* when assuming a beam with a hollow elliptic cross-section.

Taxon	a (mm)	a_1 (mm)	b (mm)	b_1 (mm)	I_z (mm^4)	Mz (N · mm)	σ_x (N · mm^2)
Dasypus kappleri	3.51	0.51	8	6	82.81	1010	48.8
Chaetophractus villosus	4.94	2.94	12	10	274.71	946	20.7

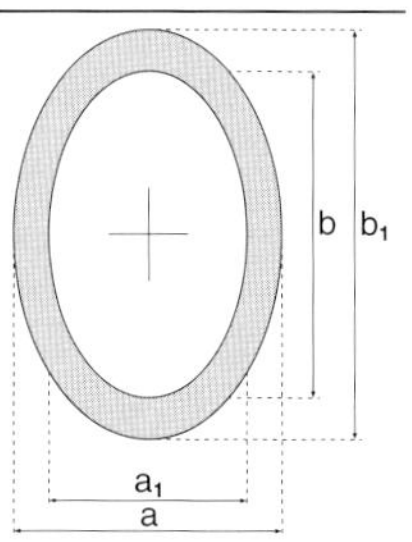

Finite element analysis in mammalian mandibles

Brief introduction to finite element analysis

The finite element analysis (FEA) is a non-invasive computer simulation technique used in engineering. It uses the numerical technique of the finite element method (FEM) to solve the equations of continuum mechanics by dividing a geometry into a finite number of discrete elements known as a mesh (Kupczik 2008) to which the equations are applied. Mechanical properties of these elements are defined in order to give the structure a realistic behavior. The model is constrained to anchor it in space and includes the external forces applied to it. FEA is used to solve the equations of the elasticity in a deformable body under the effect of external forces.

FEA and beam theory are two different approaches to solve the same mechanical problem, yet with the limitation that beam theory can be exclusively used when assuming linear geometries with cross-section, whereas FEA can be applied to any geometry. The equations of lever mechanics based on Newton's Laws are included when solving a mechanical problem in both FEA and beam theory. They are also solved when these methods are used. Therefore, beam theory and FEA can also solve reaction forces but it makes no sense to use these approaches to exclusively calculate these external forces, since they can be solved by hand using Newton's laws (Sellers et al. 2017, Snively et al. 2015).

In mammalian mandibles, FEA enables the observation of stress distribution patterns of the specimens by simulating loadings and forces involved in the masticatory function. Under equivalent loads, these stress patterns can be interpreted as a sign of the relative strength, with higher stress indicating a weaker mandible. Assuming that more robust or stronger mandibles would be needed both for processing harder food items, these mandibles should be expected to be weaker (i.e., displaying higher stress levels) than those belonging to animals feeding on harder items, such as nuts, shells or bones. On the other hand, differences in the stress distribution pattern may give a clue regarding different aspects of the feeding ecology of the analyzed species.

There are several types of FEA depending on the nature of the geometry imported. A three-dimensional geometry e.g., from a CT-scan implies the use of three-dimensional solid models; a two-dimensional geometry requires the solution of a planar model using the equations of plane elasticity. Surface geometries in a 3D space require the use of shell models, whilst geometries based on lines need the use of linear elements. The latter is the exact computational translation of beam theory.

It is usual to model mammal mandibles as three-dimensional solid models (Gill et al. 2014, Gröning et al. 2011b, Püschel et al. 2018, Tseng & Binder 2010) or to represent them as simpler planar models (Fletcher et al. 2010, Marcé-Nogué et al. 2017b, Piras et al. 2013, Serrano-Fochs et al. 2015). In continuum mechanics, plane elasticity makes reference to the study of particular solutions of the general elastic problem in bodies that are geometrically mechanical prisms (that is, an area with a constant thickness) (Mase & Mase 1999). In particular, a plane stress solution occurs in structural elements where one dimension (the thickness) is very small compared to the other two, and the stresses are negligible with respect to the smaller dimension.

Steps to perform a finite element analysis

Geometry of the model. Models of mammalian mandibles can be obtained from the digitalization of real specimens by using CT-scans, photogrammetry or similar digitalization techniques that can register morphologies in three-dimensions and in full detail (see review in Cunningham et al. 2014). Another option is to generate them from scratch based on photos or other templates in CAD software (see Rahman & Lautenschlager 2016 for the box modelling approach).

The advances in computational tools to perform virtual reconstruction and restoration techniques of the original morphology of fossils, enable the creation of accurate and realistic three-dimensional models (Lautenschlager 2016). In all cases, during the reconstruction process, irregularities in the surface due to the generation of the model from the CT scanner need to be repaired using refinement and smoothing tools (Marcé-Nogué et al. 2011).

Planar models can be created from published photographs or built from scratch. In order to take the photographs in the most consistent way, some procedures need to be carried out in order to standardize images (de Esteban-Trivigno 2011), and then the use of CAD software allows the creation of models based on splines which are traced on the picture (see Fortuny et al. 2010 and Serrano-Fochs et al. 2015 for the full methodology).

Material properties. Bone is a rigid organ that forms part of the vertebrate skeleton. The hard and dense outer layer is composed of cortical bone tissue. Trabecular bone tissue corresponds to the internal tissue of the skeletal bone, being an open cell porous network. The mechanical properties of a particular bone depend on both composition and structure, which are fundamental parameters to be considered when simulating the behavior and response of a model when loads are applied. Although composition is not constant in living tissues, it is normally assumed (for simplicity and computational costs) that the mechanical properties of cortical and trabecular bone depend on apparent density and mineral content (see Doblaré et al. 2004 for a review).

The difficulties in modeling bone material properties in FEA models have driven researchers to assume isotropic behavior and to consider that bone, like most biological materials, is elastic and fails under a ductile model of fracture (Nalla et al. 2003). Isotropy means that the properties are taken to be the same in all directions, whereas anisotropy means having different properties in different directions. Orthotropy is present when the material properties differ along the three orthogonal axes. Another common assumption of most FE analyses in the literature is to consider the mechanical behavior of bone (trabecular and cortical) as being linear, which is usually an accurate enough assumption (Doblaré et al. 2004). This means that the relationship between stress and strain is proportional and the constitutive equation of linear elasticity based on Hooke's law can be assumed. This also implies that using Young's modulus and Poisson coefficient is enough to characterize bone. Cortical bone is stiffer and, therefore, has a higher Young's modulus than trabecular bone.

One of the main problems when working with animal bones – and especially with fossil taxa – is the absence of material data to define Hooke's law (Young's modulus and coefficient of Poisson). This problem is normally tackled by using data available from closely related species, or from species with similar bone structure. The use of appropriate material properties depends on the available information from phylogenetically or ecologically related taxa. It is important to bear in mind that modelling other materials such as soft tissues, implies the application of non-linear constitutive equations, which in turn results in more computation time, (i.e., solutions mostly based on the equations of hyperelasticity). A hyperelastic material is a type of constitutive model of an ideally elastic material for which the stress-strain relationship derives from a strain energy density function that does not follow a linear relationship. It is common to use these constitutive models in FEA for ligaments (Qian et al. 2009), cartilage (Gislason et al. 2017), or arteries (Early et al. 2009) but not when analyzing osteological structures, such as mandibles.

Biting scenarios. Different biting scenarios have been used to study the chewing mechanism of mammals. According to Preuschoft & Witzel (2002), the skull of mammals can be loaded in three different ways: (1) assuming the weight of the head and prey or food acting downward on the mandible, (2) applying arbitrary forces or movements in the plane of the tooth row that create movements of the prey in relation to the head (defined as extrinsic loads by McHenry et al. (2007), and (3) applying the adduction of the mandible via the muscle forces which leads to reaction bite forces in the teeth (defined as intrinsic loads by McHenry et al. 2007). The cases with intrinsic loads can be modelled at the different tooth positions as bilateral (Püschel et al. 2018) or unilateral bites (Ledogar et al. 2016), whereas the group with extrinsic loads can be used to model axial twist, lateral shake, pullback, and dorsoventral movements (Attard et al. 2014).

The values of the forces in the different scenarios can be obtained from different information sources (e.g., literature, experiments, other modeling approaches such as MDA, etc.) or can be arbitrarily selected. When comparing different extrinsic models, arbitrary values can be a good choice (Attard et al. 2014), whereas for the intrinsic cases the values of the muscular forces can be calculated as a muscular contraction (see muscular forces in the intrinsic cases).

Several authors have used multibody dynamic analysis (MDA) to previously predict bite forces or muscular forces as a source of information for the forces applied at the FEA models (Lautenschlager et al. 2018). A multibody dynamic system consists of solid bodies that are connected to each other by joints that restrict their relative motion. This kind of studies analyzes how mechanical systems move under the influence of forces and can also be used to identify which forces are produced during the dynamic behavior of the system under analysis. This topic is, however, out of the scope of this chapter and further exposition can be found in Curtis (2011) where the application of MDA in vertebrates is reviewed.

Muscular forces in the intrinsic cases. The external loads that are acting on a body are applied as force or pressure on a surface. In a dynamic analysis, muscles are acting in combination to create a succession of individual static cases with several component contributions through time. In a static analysis (which is the one we are mostly referring to here) the muscle can be simplified as a force vector with an associated value for a specific case study. It is usual to analyze the so-called "theoretical maximum" which is when all muscles are acting together in a maximal tetanic contraction.

Muscle forces are calculated using muscular cross-sectional areas and a value for the specific muscular tension of 0.3 MPa (force per unit area), which is assumed as muscular contraction. This value, estimated by Alexander (1992), is considered as an isometric contraction of each adductor muscle. The physiological cross-sectional area (PCSA) is the area of the cross-section of a muscle perpendicular to its fibers, generally at its widest point.

$$F = P \cdot A_{PCSA} \qquad \text{(equation 10)}$$

Where F is the muscle force, P is the specific muscular tension and A_{PCSA} is the physiological cross-sectional area.

Usually, when solving models of mammalian mandibles, the muscle forces (pterygoid, masseter, and temporalis) correspond to input values that have previously been calculated, whilst bite force is the output value solved by the static equations.

Muscular forces can be applied homogeneously on the insertion areas (Püschel et al. 2018) or can be distributed tangentially over the attachment areas wrapped around the bone surface (Tseng et al. 2017). To define the direction of the muscular forces, there are several accurate methods that create frames of linear elements all over the muscle attachments that linked the upper and lower jaws, with forces applied at the center of the frame (Attard et al. 2014). The simplest approach to determine the directions of the forces is defining the lines joining the centroid of the insertion area on the skull with the centroid of the insertion areas on the mandible (Panagiotopoulou et al. 2017, Püschel et al. 2018). In fact, if we assume homogeneous muscle force all over the muscle insertion, a single equivalent load applied at the centroid of the surface can replace a homogeneously distributed load applied on a surface. However, it should be mentioned that this single equivalent force generates little difference in the output values of stress and strain in the regions immediately outside of the applied force. This is because the specific force may create an artificially high stress (for more details see the problem of the maximum peak value addressed below).

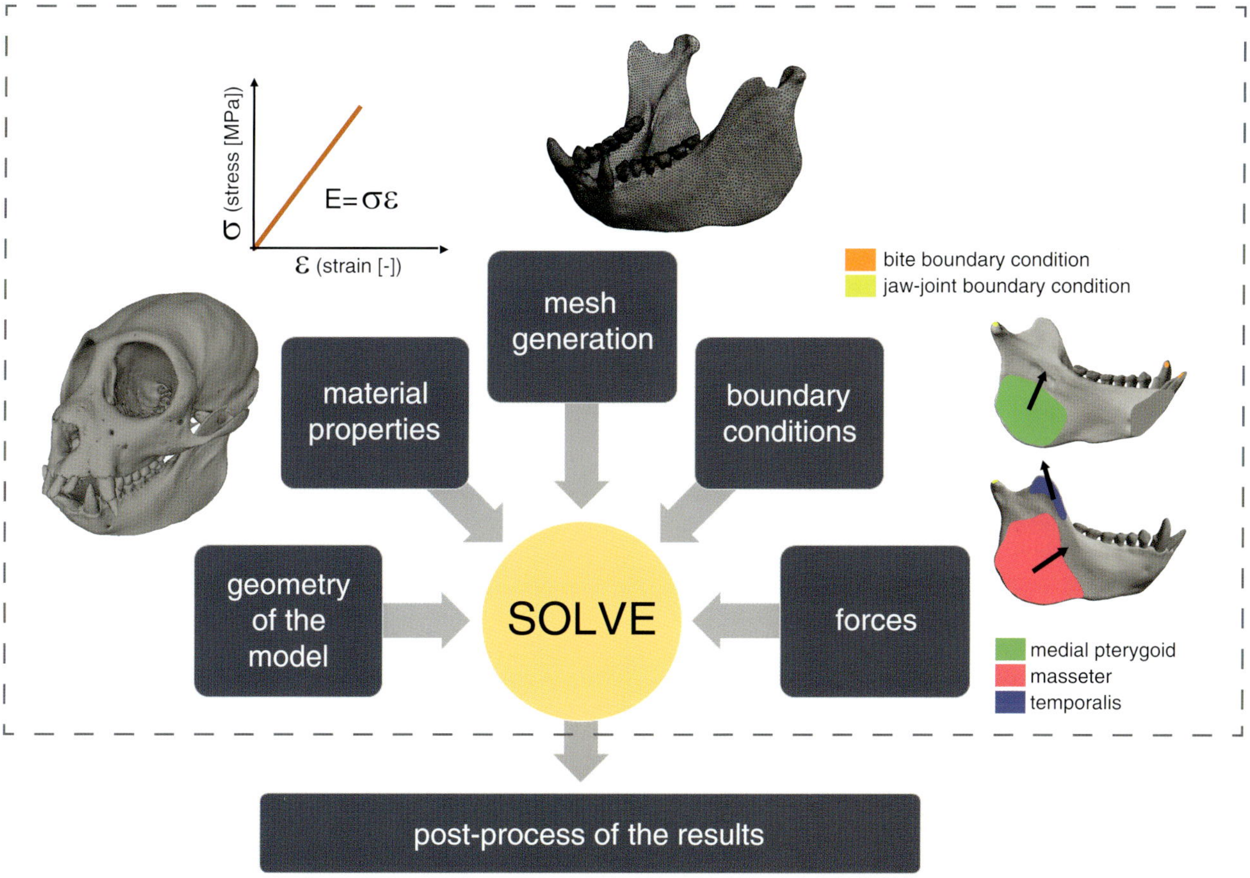

Fig. 4.5. Flow chart of a finite element analysis. All the steps involved in the post-process: Creation of the model's geometry, definition of the material properties, generation of the FEA mesh, definition of the boundary conditions and application of the forces.

Boundary conditions. Boundary conditions are defined to represent the unknown loads (for Newton's third law), displacements, and constraining anchors that the structure experiences during the action of the forces. In general, it is recommended to avoid over-constrained models and redundant supports. Redundant supports are those in excess when the equilibrium is already guaranteed. It means that the number of unknown loads is higher than the number of degrees of freedom of the body. In a FEA planar model, the degrees of freedom of the body are three (the displacements in both directions and the rotation on the plane). For example, in Marcé-Nogué et al. (2017b) the boundary conditions are fixing both directions in the jaw-joint and only the vertical displacement at the bite point. This is a total of three constraints which is exactly the number of degrees of freedom for a planar model. The degrees of freedom of the body are six when analyzing 3D models using solid elements: three displacements in each direction, and three rotations in each axis. Excessive constraints tend to add stiffness to the model, thus making it stiffer than it should be.

In addition, it is fundamental to avoid under-constrained models which could produce rigid-body motions of the model. This means that the number of unknowns is lower than the number of degrees of freedom and that we need to be sure that the model is fixed to avoid any possibility of translation or rotation.

In the case of mandibular models, boundary conditions are often defined so they represent the fixed displacements that mandibles experience during biting scenarios. The first boundary condition restrains the condyle at the level of the contact points with the mandibular fossa of the cranium. in order to represent the immobilization of the mandible, thereby constraining the translation of the jaw in all the directions.

In intrinsic cases, a second boundary condition is applied to individual dental positions by fixing the displacement in the vertical axis. They are usually applied at different dental positions in order to simulate different biting scenarios (e.g., molar or incisive bite). According to Newton's third law, the corresponding reaction force represents the unknown bite force.

When a unilateral bite is modelled, then the phenomenon of torsion has great importance because only one bite point is being considered on one side of the mandible (namely the working side), which means that there is a balancing side that is not directly acting in the biting action. In the balancing side, muscle forces can be reduced in order to remove a distractive tensile reaction force at the working side jaw to avoid being pulled apart (Greaves 1978, Spencer 1999).

In extrinsic cases, the fixed condition of the teeth is not used because instead of defining muscular forces to the posterior part of the mandible, a force is applied at the dental position. For cases in which the weight of the head and prey/food are simulated, this fixed condition of the teeth is also not considered because the weight of the prey/food is also applied to the teeth as a force.

Mesh generation. The mesh of a FEA model is a collection of interconnected elements joined at nodes, which defines the shape of the model where the physical problem will be solved using the mathematical equations of the FEM. Elements usually consist of triangles, quadrilaterals, tetrahedrals, hexahedrals or other simple convex polygons which simplify and discretize the geometry of the original model.

A problem can be solved as a two-dimensional model or as a three-dimensional model. However, different finite elements are required (see Marcé-Nogué et al. 2015 for further details about FEA meshes). When the geometry that is required to be meshed consists of surfaces or lines, it is necessary to use shell or beam elements projected in a three-dimensional space. These elements require different formulation compared to the standard solid elements used in three-dimensional models but, usually, FEM packages allow their easy implementation. Two-dimensional problems are solved using the equations of plane elasticity and the model can be meshed with planar elements, which means that it is required to consider their thickness.

Currently, the main commercial FEA packages include several methods to automatically generate the mesh of the desired geometry without human intervention. This improves the analysis, but the mathematical complexity of the FEM still means that mesh generation has to be treated with care. It is important to find a reasonable balance between the best mesh to be built and analyzed, and computational limitations. Depending on the specific package, software or the file format of the imported geometry, there are differences in the order of steps required to carry out a FEA. In some cases the mesh is generated in the first steps and then, the creation of the boundary conditions and the application of the forces are performed on the mesh (Figueirido et al. 2018) or, inversely, in other cases the inclusion of the forces and boundary conditions are performed on the geometry to then finally mesh the model (Püschel et al. 2018).

The different types of elements and the different types of mesh (structured, non-structured, uniform, non-uniform) and their pros and cons in the context of biological problems are discussed in detail in Marcé-Nogué et al. (2015). The quality assessment and the reliability of the mesh to solve FEA models is also discussed in Marcé-Nogué et al. (2015), and some examples of convergence results when the size of the elements is reduced can be found in Bright & Rayfield (2011b) and Tseng & Flynn (2015), where vertebrate species were analyzed.

Understanding the outputs of FEA

Scalar results. The solution of a FEA model generates several different outputs. A scalar is a quantity represented by a numerical value or magnitude that is independent of specific classes of coordinate systems, or one that is usually said to be described by a single real number. Strain energy or mechanical advantage are examples of scalar values. For example, they were used by Tseng et al. (2017) in FEA mandibular models to study the feeding capability in fossil and living Lutrinae. Based on their results, it was hypothesized that the extinct species of giant otter *Siamogale melilutra* exhibited feeding ecomorphological adaptations that have no living analog.

Vectorial results. FEA vectorial results are the ones represented by a vector. A vector is a quantity that has magnitude and direction and that is commonly represented by a directed line segment whose length represents the magnitude and whose orientation in space represents the direction. This can be also represented by separating the vector into two or three components depending on the dimensionality of the problem (i.e., 2D or 3D).

The force reactions derived by Newton's laws are vectorial results. In cranial mechanics, the bite force is computed as an output result when muscle forces are applied as input variables. Bite force strongly depends on the size (Wroe et al. 2005) of the models and needs to be carefully used when comparing models of dissimilar sizes. However, if analyzed properly, it can be used in a comparative setting to evaluate bite performances given different tooth positions (Figueirido et al. 2014).

Displacements are also a vectorial result providing the information from the original position of a node of the model with respect to the new position of the same node after being loaded by the input forces. These results have been used to compare the deformation between the unloaded and the loaded model (Gröning et al. 2011a).

Tensorial results. Tensors are important in physics because they provide a concise mathematical framework to formulate and solve problems of physics in areas such as stress, elasticity, fluid mechanics, and general relativity.

Stresses and strains are tensorial which means that they are represented by a tensor. Tensors are geometric objects that describe linear relations between geometric vectors, scalars, and other tensors and are represented by an organized multidimensional array of numerical values. The order (also known as degree or rank) of a tensor is the dimensionality of the array needed to represent it, or in other words, the number of indices needed to label a component of that array. For example, stresses and strains in three-dimensional models are represented by a matrix (a three-dimensional array) in a basis, and consequently correspond to a 3rd-order tensor.

In plane elasticity, tensors are kept as a third order tensor but with zero values in all the stresses related to the third direction in plane stress, and with zero values in all the strains related to the third direction in plane strain. Plane stress stays different from zero in ε_Z and plane strain σ_Z but these values are negligible with respect to the other values, and, therefore, we can assume plane elasticity as a two-dimensional solution and solve only the x-y components.

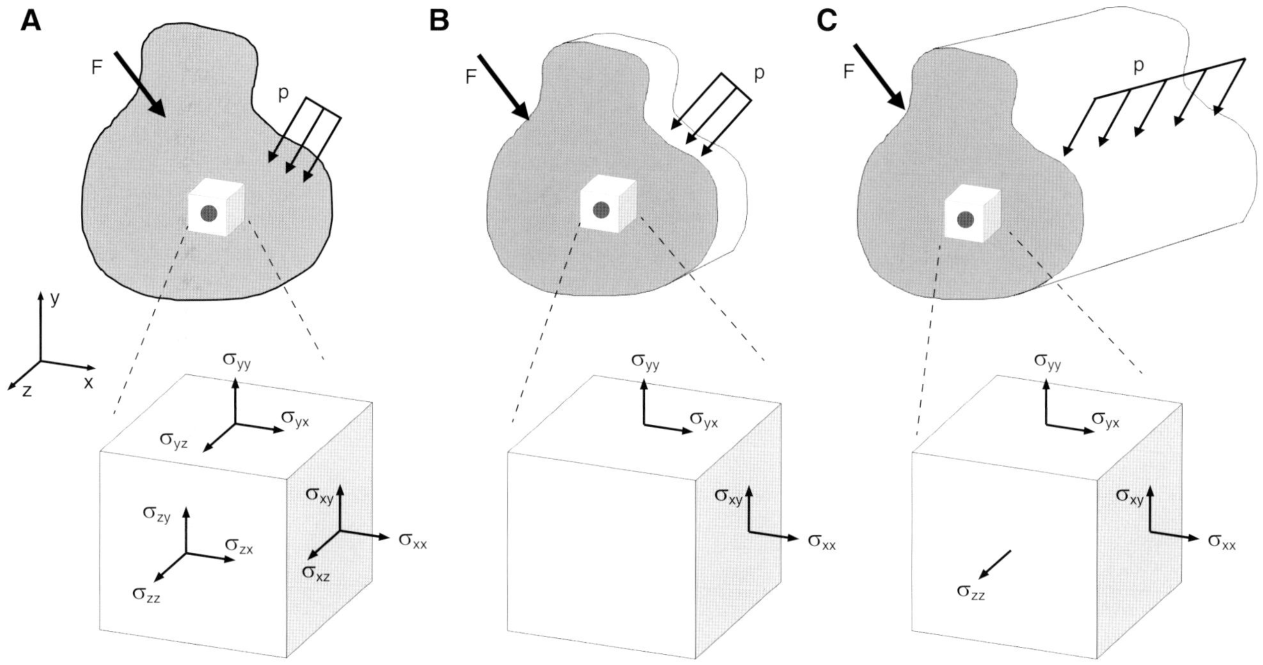

Fig. 4.6. Components of the tensor of stress in **A,** three-dimensions, **B,** plane stress and **C,** plane strain.

$$\sigma = \begin{bmatrix} \sigma_x & \sigma_{xy} & \sigma_{xz} \\ \sigma_{xy} & \sigma_y & \sigma_{yz} \\ \sigma_{xz} & \sigma_{yz} & \sigma_z \end{bmatrix} \quad \varepsilon = \begin{bmatrix} \varepsilon_x & \varepsilon_{xy} & \varepsilon_{xz} \\ \varepsilon_{xy} & \varepsilon_y & \varepsilon_{yz} \\ \varepsilon_{xz} & \varepsilon_{yz} & \varepsilon_z \end{bmatrix}$$

(equation 11 and 12, for three-dimension problems)

$$\sigma = \begin{bmatrix} \sigma_x & \sigma_{xy} & 0 \\ \sigma_{xy} & \sigma_y & 0 \\ 0 & 0 & 0 \end{bmatrix} \quad \varepsilon = \begin{bmatrix} \varepsilon_x & \varepsilon_{xy} & 0 \\ \varepsilon_{xy} & \varepsilon_y & 0 \\ 0 & 0 & \varepsilon_z \end{bmatrix}$$

(equation 13 and 14, for plane stress problems)

$$\sigma = \begin{bmatrix} \sigma_x & \sigma_{xy} & 0 \\ \sigma_{xy} & \sigma_y & 0 \\ 0 & 0 & \sigma_z \end{bmatrix} \quad \varepsilon = \begin{bmatrix} \varepsilon_x & \varepsilon_{xy} & 0 \\ \varepsilon_{xy} & \varepsilon_y & 0 \\ 0 & 0 & 0 \end{bmatrix}$$

(equation 15 and 16, for plane strain problems)

This means that, in order to define the stress state or the strain state of a FEA model, it is necessary to know all the values of the components of the respective tensors in each point of the model under analysis. FEA simplifies this statement by solving the stress and strain tensor in points inside each element of the mesh. It is important not to confuse the FEA elements with the sides of the square in Figure 4.6. The square in Figure 4.6 is not a FEA element but just the graphical representation of the Cartesian planes that are required to be known in a point when we want to know all the values of the stress/strain tensor.

Depending on the orientation of the plane under consideration, each component of the tensor represents a value of stress/strain: the component normal to the plane, called normal stress, and another component parallel to this plane, called the shearing stress. The symmetry of the stress/strain tensor simplifies the question to know only six values for defining all the tensor.

Principal and equivalent values. Principal stresses/strains are the maximum normal stress a body can have at a point of its body. It represents only normal stress and it does not have any shear stress component. Mathematically speaking, principal values are the ones in the tensor when the shear components are null. They are useful to summarize all the stress states of a FEA model for the two maximum and minimum values instead of using all the different values of the stress tensor.

$$\sigma = \begin{bmatrix} \sigma_I & 0 & 0 \\ 0 & \sigma_{II} & 0 \\ 0 & 0 & \sigma_{III} \end{bmatrix} \quad \varepsilon = \begin{bmatrix} \varepsilon_I & 0 & 0 \\ 0 & \varepsilon_{II} & 0 \\ 0 & 0 & \varepsilon_{III} \end{bmatrix}$$

Equivalent values are scalar values of stress that can be computed from the stress/strain tensor in order to summarize the stress state defined by all the components of the tensor to a unique value. In other words, an equivalent stress is the value of stress that creates a uniaxial stress state in the point to study, equivalent to the previous multiaxial stress state. The von Mises stress criterion, the Tresca criterion or the Rankine theory are different ways to reduce the multiaxial stress state to an equivalent stress. Von Mises and Tresca criteria are applicable for ductile materials whereas Rankine's theory is preferred for brittle materials (Chen & Saleeb 1994).

The von Mises criterion is the most popular criterion for predicting the yield of ductile materials. Bone can be assumed to be either brittle (Doblaré et al. 2004) or ductile (Dumont et al. 2009). However, according to Doblaré et al. (2004), when isotropic material properties are used in cortical bone, von Mises is the most accurate criterion for predicting fracture location. This probably explains why it is in fact the most common criterion used in mandibular mechanics (Panagiotopoulou et al. 2017, Piras et al. 2013, Püschel et al. 2018, Tseng et al. 2017).

Uses of stress and strain. There are several different approaches that have been applied to analyze the results derived from mandibular mechanics. While some works analyze stress (Attard et al. 2014, Lautenschlager et al. 2016, Püschel & Sellers 2016, Tseng et al. 2017), other focus on strain (Gröning et al. 2013, Oldfield et al. 2012,

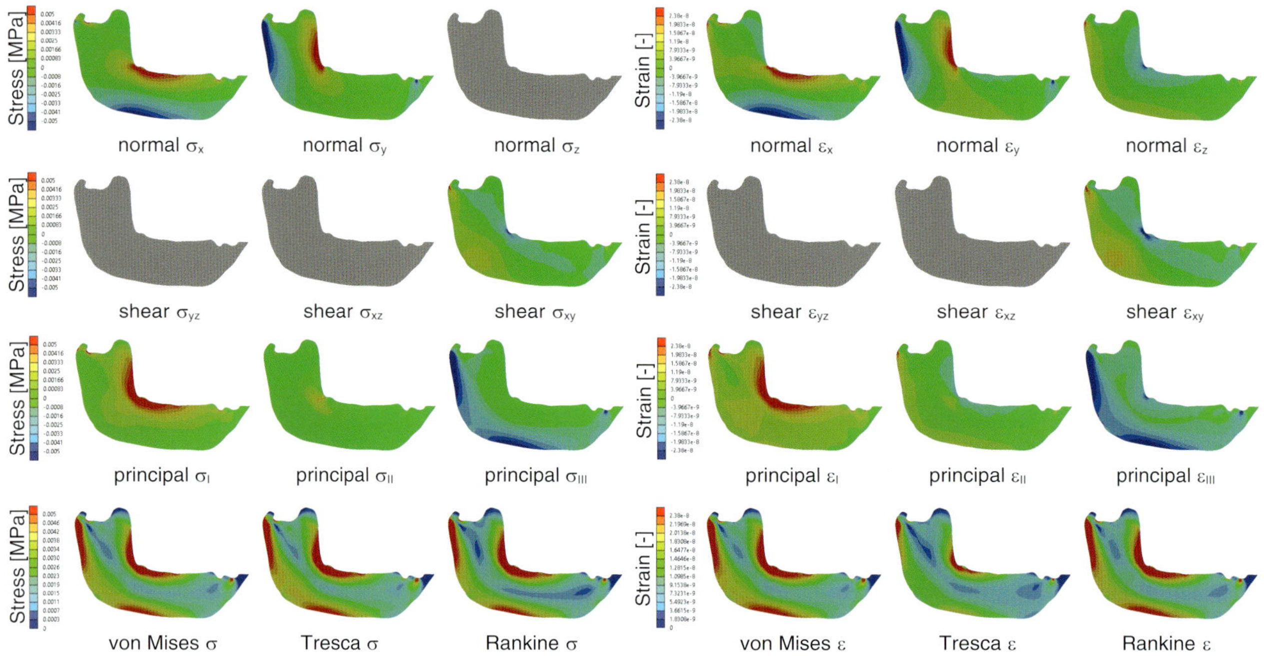

Fig. 4.7. Stress and strain distribution in a plane stress model. Considered are all the stress variables of the tensor, principal stresses and equivalent von Mises, Tresca and Rankine. Model of *Macaca fascicularis* from Marcé-Nogué et al. (2017b).

Panagiotopoulou et al. 2017). When using linear and elastic properties, it has been shown that both values are proportional and that the results of the distributions along the mandible are the same. This is due to the application of Hooke's law, which establishes a proportional relation between stress and strain.

Both stress and strain are accepted when comparing models below the fracture point, but, when we want to extrapolate these models to study fracture mechanics, a new debate is triggered: Is it stress or strain that causes failure? Probably the longest standing issue regarding failure criteria is whether they should be expressed in terms of stresses or in strains.

In engineering, the most commonly used form is the von Mises criterion expressed in terms of stress. Other criteria that are used are the maximum shear stress criterion (Tresca), the maximum normal stress criterion (Rankine) or the maximum normal strain criterion (Rankine). Despite the von Mises stress being a stress-designed criterion based on the maximum distortion of the strain energy, the stress-strain relations for isotropy can be used to convert the von Mises criterion in terms of stresses into a metric expressed in terms of strains. This transformation can be easily done using a simple mathematical procedure, but this does not obliterate the incontrovertible assumption that considers that all criteria can be switched from stress to strain (and vice versa). This is not always possible, especially when considering anisotropic material properties, but equally so for some isotropic cases. As an example, the maximum normal stress criterion and the maximum normal strain criterion take different forms when interconverted.

In this text, stress is taken as the fundamental form to be used when focusing on failure criteria. This is the direct consequence of the following conditions. Stress must be used if one wishes to have compatibility with fracture mechanics in the brittle range and with dislocation dynamics in the ductile range. Both of these classical theories require formulations in terms of stress.

On the other hand, strain has been shown valuable when comparing simulation results with experimental data (Bright & Rayfield 2011a, Gröning et al. 2012a), since strain gauges are commonly used in experimental settings. These electrical components experience changes in length as changes in resistance, and this property can be used to measure strain on a surface (e.g. bone) and compare it with FEA models.

Comparative framework

One of the most important decisions when comparing different specimens is the selection of taxa, provided that it allows to create an appropriate biological comparative framework. In addition, it is also important to build the models considering that they will be compared under equivalent loads in order to interpret the stress pattern differences of the models as a sign of relative strength. Comparing the strength allows to discern that more robust or stronger mandibles are needed to be able to process more demanding diets such as, for example, nuts, shells, or bones. A comparative framework can be defined as all the decisions we need to take into account in the post-processing of FEA models that enable the comparison between models to analyze the biomechanical behavior of different species or specimens under equivalent mechanical scenarios. Post-processing involves all the steps carried out after solving the FEA models. All these considerations are not necessary, if the purpose is to simply analyze the distribution of stress or strain of just one mandible, e.g. to compare it against real data or to study mandibular facture.

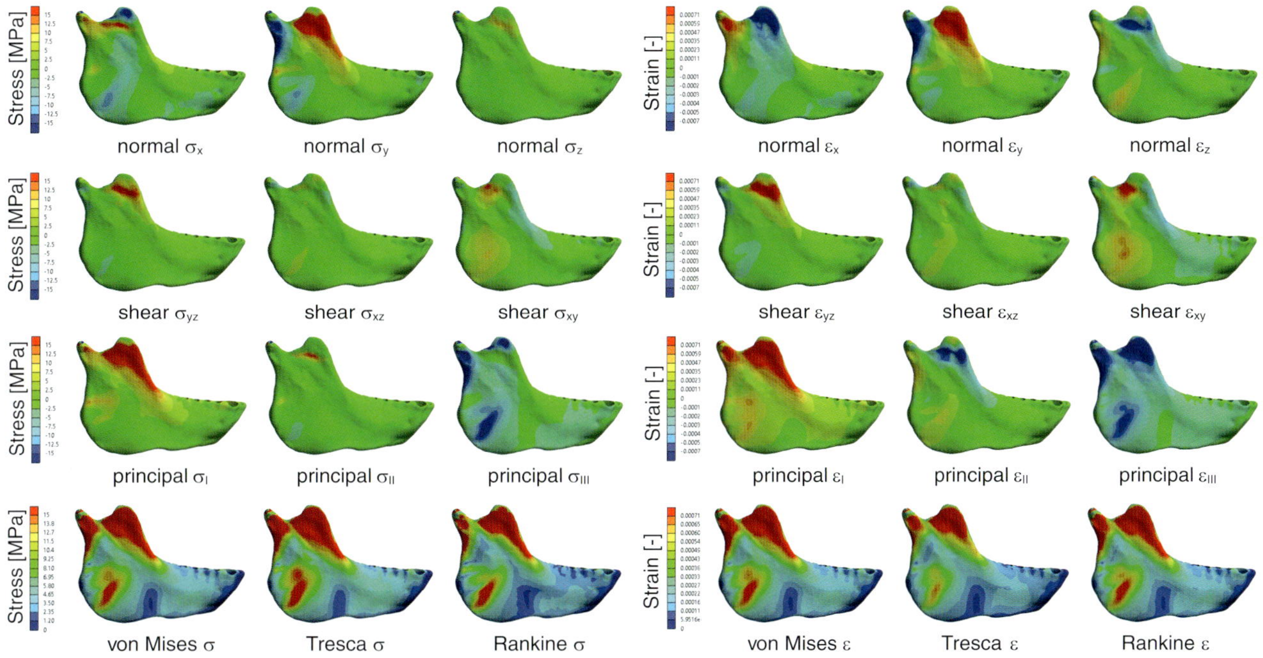

Fig. 4.8. Stress and strain distribution in a three-dimensional solid model. Considered are all the stress variables of the tensor, principal stresses and equivalent von Mises, Tresca and Rankine. Model of *Pithecia monachus* from Püschel et al. (2018).

Orientation of the models. All models must be oriented in the same way. Several options exist for mammalian mandibles, such as positioning them in an anatomical orientation, central occlusion or according to anatomical planes, such as the Frankfurt horizontal plane (Marcé-Nogué et al. 2017b, Osborn 1987). Another option is to apply algorithms such as a best-fit alignment to orient all the models with respect to a common plane (e.g., Püschel et al. 2018 applied to talar morphology). It is important to orient all models consistently because different orientations can result in different muscle force orientations, which can potentially have a significant effect in the variation of the results (Gröning et al. 2012b).

Scaling the models. When the analyzed specimens exhibit significant size differences, it is necessary to adjust the models to a comparable size. There are two main ways of answering the question of how performance can be compared so that the effects of size and shape are disentangled, in order to exclusively focus on how shape affects mechanical performance for a given loading condition (Dumont et al. 2009).

(1) Change the size of all the models to the same size and apply the same force.
(2) Keep the differences in size in the models and apply an appropriate force that generates the same effect as carrying out the procedure described in 1.

In vertebrate models, it is more common to follow the second approach, thus keeping the original size of the model. However, even though both approaches are correct, they display the results in a different context. Some recent works have discussed which equations should be used to scale the models to the same size in order to study the stress patterns or the strain energy (Dumont et al. 2009, Marcé-Nogué et al. 2013), thus proposing that size could be removed either by modifying the dimensions of the model, or the values of the muscular forces or muscular pressures applied. It is important to note that when comparing stress/strain the forces should be scaled in a different way than when comparing other values, such as displacements.

For FEA planar models, the forces applied should be scaled using – for example – the quasi-homothetic transformation proposed by Marcé-Nogué et al. (2013) to allow meaningful comparison between planar models. In this case, it is important to bear in mind if our planar models are following the equations of plane stress or plain strain. For FEA models in three-dimensions, the volume ratio should be considered for calculating the different forces. The equations that must be fulfilled between two models that are to be compared without the effect of size are shown in Table 4.4 depending on the elasticity problem used.

Some criticisms and improvements have been made in recent papers (Parr et al. 2012) regarding the widely used scaling methodology proposed by Dumont et al. (2009). In particular, the analysis of 3D complex biological structures

Table 4.4. Force equations in a scaled model B with reference to model A. A_A is the area of the reference model, A_B the area of the scaled model, t_A the thickness of the reference model, t_B the thickness of the scaled model, V_A the volume of the reference model and V_B the volume of the scaled model.

Model type	Comparing strain/stress	Comparing displacements
Plane stress	$F_B = \left(\sqrt{\frac{A_B}{A_A}}\right)\left(\frac{t_B}{t_A}\right)F_A$	$F_B = \left(\frac{t_B}{t_A}\right)F_A$
Plane strain	$F_B = \left(\sqrt{\frac{A_B}{A_A}}\right)F_A$	$F_B = F_A$
3D models	$F_B = \left(\sqrt[3]{\frac{V_B}{V_A}}\right)^2 F_A$	$F_B = \left(\sqrt[3]{\frac{V_B}{V_A}}\right)F_A$

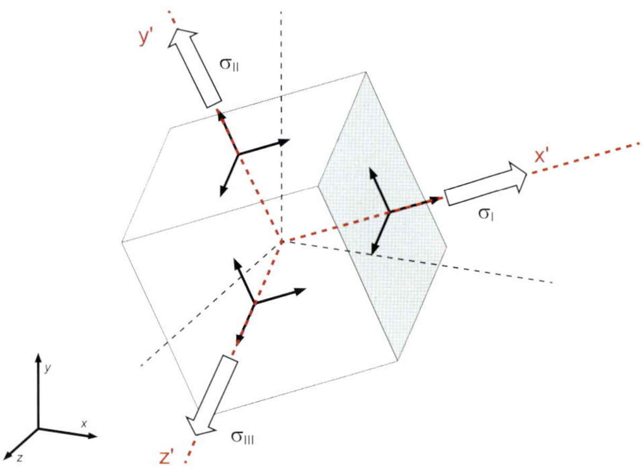

Fig. 4.9. Principal components of stress and principal directions.

has been improved, by considering complex inner cavities such as those present in skulls. In the latest literature it is possible to find several works focused on mammalian mandibles which followed the suggestions of Dumont et al. (2009) to scale their models in order to be able to compare von Mises stress via the surface ratio (e.g., Lautenschlager et al. 2016). Many other works have scaled their models using the volume ratio to then compare stress based on the allometric proportions between muscular forces and body mass (e.g. Attard et al. 2014, Tseng et al. 2017).

The main controversy regarding the scaling procedures is whether the scaling should be carried out by means of a surface ratio or via a volume ratio. Interestingly, authors such as Dumont et al. (2009) which propose the use of surface ratios, do not deny the usefulness of volume ratios when scaling models to compare von Mises stress. In fact, they define an "area proportion" which can be calculated from the areas but can also be deduced from volumes $[\text{volume}]^{2/3}$ or length $[\text{length}]^2$ which can be substituted for an area ratio. This means that both the surface ratio and the volume ratio are equivalent when the latter is raised to the $^2/_3$ power. However, exchanging ratios in the equations cannot be done when comparing models that differ in shape, because models with different shapes do not necessarily have the same surface area to volume ratios.

To clarify some of the issues in this debate (i.e., how the scaling should be done when comparing von Mises stress), Marcé-Nogué et al. (2013) published a methodological paper that was later followed by a practical application for 3D models by Fortuny et al. (2015), where it was shown that the relationship using the equations of continuum mechanics should be made by means of the volume ratios and not using surface ratios. These results agree with the fact that muscular forces and body mass/volume are related with the $^2/_3$ power (Meers 2003, Wroe et al. 2005).

Material properties. The difficulty of obtaining information on the material properties (namely Young's modulus and Poisson's coefficient) of bony structures due to the absence of data for the species under study or any close relatives, makes it very difficult to create a comparative framework with different material properties for each model. It is common to consider bone as an isotropic material for simulations, as real data for bone materials are unavailable for most species. Moreover, if the comparison is made between the same bony structures in close relatives, it makes sense to use the same material properties for all the specimens compared. This is the usual procedure when comparing mammalian mandibles (e.g., Serrano-Fochs et al. 2015, Tseng & Binder 2010), even though some authors have obtained heterogenic and orthotropic properties by carrying out experiments (Panagiotopoulou et al. 2017).

When linear and elastic material properties are assumed in a comparative analysis, the effect of the elastic modulus of the material is irrelevant for stress patterns. In the case of the strains and displacements, there is an inverse proportionality that is kept constant, between the values of the metrics and the changes in the elastic modulus. These properties allow comparative studies without considering the real elastic materials properties (Gil et al. 2015).

This evidence suggests that the use of the values that define a linear material (namely Young's modulus and coefficient of Poisson) is not crucial for the development of the analyses proposed here, because these values do not affect the results when a relative comparison of stress results between models is performed. In that sense, material properties should not be an issue when comparing bony structures.

Simplifications. Apart from the material properties discussed above, models are, by definition, a simplification of reality which means that one has to be aware of which assumptions are required, so as to keep them in mind when interpreting the results. There are at least two issues of great importance that should be considered in relation to model simplifications (when required):

(1) Knowing whether the model simplifications are not extremely affecting the sensitivity of the results.
(2) Carrying out the same simplifications in all the models under comparison in order to avoid non-comparable scenarios/results.

In that context, a sensitivity analysis will test the influence of the different parameters of the FEA models. For example, Gröning et al. (2012b) tested the sensitivity of FEA models of a human mandible with different trabecular bone defined as a solid inner material with a lower Young's modulus. For instance, this kind of analysis can give us information of how we can simplify trabecular bone without excessively changing the results of our models. Simplified modelling assumptions of material properties and muscle activation patterns may introduce analytical errors in analyses where quantitative accuracy is critical for obtaining rigorous results (Panagiotopoulou et al. 2017). For example, when comparing mammalian mandibles it is common to segment the models as solid models without including trabecular bone properties, since it has been shown that the exclusion of trabecular bone does not affect the general results of a comparative FEA (Fitton et al. 2015) and because it has been argued that simplified models can reproduce the overall stress distribution patterns in ex vivo validation experiments (Bright & Rayfield 2011a).

Additionally, it is common not to segment the periodontal ligament (PDL) despite some debate in the literature regarding the importance (or lack of it) of modelling this tissue in models of the mandible in FEA (Bright 2014). Some modelling studies of the primate mandible have suggested that the presence or absence of the PDL might affect the obtained results substantially throughout the whole structure (Gröning et al. 2011a, Marinescu et al. 2005).

Another discussed simplification in mammalian mandibles is the accuracy of the results when analyzing planar models instead of 3D models. Planar models can be useful to get insights of the main patterns of variation of stress from both a morphological and a biomechanical perspective, as it has been exemplified by Fletcher et al. (2010), Marcé-Nogué et al. (2017b), Neenan et al. (2014), Piras et al. (2013), and Serrano-Fochs et al. (2015). Planar models can be consistent with the results obtained from more complex scenarios, even though the level of detail is not the same. Probably the most important reason is that planar models study a specific biological scenario that can be simplified, and the validity of the models will be accepted if they can achieve the objective for which they were created. For example, planar models of primate mandibles can be good FEA models when distinguishing hard and soft-food eaters because the use of this simpler models allows to distinguish these categories in terms of stress values (Marcé-Nogué et al. 2017b).

Planar models are easier and faster to build than 3D models and can also be a better option in some cases, as they allow the creation of models that would not be easy to carry out in 3D when the original data are impossible to reconstruct. On one side, in the field of paleontology it is usual to rely on morphotaxa defined from scarce and fragmentary elements of few individuals, which means that it could be potentially difficult to carry out reconstructions in three dimensions, but not equally so in two dimensions. On the other side, 3D reconstructions of the fossils can be difficult to produce if no access is granted to digitize the original materials.

Using FEA of mandibles to understand diet

The use of FEA to study the mechanics of mandibles and relating it to diet has been widely used in the last years (Fletcher et al. 2010, Gill et al. 2014, Gröning et al. 2011b, Lautenschlager et al. 2018, Marcé-Nogué et al. 2017b, Neenan et al. 2014, Piras et al. 2013, Püschel et al. 2018, Serrano-Fochs et al. 2015, Tseng & Binder 2010, Zhou et al. 2019). The underlying hypothesis is that under equivalent loads, the stress patterns observed in the different FEA models can be interpreted as a sign of the relative strength, with specimens with higher stress being weaker. Therefore, the hypothesis assumes that more robust or stronger mandibles would be needed to process more demanding diets. The study of extant species has been fundamental to this approach, because the information of the dietary preferences of extant species is readily available hence facilitating the correlation of this information with the results obtained using FEA models. Some examples are presented here, showing how authors have used FEA to infer or study the dietary preferences of different mammals.

Figueirido et al. (2014) exclusively analyzed extant species when comparing the results obtained in both cranial and mandibular FEA models of red and giant pandas, with the dietary differences of each one of these species reflecting their distinct bamboo feeding preferences. Püschel et al. (2018) studied the sclerocarpy adaptations in the mandible of pitheciid primates. This was done using FEA models which demonstrated that there is indeed a relative robusticity continuum for some aspects of shape, but this gradient could be related to other factors rather than sclerocarpic specialization. Finally, Marcé-Nogué et al. (2017b) studied a large number of planar models of different primate mandibles under different loading scenarios. They found that there are significant differences in mandibular biomechanical performance depending on food categories and/or food hardness in Primates.

The understanding of dietary preferences in extant species enables a comparative framework with fossil taxa to infer dietary preferences in extinct species. This is, probably, one of the most widely applied aspects of FEA for mandible biomechanics, both using three-dimensional and planar models.

In that sense, using planar models Fletcher et al. (2010) tested, if mandibles of hindgut-fermenter ungulates are more robust than those of ruminants. They addressed the question, if extinct hindgut or foregut fermenters can be identified in the fossil record. Similarly, Serrano-Fochs et al. (2015) studied FEA mandible models of armadillos. The results of this FEA showed that omnivorous species have stronger mandibles than insectivorous species, and also that the studied fossil herbivore taxa possess very strong mandibles due to a great amount of oral processing.

Three-dimensional models of extant taxa have been used in several carnivores to study fossil species. For example, Tseng & Binder (2010) compared strain distributions during a biting scenario between two bone-cracking ecomorphologies that evolved convergently, *Crocuta crocuta* and *Canis lupus*, and the fossil *Dinocrocuta gigantea*. *Kolponomos* has been interpreted as an otter-like shell-crusher based on a similar dentition. Tseng et al. (2016) studied different feeding strategies of extant carnivores in different FEA models. They tested whether there was a convergent morphology in the Miocene *Kolponomos* and the Pleistocene sabre-tooth felid *Smilodon* by using their anterior mandibles as anchors. The same authors studied the feeding capability of the Mio-/Pliocene giant otter *Siamogale melilutra* using FEA models of the mandible of different living Lutrinae to infer that the mandibular strength in *S. melilutra* strikingly surpasses molluscivores such as the sea otter and the Cape clawless otter (Tseng et al. 2017).

In fact, the combination of both FEA models of the mandible and skull has been common in the study of fossil carnivores. Others such as Wroe et al. (2007) studied the extinct marsupial wolf *Thylacinus cynocephalus* and the placental grey wolf, which are commonly presented as an iconic example of convergence. In spite of this convergence, comparisons of stress distributions in the FEA models not only revealed considerable similarity, but also provided informative differences suggesting ecological overlap yet different prey strategies. They suggest that the grey wolf is better adapted to withstand the high extrinsic loads likely to accompany social hunting of relatively large prey. Other authors (Attard et al. 2011) studied the biomechanical performance of the skull and mandible of *T. cynocephalus* in relation to those of two extant marsupial carnivores with

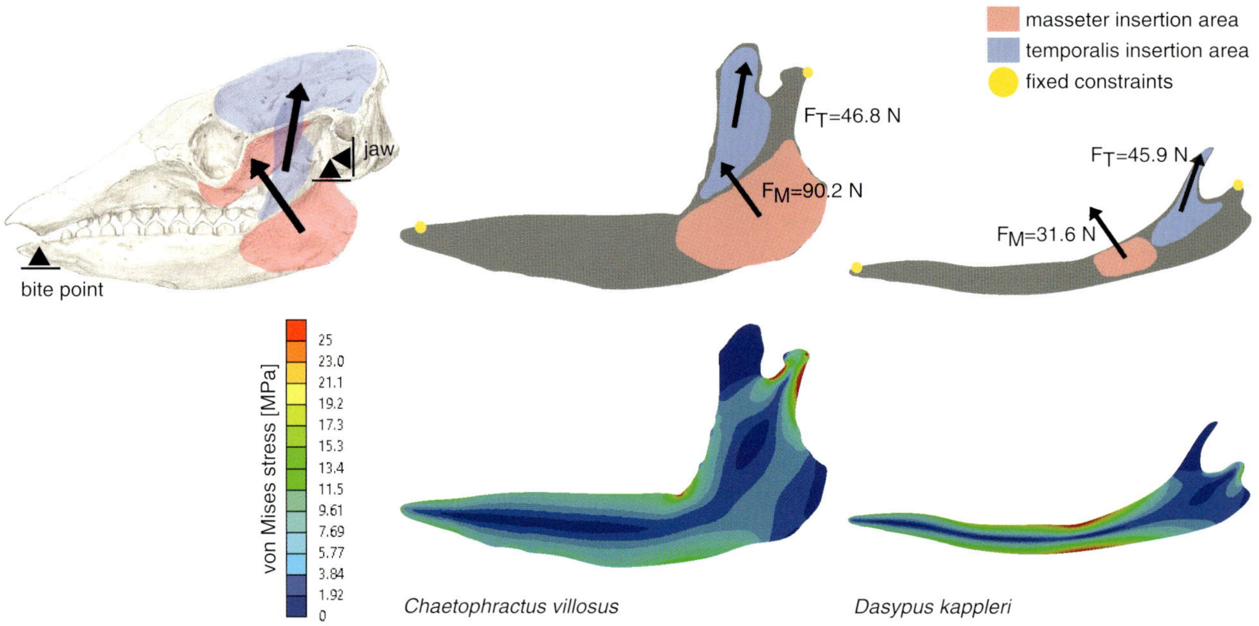

Fig. 4.10. Free-body and von Mises stress diagrams of armadillo mandibles with real muscular force values. **A,** free-body diagram of a mandibular bilateral bite using a plane stress approach and; **B,** von Mises stress from a finite element analysis of *Chaetophractus villosus* and *Dasypus kappleri* from Serrano-Fochs et al. (2015). Results using real muscular forces values.

known diets that occurred sympatrically – the Tasmanian devil, *Sarcophilus harrisii,* and the spotted-tailed quoll, *Dasyurus maculatus* – suggesting that *T. cynocephalus* is likely to have consumed smaller prey compared to its size. Oldfield et al. (2012) created several FEA models of ursids to predict the feeding behavior of the extinct *Agriotherium africanum.* According to these authors, the findings did not resolve whether the fossil was more likely a predator or a scavenger of large terrestrial vertebrates but showed that it was well adapted to resist the forces generated by either activity. Sabre-tooths have been examined by Wroe et al. (2013) using FEA models to suggest that, in many respects, the placental *Smilodon fatalis* was more similar to the marsupial *Thylacosmilus atrox* than to a conical-toothed cat. Similarly, Attard et al. (2014) used FEA models of marsupial carnivores to test whether the fossil species *Nimbacinus dicksoni* was more similar to its recently extinct relative *Thylacinus* or to several large living marsupials. This comparison in a key aspect such as feeding ecology suggested that it was a large-prey specialist.

The difficulties of comparing fossil taxa with extant species has driven a handful of works using FEA models of the mandible to study dietary preferences in extinct species only comparing them to their extinct close relatives. However, the advances in computational tools to perform virtual reconstruction and restoration techniques of the original morphology of fossils (Lautenschlager 2016) allow the creation of accurate and realistic three-dimensional models. In that line, Gill et al. (2014) studied the dietary specializations and diversity in feeding ecology of the earliest stem mammals using FEA. Three-dimensional models of the mandible of the early mammaliaforms *Morganucodon* and *Kuehneotherium* were analyzed suggesting that the lineage split during the earliest stages of mammalian evolution due to ecomorphological specialization and niche partitioning.

Table 4.5. Real muscular force values and obtained results: Bite force, mechanical advantage and strain energy of *Chaetophractus villosus* and *Dasypus kappleri* mandibles using a plane stress approach.

Taxon	Masseter force (N)	Temporalis force (N)	Bite force (N)	Mechanical advantage (MA)	Strain energy (mJ)
Dasypus kappleri	31.6	45.9	15.29	0.197	6.8667
Chaetophractus villosus	90.2	46.8	34.69	0.253	4.9075

Table 4.6. Scaled muscular force values and obtained results: Bite force, mechanical advantage and strain energy of *Chaetophractus villosus* and *Dasypus kappleri* mandibles using a plane stress approach.

Taxon	Thickness (mm)	Model area (mm^2)	Masseter force (N)	Temporalis force (N)	Bite force 1(N)	Mechanical advantage (MA)	Strain energy (mJ)
Dasypus kappleri	3.51	971.37	62.0	33.2	23.22	0.244	15.54
Chaetophractus villosus	4.94	1038.90	90.2	46.8	34.69	0.253	4.9075

Practical example B. Finite element analyses (FEA) were applied to assess the lower mandible biomechanics of *Dasypus kappleri* and *Chaetophractus villosus*. The free-body diagram of the mandible, including masseter and temporalis muscle forces, is the same as shown in previous examples (Fig. 4.10). Constraints are applied in the bite position and the jaw joint.

Reconstructed muscular force values: Firstly, we solve these FEA models using realistic values from muscular contraction in both models. The muscle forces are calculated, as explained previously, using the muscle insertion area and a value for the specific tension of 0.3 MPa (force per unit area) is assumed as muscular contraction.

Using FEA we can obtain the von Mises stress patterns of each mandible (Fig. 4.10) and the bite forces calculated by the software according to the equilibrium static laws (Tab. 4.5). As expected, the bite forces calculated here are almost identical to the ones calculated before using the equations of lever mechanics.

The von Mises stress distribution is similar for both mandibles, with lower stress magnitudes in the ramus, null values in the coronoid process, and higher values of stress in the corpus. The mandibular corpus of *D. kappleri* is narrower than the mandibular corpus of *C. villosus* which results in higher stresses appearing in the mandibular corpus of *D. kappleri* and shows that this mandible is at a higher risk of fracture during chewing. Moreover, *D. kappleri* spends more strain energy during the biting. If we consider the strain energy as a measure of structural stiffness, and higher energy meaning lower stiffness (Tseng et al. 2017), all these results are consistent with the ones from beam models also suggesting that *C. villosus* is capable to withstand higher bending moments than *D. kappleri* when chewing.

Summarizing, under the same conditions of chewing with the maximum isometric contraction of each muscle, *C. villosus* is capable of generating higher bite forces with lower stress values in the mandible. This is in line with the fact that *D. kappleri* is a generalist insectivore, whereas *C. villosus* consumes a more demanding omnivore diet.

Scaled muscle forces values: Secondly, we solve the same models by using the scaling approach for the values of the muscle forces. In order to compare the models, a scaling of the values of the forces is applied. We use the equations defined in Table 4.4, according to which we are comparing stresses in planar models using *Chaetophractus villosus* as the reference model (Tab. 4.6).

In this case, we assume both models have the exact same muscle forces applied but remove the effect of size. This is equivalent to the creation of two models with the same model area and the same values of forces. The results suggest a more fragile and weaker mandible for *D. kappleri* but the results are coherent with the same dietary conclusions.

New developments in the comparison of mammalian mandibles: the example of the armadillo mandible

Quantification of FEA results

In recent years, FEA models and the respective results have been compared in a predominantly qualitative way, by comparing the stress or strain distributions. Visual comparison using only the stress or strain distribution of contour plots can be a good option when the number of models to compare is small and the results are distinctive enough between the analyzed models. For example, in Tseng & Binder (2010), comparison of stress distributions during a biting scenario in *Crocuta crocuta*, *Canis lupus* and the fossil *Dinocrocuta gigantea* was feasible, because the differences in the graphical representation of the von Mises stress are clearly visible and distinct from one another. However, when the number of models to be compared increases, such comparison becomes difficult (Neenan et al. 2014) and more rigorous quantification of the FEA results is useful.

Quantifying stress data at specific points of a model has been useful in ecomorphological analyses (Attard et al. 2014, Fortuny et al. 2011, 2016, Piras et al. 2015, Serrano-Fochs et al. 2015), but to take into account the biomechanical behavior of the complete model, single mean values of stress were adopted as quantitative measurement to provide a metric of the relative strength of vertebrate structures. This approach involves computation of the von Mises stress averages in finite element models. In spite of this approach having been used in several paleobiological studies (Aquilina et al. 2013, Farke 2008, Figueirido et al. 2014, Fish & Stayton 2014, Neenan et al. 2014, Parr et al. 2012, Tseng 2009), we employed the recently proposed “weight-meshed values” and the “quasi-ideal meshes” to compute these values as adapted by Marcé-Nogué et al. (2016, 2017a). The latter is more robust, because it takes into account the effects caused by the different size of the individual mesh elements in a FEA model.

Similarly, the quantification and post-processing of the results of FEA models has also evolved to be combined with other morpho-functional methodologies such as geometric morphometrics (GM). FEA and GM outputs have been used in mandible models to explore questions in functional morphology, ecomorphology, macroevolution, among others, by applying standard statistical methods. These methods involve multivariate regressions, ANOVAs or PLSs (Maiorino et al. 2015, Piras et al. 2013) or, even, GM as a tool to analyze the deformations resulting from FEA analyses (Gröning et al. 2011a).

Finally, it is worth mentioning that the influence of the size of the individual elements of the FEA mesh in the results must be always taken into account by performing sensitivity analysis (Tseng & Flynn 2015). To avoid the influence of the size of the elements in the results, an ‘intervals’ method’ was proposed by Marcé-Nogué et al. (2017a), which incorporates element volumes to provide an additional way of quantifying and comparing FEA results, thus avoiding the problems with the mesh. This method could allow for considerably more effective comparisons of finite element models, and maybe more precise distinction between dietary traits.

The problem of the maximum peak value

Maximum values of stress or strain (or peak stress/strain) could be potentially useful as a value to consider when comparing the biomechanical performance of different FEA models. However, peak stress and strain should not be taken into account because the value could be a numerical singularity, which is a well-known problem in FEA (Morris 2008).

This singularity could be a consequence of a boundary condition or a perfect square within the geometry. A simply supported condition is an idealistic assumption. This condition has been applied specially in planar models, as in the examples provided by this chapter, where fixed displacements were imposed at the bite position and at the jaw joint in a simple point of the model. This condition creates a numerical effect that leads to the stress increasing without limit when the mesh is refined for both 2D and 3D elements. In 3D models of mandibles, simply supported conditions are usually applied in the bite position or in the jaw joint where the displacements are restrained in a specific area of the model. In a similar way, the perfect square condition is also an idealistic assumption where a corner with zero radius in the original geometry creates a stress concentration that also leads the value to increase without limit when the mesh is refined. To avoid artificially high stress magnitudes, it is not recommended to record the maximum stress or strain values in the whole model, because maximum values could be the result of singular high yet unrealistic magnitudes. As some authors have warned (e.g., Rayfield 2007), it may be difficult to assess the peak stresses of the model in the whole model. Some approaches regarding how to handle this problem are described in Marcé-Nogué et al. (2015). For example, this can be achieved by just studying the results obtained from an area of the model that is far enough from the artificially high magnitudes, or by adapting the mesh to avoid the visualization of this high value. Coarse meshes can avoid these extremely high stress values whereas very fine meshes can concentrate these high values in a very small region of the model. Finally, other authors have suggested to exclude elements displaying the stress or strain values within the top 5 % (Tseng & Flynn 2018).

Use of specific points to quantify data

Quantifying stress or strain data at specific points has been useful in ecomorphological analyses of mammalian mandibles. An example of this can be found in Attard et al. (2014) who selected equidistant points along the mandible to measure the distribution of von Mises stress for each loading case and then compared them with other models. This approach allows us to plot these values of stress in a graph where one of the axes is the value of stress and the other one is the position from anterior to posterior along the mandible.

This approach has also been used in combination with the values of von Mises stress in specific points with data from morphometric analysis in other vertebrate groups (Fortuny et al. 2011). Or in a different way, using the values of stress from the specific points to plot in a dispersion plot of a principal component analysis (PCA) (Fortuny et al. 2016, Marcé-Nogué et al. 2015). Quantifying data from FEA models allows the use of such data to test the significance using statistics or other methods, opening the door to new methodologies to be used in combination of FEA.

Case example: armadillos

Biomechanical problem and FEA data. Finite element analyses were applied to assess the lower jaw biomechanics of cingulate xenarthrans in Serrano-Fochs et al. (2015). The main goal of this work was to make a comparative assessment of the biomechanical capabilities of the mandible based on FEA, and to relate the obtained stress patterns with dietary preferences. To do this, a comparative framework was created following the indications of the previous section of this chapter, so to solve the biomechanical problem described in the free-body diagram of Figure 4.11. The mandible was constrained at the most posterior part and at the condyle level of the mandibular notch following the procedures described in Serrano-Fochs et al. (2015) and Marcé-Nogué et al. (2016, 2017a).

Planar models of 11 armadillo mandibles (Tab. 4.1), each one corresponding to a different species, were created. Two main masticatory muscles (i.e., temporalis and masseter) were included in the model as a vector between the centroid of the muscular attachment on the mandible and the centroid of the equivalent muscle attachment on the skull. To compare the models, a scaling of the values of the forces was applied using the equations defined in Table 4.4 in order to compare stresses in planar models, by using *Chaetophractus villosus* as reference model (Tab. 4.7).

Isotropic and linear elastic properties were assumed for the bone. In the absence of data for Cingulata or any other closer relative, and due to the absence of data for any mammalian clade with a similarly shaped mandible, the authors decided to apply the mandibular material properties of *Macaca rhesus*: E (elasticity modulus) = 21000 MPa and v (Poisson coefficient) = 0.45 (Dechow & Hylander 2000). The available properties of *Macaca rhesus* were chosen, provided that it has a wide range of habitats and diet which resembles omnivorous or generalist insectivorous diets, as observed in the analyzed armadillos (Richard et al. 1989).

The stress distribution obtained in each armadillo is shown in Figure 4.12A. These stress patterns can be interpreted as a sign of relative strength. Assuming that more robust or stronger mandibles would be needed both for processing harder food items, insect-feeding armadillos (which have none or little processing in the mouth) should be expected to have weaker mandibles (i.e. with higher stress levels) than those feeding on other items, such as herbivorous or omnivorous species.

On the other hand, differences in stress distribution patterns may provide a clue regarding different aspects of the feeding ecology of the analyzed species. The aim of the methods presented below is to show that using a comparative framework and post-processing the results (e.g., with specific points, mesh-weighted values, quasi-ideal meshes or the intervals' method) the interpretation of the results can be greatly improved.

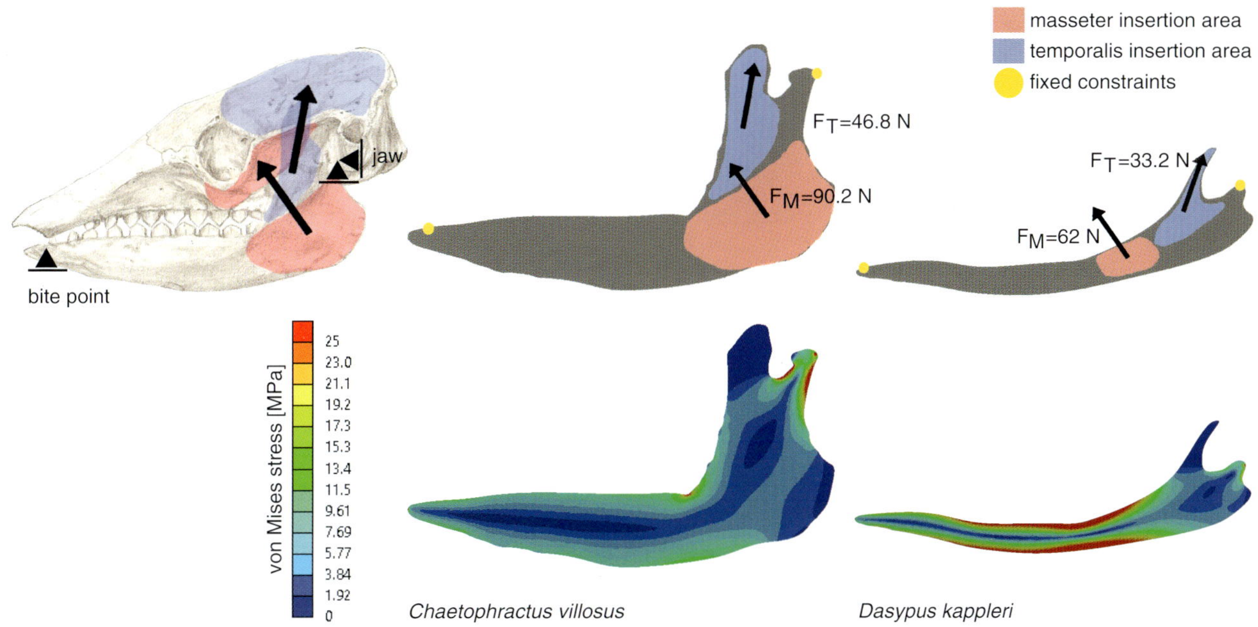

Fig. 4.11. Free-body and von Mises stress diagrams of armadillo mandibles with scaled muscular force values. **A,** Free-body diagram of a mandibular bilateral bite using a plane stress approach; **B,** von Mises stress from a finite element analysis of *Chaetophractus villosus* and *Dasypus kappleri* from Serrano-Fochs et al. (2015). Results using scaled muscular force values.

Specific points. In Serrano-Fochs et al. (2015) the strength of the mandibles was studied using the values of stress in eight specific points. The values of stress in each specific point were plotted according to the different dietary traits of the studied armadillos. They are classified as omnivores or insectivorous specialists and generalists (Fig. 4.12A) and a Kruskal-Wallis test was applied to check if there were significant differences in the results. However, despite the lack of significant results, some patterns appear in the boxplots for some specific points (Fig. 4.12B). Moreover, a principal component analysis (PCA) of the stress values for all specific points recorded was carried out using the variance-covariance matrix (Fig. 4.12C). The aim of these analyses was to evaluate the stress values in a multivariate manner and to look for diet-related patterns.

The most obvious result is that in general, insectivorous species showed a larger variance than omnivorous species, and also that omnivorous species have some coincident stress patterns. However, the absence of significant results could be due to a lack of statistical power, or possibly because the selected points represent local non-significant regions, or the specific points were hindered by the lack of information about the whole FEA model.

Mesh-weighted average values and quasi-ideal mesh. A quantitative single measurement of the relative strength of the studied structure could be preferred to summarize the strength of the whole model. The most common approach has been the computation of the average von Mises stresses of the various models considered. Even though

Table 4.7. List of the species analyzed in the present study. The classification of each species was made on the basis of the current knowledge about the ecology of armadillos (see Serrano-Fochs et al., 2015 for discussion). The geometric properties and the applied forces by the masseter and temporalis muscles are also provided. Abbreviations preceding the names of institutions are used to identify the location were the specimens are housed. AMNH, American Museum of Natural History, New York, USA; MNCN, Museo Nacional de Ciencias Naturales, Madrid, Spain; MNHN, Muséum national d'Histoire naturelle, Paris, France; ZMB, Museum für Naturkunde (Zoologisches Museum), Berlin, Germany; MLP, Museo de la Plata, La Plata, Argentina.

Taxon	Diet	Collection number	Thickness (mm)	Model area (mm^2)	Masseter area (mm^2)	Temporalis area (mm^2)	Masseter force (N)	Temporalis force (N)
Priodontes maximus	Specialist insectivore	AMNH 208104	6.41	2051.70	616.02	255.06	1.29	0.53
Cabassous unicinctus	Specialist insectivore	MNHN 1953/457	3.51	415.75	112.08	22.91	0.37	0.08
Tolypeutes matacus	Generalist insectivore	AMNH 246460	3.56	497.40	157.01	64 116.00	0.35	0.14
Dasypus kappleri	Generalist insectivore	MNHN 1995/207	3.51	971.37	105.37	153.18	0.28	0.41
Dasypus sabanicola	Generalist insectivore	ZMB 85899	2.78	527.86	150.66	71 545.00	0.27	0.13
Dasypus novemcinctus	Generalist insectivore	AMNH 133338	2.94	613.54	225.77	92 174.00	0.32	0.13
Chlamyphorus truncatus	Generalist insectivore-fossorial	ZMB 4321	2.00	113.19	16 035.00	34 006.00	0.04	0.09
Chaetophractus villosus	Omnivore/Carnivore	MNCN 2538	4.94	1038.90	300.58	156.08	0.66	0.34
Chaetophractus vellerosus	Omnivore/Carnivore	MLP 18.XI.99.9	3.68	538.80	145.04	117.03	0.30	0.24
Euphractus sexcinctus	Omnivore/Carnivore	MNHN 1917/13	5.66	1019.20	331.22	190.60	0.72	0.41
Zaedyus pichiy	Omnivore/Carnivore	MLP 9.XII.2.10	3.51	327.35	89 737.00	66 091.00	0.23	0.17

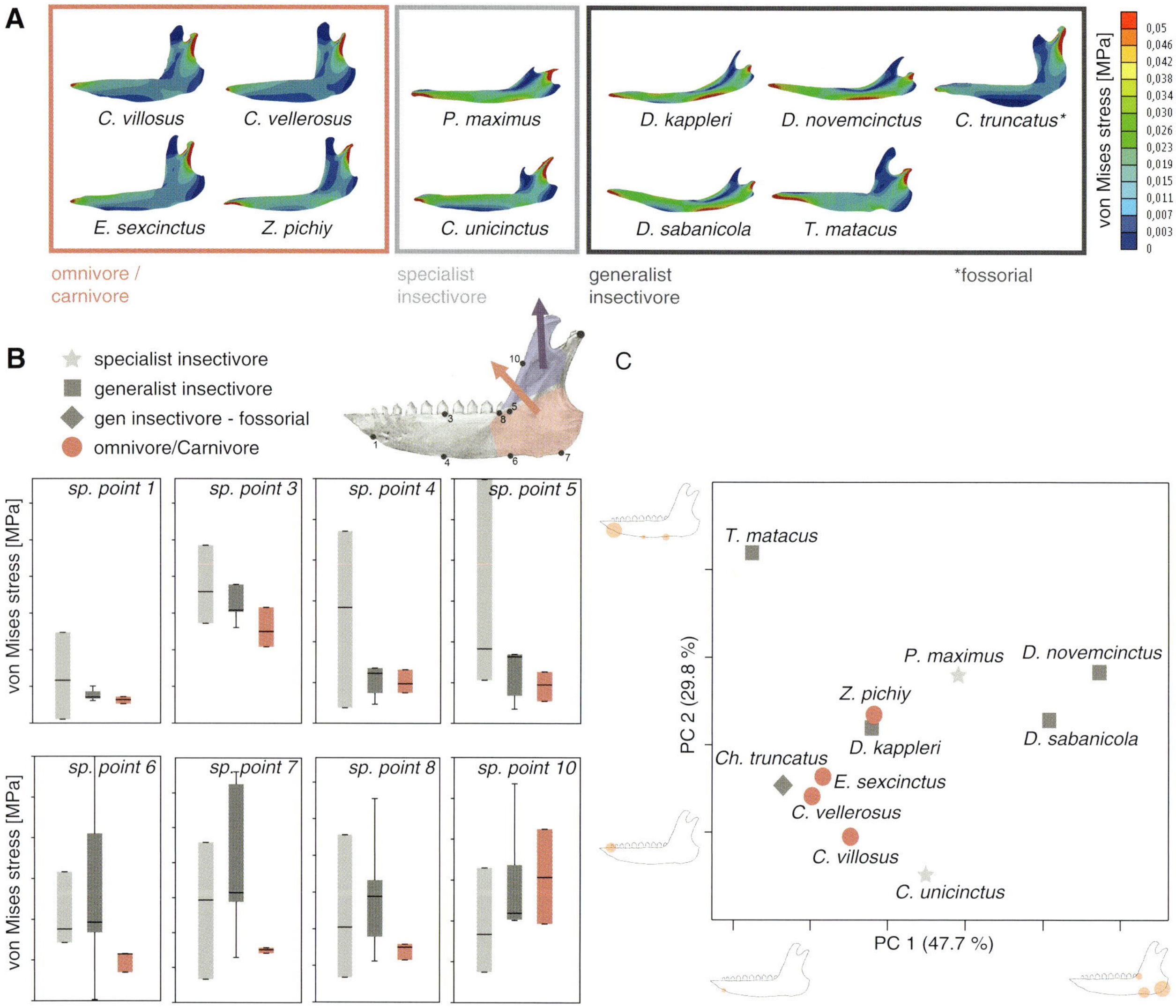

Fig. 4.12. Von Mises stress and PCA of armadillo mandibles according to diet. **A,** distribution of von Mises stress in the 11 armadillos' mandibles separated by diet: Omnivore/carnivore and specialist or generalist insectivore. **B,** boxplots of the von Mises stress values for the specific points; **C,** dispersion graph of the two first principal components. A hypothetical lower jaw is represented in each extreme of the axes, highlighting those landmarks with higher loadings in each PC (i.e., that have more importance in that PC) with red circles. The size of the circles approximates the standardized value of the loading. Modified from Serrano-Fochs et al. (2015). Abbreviations: *P. maximus*, *Priodontes maximus*; *C. unicinctus*, *Cabassous unicinctus*; *T. matacus*, *Tolypeutes matacus*; *D. kappleri*, *Dasypus kappleri*; *D. sabanicola*, *Dasypus sabanicola*; *D. novemcinctus*, *Dasypus novemcinctus*; *Ch. truncatus*, *Chlamyphorus truncatus*; *Ch. villosus*, *Chaetophractus villosus*; *Ch. vellerosus*, *Chaetophractus vellerosus*; *E. sexcinctus*, *Euphractus sexcinctus*; *Z. pichiy*, *Zaedyus pichiy*.

this approach has been used previously in paleobiological studies of mammalian mandibles (Figueirido et al. 2014, Neenan et al. 2014, Wroe et al. 2013), it represents a common error that should be avoided. When obtaining descriptive statistics from FE models, such as mean or median values, the problem of having non-uniform meshes is crucial and could lead to skewed results. For example, if by chance high stress values are obtained in larger elements, these high stress values will be under-represented because they are present in fewer elements despite the area with high values being large.

To solve this issue, Marcé-Nogué et al. (2016) proposed the so-called weight-meshed values, which take into account the type and size of the finite element mesh: mesh-weighted average mean (MWAM) and mesh-weighted median (MWM). In the same work, Marcé-Nogué et al. (2016) proposed the creation of quasi-ideal meshes (QIM) which are meshes where all the elements have practically the same size. The use of QIM avoids the use of weight-meshed corrections in the mesh, thus allowing the use of average values without correcting them. Moreover, with the use of QIM it is also proposed to use percentile values (e.g., 25th, 50th, 75th or 95th) as quantitative values, which allows to visualize the distribution of stress in each FEA model as a boxplot, facilitating the comparison between models. Moreover, and following the idea proposed by Walmsley et al. (2013), the use of the 95th percentile can be assumed as the maximum peak value to avoid the necessity to consider numerical singularities.

The arithmetic mean is calculated by summing all individual observations or items of a sample and dividing this sum by the number of items in the sample. In FEA results of stress, the arithmetic mean (AM) would be the sum of the value of the von Mises stress (σ_{VM}) of each element

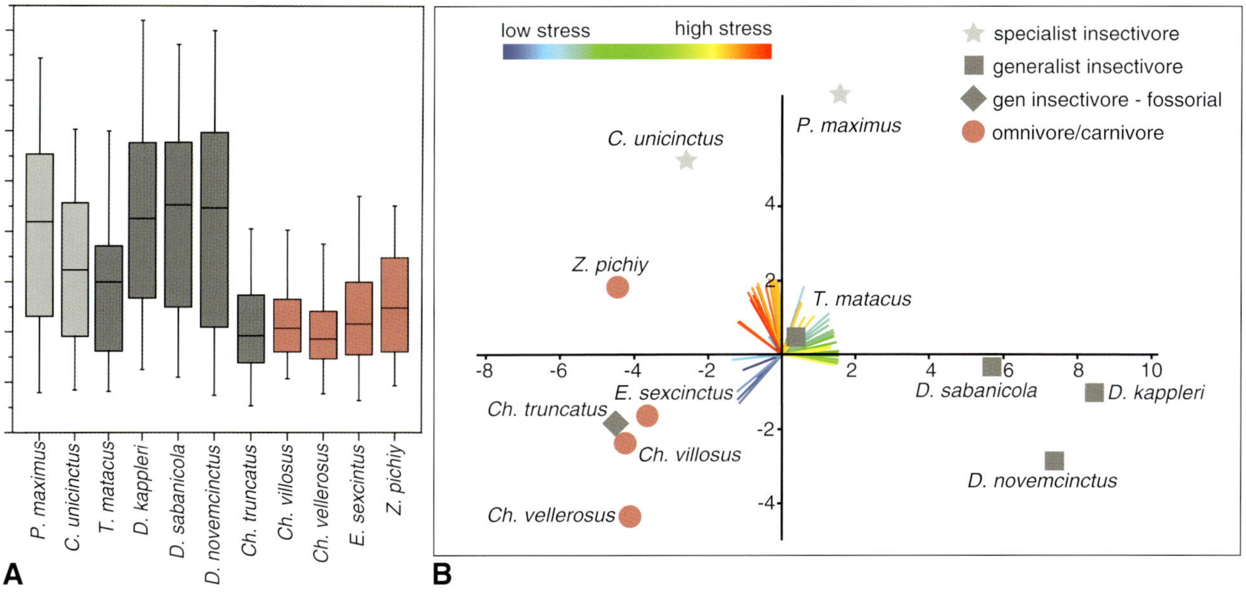

Fig. 4.13. Von Mises stress and PCA of armadillo mandibles under assumption of quasi-ideal meshes. **A,** boxplots of von Mises stress distributions when quasi-ideal meshes (QIM) are assumed for the Cingulata mandibles analyzed; **B,** PCA based on the correlation matrix. The loadings for each variable are colored according with the range of stress they represent, with reddish colors for high level of stress, and bluish for low levels. X-axis: PC1. Y-axis: PC2. Modified from Marcé-Nogué et al. (2016, 2017a). Abbreviations: *P. maximus*, *Priodontes maximus*; *C. unicinctus*, *Cabassous unicinctus*; *T. matacus*, *Tolypeutes matacus*; *D. kappleri*, *Dasypus kappleri*; *D. sabanicola*, *Dasypus sabanicola*; *D. novemcinctus*, *Dasypus novemcinctus*; *Ch. truncatus*, *Chlamyphorus truncatus*; *Ch. villosus*, *Chaetophractus villosus*; *Ch. vellerosus*, *Chaetophractus vellerosus*; *E. sexcinctus*, *Euphractus sexcinctus*; *Z. pichiy*, *Zaedyus pichiy*.

divided by the number of elements of the mesh (equation 19).

$$AM = \frac{\sum_{i=0}^{n} \sigma_{VM}}{n} \qquad \text{(equation 19)}$$

The mesh-weighted arithmetic mean (MWAM) corresponds to the sum of the value of the von Mises stress for each element multiplied by its own area (A) and divided by the total area (equation 20). This value is equivalent to the division of the arithmetic mean of the product of stress and area by the arithmetic mean of the area, which is easier to calculate and does not require correction of weight element by element.

$$MWAM = \frac{\sum_{i=0}^{n}(\sigma_{VM}^{i} \cdot A^{i})}{\sum_{i=0}^{n} A^{i}} = \frac{\frac{\sum_{i=0}^{n}(\sigma_{VM}^{i} \cdot A^{i})}{n}}{\frac{\sum_{i=0}^{n} A^{i}}{n}} = \frac{AM(\sigma_{VM}^{i} \cdot A^{i})}{AM(A^{i})} \qquad \text{(equation 20)}$$

The median is the middle measurement of any set of sorted data. In the case of FEA, the median would be the value separating the higher half from the lower half of the values of von Mises stress recorded in each element of the mesh after they have been ordered.

Here the mesh-weighted median (MWM) of stress distribution has been defined as the division of the median of the product of stress and area by the median of the area (equation 21), based on the formulation presented in equation 20.

$$MWM = \frac{median\,(\sigma_{VM}^{i} \cdot A^{i})}{median\,(A^{i})} \qquad \text{(equation 21)}$$

Two indicators were proposed to evaluate whether a mesh is uniform enough to use the raw stress data for statistical analysis: the percentage error of the arithmetic mean (PEofAM) and the percentage error of the median (PEofM). These two indicators evaluate the difference between the non-weighted value and the weighted value of mean and median (PEofAM in equation 22 and PEofM in equation 23),

$$PEofAM = \left(\frac{MWAM - AM}{MWAM}\right) \times 100 \qquad \text{(equation 22)}$$

$$PEofM = \left(\frac{MWM - M}{MWM}\right) \times 100 \qquad \text{(equation 23)}$$

If the mesh of the model is close to an ideal fine uniform mesh, the non-weighted and the weighted indicators should be equal. If the error is lower than a certain threshold, the mesh can be considered a QIM, and quantitative and statistical analysis of the FEA data can be computed without any corrections and e.g., the percentile values 25th, 50th, 75th and 95th can be computed.

When this approach was used to analyze the armadillo mandibles, the significant differences among different diets found in the statistical analyses were more conclusive than the results obtained when analyzing stress values at a few specific points (Serrano-Fochs et al. 2015). This supports the analysis of the stress values of the whole specimen rather than focusing on just a few isolated points. An alternative would be to use both sources of information as isolated points, which could give an alternative view of some specific areas.

Statistics showed significant results when comparing omnivore and insectivore species (Tab. 4.8). However, this significance should be interpreted as illustrative more than fully conclusive. This is because when a Bonferroni correction was applied, the mean values between specialized insectivores and omnivores are non-significant (most probably due to the small sample size of the specialist insectivores).

Intervals' method. In Marcé-Nogué et al. (2017a), a new method was proposed, named the intervals' method, that can be used to analyze data from finite element models in a comparative multivariate framework. As a case study, several armadillo mandibles were analyzed with success, showing that the proposed method was useful to distinguish and characterize biomechanical differences related to diet/ecomorphology.

This so-called intervals' method consists of generating a set of variables, each one defined by an interval of stress values. Each variable is expressed as a percentage of the area (in planar models) or the volume (in 3D models) of the mandible occupied by those stress values. Once all the stress values of a single specimen are obtained, they can be subdivided into different intervals, each one of them representing the amount of area in percent values of the original model having a specific range of stress values.

Afterwards these newly generated variables can be analyzed using multivariate methods. The use of this new method in the FEA models of cingulate mandibles allowed us to positively discriminate between specialist and generalist species of insectivores, which was not possible with the approaches defined above. The method described in Marcé-Nogué et al. (2017a) requires the definition of certain parameters such as a fixed upper threshold (FT_{upper}) and a specific number of intervals N. For example, in this work, $FT_{upper} = 0.1$ MPa and $N = 50$. This value is assumed based on the convergence procedure defined in Marcé-Nogué et al. (2017a), when the results of the regressions of the PCs indicate that the scores are almost completely correlated.

We carried out a PCA using the variance-covariance matrix (Fig. 4.13). The PCA of the correlation matrix successfully distinguishes the main three diets, with omnivore-carnivore species on lower-left area of the plot. PC2 separates specialist insectivores from the rest of species, while *Chlamyphorus truncatus* (i.e., a generalist insectivorous species exhibiting a very particular diet due to its completely fossorial lifestyle) is located near the omnivore/carnivore species in the negative part of the PC1 and PC2. Within this PCA, omnivore/carnivore species were characterized by very low stress values (lower-left quadrant of the plot), whilst generalist insectivores showed a proportionally larger area of intermediate stress values than the rest of the species. Specialist insectivores have proportionally larger areas of high stress.

Further usage of quantification of stress

Once we have quantified our models using specific points, average values or percentiles from a QIM or applying the intervals' method, we can use different ideas and methods to organize our data. On one hand, values of specific points, scalar values from the models (such as strain energy) and the mesh-weighted proposed averages values can be used for the comparison of models and their behavior. On the other hand, the intervals' method or the use of the stress quartiles are a good approach to check similarities between models.

Scalar values from the FEA models such as the strain energy were used to compare with other scalar values such as the mechanical advantage (Tseng et al. 2016) or used in regression with other values such as the volume of the model to check for allometry (Tseng et al. 2017).

Average values were used to compare different dietary traits and hardness of ingesta in planar FEA models of extant primates (Marcé-Nogué et al. 2017b), showing that there is a strong association between mandibular biomechanical performance, mandibular form, food hardness, and diet categories. The same work (Marcé-Nogué et al. 2017b) used average values to estimate the ancestral states for internal nodes using maximum likelihood and then by interpolating the states along the branches of the phylogeny. This approach was applied to get insights about the possible evolution of mandibular stiffness.

Finally, it is worth mentioning that the use of the different percentile values as a unique set of multivariate data, the average values or the powerful intervals' method, are all opening a window of opportunity for post-processing FEA-derived data. Using these kinds of data allows the application of multivariate analysis, and they can even be used to run machine learning algorithms in order to infer dietary preferences of fossil taxa from the data of living species. Both supervised (Marcé-Nogué et al. 2020) and unsupervised (Zhou et al. 2019) algorithms can be used to study mandibular biomechanics and its relationship with diet.

Table 4.8. Statistics for the Kruskal-Wallis test: mesh-weighted arithmetic mean (MWAM), and mesh-weighted median (MWM).

		Pairwise test p-value: Bonferroni non-corrected/corrected		
	Kruskall-Wallis p-value	Specialist insectivore vs. generalist insectivore	Specialist insectivore vs. omnivore	Generalist insectivore vs. omnivore
MWAM	0.0089	0.8197/1	0.0369/0.1107	0.0058/0.0173
MWM	0.0081	0.5676/1	0.0358/0.1073	0.0057/0.0170

Acknowledgments

The author was supported by the Deutsche Forschungsgemeinschaft (DFG, German Research Foundation, KA 1525/9-2). This research forms part of the DFG Research Unit 771 "Function and performance enhancement in the mammalian dentition – phylogenetic and ontogenetic impact on the masticatory apparatus." Thomas A. Püschel is acknowledged for his technical and scientific comments on earlier versions of this chapter and Dan Sykes for his help proof-reading this work. The author wants to thank the CERCA program (Generalitat de Catalunya) and Lluís Gil, Josep Fortuny and Soledad de Esteban-Trivigno for reading and commenting earlier versions of this work. Last but not least, the review comments by Stefan Lautenschlager and Jack Tseng greatly improved this chapter.

References

* indicates publications that originated from the DFG Research Unit 771.

Alexander, R. M. (1992): Exploring Biomechanics. Animals in Motion. New York Scientific American Library.

Alexander, R. M. (2005): Mechanics of animal movement. Current Biology 15: 616–619.

Aquilina, P., Chamoli, U., Parr, W. C. H., Clausen, P. D. & Wroe, S. (2013): Finite Element Analysis of three patterns of internal fixation of fractures of the mandibular condyle. British Journal of Oral and Maxillofacial Surgery 51: 326–331.

Attard, M. R. G., Chamoli, U., Ferrara, T. L., Rogers, T. L. & Wroe, S. (2011): Skull mechanics and implications for feeding behaviour in a large marsupial carnivore guild: The thylacine, Tasmanian devil and spotted-tailed quoll. Journal of Zoology 285: 292–300.

Attard, M. R. G., Parr, W. C. H., Wilson, L. A. B., Archer, M., Hand, S. J., Rogers, T. L. & Wroe, S. (2014): Virtual reconstruction and prey size preference in the mid Cenozoic thylacinid, *Nimbacinus dicksoni* (Thylacinidae, Marsupialia): PLoS One 9: e93088

Biknevicius, A. R. & Ruff, C. B. C. (1992): The structure of the mandibular corpus and its relationship to feeding behaviours in extant carnivorans. Journal of Zoology 228: 479–507.

Bouvier, M. (1986a): A biomechanical analysis of mandibular scaling in Old World monkeys. American Journal of Physical Anthropology 69: 473–482.

Bouvier, M. (1986b): Biomechanical scaling of mandibular dimensions in New World monkeys. International Journal of Primatology 7: 551–567.

Bright, J. A. (2014): A review of paleontological finite element models and their validity. Journal of Paleontology 88: 760–769.

Bright, J. A. & Rayfield, E. J. (2011a): Sensitivity and ex vivo validation of finite element models of the domestic pig cranium. Journal of Anatomy 219: 456–471.

Bright, J. A. & Rayfield, E. J. (2011b): The response of cranial biomechanical finite element models to variations in mesh density. Anatomical Record 294: 610–620.

Chen, W.-F. & Saleeb, A. F. (1994): Constitutive Equations for Engineering Materials: Elasticity and Modeling. Elsevier.

Christiansen, P. & Wroe, S. (2007): Bite forces and evolutionary adaptations to feeding ecology in carnivores. Ecology 88: 347–358.

Crompton, A. W. & Parkyn, D. G. (2009): On lower jaw of *Diarthrognathus* and the origin of the mammalian lower jaw. Proceedings of the Zoological Society of London 140: 697–749.

Cuff, A. R. & Rayfield, E. J. (2013): Feeding mechanics in spinosaurid theropods and extant crocodilians. PLoS One 8: e65295.

Cunningham, J. A., Rahman, I. A., Lautenschlager, S., Rayfield, E. J. & Donoghue, P. C. J. (2014): A virtual world of paleontology. Trends in Ecology and Evolution 29: 347–357.

Curtis, N. (2011): Craniofacial biomechanics: An overview of recent multibody modelling studies. Journal of Anatomy 218: 16–25.

de Esteban-Trivigno, S. (2011): Ecomorfología de xenartros extintos: Análisis de la mandíbula con métodos de morfometría geométrica. Ameghiniana 48: 381–398.

Dechow, P. C. & Hylander, W. L. (2000): Elastic properties and masticatory bone stress in the macaque mandible. American Journal of Physical Anthropology 112: 553–574.

Doblaré, M., García, J. M. & Gómez, M. J. (2004): Modelling bone tissue fracture and healing: a review. Engineering Fracture Mechanics 71: 1809–1840.

Dumont, E. R., Davis, J. L., Grosse, I. R. & Burrows, A. M. (2011): Finite Element Analysis of performance in the skulls of marmosets and tamarins. Journal of Anatomy 218: 151–162.

Dumont, E. R., Grosse, I. R. & Slater, G. J. (2009): Requirements for comparing the performance of finite element models of biological structures. Journal of Theoretical Biology 256: 96–103.

Early, M., Lally, C., Prendergast, P. J & Kelly, D. J. (2009): Stresses in peripheral arteries following stent placement: A Finite Element Analysis. Computer Methods in Biomechanics and Biomedical Engineering 12: 25–33.

Farke, A. A. (2008): Frontal sinuses and head-butting in goats: a Finite Element Analysis. Journal of Experimental Biology 211: 3085–3094.

Figueirido, B., Lautenschlager, S., Pérez-Ramos, A. & Van Valkenburgh, B. (2018): Distinct predatory behaviors in scimitar- and dirk-toothed sabertooth cats. Current Biology 28: 3260–3266

Figueirido, B., Tseng, Z. J., Serrano-Alarcon, F. J., Martin-Serra, A., Pastor, J. F. F., Serrano-Alarcón, F. J. & Martín-Serra, A. (2014): Three-dimensional computer simulations of feeding behaviour in red and giant pandas relate skull biomechanics with dietary niche partitioning. Biology Letters 10: 20140196

Fish, J. F. & Stayton, C. T. (2014): Morphological and mechanical changes in juvenile red-eared slider turtle (*Trachemys scripta elegans*) shells during ontogeny. Journal of Morphology 275: 391–397.

Fitton, L. C., Prôa, M., Rowland, C., Toro-Ibacache, V. & O'Higgins, P. (2015): The impact of simplifications on the performance of a Finite Element model of a *Macaca fascicularis* cranium. Anatomical Record 298: 107–121.

Fletcher, T. M., Janis, C. M. & Rayfield, E. J. (2010): Finite Element Analysis of ungulate jaws: Can mode of digestive physiology be determined? Palaeontologica Electronia 13, 3; 21A: 15 p. http://palaeo-electronica.org/2010_3/234/index.html

Fortuny, J., Marcé-Nogué, J., de Esteban-Trivigno, S., Gil, L. & Galobart, Á. (2011): Temnospondyli bite club: ecomorphological patterns of the most diverse group of early tetrapods. Journal of Evolutionary Biology 24: 2040–2054.

Fortuny, J., Marcé-Nogué, J., Gil, L. L. L & Galobart, À. (2012): Skull mechanics and the evolutionary patterns of the otic notch closure in capitosaurs (Amphibia: Temnospondyli). Anatomical Record 295: 1134–1146.

Fortuny, J., Marcé-Nogué, J., Heiss, E., Sanchez, M., Gil, L. & Galobart, À. (2015): 3D bite modeling and feeding mechanics of the largest living amphibian, the Chinese giant salamander *Andrias davidianus* (Amphibia: Urodela): PLoS One 10: e0121885.

Fortuny, J., Marcé-Nogué, J., Steyer, J.-S., de Esteban-Trivigno, S., Mujal, E. & Gil, L. (2016): Comparative 3D analyses and palaeoecology of giant early amphibians (Temnospondyli: Stereospondyli): Scientific Reports 6: 30387.

Freeman, P. W. (1979): Specialized insectivory: Beetle-eating and moth-eating molossid bats. Journal of Mammalogy 60: 467–479.

Freeman, P. W. (1981): Correspondence of food habits and morphology in insectivorous bats. Journal of Mammalogy 62: 166–173.

Freeman, P. W. (1984): Functional cranial analysis of large animalivorous bats (Microchiroptera). Biological Journal of the Linnean Society 21: 387–408.

Freeman, P. W. (1988): Frugivorous and animalivorous bats (Microchiroptera): Dental and cranial adaptations. Biological Journal of the Linnean Society 33: 249–272.
Freeman, P. W. (2000): Macroevolution in Microchiropteria: Recoupling morphology and ecology with phylogeny. Evolutionary Ecology Research 2: 317–335.
Gil, L., Marcé-Nogué, J. & Sánchez, M. (2015): Insights into the controversy over materials data for the comparison of biomechanical performance in vertebrates. Palaeontologia Electronica 18.1.10A: 1-24. https://doi.org/10.26879/509
Gill, P. G., Purnell, M. A., Crumpton, N., Brown, K. R., Gostling, N. J., Stampanoni, M. & Rayfield, E. J. (2014): Dietary specializations and diversity in feeding ecology of the earliest stem mammals. Nature 512: 303–305.
Gislason, M. K., Foster, E., Bransby-Zachary, M. & Nash, D. H. (2017): Biomechanical analysis of the Universal 2 implant in total wrist arthroplasty: A Finite Element study. Computer Methods in Biomechanics and Biomedical Engineering 20: 1113–1121.
Greaves, W. S. (1978): The jaw lever system in ungulates: a new model. Journal of Zoology 184: 271–285.
Gröning, F., Bright, J. A., Fagan, M. J. & O'Higgins, P. (2012a): Improving the validation of finite element models with quantitative full-field strain comparisons. Journal of Biomechanics 45: 1498–1506.
Gröning, F., Fagan, M. J. & O'Higgins, P. (2011a): The effects of the periodontal ligament on mandibular stiffness: A study combining Finite Element Analysis and Geometric Morphometrics. Journal of Biomechanics 44: 1304–1312.
Gröning, F., Fagan, M. & O'Higgins, P. (2012b): Modelling the human mandible under masticatory loads: Which input variables are important? Anatomical Record 295: 853–863.
Gröning, F., Fagan, M. J. & O'Higgins, P. (2013): Comparing the distribution of strains with the distribution of bone tissue in a human mandible: A Finite Element study. Anatomical Record 296: 9–18.
Gröning, F., Liu, J., Fagan, M. J. & O'Higgins, P. (2011b): Why do humans have chins? Testing the mechanical significance of modern human symphyseal morphology with Finite Element Analysis. American Journal of Physical Anthropology 144: 593–606.
Hogue, A. S. (2008): Mandibular corpus form and its functional significance: Evidence from marsupials. In: Vinyard, C., Ravosa, M. & Wall, C. (eds.). Primate Craniofacial Function and Biology. Springer Academic Publishers, New York: 329–356
Hogue, A. S. & Ravosa, M. J. (2001): Transverse masticatory movements, occlusal orientation, and symphyseal fusion in selenodont artiodactyls. Journal of Morphology 249: 221–241.
Kardong, K. V. (2014): Vertebrates: Comparative Anatomy, Function, Evolution. Open University Press, London.
Kupczik, K. (2008): Virtual biomechanics: Basic concepts and technical aspects of Finite Element Analysis in vertebrate morphology. Journal of Anthropological Sciences 86: 193–198.
Lautenschlager, S. (2016): Reconstructing the past: Methods and techniques for the digital restoration of fossils. Royal Society Open Science 3: 160342.
Lautenschlager, S., Brassey, C. A., Button, D. J. & Barrett, P. M. (2016): Decoupled form and function in disparate herbivorous dinosaur clades. Scientific Reports 6: 26495.
Lautenschlager, S., Gill, P. G., Luo, Z.-X., Fagan, M. J. & Rayfield, E. J. (2018): The role of miniaturization in the evolution of the mammalian jaw and middle ear. Nature 561: 533–537.
Ledogar, J. A., Smith, A. L., Benazzi, S., Weber, G. W., Spencer, M. A., Carlson, K. B., McNulty, K. P., Dechow, P. C., Grosse, I. R., Ross, C. F., Richmond, B. G., Wright, B. W., Wang, Q., Byron, C., Carlson, K. J., de Ruiter, D. J., Berger, L. R., Tamvada, K., Pryor, L. C., Berthaume, M. A. & Strait, D. S. (2016): Mechanical evidence that *Australopithecus sediba* was limited in its ability to eat hard foods. Nature Communications 7: 10596.
Maiorino, L., Farke, A. A., Kotsakis, T., Teresi, L. & Piras, P. (2015): Variation in the shape and mechanical performance of the lower jaws in ceratopsid dinosaurs (Ornithischia, Ceratopsia): Journal of Anatomy 227: 631–646.
Marcé-Nogué, J., de Esteban-Trivigno, S., Escrig, C. & Gil, L. (2016): Accounting for differences in element size and homogeneity when comparing Finite Element models: Armadillos as a case study. Palaeontologia Electronica 19: 1–22.
*Marcé-Nogué, J., de Esteban-Trivigno, S., Püschel, T. A. & Fortuny, J. (2017a): The intervals method: a new approach to analyse finite element outputs using multivariate statistics. PeerJ 5: e3793.
Marcé-Nogué, J., DeMiguel, D., Fortuny, J., de Esteban-Trivigno, S. & Gil, L. (2013): Quasi-homothetic transformation for comparing the mechanical performance of planar models in biological research. Palaeontologia Electronica 16, 3: 1–15. https://doi.10.26879/365
Marcé-Nogué, J., Fortuny, J., de Esteban-Trivigno, S., Sánchez, M., Gil, L .& Galobart, À. (2015): 3D computational mechanics elucidate the evolutionary implications of orbit position and size diversity of early amphibians. PLoS One 10: e0131320.
Marcé-Nogué, J., Fortuny, J., Gil, L. & Galobart, A. (2011): Using reverse engineering to reconstruct tetrapod skulls and analyse its feeding behaviour. In: Topping, B. H. V. & Tsompanakis, Y. (eds.). Proceedings of the 13th International Conference on Civil, Structural and Environmental Engineering Computing. Civil-Comp Press, Stirlingshire. https://doi.10.4203/ccp.96.237
Marcé-Nogué, J., Fortuny, J., Gil, L. & Sánchez, M. (2015): Improving mesh generation in Finite Element Analysis for functional morphology approaches. Spanish Journal of Palaeontology 31: 117–132.
*Marcé-Nogué, J., Püschel, T. A, Daasch, A. & Kaiser, T. M. (2020): Broad-scale morpho-functional traits of the mandible suggest no hard food adaptation in the hominin lineage. Scientific Reports 10: 6793.
*Marcé-Nogué, J., Püschel, T. A. & Kaiser, T. M. (2017b): A biomechanical approach to understand the ecomorphological relationship between primate mandibles and diet. Scientific Reports 7: 8364.
Marinescu, R., Daegling, D. J. & Rapoff, A. J. (2005): Finite-element modeling of the anthropoid mandible: The effects of altered boundary conditions. Anatomical Record 283: 300–309.
Mase, G. E. & Mase, G. T. (1999): Continuum Mechanics for Engineers. CRC Press, New York.
Maynard-Smith, J. & Savage, R. (1959): The mechanics of mammalian jaws. School Science Review 289–301.
McHenry, C. R., Wroe, S., Clausen, P. D., Moreno, K. & Cunningham, E. (2007): Supermodeled sabercat, predatory behavior in *Smilodon fatalis* revealed by high-resolution 3D computer simulation. Proceedings of the National Academy of Sciences of the United States of America 104: 16010–16015.
Meers, M. B. (2003): Maximum bite force and prey size of *Tyrannosaurus rex* and their relationships to the inference of feeding behavior. Historical Biology 16: 1–12.
Morris, A. (2008): A Practical Guide to Reliable Finite Element Modelling. John Wiley & Sons, Chichester.
Nalla, R. K., Kinney, J. H. & Ritchie, R. O. (2003): Mechanistic fracture criteria for the failure of human cortical bone. Nature Materials 2: 164–168.
Neenan, J. M., Ruta, M., Clack, J. A. & Rayfield, E. J. (2014): Feeding biomechanics in *Acanthostega* and across the fish-tetrapod transition. Proceedings of the Royal Society B 281: 20132689
Oldfield, C. C., McHenry, C. R., Clausen, P. D., Chamoli, U., Parr, W. C. H., Stynder, D. D. & Wroe, S. (2012): Finite Element Analysis of ursid cranial mechanics and the prediction of feeding behaviour in the extinct giant *Agriotherium africanum*. Journal of Zoology 286: 171–179.
Osborn, J. W. (1987): Relationship between the mandibular condyle and the occlusal plane during hominid evolution: Some of its effects on jaw mechanics. American Journal of Physical Anthropology 73: 193–207.
Panagiotopoulou, O., Iriarte-Diaz, J., Wilshin, S., Dechow, P. C., Taylor, A. B., Mehari Abraha, H., Aljunid, S .F. & Ross, C. F. (2017): In vivo bone strain and finite element modeling of a rhesus macaque mandible during mastication. Zoology 124: 13–29.

Parr, W. C. H., Wroe, S., Chamoli, U., Richards, H. S., McCurry, M. R., Clausen, P. D & McHenry, C. R. (2012): Toward integration of Geometric Morphometrics and computational biomechanics: New methods for 3D virtual reconstruction and quantitative analysis of Finite Element models. Journal of Theoretical Biology 301: 1–14.

Piras, P., Maiorino, L., Teresi, L., Meloro, C., Lucci, F., Kotsakis, T. & Raia, P. (2013): Bite of the cats: Relationships between functional integration and mechanical performance as revealed by mandible geometry. Systematic Biology 62: 878–900.

Piras, P., Sansalone, G., Teresi, L., Moscato, M., Profico, A., Eng, R., Kotsakis, T. (2015): Digging adaptation in insectivorous subterranean eutherians. The enigma of *Mesoscalops montanensis* unveiled by Geometric Morphometrics and Finite Element Analysis. Journal of Morphology 276: 1157–1171.

Porro, L. B., Holliday, C. M., Anapol, F., Ontiveros, L. C. L. T. & Ross, C. F. (2011): Free body analysis, beam mechanics, and finite element modeling of the mandible of *Alligator mississippiensis*. Journal of Morphology 272: 910–937.

Preuschoft, H. & Witzel, U. (2002): Biomechanical investigations on the skulls of reptiles and mammals. Senckenbergiana lethaea 82: 207–222.

Püschel, T. A. & Sellers, W. I. (2016): Standing on the shoulders of apes: Analyzing the form and function of the hominoid scapula using Geometric Morphometrics and Finite Element Analysis. American Journal of Physical Anthropology 159: 325–341.

Püschel, T. A., Marcé-Nogué, J., Gladman, J. T., Bobe, R. & Sellers, W. I. (2018): Inferring locomotor behaviours in Miocene New World monkeys using Finite Element Analysis, Geometric Morphometrics and machine-learning classification techniques applied to talar morphology. Journal of The Royal Society Interface 15: 20180520.

*Püschel, T. A., Marcé-Nogué, J., Kaiser, T. M., Brocklehurst, R. J. & Sellers, W. I. (2018): Analyzing the sclerocarpy adaptations of the Pitheciidae mandible. American Journal of Primatology 80: e22759.

Qian, L., Todo, M., Morita, Y., Matsushita, Y. & Koyano, K. (2009): Deformation analysis of the periodontium considering the viscoelasticity of the periodontal ligament. Dental Materials 25: 1285–1292.

Radinsky, L. B. (1981a): Evolution of skull shape in carnivores: 2. Additional modern carnivores. Biological Journal of the Linnean Society 16: 337–355.

Radinsky, L. B. (1981b): Evolution of skull shape in carnivores 1. Representative modern carnivores. Biological Journal of the Linnean Society 15: 369–388.

Rahman, I. A. & Lautenschlager, S. (2016): Applications of three-dimensional box modeling to paleontological functional analysis. The Paleontological Society Papers 22: 119–132.

Rayfield, E. J. (2007): Finite Element analysis and understanding the biomechanics and evolution of living and fossil organisms. Annual Review of Earth and Planetary Sciences 35: 541–576.

Richard, A. F., Goldstein, S. J. & Dewar, R. E. (1989): Weed macaques: The evolutionary implications of macaque feeding ecology. International Journal of Primatology 10: 569–594.

Ross, C. F. & Iriarte-Diaz, J. (2014): What does feeding system morphology tell us about feeding? Evolutionary Anthropology 23: 105–120.

Ross, C. F., Iriarte-Diaz, J. & Nunn, C. L. (2012): Innovative approaches to the relationship between diet and mandibular morphology in primates. International Journal of Primatology 33: 632–660.

Sella-Tunis, T., Pokhojaev, A., Sarig, R., O'Higgins, P. & May, H. (2018): Human mandibular shape is associated with masticatory muscle force. Scientific Reports 8: 1–10.

Sellers, K. C., Middleton, K. M., Davis, J. L. & Holliday, C. M. (2017): Ontogeny of bite force in a validated biomechanical model of the American alligator. Journal of Experimental Biology 220: 2036–2046.

Serrano-Fochs, S., de Esteban-Trivigno, S., Marcé-Nogué, J., Fortuny, J. & Fariña, R. A. (2015): Finite Element Analysis of the Cingulata jaw: An ecomorphological approach to armadillo's diets. PLoS One 10: e0120653.

Snively, E., Fahlke, J. M. & Welsh, R. C. (2015): Bone-breaking bite force of *Basilosaurus isis* (Mammalia, Cetacea) from the Late Eocene of Egypt estimated by Finite Element Analysis. PLoS One 10: e0118380.

Spencer, L. M. (1995): Morphological correlates of dietary resource partitioning in the African Bovidae. Journal of Mammalogy 76: 448–471.

Spencer, M. A. (1999): Constraints on masticatory system evolution in anthropoid primates. American Journal of Physical Anthropology 108: 483–506.

Strait, D. S., Constantino, P. J., Lucas, P. W., Richmond, B. G., Spencer, M. A., Dechow, P. C., Ledogar, J. A. (2013): Viewpoints: Diet and dietary adaptations in early hominins: The hard food perspective. American Journal of Physical Anthropology 151: 339–355.

Taylor, A. B. (2006): Feeding behavior, diet, and the functional consequences of jaw form in orangutans, with implications for the evolution of *Pongo*. Journal of Human Evolution 50: 377-393.

Thomason, J. J. (1991): Cranial strength in relation to estimated biting forces in some mammals. Canadian Journal of Zoology 69: 2326–2333.

Timoshenko, S. (1955): Strength of Materials. Van Nostrand, New York.

Tseng, Z. J. (2009): Cranial function in a late Miocene *Dinocrocuta gigantea* (Mammalia: Carnivora) revealed by comparative Finite Element Analysis. Biological Journal of the Linnean Society 96: 51–67.

Tseng, Z. J. (2013): Testing adaptive hypotheses of convergence with functional landscapes: A case study of bone-cracking hypercarnivores. PLoS One 8: e65305.

Tseng, Z. J. & Binder, W. J. (2010): Mandibular biomechanics of *Crocuta crocuta*, *Canis lupus*, and the late Miocene *Dinocrocuta gigantea* (Carnivora, Mammalia): Zoological Journal of the Linnean Society 158: 683–696.

Tseng, Z. J . & Flynn, J. J. (2015): Convergence analysis of a Finite Element skull model of *Herpestes javanicus* (Carnivora, Mammalia): Implications for robust comparative inferences of biomechanical function. Journal of Theoretical Biology 365: 112–148.

Tseng, Z. J. & Flynn, J. J. (2018): Structure-function covariation with nonfeeding ecological variables influences evolution of feeding specialization in Carnivora. Science Advances: 4: eaao5441.

Tseng, Z. J., Grohé, C. & Flynn, J. J. (2016): A unique feeding strategy of the extinct marine mammal *Kolponomos*: convergence on sabretooths and sea otters. Proceedings of the Royal Society B 283: 20160044.

Tseng, Z. J., Su, D. F., Wang, X., White, S. C. & Ji, X. (2017): Feeding capability in the extinct giant *Siamogale melilutra* and comparative mandibular biomechanics of living Lutrinae. Scientific Reports 7: 15225.

van Eijden, T. M. (2000): Biomechanics of the mandible. Critical Reviews in Oral Biology & Medicine 11: 123–136.

Varela, L. & Fariña, R. A. (2015): Masseter moment arm as a dietary proxy in herbivorous ungulates. Journal of Zoology 296: 295–304.

Veitschegger, K., Wilson, L. A. B., Nussberger, B., Camenisch, G., Keller, L. F., Wroe, S. & Sánchez-Villagra, M. R. (2018): Resurrecting Darwin's Niata – anatomical, biomechanical, genetic, and morphometric studies of morphological novelty in cattle. Scientific Reports 8: 9129.

Vizcaíno, S. F., Bargo, M. S., Cassini, G. H. & Toledo, N. (2016): Forma y función en paleobiología de vertebrados. Editorial de la Universidad Nacional de La Plata, La Plata.

Walmsley, C. W., Smits, P. D., Quayle, M. R., McCurry, M. R., Richards, H. S., Oldfield, C. C. & McHenry, C. R. (2013): Why the long face? The mechanics of mandibular symphysis proportions in crocodiles. PLoS One 8: e53873.

Wroe, S., Chamoli, U., Parr, W. C. H., Clausen, P. D., Ridgely, R. & Witmer, L. M. (2013): Comparative biomechanical modeling of metatherian and placental saber-tooths: A different kind of bite for an extreme pouched predator. PLoS One 8: e66888

Wroe, S., Clausen, P. D., McHenry, C. R., Moreno, K. & Cunningham, E. (2007): Computer simulation of feeding behaviour in the thylacine and dingo as a novel test for convergence and niche overlap. Proceedings of the Royal Society B 274: 2819–2828.

Wroe, S., Ferrara, T. L., McHenry, C. R., Curnoe, D. & Chamoli, U. (2010): The craniomandibular mechanics of being human. Proceedings of the Royal Society B 277: 3579–3586.

Wroe, S., McHenry, C. R & Thomason, J. J. (2005): Bite club: comparative bite force in big biting mammals and the prediction of predatory behaviour in fossil taxa. Proceedings of the Royal Society B 272: 619–625.

*Zhou, Z., Winkler, D. E., Fortuny, J., Kaiser, T. M. & Marcé-Nogué, J. (2019): Why ruminating ungulates chew sloppily: Biomechanics discern a phylogenetic pattern. PLoS One 14: e0214510.

Zienkiewicz, O. C. & Taylor, R. L. (2000): Finite Element Method – The Basis (Vol. 1). Butterworth-Heinemann, Oxford.

Interaction between food, saliva, and tooth surface

Marcus Mau, Achim Johann, Thomas M. Kaiser, and Karl-Heinz Südekum

Introduction

Saliva is an outstanding fluid, especially in terms of research and diagnostic possibilities. Its composition – namely electrolytes, hormones and especially its proteome – contains valuable information about feeding status, nutritional requirements as well as adaptation to diet and environment. The biggest advantage of saliva as a 'research tool', however, is the possibility to collect it on a non-invasive basis; and there is almost no need for special training. Therefore, the results of our analysis of salivary proteomes from five different herbivorous species (camel, cattle, gelada baboon, goat, hamadryas baboon) already ignited major interest in salivary research, with the future goal to maintain and improve livestock productivity on one hand and certainly zoo animal welfare on the other. Moreover, the comprehensive analysis and identification of salivary proteins is a necessary pre-requisite to better understand animal physiology and diet adaptation.

How to address food adaptation in animal saliva?

When starting to work on animal saliva, one soon realises the great variety of saliva within mammals. What we know about human saliva and its use in measuring hormone levels, or its ability to pre-digest high-molecular carbohydrates such as starch, shows only the limits of our current understanding of salivary functions. For example, how do we explain the huge amount of saliva produced by cattle – up to 180 litres per day? Bovine saliva lacks amylase or comparable digestive enzymes. On the other hand, free-ranging cattle feed on fibre-rich, "difficult-to-digest" plant diets all their lives. So, how do they manage? What role does saliva play in these herbivores?

The aim of this research project was to characterise salivary proteins in herbivores, and to find certain interactions between them and tooth enamel as well as plant phytoliths or grit, in order to protect teeth from abrasion. The answers to the following five questions produced the highest practical interest:

1. Does saliva composition correlate with feeding traits in herbivorous mammals, and is salivary composition adaptive to dietary changes?
2. Which saliva components interact specifically with tooth enamel and/or dentin and thus might play a certain role in the reduction of diet abrasiveness?
3. Do various herbivorous species, such as grazers, show similar salivary protein patterns and binding capabilities in adaptation to their fiber-rich and highly abrasive diet?
4. Are salivary proteins of grazing mammals able to bind and enclose plant phytoliths?
5. Which other, so-called accessory functions does saliva of herbivorous species provide?

Animal saliva sampling

Work on animal salivary proteins demands the collection of larger amounts of saliva and adequate frozen storage. However, full saliva – as it is obtained by using cotton swabs – always represents a mixture of oral secretions and proteins originating from different salivary glands as well as from gingival fluid. Saliva further contains cellular debris and food particles. Furthermore, salivary components might be altered or partially degraded by the oral microbiome.

Animal saliva sampling can be adapted from collecting human saliva, using cotton pads and centrifuge tubes. While goat and cattle can be easily handled during the procedure, camels turned out to be slightly more laborious. After the first tries ending up in the researcher being covered with camel saliva, well trained camels could be sampled like cows.

For the collection of whole saliva from goats, camels and cattle, a cotton pad was held into the animal's mouth, and allowing the animal to chew on the cotton wool to obtain a saliva sample.

Sampling in the primate species (hamadryas baboons, gelada baboons) was performed during routine checks or operations in zoological gardens, while the animals were unconscious.

All saliva samples were kept on ice and centrifuged according to a standard protocol, described in Mau et al. (2006), to precipitate food debris and other solid particles in the sample. The samples were kept frozen at −80 °C.

T. Martin & W. v. Koenigswald (eds.): Mammalian Teeth – Form and Function. Pp. 81–86, 3 figs., 2 tabs.

DOI: http://doi.org/10.23788/mammteeth.05 Published 22 December 2020

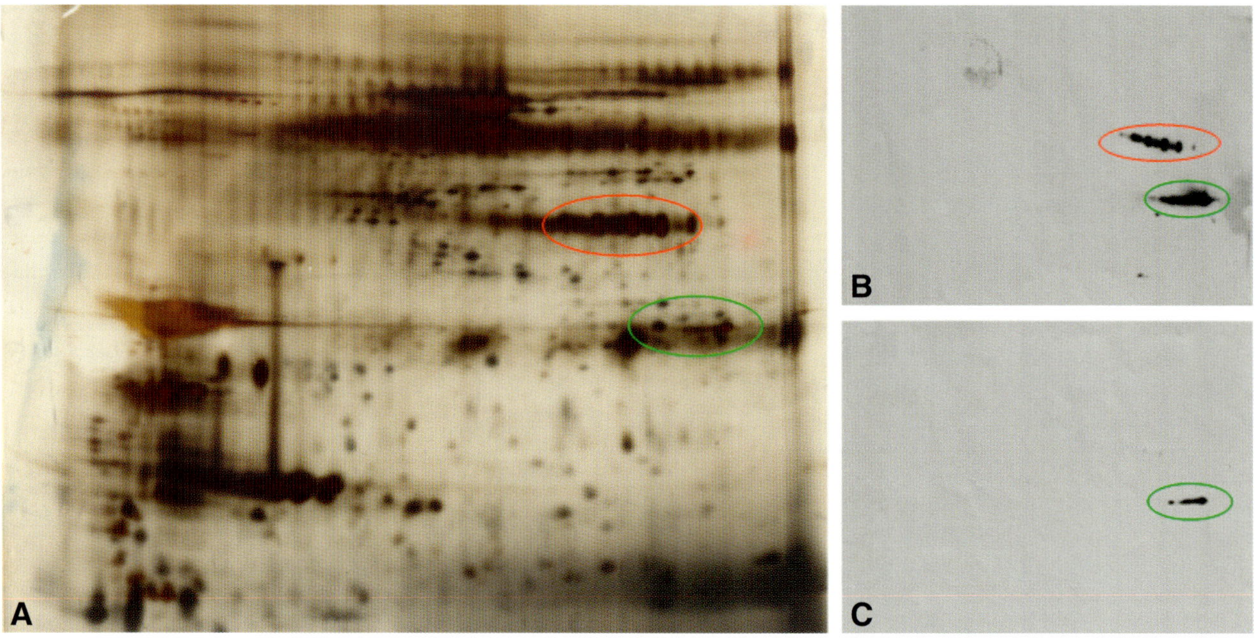

Fig. 5.1. **A,** 2D gel electrophoresis of an exemplary bovine saliva sample. The upper red marking shows the position of CA VI in the 2D gel. The lower green mark describes CA II spots. **B,** the anti-bovine CA antibody reacted with CA VI (upper red mark) and CA II (lower green mark). **C,** a second used anti-human CA II antibody did only cross-react with bovine CA II (green marking) in the same saliva sample. From Mau et al. (2010).

Saliva composition

To identify and also to quantify common salivary proteins from whole animal saliva of different herbivorous mammals and primate species, immunoblots were performed using standard Coomassie R250 staining as well as commercially available anti-bovine carbonic anhydrase VI (CA VI) and anti-human CA II antibodies. The anti-bovine CA VI antibody was found to further cross-react with bovine CA II as described in detail in Mau et al. (2009).

Furthermore, the detection protocols were improved using bovine saliva for the first time in 2D gel electrophoresis as shown by Mau et al. (2010). This enabled us to reproduce certain species-specific salivary protein patterns and to isolate and identify diverse protein spots, such as CA VI, in mass spectrometry (Fig. 5.1).

Salivary protein patterns of various grass-eating mammals

For the first time saliva samples originating from various ruminating mammals and primates were collected and analysed to find characteristic salivary protein patterns discriminating the various feeding traits in mammals.

In this experiment saliva samples of four grazing species were compared using 1D gel electrophoresis in order to describe common salivary proteins as well as distinct differences in the protein patterns. The mean protein concentrations of bovine, camel, goat and gelada saliva are presented in Table 5.1.

The lowest protein concentration was observed in bovine saliva, while gelada baboons presented saliva secretions very rich in protein. Camels and goats showed intermediate salivary protein concentrations.

Protein bands in 1D gel electrophoresis were visualised using Coomassie R250 staining. Protein masses ranged from 140 kD to 10 kD showing distinct species-specific patterns with differences not only in the range of the molecular weights, but also in the number of protein bands. Bovine and goat saliva consisted of 10 to 14 visible protein

Table 5.1. Mean salivary protein concentrations of four herbivorous mammal species. From Mau et al. (2009).

Species	Feeding trait	Digestive tract	n	Protein (mg/ml)
Bos primigenius f. *taurus*	grazer	foregut-fermenting	4	1.18 ± 0.202
Camelus ferus f. *bactrianus*	intermediate	foregut-fermenting	3	3.20 ± 0.929
Capra aegagrus f. *hircus*	intermediate	foregut-fermenting	11	3.20 ± 2.294
Theropithecus gelada	grazer	hindgut-fermenting	13	6.71 ± 2.693

bands, whereas camel saliva showed 12 distinct bands. Gelada saliva samples, on the other hand, contained up to 29 protein bands.

Although both species are grass-eaters, cattle and geladas had very distinct salivary protein patterns, with geladas having up to three times more protein bands in their saliva. This difference is apparently linked to the evolutionary origins of the two species, with the gelada baboon being a primate and the cattle an artiodactyl. Interestingly, in all four species there were two prominent proteins at 42 kD and 29 kD, identifying again as CA II and VI by immunoblotting as well as in mass spectrometry.

Carbonic anhydrases had been described before as important enzymes that guarantee oral homeostasis. With the help of the established immunoblot for CA II and CA VI, we detected CA II in saliva of different mammals such as cattle, goat, camel and gelada baboons. In comparison, human saliva only contained very small quantities of CA II. As a primarily cytoplasmatic enzyme, the CA II in ruminant saliva might be actively secreted by apocrine secretion from parotid glands as described earlier in cattle by Stolte & Ito (1996).

The presence of two highly active CA isoenzymes (CA II and CA VI) in ruminant saliva might indicate a co-operation of the two in regulating oral pH, thus maintaining constant conditions for the fermentation of grass matter in the rumen. Based on their proposed function, a decline in either CA II or CA VI salivary concentrations could result in impaired digestion or might increase the risk for acidosis in ruminants. Could CA II in ruminant saliva therefore act as a potential marker for acidosis risk in livestock animals?

Furthermore, CA II concentrations in primate and also human saliva seemed comparably marginal. However, based on our data we postulated a descriptive inverse correlation between salivary CA II concentrations and CA II levels in the brains of various taxa, such as cattle, primates and humans. According to the expensive tissue hypothesis this means, the higher the CA II level in saliva of an animal species, the lower is the CA II level in the brain of the same species and vice versa as published by Mau et al. (2009b). This finding might demonstrate the "evolutionary balance" between two otherwise energetically expensive tissues (here: brain versus gut).

Visualising feeding adaptations in animal saliva

Tannin-binding capacity

Additionally, we performed direct electrophoretic comparisons between gelada saliva and saliva samples of various primate species (*Papio hamadryas*, *Pan troglodytes*, *Homo sapiens sapiens*) with distinct feeding traits. It was hypothesised that gelada baboons might lack certain proteins, not necessary for grass diet. Other primate species were chosen for comparison, because primates are a diverse group with various feeding adaptations.

However, animal nutrition and feeding are only half of the evolutionary story. Many plants "learned" to prevent excessive feeding by the development of certain counter-measures. In grasses one will find so-called phytoliths embedded into the cell structure of the leaves to wear down animal teeth. Many trees and herbs, on the other hand, protect themselves by producing tannins, which precipitate proteins and enzymes, thereby influencing digestion.

Answering to these counter-measures, primates developed different adaptations to circumvent pytoliths or tannins. For example, in adaptation to a highly abrasive grass diet, the gelada baboon developed high-crowned molar teeth. *Papio hamadryas*, a primate species feeding primarily on tannin-rich foliage, possibly developed specialised proline-rich, tannin-binding proteins to bind and to neutralise plant tannins.

In order to investigate, whether such feeding adaptations are present in saliva of hamadryas and gelada baboons, 1D gel electrophoresis of native saliva was performed. The results supported the idea of special salivary proteins, which are characteristic for certain dietary niches in primate species. While saliva of hamadryas baboons showed a rich pink staining with Coomassie – indicative for proline-rich proteins (PRP), which also include the so-called tannin-binding proteins (TBP) – these bands were completely absent in gelada saliva (Mau et al., 2009a; Fig. 5.2). Protein identification as hamadryas baboon PRP and TBP was subsequently managed using mass spectrometry (Mau et al., 2011).

The absence of salivary TBP is common in grazing mammals, such as gelada baboons and cattle. Therefore it seems that the lack of TBP demonstrates a direct adaptation of a species to a diet based on grasses, which do not contain tannins. Diet obviously influences saliva composition.

In order to further test this interaction hypothesis, we analysed the amylase protein expression in primate saliva. Earlier studies have shown that the expression and activity of salivary amylase in primates is correlated with the level of starch present in a certain diet.

We compared amylase protein expression in saliva samples of humans, gelada baboons, hamadryas baboons and additionally in a few chimpanzee samples. The two baboon species showed a higher expression level than the hominid primate species or humans (Mau et al., 2010a).

Chimpanzees primarily feed on fruits, which are poor in starch. On the other hand, baboons and geladas seasonally refer to seeds and plant roots that are naturally rich in starch, which explains the different expression levels of salivary amylase.

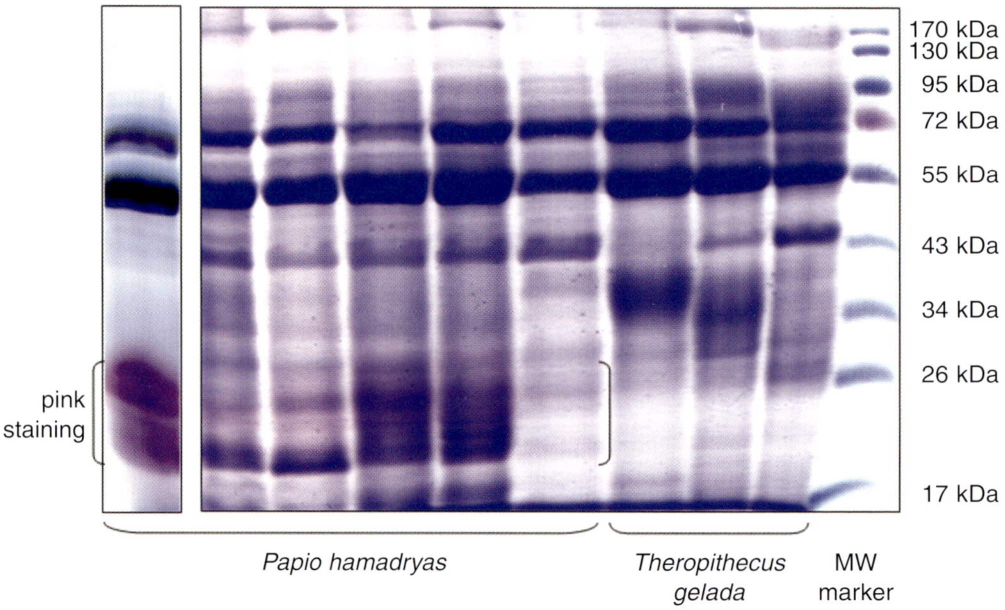

Fig. 5.2. Salivary protein patterns of hamadryas baboons (*Papio hamadryas*) und gelada baboons (*Theropithecus gelada*) stained with Coomassie R250. The original protocol to identify proline-rich proteins (PRP) in saliva was published by Beeley et al. (1991). All potential PRP shown as pink bands in hamadryas saliva. Gelada saliva totally lacks the PRP as demonstrated by the absence of any pink staining protein bands. The protein group of proline-rich proteins amongst others includes the tannin-binding proteins (TBP) to neutralise dietary tannins. From Mau et al. (2009a).

Table 5.2: Use of mass spectrometry to identify proteins from bovine saliva that bound to phytoliths, bovine dental enamel or dentin. From Mau et al. (2013).

Protein name	Matching species	Protein accession no. (NCBI)	Total protein score	Peptides matched in MS/MS	Molecular weight [kDa]	Sequence coverage [%]
Lactoperoxidase	*Bos taurus*	P80025	171	DSLQKVSFSR EQINAVTSFLDASLVYGSEPSLASR FGHMEVPSTVSR IIKDGGIDPLVR IVGYLDEEGVLDQNR TPDNIDIWIGGNAEPMVER	80.59	13
carbonic anhydrase VI (CA-VI)	*Bos taurus*	P18915	175	DLDIQDMLPGDLR DYAENTYYSNFISHLEDIR QGEFPMTNNGHTVQISLPSSMR VVEANFVSHPHQEYTLGSK YGSYEEAQNEPDGLAVLAALVEVK	36.98	30
BSP30	*Bos taurus*	P79124	260	GLGTFDSTIEIIQNLK KLIGEPQVTTQQEI LDLNVDLQTSVSIETDAETGDSR LIGEPQVTTQQEI RLISGLER RPGLLNDVVDFGVNLVR	26.37	33
odorant-binding protein (OBP)	*Bos taurus*	P07435	555	AQEEEAEQNLSELSGPWR ATKQDDTYVADYEGQNVFK ELVFDDEKGTVDFYFSVK LNVEDEDLEK LNVEDEDLEKFNK NVVNFLENEDHPHPE QDDGTYVADYEGQNVFK THLVAHNINVDK TVYIGSTNPEK	18.49	67
Haemoglobin β	*Bos taurus*	P02070	354	AAVTAFWGK EFTPVLQADFQK FFESFGDLSTADAVMNNPK LLVVYPWTQR NFGKEFTPVLQADFQK VKVDEVGGEALGR VLDSFSNGMK VVAGVANALAHR	15.94	61

Fig. 5.3. Microscopic evidence for the presence of phytoliths in grass preparations. **A, B,** showing the characteristic bilobate short cell-form of phytolith; **C,** a single phytolith in comparison to the irregular appearance of environmental silicates. The mean length of the phytoliths prepared was 22.8 µm. Scale bar = 100 µm. From Mau et al. (2013).

From saliva proteins to feeding guidelines

As a practical outcome of our research, we developed new feeding guidelines for captive gelada baboons that are bred in accordance with the worldwide gelada breeding programme in zoological gardens. Saliva was collected from various zoo animals to determine certain dietary adaptations on enzymatic and physiological levels that might change our current understanding of animal nutrition and the feeding in zoo environments. The interest was primarily focused on species with specialized nutritional needs or fermentation strategies such as geladas, chimpanzees and camels.

The work produced highly valuable data especially on geladas, which, despite their adaptation to grass diet (= main food source), revealed to have the enzymatic precondition (namely salivary amylase) to use starch-rich food items very effectively, especially in seasons, when fresh grass is lacking in the wild. Therefore, in order to prevent obesity in captive geladas, the reduction of starchy food in their daily rations is mandatory (Mau & Johann 2011a).

Accessory salivary proteins

Furthermore, distinct salivary proteins were hypothesized to prevent enamel abrasion and adverse chemical effects on teeth of herbivorous mammals. To detect potential candidates for protecting proteins, whole bovine and goat saliva was incubated with dental enamel or glass powder. Salivary proteins, bound to dental enamel and glass, were analysed by 1D gel electrophoresis. Two salivary proteins were found to adhere to bovine enamel and were isolated by MS/MS mass spectrometry and subsequently identified by protein sequencing. Both the bovine odorant-binding protein (bOBP) and the CA VI were present in these protein layers.

The bovine CA VI adhering to teeth may serve as a local pH-regulator to protect bovine enamel and dentin from bacterial acids. The bOBP on the other hand may bind odorant particles right in the oral cavity and thus could enhance their transmission to the vomeronasal organ (VNO), as for example during flehmen. The signals of the VNO are important determinants in sexual behaviour of mammals. These results suggested that salivary proteins may support olfaction and protect enamel of ruminants from mechanical and chemical destruction as postulated by Mau et al. (2006).

Furthermore, in a pilot study phytoliths from common feeding plants were isolated for in vitro incubation with bovine saliva samples. One gram of commercially available grass pellets was incinerated according to the method originally published by Parr et al. (2001). Acid-insoluble ash contained microscopic environmental grit (dust and soil particles) and biogenic, dumbbell-shaped phytoliths (medium length: 22.8 µm; Fig. 5.3). To detect potential candidates for silicate-binding proteins, bovine whole saliva was then incubated with isolated grass-derived phytoliths and silicates. Interactions of salivary proteins with pulverized bovine dental enamel and dentin were additionally analysed. After intense washing, the powder fractions were loaded onto 1D-polyacrylamide gels, most prominent adhesive protein bands were cut out and proteins were identified by mass spectrometry within three independent replicates (Table 5.2). All materials were mainly interacting with bovine OBP, BSP-30 and CA VI. The phytolith/silicate fraction showed additionally stronger interactions with haemoglobin β and lactoperoxidase (Mau et al., 2013).

Conclusions

Saliva sampling is a cost-effective and non-invasive procedure, the use of which is both practical and attractive in livestock research and also in a zoo environment. We demonstrated that saliva proteins might fulfil a great number of different tasks in oral homeostasis and may further protect teeth from abrasion in correlation to distinct feeding adaptations. In the future, the emerging field of salivary research will give birth to new applications of saliva as a tool to detect biomarkers of physiology and diseases, not only in animal husbandry. It might further help to develop therapeutic methods which regulate pH-levels and/or abrasion at the occlusal surface.

Acknowledgments

We thank the reviewers Marcus Clauss (Zurich) and Carsten Staszyk (Gießen) for their constructive review comments and Aditya Kombra for checking the English.

References

Beeley, J. A., Sweeney, D., Lindsay, J. C. B., Buchanan, M. L., Sarna, L. & Khoo, K. S. (1991): Sodium dodecyl sulphate-polyacrylamide gel electrophoresis of human parotid salivary proteins. Electrophoresis 12: 1032–1041.

Mau, M. & Johann, A. (2011): Saliva research in zoos – An aid for appropriate feeding of nutritional specialists. Zoologischer Garten N.F. 80: 262–270. [Article in German]

Mau, M., De Almeida, A. M., Coelho, A. V. & Südekum, K.-H. (2011): First identification of tannin binding proteins in saliva of *Papio hamadryas* using MS/MS mass spectrometry. American Journal of Primatology 73: 896–902.

Mau, M., Kaiser, T. M. & Südekum, K.-H. (2009): Evidence for the presence of carbonic anhydrase 29-kDa-isoenzyme in salivary secretions of three ruminating species and the gelada baboon. Archives of Oral Biology 54: 354–360.

Mau, M., Kaiser, T. M. & Südekum, K.-H. (2010): Carbonic anhydrase II is secreted from bovine parotid glands. Histology and Histopathology 25: 321–329.

Mau, M., Kaiser, T. M. & Südekum, K.-H. (2013): Pilot study on binding of bovine salivary proteins to grit silicates and plant phytoliths. Zoological Research 34 (E3): E87–E92.

Mau, M., Müller, C., Langbein, J., Rehfeldt, C., Hildebrandt, J. P. & Kaiser, T. M. (2006): Adhesion of bovine and goat salivary proteins to dental enamel and silicate. Archives of Animal Breeding 49: 439–446.

Mau, M., Südekum, K.-H., Johann, A., Sliwa, A. & Kaiser, T. M. (2009a): Saliva of the graminivorous *Theropithecus gelada* lacks proline-rich proteins and tannin-binding capacity. American Journal of Primatology 71: 663–669.

Mau, M., Südekum, K.-H., Johann, A., Sliwa, A. & Kaiser, T. M. (2010a): Indication of higher salivary α-amylase expression in hamadryas baboons and geladas compared to chimpanzees and humans. Journal of Medical Primatology. https://doi:10.1111/j.1600-0684.2010.00407.x

Mau, M., Südekum, K.-H. & Kaiser, T. M. (2009b): Why cattle feed much and humans think much – new approach to confirm the expensive tissue hypothesis by molecular data. Bioscience Hypotheses 2: 205–208.

Parr, J. F., Lentfer, C. J. & Boyd, W. E. (2001): A comparative analysis of wet and dry ashing techniques for the extraction of phytoliths from plant material. Journal of Archaeological Science: Reports 28: 875–886.

Stolte, M. & Ito, S. (1996): A comparative ultrastructural study of the parotid gland acinar cells of nine wild ruminant species (Mammalia, Artiodactyla). European Journal of Morphology 34: 79–85.

Aspects of food comminution in ungulates and their consequences for energy budget

Jürgen Hummel, Marcus Clauss, and Karl-Heinz Südekum

Introduction

While green plant material represents the most commonly available food resource in terms of quantity, its nutritional quality poses considerable challenges. For chemical digestion, the lack of endogenous enzymes to digest the fibrous parts of plants (basically cell wall) is overcome in all major vertebrate herbivore lines by the development of large fermentation chambers within the digestive tract hosting symbiotic microbes (Stevens & Hume 1995). However, physical resistance of the material ingested as grass or browse represents another, equally relevant challenge herbivores are confronted with (Welch 1982). Reduction of particle size and disintegration of plant material in terrestrial herbivores is largely depending on the chewing process (Poppi et al. 1980, Murphy & Nicoletti 1984), which has been described to be responsible for about 80% of particle size reduction in forage like ryegrass or alfalfa in cattle (McLeod & Minson 1988). Comminution can be considered to be important because smaller ingesta particles will allow a denser packing of the digesta and therefore a higher intake. Even more important, it is of crucial significance for the speed of fermentation (Bjorndal et al. 1990); in contrast to cells without a cell wall, access to nutrients (cell solubles and cell wall itself) for microbes and consequently the host is depending on mechanical disruption of the fibrous cell wall as the major barrier (Fig. 6.1). Because smaller particles are fermented faster, animals that achieve a higher particle size reduction can afford shorter digesta retention times. This is evident in a comparison of reptiles and mammals (Fritz et al. 2010), but also within mammalian herbivores (Clauss et al. 2009b).

Degree of comminution can be regarded as an indicator for the elaborateness of the food processing system and also the level of metabolism of herbivores. Again, this is most obvious in herbivores with a low level of energy metabolism (brachymetabolic) like tortoises and herbivorous lizards, which show little to no particle size reduction beyond bite size (Fritz et al. 2010). At least some herbivorous dinosaurs, for which level of metabolism is a matter of discussion, were capable of some food comminution, e.g. ceratopsids and some ornithopods (Norman & Weishampel 1985, Weishampel & Jianu 2000). In birds, the gizzard is capable of comprehensive particle disintegration (Moore 1999, Fritz et al. 2011), while among extant mammals, a more or less comprehensive comminution by elaborated teeth is present in all lineages specialized on herbivory. Mammalian dentitions and chewing apparatus must surely be considered key innovations, contributing significantly to mammalian success in the large herbivore niche observed today (Ungar 2010, Berkovitz & Shellis 2018).

While in general all mammalian herbivores have developed an elaborate food comminution system, relevant differences become obvious when looking at it in more detail; important variables like the species-specific level of intake, retention time of food in the gut, digestion and metabolism of mammals influence each other, and food comminution is a crucial variable in this concert (Clauss et al. 2013). Differences between taxonomic groups are obvious and have triggered comparative research on their performance (Udén & Van Soest 1982, Fujikura et al. 1989, Fritz et al. 2009). Besides phylogenetic relationships, feeding preference (browse vs. grass) also takes considerable influence on the process of food comminution (Clauss et al. 2008, Codron et al. 2019).

In this contribution the focus will be on the process of food comminution in its effect on energy digestion and energy budget. While overall efficiency of the food comminution process of mammalian herbivores must be considered to be influenced by a concert of different factors, including, for example, the number of erupted teeth in juvenile animals (Grandl et al. 2018), effectiveness of

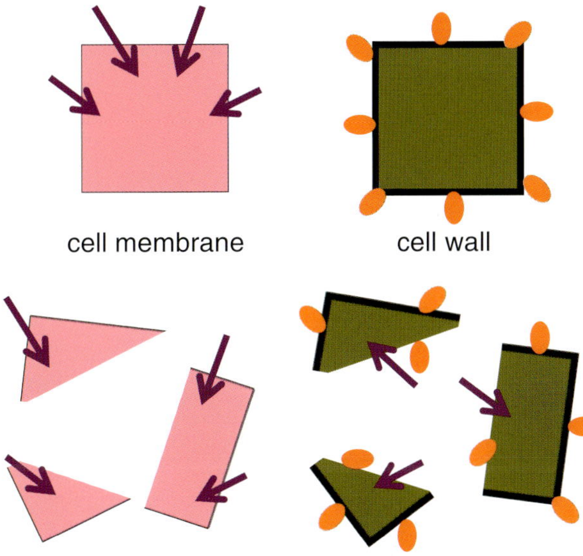

Fig. 6.1. Different digestibility of animal and plant cells. An animal cell without a cell wall can be directly digested by enzymes (upper left). In plant cells, the cell wall represents a barrier to digestive enzymes produced by vertebrates (upper right). Only cell-wall digesting microbes (orange ovals) can disrupt the cell wall so that enzymes can reach the inside of the cell (lower left). Disrupting cells mechanically will hence dramatically increase the rate of digestion (lower right.)

T. Martin & W. v. Koenigswald (eds.): Mammalian Teeth – Form and Function. Pp. 87–101, 9 figs., 3 tabs.

DOI: http://doi.org/10.23788/mammteeth.06 Published 22 December 2020

dentition (e.g. occlusal area or enamel ridge length) and chewing behavior are considered to be major factors for chewing efficiency (Péréz-Barbería & Gordon 1998). After discussing relevant characteristics of food plants, a closer look on strategies of food comminution and on influencing factors on the process will be taken. Finally, major variables of the food comminution process will be summarized and compared for equids and ruminants as examples for two major and distinct herbivore lineages.

Forage characteristics

When discussing adaptations and characteristics of herbivores, a look at the particularities of green plant material as food represents a logic starting point. Any adaptations seen in the feeding process of herbivores should have been triggered by characteristics of food. Among the factors to consider here are comminution resistance, particle fracture patterns in the comminution process and consequences of particle disruption for the digestion process. The dependence of particle disintegration on chewing is a consequence of lignification, since lignin represents the major material not degradable even by gut microbes. This can be easily visualized by investigating a *Ficus* leaf after the gut passage in *Iguana*, a well-fermenting but non-chewing herbivore: After gut passage, the lignified plant vascular system stays rather intact, while the rest of the tissue has been digested (Fig. 6.2). On the other hand, the less lignified a plant is, the more particle disintegration is possible simply due to gut microbial fermentation or peristalsis. Correspondingly, ingesta particle size in virtually non-lignified seagrasses (e.g. *Halophila ovalis*) has been shown to be continuously decreasing during gut passage in dugongs, most likely due to microbial activity on the available material (Lanyon & Sanson 2006).

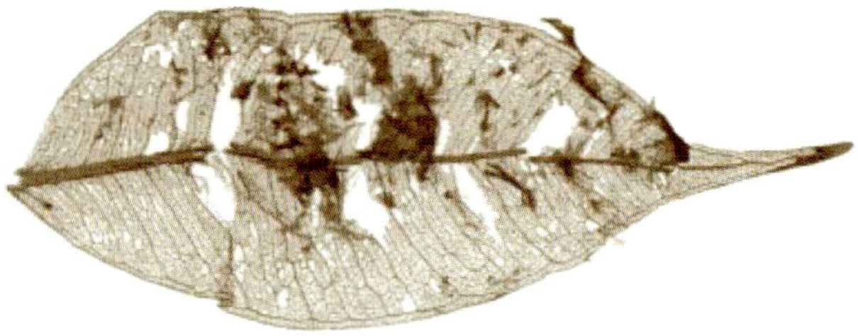

Fig. 6.2. *Ficus* leaf after gut passage in an iguana. Note that while much of the leaf parenchym was digested, the shape of the leaf stays intact. Photo: J. Fritz.

Comminution resistance

Mechanical properties of green plant parts in the comminution process are among the first factors when characterizing the material as food. An evaluation of the fiber content (e.g. crude fiber, neutral-detergent fiber or acid-detergent fiber) is the typical approach of nutritionists. Neutral-detergent fiber (NDF) represents the sum of hemicelluloses + cellulose + lignin, acid-detergent fiber (ADF) that of cellulose + lignin, and crude fiber (CF) a less defined conglomerate of fractions of hemicellulose, cellulose, pectin and lignin. All these fiber fractions are closely associated with chewing efforts. For example, a relatively constant value of 3 h of chewing (rumination) has been assigned to 1 kg of CF (Piatkowski et al. 1990), and NDF content of the diet is regarded the most significant single predictor of feed intake and chewing duration in dairy cows (Mertens 1987, Nørgaard et al. 2010, Jensen et al. 2016).

However, for deeper insight, material science-orientated approaches may be preferable, using the energy necessary for comminution (comminution resistance) to rank and characterize materials (Sanson et al. 2001, Sanson 2006). Different approaches have been tried to arrive at such a quantification of the energy necessary to comminute a given amount of food. They range from technical approaches using material testing machines on individual plant parts (Sanson et al. 2001) or measuring the energy necessary to comminute a larger sample of the material in a mill (Laredo & Minson 1973, Paul & Mika 1981, Paul et al. 1981, Paul & Schild 1982, Paul 1984, Baron 1994, Blümmel et al. 1996) to even direct measurement of the metabolism of animals in relation to chewing activities (Susenbeth et al. 1998, 2004). For ruminants, energetic costs (measured as metabolizable energy ME) for chewing (eating + rumination) are estimated to be 10 % of maintenance energy requirements for high quality forage and up to 30 % (!) of maintenance energy requirements for low quality forage (straw); the cost for eating approaches values of 35 J/(min · kg BM) (BM = body mass) for species like cattle, sheep and horse, while costs for rumination are assumed to be only 27 % of those for eating (Susenbeth et al. 1998). This latter finding is best explained by the fact that material regurgitated for rumination has been washed, moistened, pre-fermented and softened by its previous stay in the rumen (Kennedy et al. 1993, Hatt et al. 2019, 2020).

Although less investigated, further characteristics of forages besides fibrousness may also influence the effort necessary for oral food comminution. Dry matter (DM) content is relevant in the nutrition of domestic ruminants, where grass based forages can be fed in ways differing considerably in DM. Fresh grass (DM: 230 g/kg) has been found to require considerably more energy effort for ingestive comminution (527 kJ/kg DM) than grass silage (DM: 310 g/kg; 355 kJ/kg DM) or grass hay (DM: 868 g/kg; 342 kJ/kg DM) (Susenbeth et al. 2004). These differences were not explained by fiber content (NDF contents: fresh grass 535 g/kg DM; grass silage 566 g/kg DM, grass hay 502 g/kg DM); instead, it can be speculated that variation in brittleness or in volume of identical amounts of DM are factors influencing the effort required for comminution.

Fracture patterns

Somehow related to this topic, plant groups like grasses (monocot) and browse and forbs (dicot) appear to differ systematically in their fracture patterns: While grass is

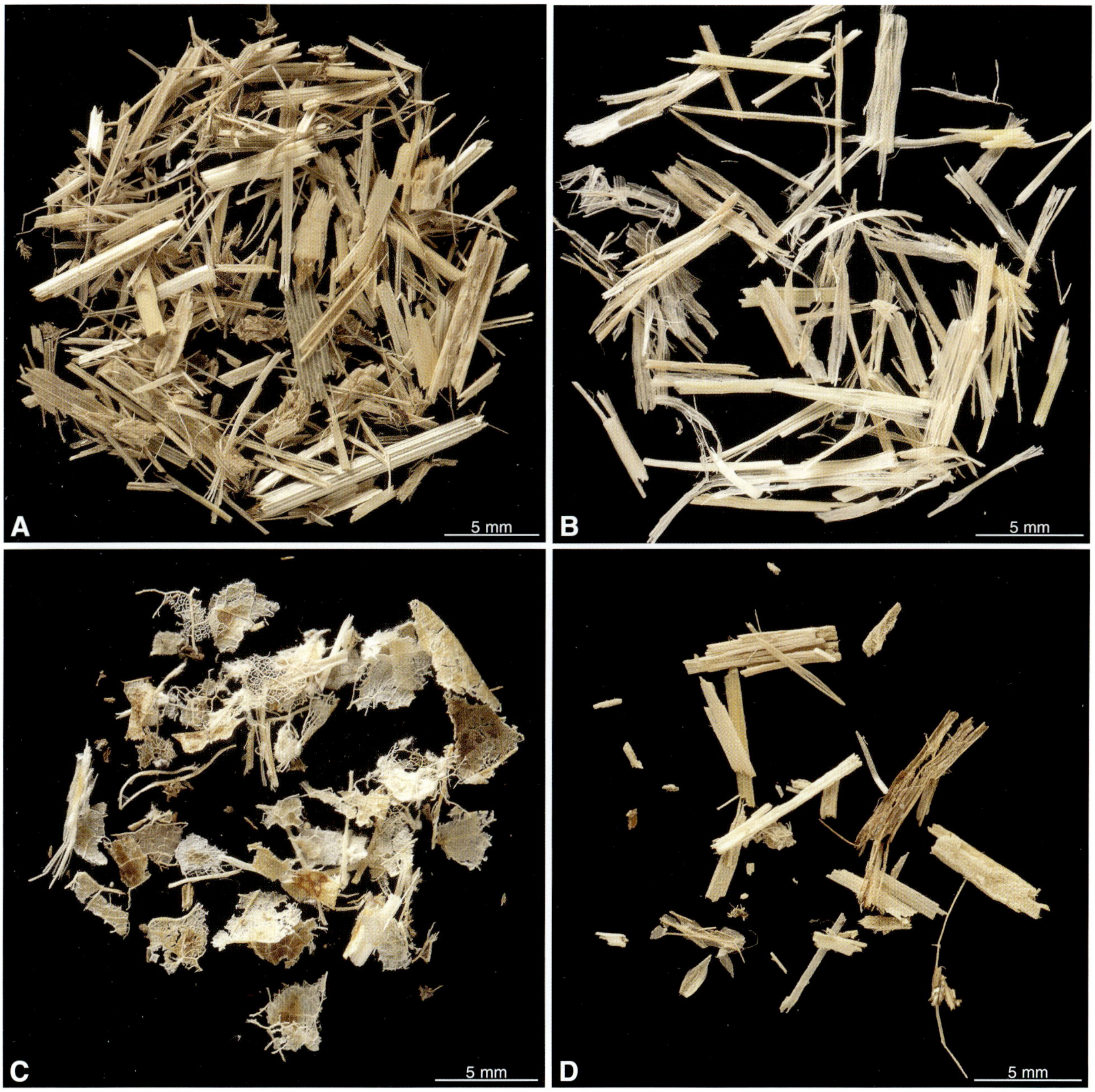

Fig. 6.3. Fecal particles retained on 2 mm sieve after wet sieving. **A,** grass 1; **B,** grass 2; **C,** browse; **D,** alfalfa. Photos: G. Oleschinski.

comminuted into longish particles, particles developing from browse are of a more polyedric shape (Fig. 6.3) (Spalinger et al. 1986, Clauss et al. 2011). Such particle shapes potentially can have consequences e.g. for rumen stratification, which is far less present in browsing ruminants (Clauss et al. 2009a, Lechner et al. 2010), or for gut fill/intake, as it has been suggested for alfalfa as an important dicot forage plant (Troelsen & Campbell 1968).

Particle size and fermentation

Besides influencing physical behavior of food particles in the gut, comminution will also improve the energy budget of herbivores by accelerating the process of microbial fermentation substantially. The plant cuticle represents a largely impenetrable barrier to microbial access, which would only be possible e.g. via stomata without intensive comminution (see Fig. 6.1). In consequence, particle size is negatively correlated to degradation rate (Tab. 6.1). Only fungi, present to some degree in the gastro-intestinal tract of ruminants, have been described to have the capacity to penetrate this plant cuticle barrier and to facilitate bacterial access to fermentative material in this way (Van Soest 1994). Cellulose digestion is inherently a relatively slow process compared to polysaccharides like starch or pectin (Weimer 1996) and will always require relative long retention times to be effective. However, food comminution has the potential to speed up this process at least to some degree; this is reflected in the mean retention times of food in the gut of the respective herbivores, which are at the level of 90–250 h in herbivorous reptiles like tortoises, while those in ruminants are at the level of 48 h (Franz et al. 2011). The degree to which particle size influences

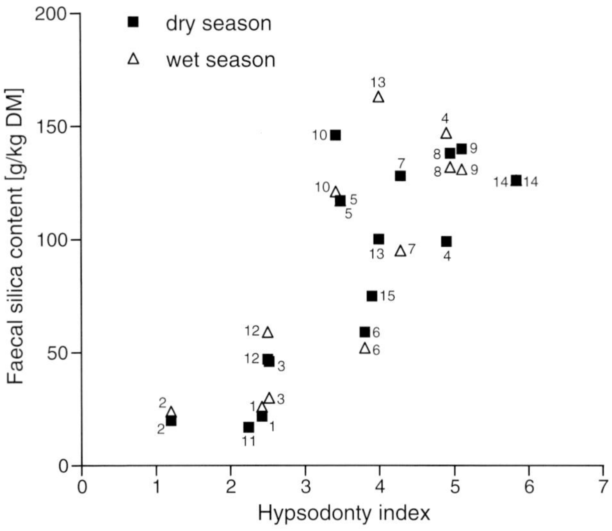

Fig. 6.4. Relation of fecal silica contents and hypsodonty index (degree of high-crownedness of teeth, estimated as ratio of tooth height/width of the 3rd molar). 1, greater kudu; 2, giraffe; 3, nyala; 4, impala; 5, waterbuck; 6, sable antelope; 7, roan antelope; 8, blue wildebeest; 9, tsessebe; 10, african buffalo; 11, black rhino; 12, african elephant; 13, warthog; 14, plains zebra; 15, white rhino. From Hummel et al. (2011).

fermentation rate may vary between plants, however. This was investigated in a sample of nine plants analyzed in an artificial fermentation system, the Hohenheim gas test. The samples were ranging from monocots (maize leaves; honey grass) to herbs (dandelion, ribwort) and browse (mulberry, hazel, elder, willow) and also included a leafy vegetable (endive). A significant influence of particle size (50, 20, 10, 6 and 2 mm, plus milled through a 1 mm sieve, standard size for chemical analyses) was obvious for most samples. The magnitude of the influence of particle size can be quantified as the relative increase of (fractional) fermentation rates with decreasing particle size, using a simple mono-exponential model. A high value indicates a high influence of particle size on fermentation rate, therefore. The effect was higher for the grass (14 %/mm) and maize (28 %/mm) samples (both monocots) compared to the 6 herb/browse samples with an average change of fractional fermentation rate of 7.3±3.9 %/mm (dicots). The high standard deviation indicates that some dicots (e.g. mulberry) also reached a high level of ~15 %/mm. Differences in leaf morphology like structure of cuticle and lignification must be regarded as most likely reasons for such variation. No clear influence of particle size was seen for endive, as leafy vegetable the only sample bred for human consumption and therefore for low lignin content.

Abrasiveness

The presence of abrasives in ingesta leads to increased tooth wear, up to the level of a serious loss of function in old animals. Abrasiveness associated with forage intake, either internally via phytoliths or externally via dust and sand, has been considered a major driver of the evolution of hypsodont teeth among herbivores. Since silica is the hardest material in ingested food, it has the highest potential to induce wear even on a material as tooth enamel (Williams & Kay 2001), generally considered to be the most abrasion-resistant animal tissue. While this relation appears straightforward and has been generally accepted, few attempts have been made to directly test this hypothesis. Besides its exceptional mechanical resistance, silica is also known to be chemically indigestible and is therefore used as a marker substance in digestion studies (Porter 1987). In consequence, fecal levels can be considered a reasonable proxy for the ingested amounts of silica, obviously slightly hampered by differences in overall digestibility of feeds. Since the residual ash of acid-detergent fiber (ADF) analysis recovers all silica in a sample (phytoliths + dust/sand) (Van Soest et al. 1991), it can be considered a valid and valuable methodological approach, allowing the quantification of larger sample sizes. In a study including 15 large herbivore species, a clear correlation between fecal silica levels (~ silica intake) and hypsodonty was found (Hummel et al. 2011) (Fig. 6.4). Browsing species had fecal silica contents of 17–46 g/kg DM, while grazing taxa had levels of 52–146 g/kg DM, with no difference found between wet and dry season. While the study demonstrates a clear

Table 6.1. Influence of food particle size on fermentation (DM, dry matter; NDF, Neutral-detergent fiber = hemicellulose + cellulose + lignin).

Plant	Particle size	Fermentation rate	Degradation	Source
Grass		NDF:	NDF (72 h)	4)
(*Tripsacum dactyloides*, *Dactylis glomerata*),	<2 mm 1)	0.066 h^{-1}	54.2 %	
Legume (*Trifolium pratense*)	<5 mm 1)	0.048 h^{-1}	57.4 %	
Meadow hay	0.1–0.4 2)	6.43 ml gas/(g DM × h)		5)
	0.4–1.0 2)	5.10 ml gas/(g DM × h)		
	1.0–2.0 2)	4.51 ml gas/(g DM × h)		
Grass hay			DM (24 h)	6)
	5 mm 1)		21.6 %	
	20 mm 1)		20.1 %	
	40 mm 1)		17.2 %	
Grass (*Pennisetum purpureum*)		DM:	DM (140 h)	7)
	3 mm 3)	0.05 h^{-1}	83 %	
	30 mm 3)	0.03 h^{-1}	85 %	

1) Particles passing the respective sieves
2) Particles passing the larger sieve size but retained on the following smaller sieve size
3) Exactly cut this size
4) Bowman & Firkins (1993)
5) Gerson et al. (1988)
6) Fadlalla et al. (1987)
7) Bjorndal et al. (1990)

relation of silica and development of high-crowned teeth, it could not answer the question which part of total silica (biogenic phytoliths or dust/sand) is responsible for most of the observed tooth wear. This question can be addressed with an experimental approach using diets varying the types of abrasives in a systematic way (Müller et al. 2014, Karme et al. 2016, Merceron et al. 2016, Ackermans et al. 2019, Martin et al. 2019). An alternative approach to the problem would be silica analysis methods that allow some differentiation; e.g., methods have been suggested to distinguish organic, soluble mineral and polymeric silica (Kolesnikov & Abaturov 1997, Kolesnikov & Gins 2001).

Food comminution in herbivores

Fecal particle sizes as a proxy for effectiveness of food comminution

The fact that digesta particle size is changed little after oral processing provides the opportunity to quantify chewing performance of different herbivores non-invasively via fecal particle size. It has been quantified using the wet sieving method mostly (Udén & Van Soest 1982, Kennedy 1984, Lechner-Doll 1986) (Fig. 6.5). While this is an established method, optical analysis is considered a valid alternative today (Luginbuhl et al. 1991, Péréz-Barbería & Gordon 1998, Nørgaard 2006, Kornfelt et al. 2013a,b). When applying wet sieving it should be kept in mind that the sizes given represent the sieve sizes and not the particles themselves. Especially the longish particles produced from grass may, by 'toppling over', be retained by sieves of a size much shorter than the actual particle length (Lechner-Doll 1986). While optical analysis is surely a valid practice, it obviously depends on an image of the sample without overlapping particles (which would not be recognized as separate entities). This requires protracted sample preparation, and thus puts some limitation on sample sizes, which influences representativeness. The quantification of very fine particles is another aspect where optical analysis should receive particular concern, since below certain particle sizes, resolution and particle clogging may become limiting.

Considering the fate of particles of different sizes in the digestive tract, various concepts have been put forward in the literature. For ruminants and camelids, it is well-established that larger particles are retained for a longer time than smaller particles, due to a sorting mechanism in the forestomach (Lechner et al. 2010, Clauss et al. 2011, Dittmann et al. 2015). This mechanism ensures that larger particles are re-submitted to rumination and hence ultimately reduced in size. Therefore, fecal particles reflect the effect of the sum of ingestive and rumination mastication. Since repeated rumination can compensate for a lower performance of the initial chewing process, reflections on the performance of tooth surfaces appear less direct in ruminants than in non-ruminating animals. A longer rumination time has the potential to compensate for less efficient teeth to a higher degree than the duration of ingestive mastication does in non-ruminating animals. Nevertheless, fecal particles still represent a valid estimate for the final result of total daily chewing efforts in ruminants.

In other large herbivores with a forestomach, but without the physiological adaptation of rumination, current evidence suggests that there is little differential passage of particles of different sizes (Clauss et al. 2004, Schwarm et al. 2009). For horses, in which such a differential passage had sometimes been postulated, it was recently demonstrated conclusively that large and small particles move through their digestive tract together (Hummel et al. 2018).

Herbivore classes and influence of body mass

While comparative aspects of digestive physiology related to nutritional strategies like retention times have been in the focus of research for a longer time (Foose 1982, Duncan et al. 1990), comparative food comminution moved into the center of interest only more recently. Differences between large herbivores like elephants, rhinos, equids, tapirs, ruminants, camelids, pigs and hippos have been investigated in a comprehensive overview for animals on zoo diets (Fritz 2007, Fritz et al. 2009). In another data set, fecal particle size has been evaluated for 16 large herbivores (African

Fig. 6.5. Sieving machine as used for quantifying fecal particle sizes of herbivores. Photo: J. Hummel.

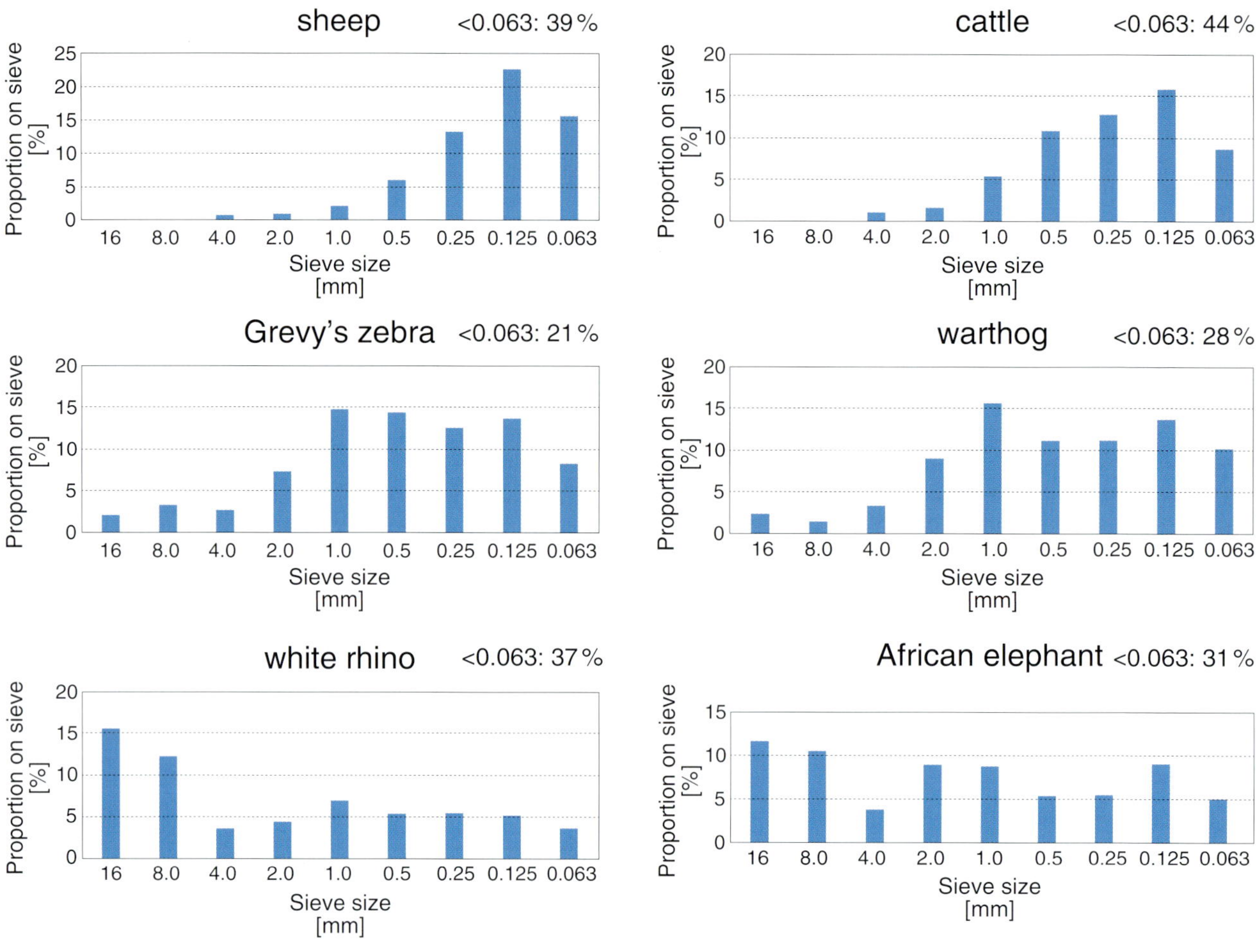

Fig. 6.6. Fecal particle size distributions of different herbivores. Based on data from Clauss et al. (2015); numbers in boxes represent the weight difference between initial sample and the sum of particles retained on the sieves.

elephant *Loxodonta africana*, white rhino *Ceratotherium simum*, Grevy's zebra *Equus grevyi*, Przewalski horse *Equus ferus przewalskii*, domestic horse and Shetland pony *Equus ferus caballus*, warthog *Phacochoerus africanus*, domestic cattle *Bos taurus primigenius*, forest buffalo *Synceros caffer nanus*, blue wildebeest *Connochaetes taurinus*, sable antelope *Hippotragus niger*, gemsbok *Oryx gazella*, waterbuck *Kobus ellipsiprymnus*, domestic sheep *Ovis orientalis aries*, domestic goat *Capra aegagrus hircus* and Bactrian camel *Camelus bactrianus*), all on a diet of ad libitum intake of meadow hay (Clauss et al. 2015) (Fig. 6.6). The hay had an NDFom (neutral-detergent fiber, ash-corrected) content of 720 g/kg OM and an ADL/NDFom ratio of 0.07. In both studies, animals of very large body mass (>1000 kg) like elephants, rhinos and hippos where characterized by particularly large fecal particle sizes, with averages of 5.1 and 5.0 mm for white rhinos and African elephants, respectively (reaching sizes of several centimeters for longest particles). It is an interesting aside that this is true although they differ considerably in their way of processing and chewing food, which becomes obvious e.g. when comparing fecal particles of browsing black rhinos and elephants in the same habitat. While ingested twigs will be present as clearly cut pieces in rhino feces, they will appear as rather "squeezed", but less sharply cut pieces in elephant dung (J. Hummel. pers. obs.).

Equids were intermediate (∅ 1.24 ± 0.22 mm), while ruminants and camels invariably had the by far smallest particle sizes (∅ 0.39 ± 0.10 mm) (Fig. 6.6). The warthog as artiodactyl hindgut fermenter had a fecal particle size similar to that seen in equids. In the ruminant data set, no major influence of body size was evident: Cattle were on a rather comparable level than e.g. sheep. Interestingly, the Bactrian camel as the only camelid in the sample plots with ruminants in principle, but has slightly larger fecal particles than any other ruminating animal (all Ruminantia). Functionally, camels are ruminants, achieving a comparable digestive efficiency, but are characterized by a slightly lower metabolism and food intake than true ruminants (Jalali et al. 2012, Dittmann et al. 2014). While sharing dental characteristics like selenodont teeth and missing upper incisors, camels differ in details of their chewing behavior, in particular in chewing each individual bolus on both sides alternating. This is in contrast to (true) Ruminantia which invariably chew a single bolus on one side of the jaw only, and change to the other side for the next bolus (Hendrichs 1965).

Given the large particle sizes found in many megaherbivores, the influence of body mass on fecal particle size became of interest. A scaling of fecal particle size with $BM^{0.14}$ has been proposed (Péréz-Barbería & Gordon 1998) for the data set of Udén (1978). It soon becomes obvious that, as outlined by Fortelius (1985), allometry among lines of considerable phylogenetic distance (e.g. elephants, rhinos, equids, hippos, ruminants) is different and generally much more pronounced from what is seen in closely related taxa or even breeds (e.g. equids reaching from a riding horse of 600 kg BM over a Przewalski horse of around 300–400 kg BM to Shetland ponies of a BM of 100 kg). If at all, only much smaller differences appear to be found in the latter case where animals of considerably different size, but uniform *bauplan* are compared. Among lineages of uniformly very large body size (elephants, rhinos, hippos), there appears to be a "megaherbivore" syndrome in the area of food comminution: Their very large fecal particle sizes render those groups distinctively different from e.g. equids or ruminants (the latter with the particularities in the food comminution outlined above). Potential explanations for the pattern could be the limited selectivity and in consequence diet quality in megaherbivores, or a higher gut fill in virtually predation-free taxa, compensating for less intense food processing.

Between breeds of a species or taxa of a distinct phylogenetic group, BM can largely vary, but virtually no effect of BM on fecal particle size is found; this is true within ruminants or equids. It has been described for horses that tooth size (and therefore also determinants of dentition performance like length of enamel cutting edges or occlusal surface) is decreasing disproportionately with BM, providing smaller breeds with a relatively larger tooth area (Radinsky 1984). In ruminants, in fact extensive breeds with a higher potential to use fiber-rich forage are often characterized by relatively small body mass (e.g. sheep: German grey heath, Skudde; cattle: Galloway, highland cattle), providing them – assuming this line of thought – with superior chewing capacity in relation to BM.

In this context, an example for differences in the food comminution process well-known from practical feeding of domestic ruminants becomes interesting. It is well established that digestibility of diets including kernels of maize and also other cereal grains benefits from a more thorough pre-feeding crushing when being fed to cattle, but not or far less when fed to sheep. In fact, whole maize kernels appear in feces of cattle (and in consequence lead to a reduced dry matter digestibility) if kernels/their hulls have not been cracked in some way prior to feeding (Wilson et al. 1973). While the practical evidence is convincing, it does not appear to be clear to which degree this is due to differences in rumen retention characteristics (leading to a faster outflow/preventing rumination of kernels) or due to closer enamel ridges in the smaller animal, avoiding escape of any undamaged kernels from the chewing process.

Feeding types (browsers and grazers)

Many parts of the foraging process of herbivores can be influenced by the specialisation on certain forage plant types (Codron et al. 2019). While the dichotomy of grazer and browser may appear to be over-simplified, morphological patterns seen among herbivores largely follow their preference for the respective plant group (Figs. 6.7, 6.8). Among the obvious differences seen in tooth morphology are hypsodonty in grazing species and differences in presence of enamel cutting edges (Kaiser et al. 2010). For ruminants, a meta-analysis has described a shift of chewing from rumination to ingestion-chewing for goats compared to sheep (Dulphy et al. 1995). This observation is in line with the idea of rumination decreasing the burden of strong abrasiveness of a grass diet due to the intake of solid or grit (Hatt et al. 2019, 2020), since goats are characterized as intermediate feeding type with a strong preference for browse, and can be expected to ingest a far less abrasive diet when given the choice under free-ranging conditions. As mentioned, the difference between the feeding types in ingesta particle dimensions (longish in grass, polyedric in browse) is largely a characteristic of the respective forage (and not tooth morphology), but it is expected to have the potential to influence intake capacity of the respective herbivore (Troelsen & Campbell 1968) with a potentially higher intake capacity, at the same gut volume, in browsers.

Investigations on fecal particle sizes in feeding types have resulted in larger particles for browsing ruminant species on their respective zoo diets (Clauss et al. 2002), but not on the diets in their natural habitats (Hummel et al. 2008, Lechner et al. 2010), which suggests that either captive grazers are fed diets more similar to their natural ones, or that teeth of grazers are more universally efficient than those of browsers. In a study on feeding types in kangaroos, larger particles were found in the forestomach of three browsing compared to one grazing species (Lentle et al. 2003). In contrast to this, in rhinoceroses, larger particles were found in the grazer (white rhino) than in the browser (black rhino) when animals were fed a grass hay diet (Steuer et al. 2010). It should be added here that results of this study may have been influenced by the age of the animals, since at least two of the white rhinoceroses (*Ceratotherium simum*) were considerably older than the four browsing black rhinos (*Diceros bicornis*). It remains an open question to which degree differences in tooth morphology must be interpreted as adaptations resulting in improved comminution capacity for different forages, or if they just represent simplifications in browsers as herbivores with a diet less demanding in several aspects of comminution.

Less information is available for other herbivore lineages including taxa with a more browsing or grazing food selection, e.g. proboscideans. While both extant species are considered intermediate feeders, food choice of the African elephant could be demonstrated to be shifted towards browse compared to the Asian elephant (Cerling et al. 1999), which is also reflected in variables of digestive physiology like retention time (Hackenberger 1987). However, no systematic difference in fecal particle sizes was obvious in the study of Fritz (2007) and in a separate evaluation of four Asian and eight African zoo specimens (Tab. 6.2).

Practical feeding of specialized browsers in captivity has been receiving considerable attention for a long time (Clauss & Dierenfeld 2008); their capacity to comminute and ingest particular plant material is of relevance for their potential intake of different forages. Interestingly, considerably larger fecal particle sizes have been described for captive compared to free-ranging giraffe (Hummel et al. 2008), a result that was recently confirmed (Schüßler et al. 2017). The fact that in the latter study fecal particle size decreased from alfalfa hay to browse for identical individuals points to an interaction between tooth structure and forage type (and rather excludes an explanation assuming

Fig. 6.7. Grazing and browsing herbivores. Photos: Forages and giraffe J. Hummel, Grevy's zebra P. Steuer.

more strongly abraded teeth in a brachydont (low-crowned) browsing species like giraffe).

Table 6.2. Mean particles sizes in feces of zoo elephants (all diets were based on grass hay, in zoos B and C supplemented with some straw, tree branches and limited concentrates).[1]

	Mean particle size [mm]	Diet	Age [years]
African elephant	4.2	Zoo A	23
	5.5	Zoo A	23
	7.6	Zoo A	23
	5.8	Zoo A	23
	5.5	Zoo A	23
	5.6	Zoo A	29
	8.0	Zoo B	21
	5.6	Zoo B	26
Asian elephant	6.1	Zoo B	55
	7.9	Zoo B	40
	7.1	Zoo C	42
	12.9	Zoo C	28

1 Two other results of the latter data set deserve attention: (1) The oldest individual in this sample (55 years) was not characterized by a deviating fecal particle size, demonstrating the effectiveness of the lifelong tooth replacement mechanism in elephants. (2) The elephant characterized by conspicuously large fecal particles (12.9 mm) was an individual with a known history of tooth problems, potentially resulting in less effective chewing.

Feed intake during lactation

While food comminution must be sufficient to meet metabolic requirements, its highest performance is necessary in times of significant additional nutrient requirements like during lactation. The latter can be considered to be the period of highest energy requirements and of highest food intake in mammals. Changes like increased rumination investment, increased digesta particle sizes, increased gut fill and increased passage rate are potential consequences of increased food intake (Welch & Smith 1969). In sheep, increasing intake from 0.5 to 1.0 kg DM decreased time spent ruminating per kg DM for approximately 20 %, while chewing during feeding was influenced very little (Grimaud 1999, as cited in Baumont et al. 2006).

Changes can be expected to be most extreme in the case of domestic animals bred for milk production like high yielding dairy cows, well known for a particularly high increase in energy requirements during lactation. In fact, increasing fecal particle sizes are reported by some authors for ruminants with increased feed intake (Luginbuhl et al. 1990, Okine & Matthison 1991), while other studies did not see an increase (Kovács et al. 1997a, b, 1998). In dairy goats, a triplication of feed intake (from 0.9 to 2.8 kg DM; maintaining a constant ratio of 50 % concentrate/50 % grass hay) during lactation led to considerable changes in mean retention time of particles ($MRT_{particle}$) (from 71 ± 8.9 h to 31 ± 2.3 h) and in organic matter digestibility (from 68 ± 2.1 % to 59 ± 2.1 %), but only to minor differences in

T. Martin & W. v. Koenigswald: Mammalian Teeth – Form and Function. – München (Pfeil) 2020 – ISBN 978-3-89937-266-3

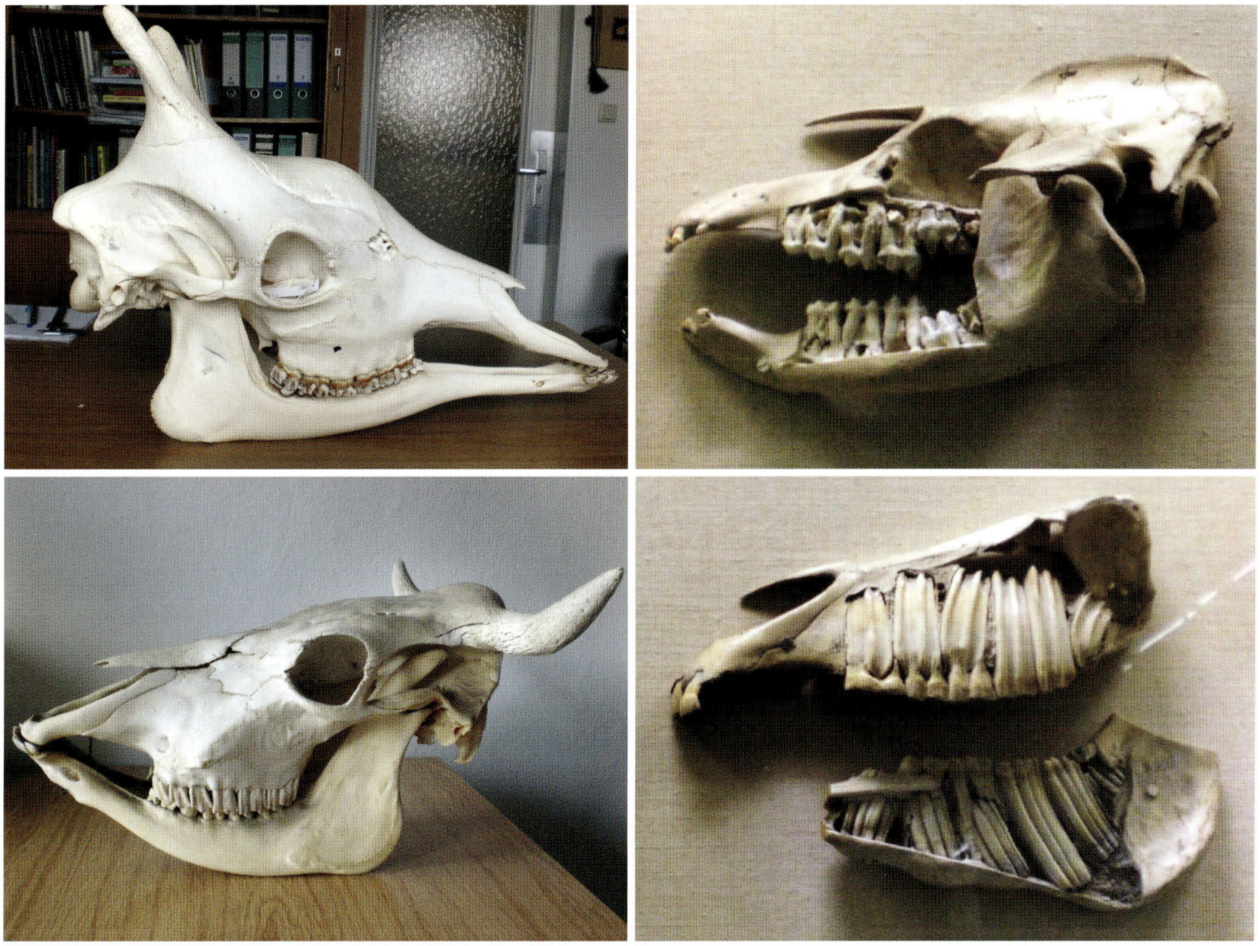

Fig. 6.8. Comparative head morphology of browsers and grazers. Artiodactyls/ruminants: brachydont (low-crowned) giraffe (*Giraffa camelopardalis*, upper left) and hypsodont (high-crowned) cattle (*Bos taurus taurinus*, lower left); perissodactyls: brachydont tapir (*Tapirus* sp., upper right) and hypsodont domestic horse (*Equus caballus*, lower right). Note that hypsodont species must shift the eye socket to the rear due to the space taken by maxillary molars. Photos: J. Hummel.

fecal particle sizes (from 0.53 ± 0.02 mm to 0.59 ± 0.02 mm for mean particle size) (Findeisen 2012).

While it can be expected that results on fecal particle sizes are influenced by the selective retention mechanism in ruminants, an effect of increased intake should be more direct in non-ruminating animals. Only limited data are available in this respect. However, in a study on rabbits, no difference in fecal particle size was present if feed intake was increased twofold during lactation from 110 g DM/day to 220 g DM/day (again with a constant ratio of 50 % grass hay/50 % pelleted concentrate) (Findeisen 2012). While retention time was influenced again from 31 ± 2.5 h to 19 ± 3.1 h for $MRT_{particle}$, no change in mean fecal particle size was present (0.56 ± 0.02 mm and 0.59 ± 0.02 mm), suggesting that the animals maintained chewing intensity (per unit food), and thus most likely increased their daily chewing time with increasing intake.

Tentative calculations on the energy budget of food comminution in herbivores

Several interactions of food comminution and energy budget have been outlined in the sections above and are summarized in Figure 6.9. A comparison of performance of herbivorous reptiles and mammals clearly demonstrates the advantageous consequences of food comminution.

Particle size reduction by chewing requires energy investment by the animal. This becomes evident from studies in which the energy required for eating forages of different particle size is quantified – typically, the same forage offered in a size-reduced form (chopped, or ground) facilitates intake at lower energetic costs (Osuji et al. 1975, Susenbeth et al. 1997).

When comparing the comminution and forage energy processing capacity of herbivores, a set of variables becomes interesting for a complete description of food processing strategy. These include:

(1) Intake and particle size of ingested food;
(2) fecal particle sizes, food digestibility and retention time;

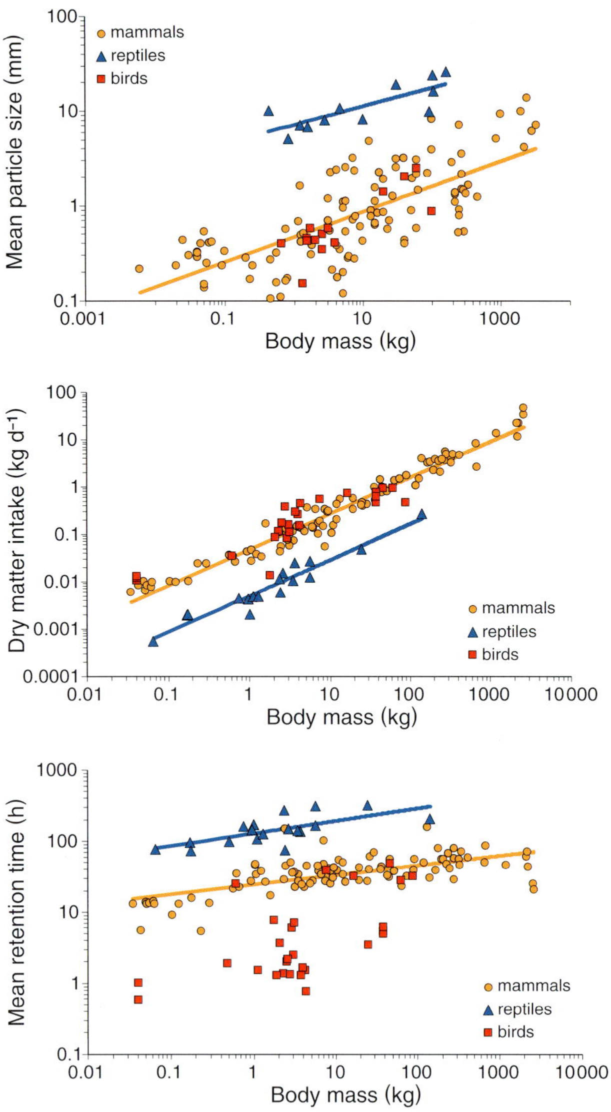

Fig. 6.9. Aspects of digestive physiology in herbivores. Data from Clauss et al. (2007), Fritz et al. (2009, 2010, 2012), Franz et al. (2011).

(3) characteristics of dentition influencing particle comminution per chew (total length of cutting edges) (in the short term invariable part of the chewing process);
(4) duration and frequency of chewing during feeding and rumination (how many chews per day and per unit of feed, amplitude of chewing strokes (= number of cutting edge contacts of a single stroke), plus possibly chewing force);
(5) energy requirement for chewing (energy effort per chew or per unit of feed intake).

Available variables have been summarized in Table 6.3 for two extant herbivores, equids represented by the horse and ruminants by cattle. The differences in fecal particle sizes are obvious and have been quantified as mean particle sizes of 1.24 and 0.38 mm for horse and cattle, respectively. The energy needed for a single feeding chewing stroke can be considered to be higher in horses, which have been described to have three times the chewing muscle mass of a comparably or even larger ruminant (bison) (Becht 1953, as cited in Fortelius 1985). Data from Osuji et al. (1975) and Vernet et al. (1995) support this assumption in showing that ingestive activity represents a higher proportion of overall maintenance energy intake in horses compared to ruminants. In addition, ruminants shift much of their chewing efforts to pre-digested material, which appears to lower the burden of comminution energy. Considerably less energy necessary for grinding has been found in pre-fermented compared to untreated feed (Kennedy et al. 1993). This is reflected in the values published by Susenbeth et al. (1998), indicating a threefold higher energy requirement for chewing during feeding compared to rumination, and supported in other studies on sheep and cattle (Suzuki et al. 2008, 2014).

Using the different published data to make more detailed, quantitative comparisons includes a large number of uncertainties and potential sources for mistakes. Experiments in which all measures – body mass, food intake, chewing behavior, particle size reduction, and energy required for chewing – have been recorded on the same animals, on the same diets, do not exist. But we consider such calculations an informative exercise nevertheless, and present two different sets of calculations.

For the first, we assume that chewing energy in ingestive chewing is the same for horses and cattle, and simply use data measured in cattle by Susenbeth et al. (1998) for horses and cattle. For the second, we use data from Vernet et al. (1995) to calculate the energy required for ingestion of a given amount of dry matter for cattle and horses, and assume that the proportion of energy required for rumination in cattle is the same as that in the data of Susenbeth et al. (1998). One of the many differences in the approaches between the two studies is the use of animals with empty rumens by Susenbeth et al. (1998), which ensures that only costs due to chewing but not energy due to initiated digestion is recorded. Correspondingly, values using the Vernet et al. (1995) data are generally of a higher magnitude. This is also the reason why data from the two studies cannot be compared directly.

We assume two similar-sized animals of 600 kg with typical values for intake (Table 6.3). Total ME requirements for maintenance would be around 60 MJ/day for a 600 kg animal, assuming an energy requirement of 0.5 MJ/kg $BM^{0.75}$ (GfE 1995, 2014). The first approach using the Susenbeth et al. (1998) data suggests a daily energy expenditure for chewing of 1.38 MJ ME for the horse compared to 1.08 MJ ME for cattle. The value for cattle is composed of 0.70 MJ due to ingestion and only 0.38 MJ due to rumination. The cost of rumination chewing would therefore be only about half the value of that for chewing during feeding, irrespective of the higher amount of chews spent during rumination. The second approach using the Vernet et al. (1995) data yields a daily energy expenditure for food processing of 10.05 MJ ME for the horse and 4.21 MJ ME for cattle, suggesting a more distinct difference.

Assuming a uniform particle size of 50 mm in the diet of horse and cattle, the difference to the mean fecal particle size can be calculated (representing the reduction of particle size; 49.62 mm for cattle and 48.76 mm for the horse), weighted by the daily intake (7.5 kg DM for a 600 kg cow and 12.6 kg DM for a 600 kg horse) and divided by the daily amount of chews to arrive at a number quantifying the "particle comminution capacity" of a single chewing stroke. Thus, the particle comminution capacity of cattle is 49.62 mm · 7.5 kg DM/34 074 chews = 0.011 mm · kg/chew; whereas that of horses is 48.76 mm · 12.6 kg DM/27 625 chews = 0.022 mm · kg/chew. In other words, a single chewing stroke in the horse is about twice as efficient as that of cattle. Most likely, this is achieved by a combination of an extended chewing surface due to the molarisation of horse premolars (whereas the premolars of cattle are, on the opposite, reduced), and the most intricate set of enamel folds found in large mammal molars to date (plus possibly the more elaborate chewing muscles in horses). The calculation thus is evidence for the putative power of morphological adaptations to achieve a high chewing efficiency.

When putting this particle comminution capacity in perspective with the required daily energy expenditure for chewing, our two different approaches lead to diverging results. In the first approach, horses achieve 0.022 mm · kg/chew using 1.38 MJ ME, or 0.016 mm · kg/chew per MJ ME. Cattle achieve 0.011 mm · kg/chew using 1.08 MJ ME, or 0.010 mm · kg/chew per MJ ME. Based on this approach, the ruminant system partially, but not completely compensates for the difference in chewing efficiency based on the effect per chew. In the second approach, with a much larger difference in energy used for chewing, horses achieve an energetic efficiency of chewing of 0.0022 mm · kg/chew per MJ, and cattle are even superior at 0.0026 mm · kg/chew per MJ. Based on the many assumptions necessary to derive the figures, these results should not be used to construe an energetic advantage of ruminants based on chewing only, but rather underline that very different organismal solutions – the equid and the ruminant system – arrive at similar efficiencies, where differences may or may not await detailed investigation. While it may appear surprising that the ruminant system is not distinctively superior in these calculations, one needs to consider in addition the higher digestive efficiency achieved by ruminants, which is also due to their higher digesta particle comminution. In evolutionary terms, animals should not be conceptualized at excelling at individual feats, such as chewing efficiency, but in overall traits such as energetic efficiency that, ultimately, results in an efficiency of turning environmental resources into offspring (Clauss et al. 2019).

Conclusion and final remarks

- Efficient food comminution is a key requirement for evolutionary successful herbivores.
- Physical processing and processing characteristics of animal food represent interesting fields of research, where we still have much to learn.
- In most plants, an increasing effect of particle comminution on fermentation rate is present; on average, it appears to be higher in grass than in browse.
- In the comparison of grazer vs. browser, all available data on differences on the level of physical characteristics of forage and its interaction with morphology/physiology of the animal (particle shapes, rumen

Table 6.3. Variables relevant for the energy budget of food comminution of ruminants and horses.

		Cattle	Horse	
Body mass (kg)		600	600	
Intake (g DM/kg $BM^{0.75}$)		62	104	Foose (1982)
(kg DM/day)		7.5	12.6	
Chewing frequency (chews/min)	Ingestion:	75 (65–85)[1], 73[2]	77[2]	[1] Jarrige et al. (1995),
	Rumination:	60 (55–65)[1]		[2] Dittmann et al. (2017)
Total number of chews per day	Ingestion:	13 942[3]	27 625[4]	[3] Piatkowski et al. (1977)
	Rumination:	20 132[3]		[4] Porzig & Sambraus (1991)
	Total:	34 074	27 625	
Total duration (min/d)	Ingestion:	185	359	
	Rumination:	336		
	Total:	521	359	
Energy requirement chewing (J/(min · $BM^{0.75}$)	Ingestion:	30	30	Susenbeth et al. (1998)
	Rumination	8.9		
Daily energy for chewing (MJ)	Ingestion:	0.70	1.38	Susenbeth et al. (1998)
	Rumination:	0.38		
	Total:	1.08	1.38	
	Ingestion:	2.73	10.05	using data from Vernet et al. (1995) for
	Rumination:	1.48		ingestion, and assuming the proportions
	Total:	4.21	10.05	of energy for ingestion and rumination as
				Susenbeth et al. (1998)
Fecal particle size (mean size, mm)		0.38	1.24	Clauss et al. (2015)
Chewing efficiency (mm · kg/chew)		0.011	0.022	
Energy efficiency for comminution		0.010	0.016	using energy from Susenbeth et al. (1998)
(mm · kg/chew · MJ)		0.0026	0.0022	using energy from Vernet et al. (1995)
Change of fermentation rate with particle size		15 %/mm particle size (grass)		

stratification, abrasiveness) point to differences at least as large as those on the chemical level.

- Body size appears to be a relevant factor for food comminution when observing it on the level of larger phylogenetic groups like ruminants, elephants, rhinos or equids, while within these groups, the influence of body size is far less pronounced. This opens the possibility that larger forms are not optimized in the particular trait under consideration (here: chewing efficiency), but that they have survived for other reasons that prevent the superiority of smaller forms to take effect on niche occupation.
- On a comparative level, data that facilitate a comparison of species in terms of chewing efficiency (accounting for the energy required to achieve a specific degree of particle size reduction by chewing, and weighting by the effects on intake capacity and rate of digestion) are still largely missing. Tentative comparisons between horses and ruminants illustrate that different morphophysiological solutions for a biological challenge exist. The degree by which such differences in a single system (here: digestive physiology) can account for perceived differences on the scale of 'evolutionary success' remains speculative.

Acknowledgments

We would like to sincerely thank Petra Jacquemien and Nadja Wahl for sophisticated lab analyses. Further, we are grateful to invaluable help of cooperating zoos, in particular Safari Beekse Bergen, Zoo Köln and Zoo Augsburg. Reviewers Peder Nørgaard (Copenhagen) and Angela Schwarm (Ås) provided constructive criticism and useful suggestions.

References

* indicates publications that originated from the DFG Research Unit 771.

Ackermans, N. L., Clauss, M., Winkler, D. E., Schulz-Kornas, E., Kaiser, T. M., Müller, D. W. H., Kircher, P. R., Hummel, J. & Hatt, J.-M. (2019): Root growth compensates for molar wear in adult goats (*Capra aegagrus hircus*). Journal of Experimental Zoology A 331: 139–148.

Baron, J.-P. (1994): Zur Eignung der Mahlwiderstandsmethode als Schnellverfahren für die Bestimmung von Futterqualitätsparametern in tropischen Gebieten. Das Wirtschaftseigene Futter 40: 59–73.

Baumont, R., Doreau, M., Ingrand, S. & Veissier, I. (2006): Feeding and mastication behaviour in ruminants. In: Bels, V. (ed.). Feeding in Domestic Vertebrates. CABI Publishing, Wallingford: 241–262.

Becht, G. (1953): Comparative biological-anatomical researches on mastication in some mammals. Proceedings Nederlands Akademie Wetenschapen C 14: 508–527.

Berkovitz, B. & Shellis, P. (2018): The Teeth of Mammalian Vertebrates. William Andrew Publishing, London.

Bjorndal, K. A., Bolton, A. B. & Moore, J. E. (1990): Digestive fermentation in herbivores: Effect of food particle size. Physiological Zoology 63: 710–721.

Blümmel, M., Paul, C., Goodchild, A. & Becker, K. (1996): Grinding energy and in vitro gas technique for the assessment of Syrian barley straws: physical and microbial degradation and voluntary feed intake by sheep. Journal of Animal Physiology and Animal Nutrition 76: 132–140.

Bowman, J. G. P. & Firkins, J. L. (1993): Effects of forage species and particle size on bacterial cellulolytic activity and colonization in situ. Journal of Animal Science 71: 1623–1633.

Cerling, T. E., Harris, J. M. & Leakey, M. G. (1999): Browsing and grazing in elephants: the isotope record of modern and fossil proboscideans. Oecologia 120: 364–374.

Clauss, M. & Dierenfeld, E. (2008): The nutrition of "browsers". In: Fowler, M. E. & Miller, R. E. (eds.). Zoo and Wild Animal Medicine. Saunders, St. Louis: 444–454.

Clauss, M., Fritz, J., Bayer, D., Nygren, K., Hammer, S., Hatt, J.-M., Südekum, K.-H. & Hummel, J. (2009a): Physical characteristics of rumen contents in four large ruminants of different feeding type, the addax (*Addax nasomaculatus*), bison (*Bison bison*), red deer (*Cervus elaphus*) and moose (*Alces alces*). Comparative Biochemistry and Physiology A 152: 398–406.

Clauss, M., Kaiser, T. & Hummel, J. (2008): The morphophysiological adaptations of browsing and grazing mammals. In: Gordon, I. J. & Prins, H. H. T. (eds.). The Ecology of Large Mammalian Herbivore Browsing and Grazing. Springer, Berlin: 47–88.

Clauss, M., Lechner-Doll, M. & Streich, J. W. (2002): Faecal particle size distribution in captive wild ruminants: An approach to the browser/grazer dichotomy from the other end. Oecologia 131: 343–349.

Clauss, M., Lechner, I., Barboza, P., Collins, W., Tervoort, T. A., Südekum, K.-H., Codron, D. & Hummel, J. (2011): The effect of size and density on the mean retention time of particles in the reticulorumen of cattle (*Bos primigenius* f. *taurus*), muskoxen (*Ovibos moschatus*) and moose (*Alces alces*). British Journal of Nutrition 105: 634–644.

Clauss, M., Müller, D. W. H. & Codron, D. (2019): Within-niche pace of life acceleration as a fundamental evolutionary principle: a mammal pilot test case. Evolutionary Ecology Research 20: 385–401.

Clauss, M., Nunn, C., Fritz, J. & Hummel, J. (2009b): Evidence for a tradeoff between retention time and chewing efficiency in large mammalian herbivores. Comparative Biochemistry and Physiology A 154: 376–382.

Clauss, M., Schwarm, A., Ortmann, S., Alber, D., Flach, E. J., Kühne, R., Hummel, J., Streich, W. J. & Hofer, H. (2004): Intake, ingesta retention, particle size distribution and digestibility in the Hippopotamidae. Comparative Biochemistry and Physiology A 139: 449–459.

Clauss, M., Schwarm, A., Ortmann, S., Streich, W. J. & Hummel, J. (2007): A case of non-scaling in mammalian physiology? Body size, digestive capacity, food intake, and ingesta passage in mammalian herbivores. Comparative Biochemistry and Physiology A 148: 249–265.

*Clauss, M., Steuer, P., Erlinghagen-Lückerath, K., Kaandorp, J., Fritz, J., Südekum, K.-H. & Hummel, J. (2015): Faecal particle size: Digestive physiology meets herbivore diversity. Comparative Biochemistry and Physiology A 179: 182–191.

Clauss, M., Steuer, P., Müller, D. W. H., Codron, D. & Hummel, J. (2013): Herbivory and body size: allometries of diet quality and gastrointestinal physiology, and implications for herbivore ecology and dinosaur gigantism. PLoS One 8: e68714.

Codron, D., Hofmann, R. R. & Clauss, M. (2019): Morphological and physiological adaptations for browsing and grazing. In: Gordon, I. J. & Prins, H. H. T. (eds.). The Ecology of Grazing and Browsing II. Springer Nature Switzerland: 81–125.

Dittmann, M. T., Hummel, J., Runge, U., Galeffi, C., Kreuzer, M. & Clauss, M. (2014): Characterising an artiodactyl family inhabiting arid habitats by its metabolism: low metabolism and maintenance requirements in camelids. Journal of Arid Environments 107: 41–48.

Dittmann, M. T., Kreuzer, M., Runge, U. & Clauss, M. (2017): Ingestive mastication in horses resembles rumination but not ingestive mastication in cattle and camels. Journal of Experimental Zoology 327: 98–107.

Dittmann, M. T., Runge, U., Ortmann, S., Lang, R. A., Moser, D., Galeffi, C., Schwarm, A., Kreuzer, M. & Clauss, M. (2015): Digesta retention patterns of solutes and different-sized particles in camelids compared with ruminants and other foregut fermenters. Journal of Comparative Physiology B 185: 559–573.

Dulphy, J. P., Balch, C. C. & Doreau, M. (1995): Adaptation des espèces domestiques à la digestion des aliments lignocellulosiques. In: Jarrige, R., Ruckebusch, Y., Demarquilly, C., Farce, M. H., & Journet, M. (eds.). Nutrition des Ruminants Domestiques. INRA Editions, Paris: 759–803.

Duncan, P., Foose, T. J., Gordon, I. J., Gakahu, C. G. & Lloyd, M. (1990): Comparative nutrient extraction from forages by grazing bovids and equids: a test of the nutritional model of equid/bovid competition and coexistence. Oecologia 84: 411–418.

Fadlalla, B., Kay, R. N. B. & Goodall, E. (1987): Effects of particle size on digestion of hay by sheep. Journal of Agricultural Science 109: 551–561.

*Findeisen, E. (2012): Food particle breakdown in herbivores – influence of food intake level on energy yield. Doctoral dissertation, Veterinary Medicine, University of Gießen.

Foose, T. J. (1982): Trophic strategies of ruminant versus nonruminant ungulates. PhD thesis University of Chicago, Chicago.

Fortelius, M. (1985): Ungulate cheek teeth: development, functional and evolutionary interactions. Acta Zoologica Fennica 180: 1–76.

Franz, R., Hummel, J., Müller, D. W. H., Bauert, M., Hatt, J.-M. & Clauss, M. (2011): Herbivorous reptiles and body mass: effects on food intake, digesta retention, digestibility and gut capacity, and a comparison with mammals. Comparative Biochemistry and Physiology A 158: 94–101.

Fritz, J. (2007): Allometrie der Kotpartikelgröße von pflanzenfressenden Säugern, Reptilien und Vögeln. Dissertation, Tierärztliche Fakultät LMU München.

Fritz, J., Hammer, S., Hebel, C., Arif, A., Michalkec, B., Dittmann, M. T., Müller, D. W. H. & Clauss, M. (2012): Retention of solutes and different-sized particles in the digestive tract of the ostrich (*Struthio camelus massaicus*), and a comparison with mammals and reptiles. Comparative Biochemistry and Physiology A 163: 56–65.

Fritz, J., Hummel, J., Kienzle, E., Arnold, C., Nunn, C. & Clauss, M. (2009): Comparative chewing efficiency in mammalian herbivores. Oikos 118: 1623–1632.

Fritz, J., Hummel, J., Kienzle, E., Streich, W. J. & Clauss, M. (2010): To chew of not to chew: Fecal particle size in herbivorous reptiles and mammals. Journal of Experimental Zoology 313A: 579–586.

Fritz, J., Hummel, J., Kienzle, E., Wings, O., Streich, W. J. & Clauss, M. (2011): Gizzard vs. teeth, it's a tie: Food-processing efficiency in herbivorous birds and mammals and implications for dinosaur feeding strategies. Paleobiology 37: 577–586.

Fujikura, T., Oura, R. & Sekine, J. (1989): Comparative morphological studies on digestion physiology of herbivores 1. Digestibility and particle distribution of digesta and faeces of domestic and feral animals. Journal Faculty of Agriculture Tottori University 25: 87–93.

Gerson, T., King, A. S. D., Kelly, K. E. & Kelly, W. J. (1988): Influence of particle size and surface area on in vitro rates of gas production, lipolysis of triacylglycerol and hydrogenation of linoleic acid by sheep rumen digesta or *Ruminococcus flavefaciens*. Journal of Agricultural Science 110: 31–37.

GfE (1995): Empfehlungen zur Energie- und Nährstoffversorgung der Mastrinder. DLG-Verlag, Frankfurt a.M.

GfE (2014): Empfehlungen zur Energie- und Nährstoffversorgung von Pferden. DLG-Verlag, Frankfurt a.M.

Grandl, F., Schwarm, A., Ortmann, S., Furger, M., Kreuzer, M. & Clauss, M. (2018): Kinetics of solutes and particles of different size in the digestive tract of cattle of 0.5 to 10 years of age, and relationships with methane production. Journal of Animal Physiology and Animal Nutrition 102: 639–651.

Grimaud, P. (1999): Effets de la sous-alimentation énergétiques sur la digestion ruminale chez les bovins et les ovins. Thèse de doctorat, Université de Rennes.

Hackenberger, M. K. (1987): Diet digestibilities and ingesta transit times of captive Asian (*Elephas maximus*) and African (*Loxodonta africana*) elephants. MSc thesis. Faculty of Graduate Studies. University of Guelph, Guelph.

Hatt, J.-M., Codron, D., Ackermans, N. L., Martin, L. F., Richter, H., Kircher, P. R., Gerspach, C., Hummel, J. & Clauss, M. (2020): Differences in the rumen washing mechanism in sheep fed diets with abrasives of various concentrations and sizes. Palaeogeography, Palaeoclimatology, Palaeoecology 550: 109728.

Hatt, J.-M., Codron, D., Müller, D. W. H., Ackermans, N. L., Martin, L. F., Kircher, P. M., Hummel, J. & Clauss, M. (2019): The rumen washes off abrasives before heavy-duty chewing in ruminants. Mammalian Biology 97: 104–111.

Hendrichs, H. (1965): Vergleichende Untersuchung des Wiederkauverhaltens. Biologisches Zentralblatt 84: 681–751.

*Hummel, J., Findeisen, E., Südekum, K.-H., Ruf, I., Kaiser, T. M., Bucher, M., Clauss, M. & Codron, D. (2011): Another one bites the dust: faecal silica levels in large herbivores correlate with high-crowned teeth. Proceedings of the Royal Society B 278: 1742–1747.

Hummel, J., Fritz, J., Kienzle, E., Medici, E. P., Lang, S., Zimmermann, W., Streich, W. J. & Clauss, M. (2008): Differences in fecal particle size between free-ranging and captive individuals of two browser species. Zoo Biology 27: 70–77.

Hummel, J., Scheurich, F., Ortmann, S., Crompton, L., Gerken, M. & Clauss, M. (2018): Comparative selective retention of particle size classes in the gastrointestinal tract of ponies and goats. Journal of Animal Physiology and Animal Nutrition 102: 429–439.

Jalali, A. R., Nørgaard, P., Weisbjerg, M. R. & Nielsen, M. O. (2012): Effect of forage quality on intake, chewing activity, faecal particle size distribution, and digestibility of neutral detergent fibre in sheep, goats, and llamas. Small Ruminant Research 103: 143–151.

Jarrige, R., Dulphy, J. P., Faverdin, P., Baumont, R. & Demarquilly, C. (1995): Activités d'ingestion et de rumination. In: Jarrige, R., Ruckebusch, Y., Demarquilly, C., Farce, M. H., & Journet, M. (eds.). Nutrition des Ruminants Domestiques. INRA Editions, Paris: 123–182.

Jensen, L. M., Markussen, B., Nielsen, N. I., Nadeau, E., Weisbjerg, M. R. & Nørgaard, P. (2016): Description and evaluation of a net energy intake model as a function of dietary chewing index. Journal of Dairy Science 99: 8699–8715.

*Kaiser, T., Fickel, J., Streich, W. J., Hummel, J. & Clauss, M. (2010): Enamel ridge alignment in upper molars of ruminants in relation to their natural diet. Journal of Zoology 281: 12–25.

Karme, A., Rannikko, J., Kallonen, A., Clauss, M. & Fortelius, M. (2016): Mechanical modelling of tooth wear. Journal of the Royal Society Interface 13: 20160399.

Kennedy, P. M. (1984): Techniques in Particle Size Analysis of Feed and Digesta in Ruminants. Canadian Society of Animal Science, Edmonton.

Kennedy, P. M., Gordon, G. L. R. & Hogan, H. P. (1993): Nutritional comparisons between cattle nd buffaloes and implications for drought and animal power. In: Pryor, W. J. (ed.). Draught Animal Power in the Asian-Australasian Region. Australian Centre for International Agricultural Research, Canberra: 66–71

Kolesnikov, M. P. & Abaturov, B. D. (1997): Silicon forms in plants and their quantitative determination (in Russian). Uspekhi sovremennoĭ biologii 117: 534–548.

Kolesnikov, M. P. & Gins, V. K. (2001): Forms of silicon in medicinal plants. Applied Biochemistry and Microbiology 37: 524–527.

Kornfelt, L. F., Nørgaard, P. & Weisbjerg, M. R. (2013a): Effect of harvest time of red and white clover silage on chewing activity and particle size distribution in boli, rumen content and faeces in cows. Animal 7: 909–919.

Kornfelt, L. F., Weisbjerg, M. R. & Nørgaard, P. (2013b): Effect of harvest time and physical form of alfalfa silage on chewing time and particle size distribution in boli, rumen content and faeces. Animal 7: 232–244.

Kovács, P. L., Südekum, K.-H. & Stangassinger, M. (1997a): Rumen contents and ruminal and faecal particle size distribution in steers fed a mixed diet at three amounts of intake. Animal Feed Science and Technology 64: 143–154.

Kovács, P. L., Südekum, K. H. & Stangassinger, M. (1997b): Effects of intake level of a mixed diet on chewing activity and on particle size of ruminated boli, ruminal digesta fractions and faeces of steers. Reproduction, Nutrition, Development 37: 517–528.

Kovács, P. L., Südekum, K.-H. & Stangassinger, M. (1998): Effects of intake of a mixed diet and time postfeeding on amount and fibre composition of ruminal and faecal particles and on digesta passage from the reticulo-rumen of steers. Animal Feed Science and Technology 71: 325–340.

Lanyon, J. M. & Sanson, G. D. (2006): Mechanical disruption of seagrass in the digestive tract of the dugong. Journal of Zoology 270: 277–289.

Laredo, M. A. & Minson, D. J. (1973): The voluntary intake, digestibility, and retention time by sheep of leaf and stem fractions of five grasses. Australian Journal of Agricultural Research 24: 875–888.

Lechner, I., Barboza, P., Collins, W., Fritz, J., Günther, D., Hattendorf, B., Hummel, J., Südekum, K.-H. & Clauss, M. (2010): Differential passage of fluids and different-sized particles in fistulated oxen (*Bos primigenius* f. *taurus*), muskoxen (*Ovibos moschatus*), reindeer (*Rangifer tarandus*) and moose (*Alces alces*): Rumen particle size discrimination is independent from contents stratification. Comparative Biochemistry and Physiology A 155: 211–222.

Lechner-Doll, M. (1986): Selektive Retention von Futterpartikeln verschiedener Größe im Magen-Darmkanal von Kamelen im Vergleich mit Rindern und Schafen im Sudan. Dissertation Tierärztliche Hochschule Hannover.

Lentle, R. G., Hume, I. D., Stafford, K. J., Kennedy, M., Springett, B. P. & Haslett, S. (2003): Observations on fresh forage intake, ingesta particle size and nutrient digestibility in four species of macropod. Australian Journal of Zoology 51: 627-636.

Luginbuhl, J. M., Fisher, D. S., Pond, K. R. & RBurns, J. C. (1991): Image analysis and nonlinear modeling tp determine dimensions of wet-sieved, masticated forage particles. Journal of Animal Science 69: 3807–3816.

Luginbuhl, J. M., Pond, K. R. & Burns, J. C. (1990): Changes in ruminal and faecal particle weight distribution of steers fed coastal bermudagrass hay at four levels. Journal of Animal Science 68: 2864–2873.

Martin, L. F., Winkler, D., Tütken, T., Codron, D., De Cuyper, A., Hatt, J.-M. & Clauss, M. (2019): The way wear goes – phytolith-based wear on the dentine-enamel system in guinea pigs (*Cavia porcellus*). Proceedings of the Royal Society B 286: 20191921.

McLeod, M. N. & Minson, D. J. (1988): Large particle breakdown by cattle eating ryegrass and alfalfa. Journal of Animal Science 66: 992–999.

Merceron, G., Ramdarshan, A., Blondel, C., Boisserie, J. R., Brunetiere, N., Francisco, A., Gautier, D., Milhet, X., Novello, A. & Pret, D. (2016): Untangling the environmental from the dietary: dust does not matter. Proceedings of the Royal Society B 283: 20161032.

Mertens, D. R. (1987): Predicting intake and digestibility using mathematical models of ruminal function. Journal of Animal Science 64: 1548–1558.

Moore, S. J. (1999): Food breakdown in an avian herbivore: who needs teeth? Australian Journal of Zoology 47: 625–632.

Müller, J., Clauss, M., Codron, D., Schulz, E., Hummel, J., Fortelius, M., Kircher, P. & Hatt, J.-M. (2014): Growth and wear of incisor and cheek teeth in domestic rabbits (*Oryctolagus cuniculus*) fed diets of different abrasiveness. Journal of Experimental Zoology A 321: 283–298.

Murphy, M. R. & Nicoletti, J. M. (1984): Potential reduction of forage and rumen digesta particle-size by microbial action. Journal of Dairy Science 67: 1221–1226.

Nørgaard, P. (2006): Use of image analysis for measuring particle size in feed, digesta and faeces. In: Sejrsen, K., Hvelplund, T., & Nielsen, M. O. (eds.). Ruminant Physiology Digestion, Metabolism and Impact of Nutrition on Gene Expression, Immunology and Stress. Wageningen Academic Press, Wageningen: 579–585.

Nørgaard, P., Nadeau, E. & Randby, A. T. (2010): A new Nordic structure evaluation system for diets fed to dairy cows: a meta analysis. In: Sauvant, D., Van Milgen, J., Faverdin, P., & Figgens, N. (eds.). Modelling Nutrient Digestion and Utilisation in Farm Animals. Wageningen Academic Publishers, Wageningen: 112–120.

Norman, D. A. & Weishampel, D. B. (1985): Ornithopod feeding mechanisms: Their bearing on the evolution of herbivory. American Naturalist 126: 151–164.

Okine, E. K. & Matthison, G. W. (1991): Effects of feed intake on particle distribution, passage of digesta, and extent of digestion in the gastrointestinal tract of cattle. Journal of Animal Science 69: 3435–3445.

Osuji, P. O., Gordon, J. G. & Webster, A. J. F. (1975): Energy exchanges associated with eating and rumination in sheep given grass diets of different physical forms. British Journal of Nutrition 34: 59–71.

Paul, C. (1984): Möglichkeiten und Grenzen des Einsatzes der Mahlwiderstandsmessung zur Schätzung von Futterwertparametern an Futterpflanzen. Landwirtschaftliche Forschung 37: 127–132.

Paul, C. & Mika, V. (1981): Mahlwiderstandsmessungen an Rauhfutter. II. Beziehungen zwischen Mahlwiderstand und Futterwertmerkmalen. Landbauforschung Völkenrode 31: 163–169.

Paul, C. & Schild, G.-J. (1982): Mahlwiderstandsmessungen an Rauhfutter. III. Tischrechnergesteuerte Meßwertaufnahme über Digitalwattmeter. Landbauforschung Völkenrode 32: 169–172.

Paul, C., Schild, G.-J. & Honig, H. (1981): Mahlwiderstandsmessungen an Rauhfutter. I. Apparative Voraussetzungen. Landbauforschung Völkenrode 31: 7–10.

Péréz-Barbería, F. J. & Gordon, I. J. (1998): Factors affecting food comminution during chewing in ruminants: A review. Biological Journal of the Linnean Society 63: 233–256.

Piatkowski, B., Gürtler, H. & Voigt, J. (1990): Grundzüge der Wiederkäuerernährung. Gustav Fischer Verlag, Jena.

Piatkowski, B., Nagel, S. & Bergner, E. (1977): Das Wiederkauverhalten von Kühen bei unterschiedlicher Trockensubstanzaufnahme und verschiedener physikalischer Form von Grasheu. Archives of Animal Nutrition 27: 563–569.

Poppi, D. P., Minson, D. J. & Ternouth, J. H. (1980): Studies of cattle and sheep eating leaf and stem fractions of grasses. I.The voluntary intake, digestibility and retention time in the reticulo-rumen. Australian Journal of Agricultural Research 32: 99–108.

Porter, P. A. (1987): The acid detergent insoluble ash digestibility marker and its use in lactating dairy cows. PhD dissertation, Cornell University, Ithaca, New York.

Porzig, E. & Sambraus, H. H. (1991): Nahrungsaufnahmeverhalten landwirtschaftlicher Nutztiere. Deutscher Landwirtschaftsverlag, Berlin.

Radinsky, L. (1984): Ontogeny and phylogeny in horse skull evolution. Evolution 38: 1–15.

Sanson, G. (2006): The biomechanics of browsing and grazing. American Journal of Botany 93: 1531–1545.

Sanson, G. D., Read, J., Aranwela, N., Clissold, F. & Peeters, P. (2001): Measurement of leaf biomechanical properties in studies of herbivory: Opportunities, problems and procedures. Austral Ecology 26: 535–546.

Schüßler, D., Gürtler, W.-D., Keyser, S. & Greven, H. (2017): Einflüsse des Nahrungsangebotes auf Aktivitätsbudgets und mittlere Partikelgrößen im Kot von Rothschildgiraffen (*Giraffa camelopardalis rothschildi*, Lydekker 1903) in der ZOOM Erlebniswelt Gelsenkirchen. Der Zoologische Garten 86: 167–175.

Schwarm, A., Ortmann, S., Wolf, C., Streich, W. J. & Clauss, M. (2009): Passage marker excretion in red kangoroo (*Macropus rufus*), collared peccary (*Pecari tajacu*) and colobine monkeys (*Colobus angolensis*, *C. polykomos*, *Trachypithecus johnii*). Journal of Experimental Zoology A 311: 647–661.

Spalinger, D. E., Robbins, C. T. & Hanley, T. A. (1986): The assessment of handling time in ruminants: the effect of plant chemical and physical structure on the rate of breakdown of plant particles in the rumen of mule deer and elk. Canadian Journal of Zoology 64: 312–321.

*Steuer, P., Clauss, M., Südekum, K.-H., Hatt, J.-M., Silinski, S., Klomburg, S., Zimmermann, W., Fickel, J., Streich, W. J. & Hummel, J. (2010): Comparative investigations on digestion in grazing (*Ceratotherium simum*) and browsing (*Diceros bicornis*) rhinoceroses. Comparative Biochemistry and Physiology A 156: 380–388.

Stevens, C. E. & Hume, I. A. (1995): Comparative Physiology of the Vertebrate Digestive System. Cambridge University Press, Cambridge.

Susenbeth, A., Dickel, T., Südekum, K.-H., Drochner, W. & Steingaß, H. (2004): Energy requirements of cattle for standing and for ingestion, estimated by a ruminal emptying technique. Journal of Animal Science 82: 129–136.

Susenbeth, A., Mayer, R., Koehler, B. & Neumann, O. (1998): Energy requirement for eating in cattle. Journal of Animal Science 76: 2701–2705.

Susenbeth, A., Mayer, R. & Lang, G. U. (1997): The effect of reducing particle size of wheat straw on feed intake, digestibility and energy requirement for chewing in cattle. Zeitschrift für Ernährungswissenschaft 36: 328–329.

Suzuki, T., Kamiya, Y., Tanaka, M., Hattori, I., Sakaigaichi, T., Terauchi, T., Nonaka, I. & Terada, F. (2014): Effect of fiber content of roughage on energy cost of eating and rumination in Holstein cows. Animal Feed Science and Technology 196: 42–49.

Suzuki, T., Takusari, N., Higuchi, K., Kurihara, M. & Terada, F. (2008): Energy expenditure for chewing in sheep fed timothy or sudangrass hay at the same intake level. Animal Science Journal 79: 590–596.

Troelsen, J. E. & Campbell, J. B. (1968): Voluntary consumption of forage by sheep and its relation to the size and shape of particles in the digestive tract. Animal Production 10: 289–296.

Udén, P. (1978): Comparative studies on rat of passage, particle size and rate of digestion in ruminants, equines, rabbits and man. PhD dissertation Cornell University, Ithaca.

Udén, P. & Van Soest, P. J. (1982): The determination of digesta particle size in some herbivores. Animal Feed Science and Technology 7: 35–44.

Ungar, P. S. (2010): Mammal Teeth – Origin, Evolution, Diversity. The Hopkins University Press, Baltimore.

Van Soest, P. J. (1994): Nutritional Ecology of the Ruminant. 2nd ed. Cornell University Press, Ithaca.

Van Soest, P. J., Robertson, J. B. & Lewis, B. A. (1991): Methods for dietary fiber, neutral detergent fiber, and nonstarch polysaccharides in relation to animal nutrition. Journal of Dairy Science 74: 3583–3597.

Vernet, J., Vermorel, M. & Martin-Rosset, W. (1995): Energy cost of eating long hay, straw and pelleted food in sport horses. Animal Science 61: 581–588.

Weimer, P. J. (1996): Why don't ruminal bacteria digest cellulose faster? Journal of Dairy Science 79: 1496–1502.

Weishampel, D. B. & Jianu, C.-M. (2000): Plant-eaters and ghost lineages: dinosaurian herbivory. In: Sues, H.-D. (ed.). Evolution of Herbivory in Terrestrial Vertebrates. Cambridge University Press, Cambridge: 123–143.

Welch, J. G. & Smith, A. M. (1969): Effect of varying amounts of forage intake on rumination. Journal of Animal Science 28: 827–830.

Welch, J. G. (1982): Rumination, particle size and passage from the rumen. Journal of Animal Science 54: 885–894.

Williams, S. H. & Kay, R. F. (2001): A comparative test of adaptive explanations for hypsodonty in ungulates and rodents. Journal of Mammalian Evolution 8: 207–229.

Wilson, G. F., Adeeb, N. N. & Campling, R. C. (1973): The apparent digestibility of maize grain when given in various physical forms to adult sheep and cattle. Journal of Agricultural Science 80: 259–267.

Case studies on functional aspects and constraints in early and late tooth ontogeny

Irina Ruf, Anne M. Schubert, and Wighart v. Koenigswald

Introduction

Despite the immense diversity of mammalian dentitions in morphology and function, the great majority of teeth share the same basic elements in their ontogeny. The term "ontogeny" covers the entire life history from the fertilized egg to the individual's death (Haeckel 1866). Haeckel applied the term to individuals as well as individual organs. We differentiate two periods in the life history of teeth: an early ontogeny when teeth are formed deep in the crypt, and a late ontogeny when teeth are erupted, and exposed to wear. These phases are deduced from the tooth morphology.

The mineralization of tooth material starts at the cusps of the occlusal surface and proceeds continuously towards the base of the crown (Alt & Türp 1997). This orientation is elementary to understand the ontogenetic constraints of complex teeth. Similar to the uniform formation of teeth the subsequent life history passes through similar stages during the wear process, when teeth are incorporated into the tooth row and start with their function e.g., comminution of food objects by cheek teeth. This function continues until the deciduous tooth is replaced by a permanent one, or the tooth is totally worn off, or the animal dies. During late ontogeny wear causes loss of material and gradual destruction of the tooth. This basic pattern of ontogeny occurs in most teeth although modified in various ways during ontogeny. Heterochronies are the most common way to alter this general pattern as demonstrated by the ontogenetic phases and types of hypsodonty (high-crowned teeth) in herbivores (see Koenigswald 2020, this volume). Only few teeth do not share this life history, especially those that are in specific tooth positions. For instance, proboscidean tusks are one of the rare exceptions, where wear is minimal and the continuously growing dentin tooth documents the entire life history of the animal (Fisher et al. 2014).

Herbivorous mammals have to cope with intensive abrasion of tooth material due to phytoliths and grit intake and have evolved certain adaptations in their dentition; however, the extent to which phytoliths or grit contribute to tooth abrasion is still discussed (e.g., Fortelius 1985, Williams & Kay 2001, Strömberg 2006, Mendoza & Palmqvist 2008, Hummel et al. 2011, Merceron et al. 2016, Martin et al. 2019). Some herbivorous species that retain brachydont (low-crowned) cheek teeth show a complex occlusal surface adapted to comminution and abrasion. Another strategy to cope with high abrasion is the evolution of hypsodont or euhypsodont (ever-growing) teeth with more or less complex occlusal surface e.g., rodent incisors, equid and rodent check teeth; in high-crowned teeth growth and abrasion are balanced (Thenius 1989, Schmidt-Kittler 2002, Koenigswald 2016, Martin et al. 2019). Variations in enamel thickness in the tooth crown can be associated with functional adaptations. For instance, in *Equus quagga* thinner coronal enamel areas allow fast exposition of enamel ridges, whereas in primates the relative thick enamel may provide a greater resistance to abrasion (Janis & Fortelius 1988, Winkler & Kaiser 2015).

Function and performance enhancement in mammalian dentitions, the major task of the DFG Research Unit 771, can only be elucidated by a deeper understanding of dental adaptations and constraints in extant as well as fossil species. One major key to this is a deeper understanding of dental ontogeny – from early development of dental tissues via timing of tooth eruption and replacement to modifications of the occlusal surface caused by wear. In this chapter, first the general framework of tooth ontogeny is outlined followed by the presentation of three specific aspects of dental ontogeny and function that represent different approaches. Each topic is exemplified by a case study that originated in the DFG Research Unit 771. The first case study deals with the enamel islets in rodent molars, a specialization of the occlusal surface formed by deep indentations that resemble fingers of a glove, and the limitations to the way they are formed. The second case study discusses perinatal tooth eruption in precocial versus altricial Rodentia and Lagomorpha. The third case study specifies individual wear stages in placental mammals that allow comparison of ontogenetic stages across species boundaries. Thus, the presented research topics concern individual teeth as well as the functional unit of the entire postcanine dentition including the morphological and functional changes of the occlusal structures during wear and life.

Ontogeny of individual teeth

In the last two decades the investigation of early development and genetic control of teeth increased our knowledge of dental patterns and evolution significantly (e.g., Aberg et al. 1997, Jernvall et al. 2000, Jernvall & Thesleff 2000, Tummers & Thesleff 2003, Harada & Ohshima 2004, Kavanagh et al. 2007, Harjunmaa et al. 2014). As the process of tooth formation and tissue genesis (amelogenesis, dentinogenesis, cementogenesis) as well as the morphology

T. Martin & W. v. Koenigswald (eds.): Mammalian Teeth – Form and Function. Pp. 102–124, 18 figs.
 – ISBN 978-3-89937-266-3
DOI: http://doi.org/10.23788/mammteeth.07 Published 22 December 2020

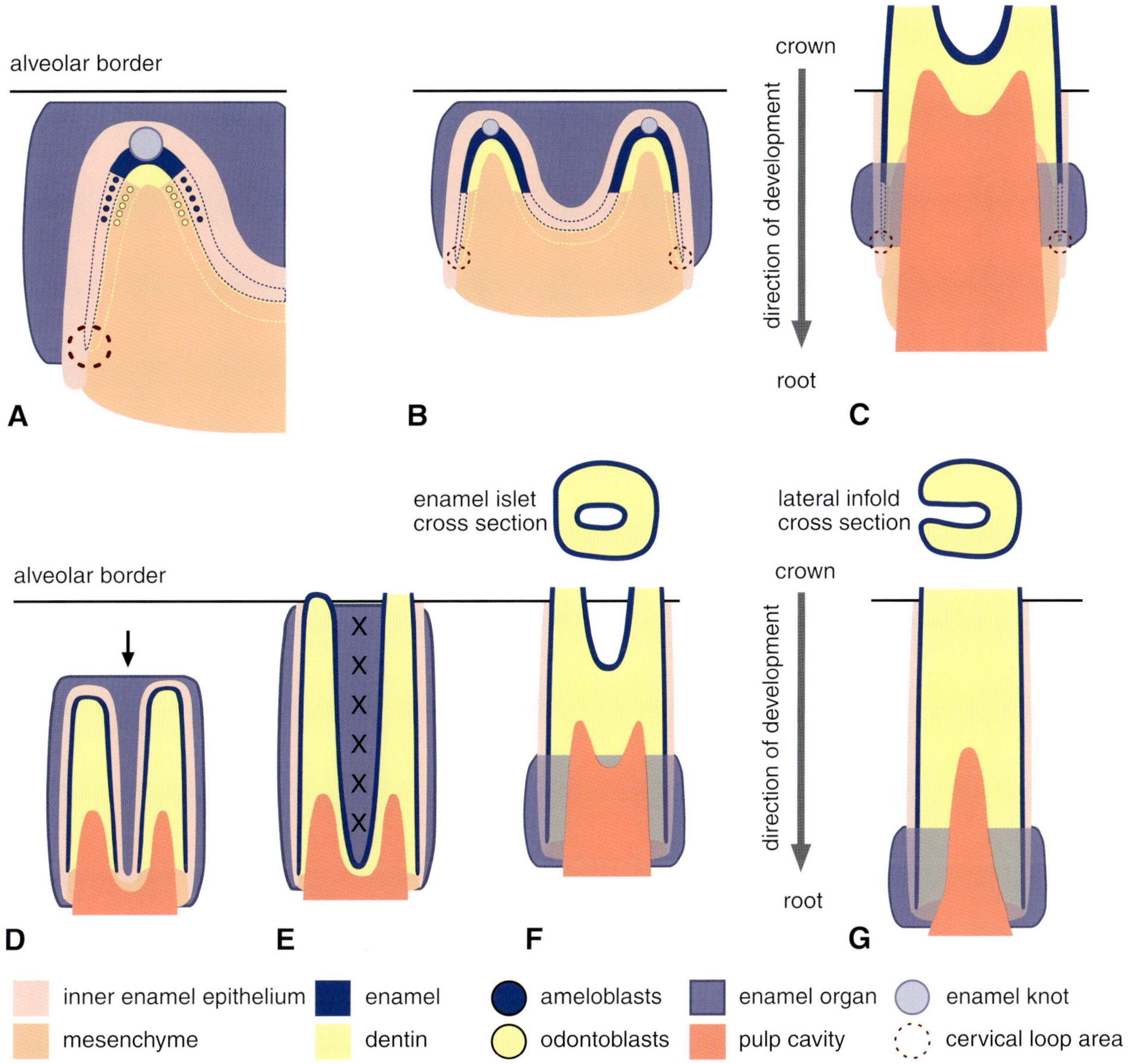

Fig. 7.1. Interdependence between tooth formation, enamel islets, lateral infolds, and abrasion. **A,** detailed aspect of a developing tooth: the enamel is formed in the crypt by the inner epithelium, whereas the dentin is formed by the mesenchyme. **B,** tooth formation and mineralization start at the top of the tooth and move continuously in apical direction. **C,** in hypsodont or euhypsodont teeth the enamel organ (cervical loop area) is still active after eruption and can add outer enamel. **D,** several hypsodont teeth have deep depressions of enamel from the coronal to the apical part of the tooth (small arrow). These are formed simultaneously with the external walls of the crown by a part of the enamel organ that sticks into the depressions. **E,** when the tooth erupts the connection between the surrounding enamel organ and the part in the depression gets lost and the enamel formation within the depression is stopped (X). **F,** the former depression becomes an enamel islet by abrasion. **G,** the connection between the enamel organ and the part in the depression is not necessarily interrupted if the depression is open to one side (lateral infold). As long as a connection to the enamel organ is preserved – even through a narrow slit – the enamel formation and the deepening of the lateral infold can continue.

of mammalian teeth is presented in many textbooks and publications in detail (e.g., Teaford et al. 2009, Ungar 2010), we selected those structures and events that are related to the presented case studies for further discussion. In order to evaluate the complex tooth morphology of herbivorous mammals, that is often characterized by enamel islets and infoldings as well as hypsodonty or euhypsodonty, a deeper understanding of the early tooth formation is demanded.

The tooth morphogenesis from bud stage to bell stage comprises the development of the dental follicle, enamel organ, and dental papilla. The ameloblasts, cells of the enamel organ, develop in the inner enamel epithelium from oral epithelial cells and produce the enamel matrix along the basal membrane, the later enamel-dentin junction (EDJ); the ameloblasts induce the maturation of the enamel matrix to form enamel. This is triggered by enamel knots (epithelial signaling centers) along the inner enamel epithelium that determine the formation of later cusps; mechanical forces and constraints during growth of the cell layers may also affect tooth morphogenesis. In parallel, the odontoblasts, that are derived from ectomesenchymal cells, are creating dentin along the EDJ inside the dental follicle (Fig. 7.1 A, B). The complex process of tooth formation is controlled by so-called bone morphogenetic proteins (BMPs) that are multi-functional growth factors playing a crucial role during the entire ontogeny; BMPs are not restricted to the development of teeth but are also involved in the development of many further structures and organs e.g., cartilage, bone, heart, neural structures, and ectodermal appendages. Especially BMP7 acts as a central mediator of epithelial-mesenchymal interactions in tooth development. BMP signal cascades as well as expression of antagonists are involved in several stages of the entire tooth development including the development of enamel, dentin, roots, pulp cavity, and periodontal ligaments. For instance, in later stages BMP expression in the tooth epithelium induces enamel formation whereas in enamel-free areas antagonistic genes are expressed (Aberg et al. 1997, Chen et al. 2004, Osborn 2008, Zouvelou et al. 2009, Graf & Mitsiadis 2013). Msx2 from the muscle segment homeobox gene family plays a major role in enamel formation by controlling proliferation and differentiation of ameloblasts. This gene regulates laminin5α3 expression in enamel knots and therefore is also involved in cusp formation (Bei et al. 2004).

The notch signaling pathway genes FGF 10 control the crown formation and the transition from the crown to the root. The formation of roots is delayed in hypsodont teeth and skipped totally in euhypsodont teeth (Tummers & Thesleff 2003). Absence of this signaling induces growth stop of the tooth and the development of roots. At the cellular level the area where inner and outer enamel epithelium join, the so-called cervical loop, is responsible for apical growth of the tooth crown (Fig. 7.1A–C). In brachydont teeth the stellate reticulum of the cervical loop is lost and the respective tissue switches to root fate. In contrast, in hypsodont and euhypsodont teeth the production of ameloblasts from the stellate reticulum of the cervical loop is maintained and thus allows prolonged or even permanent growth of the tooth (Tummers & Thesleff 2003). Ontogenetically, the evolution of hypsodont and euhypsodont teeth is regarded to be a heterochronic effect (see Koenigswald 1982, 2020, this volume).

The differentiation and mineralization of the tooth occurs from the tip of the cusps to the roots, and after eruption the tooth becomes part of the functional unit of the tooth row. Therefore, the formation of specific occlusal structures is limited as they need to be completely mineralized before tooth eruption and consequent interruption of amelogenesis as well as dentinogenesis after root formation (Fig. 7.1C, E) (Janis & Fortelius 1988, Koenigswald 2011). However, heterochronic events can compensate these restrictions (see Koenigswald 2020, this volume). Generally, occlusal morphology is determined by the shape of the enamel epithelium and mesenchyme that show specific folding during tooth development, and activators and inhibitors within these tissues control the cusp size and tooth size; the number of cusps is also co-regulated by the tooth size and cusp size as demonstrated by a tooth patterning study in rats and mice (Cai et al. 2007). A recent study on *Myodes glareolus* demonstrates that mechanical constraints from the growing jaw bones also influence molar tooth morphogenesis (Renvoisé et al. 2017). From an evolutionary perspective cusp size and number can be also explained by linking tooth development and population-level variation as a patterning cascade mode of cusp development as proposed for the postcanine dentition of seals (Jernvall 2000).

On the genetic level, expression patterns are obviously correlated with specific tooth phenotypes and heterotopic and iterative processes may be major drivers of the diversity of mammalian molar patterns. Asymmetric BMP expression affects the shape of the tooth itself and patterns of Fgf4, Lef1, p21, and Shh genes are correlated with lateral cusp topography (Aberg et al. 1997, Jernvall et al. 2000, Chen et al. 2004, Zouvelou et al. 2009, Graf & Mitsiadis 2013).

Rasp facets in herbivore dentitions

The above described developmental patterns impose limitations for adaptations in the teeth of herbivores. The physical disaggregation of tough plant material by mastication requires enamel ridges working as shearing edges on the occlusal surface of the postcanine dentition that Koenigswald (2016) defined as rasp facets. These grinding facets have evolved independently from various different tooth morphologies in ancestral forms. Several herbivorous taxa have evolved hypsodont or even euhypsodont teeth e.g., various rodent families, Rhinocerotidae, Equidae, Proboscidea and selenodont Artiodactyla such as Cervidae and Bovidae (Thenius 1989, Koenigswald 2016). Most of these dentitions are characterized by rasp facets that are flat grinding surfaces with projecting enamel crests between dentin fields. To achieve such a fully-functioning occlusal surface these teeth need to be exposed to wear in order to remove the covering enamel of the primary occlusal surface and sharpen the enamel crests in the secondary occlusal surface. The dentin between the enamel structures is widely exposed and gaps may be filled with cementum. The simplest pattern comprises an enamel crest that is formed by the surrounding enamel band only e.g., in cylindrodont rodents. This basic pattern may be optimized by additional enamel crests that occur as lateral infolds (flexus, flexids) or enamel islets (fossettes, fossettids). They increase the mechanical performance of the rasp-facet (Koenigswald 2016). A combination of enamel islets and lateral folds is common because many lateral infolds are transformed sooner or later into islets when wear continues (e.g., Koenigswald et al. 1999 and see below). Examples of such complex teeth are manyfold in Rodentia, Rhinocerotidae, Equidae, and selenodont ruminants (Fig. 7.2) (Rensberger 1973, Janis & Fortelius 1988, Thenius 1989, Koenigswald et al. 1994).

A specific ontogenetic limitation occurs when during evolution teeth become hypsodont or euhypsodont. The lower part of the erupting tooth may laterally still be covered with enamel by the surrounding enamel organ while the upper penetrates the gingiva and is already functioning (Fig. 7.1C–G). At this stage lateral infolds can be continuously formed as the cervical loop is still active. Enamel

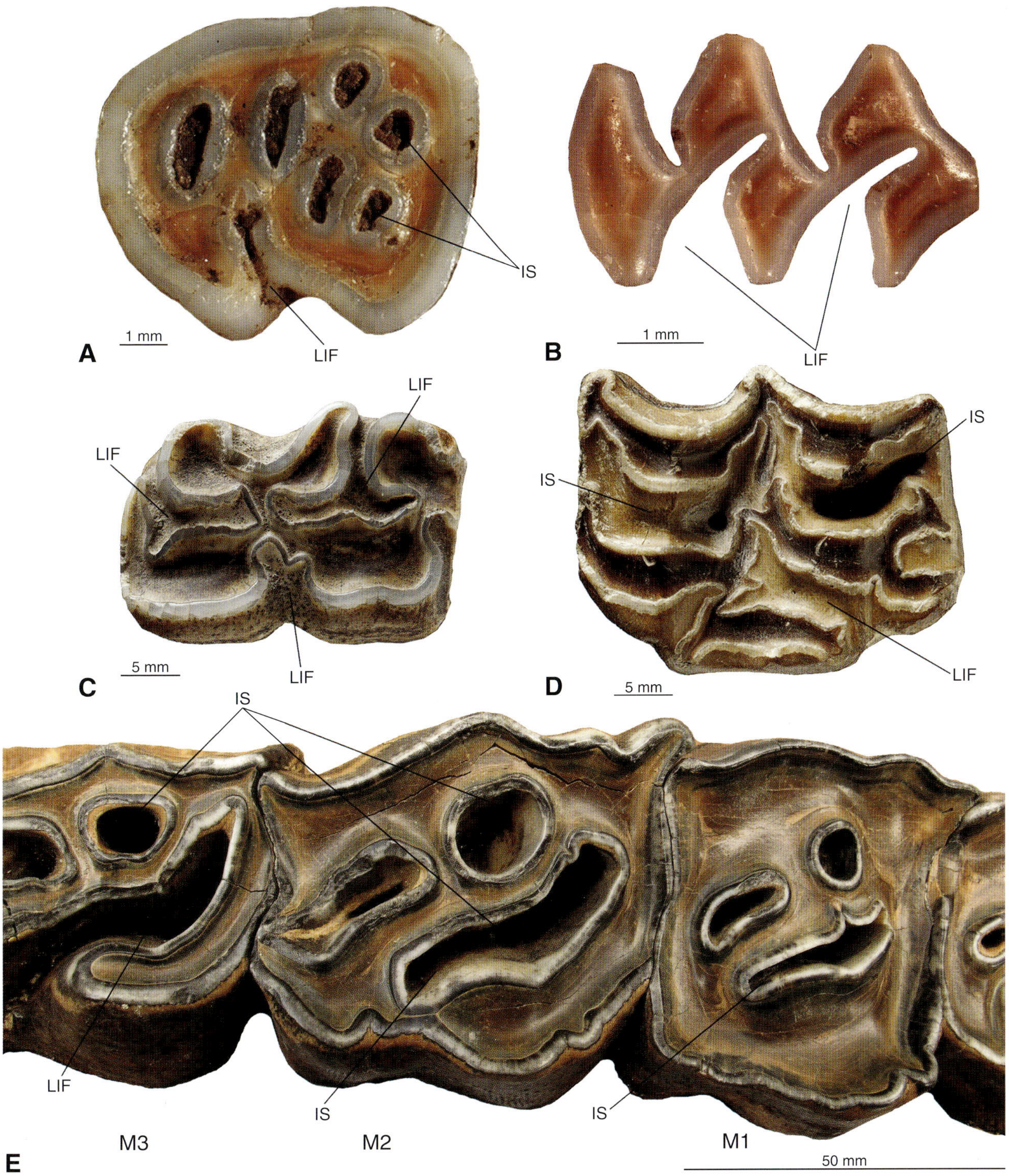

Fig. 7.2. Enamel islets and lateral infolds in selected extant and fossil species. **A,** enamel islets in the P4 of *Hystrix cristata* (Rodentia), Recent, Mt. Elgin, Kenya (IGPB KOE 25); **B,** lateral infolds in the M1 of *Ondatra zibethicus* (Rodentia), Recent (IGPB, no number); **C,** lower molar of *Equus caballus* (Perissodactyla) with lateral infolds, subfossil, Rhineland (IGPB, no number); **D,** upper molar of *Equus caballus* with enamel islets, subfossil, Rhineland (IGPB, no number); **E,** upper molars in the Late Pleistocene *Coelodonta antiquitatis* (Perissodactyla), Groß Rohrheim, Germany (SLGM, no number), showing how the lateral infold (median sulcus, still present in M3) forms an enamel islet (mediofosset, present in M2 and M1) due to the progressive wear. Abbreviations: IGPB, Institute of Geosciences, Section Paleontology, Universität Bonn; IS, enamel islet; LIF, lateral infold; M1–3, upper molar 1–3; SLGM, collection Frank Menger, Groß Rohrheim, Germany.

islets which originate from the occlusal surface, however, cannot be deepened and produced anymore, because the enamel epithelium responsible for the enamel islets is cut-off from tissue supply, as soon as the surrounding enamel is incorporated into the functional occlusal surface (Fig. 7.1 E, F). Thus, the growth of enamel islets is definitely restricted to the developmental phase before the tooth erupts (Janis & Fortelius 1988, Tummers & Thesleff 2003, Koenigswald 2011, Koenigswald 2020, this volume).

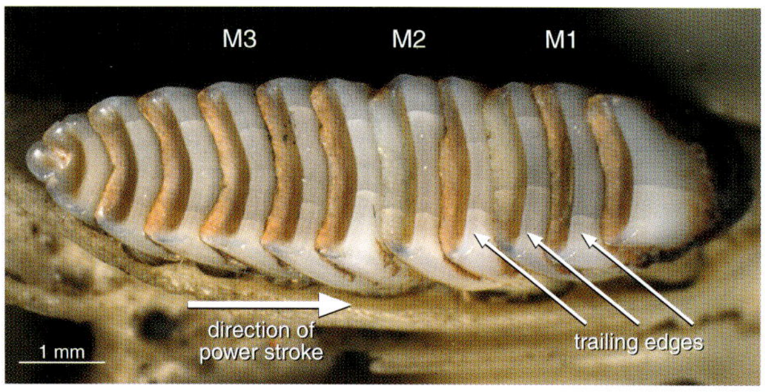

Fig. 7.3. Upper molars of the extant muroid rodent *Otomys* sp., Recent, Makapansgat, South Afrika (IGPB KOE 3355). The occlusal surface is made up by transverse enamel ridges on the trailing side, whereas those of the leading side are reduced asymmetrically. Abbreviations: IGPB, Institute of Geosciences, Section Paleontology, Rheinische Friedrich-Wilhelms-Universität Bonn; M1-3, upper molar 1-3.

Enamel islets can be found in brachydont and hypsodont teeth; in euhypsodont teeth they are restricted to juvenile stages. However, the height of hypsodont dentitions with enamel islets is limited to the space available in the jaw because these structures need to be mineralized within the crypt before tooth eruption (Koenigswald 2011). The upper and lower premolars and molars of *Equus caballus* are hypsodont and provide an interesting difference: lower cheek teeth have lateral infolds, uppers islets (Fig. 7.2C, D). Thus, the lower teeth would allow a transition to a euhypsodont architecture, but the upper teeth not. That might have stopped the increasing hypsodonty in horses. Rhinocerotidae provide another informative example as enamel islets are common in the upper molars as observed in *Coelodonta antiquitatis* (Fig. 7.2E). In *Ceratotherium simum* hypsodont molars evolved with deep lateral infolds. These folds end up as enamel islets when teeth are almost worn out. The Pleistocene *Elasmotherium* from Siberia has evolved euhypsodont molars (Brandt 1864). This was possible due to the lack of enamel islets in the flat grinding surface of its teeth, but the lateral infolds are highly complicated in order to increase the rasp function.

In contrast, enamel islets are rare in euhypsodont dentitions, and if present they are restricted to the coronal part of the teeth. In the molars of arvicolid rodents, the loss of enamel islets in combination with the accomplishment of euhypsodonty can be observed. In the hypsodont but still rooted molars of *Mimomys savini* an enamel islet is still present in m1, but in the euhypsodont m1 of *Arvicola* it is lost (Koenigswald 1982, Koenigswald & Kolfschoten 1996).

In grinding surfaces, the length of the enamel ridges is directly related to the efficiency of the mastication process, as a high amount of enamel ridges allows a faster disintegration of food particles with fewer power strokes during mastication (Rensberger 1973, Kay 1975). Teeth with hypsodont lophs show enamel ridges that are almost equally spaced either by dentin or cementum e.g., in Elephantidae, especially in *Mammuthus primigenius.* Even in teeth with a combination of islets and crests (e.g., *Equus*, *Bos*) the distances between the main functional enamel ridges tend to be similar. Too dense enamel ridges on the occlusal surface and thus a decrease in possible mastication force of the rasp facets would lead to insufficient disintegration of food particles (Schmidt-Kittler 1984). In these teeth the thickness of the enamel does not vary significantly. In some species, mainly in rodents, however, the thickness of the enamel band may be significantly reduced on one side of the lophs and lophids resulting in functional units that are correlated with the direction of jaw movement, the so-called leading and trailing edges (Greaves 1973). This asymmetry is mainly related to the position of the enamel at the leading or trailing side of the dentin, because it occurs especially in those teeth where the enamel crests are transverse to the chewing direction. This pattern evolved independently in several unrelated rodent taxa such as the archaeomyid *Archaeomys* from the Oligocene in Europe, the extant African murid *Otomys*, and fossil and extant South American caviomorphs like *Chinchilla* and *Lagostomus* (Koenigswald 1980, 2016, Rinderknecht et al. 2011, Ubilla & Rinderknecht 2016). In these teeth, the lophs are slightly inclined towards the chewing direction. Mostly, the leading side is reduced in thickness or totally missing (Fig. 7.3). The gap to the next loph is filled by cementum. As only part of the enamel band is functioning in the grinding process, it might reduce the effectiveness of the tooth. However, as the rhythm of the mastication is another component determining the efficiency of the teeth it could serve the compensation of effectiveness reduction. A similar phenomenon of an asymmetrical reduction of the enamel has been observed in the lingual infold of the upper molars of extant and fossil Lagomorpha by Koenigswald et al. (2010).

Case study 1: Ontogeny and function of enamel islets in rodents

The occlusal surface of a tooth represents only a snapshot in the ontogeny of the individual life history, because wear modifies the occlusal surface continuously. It may be assumed, that selection favors occlusion surface patterns that stay functional at least until the end of the reproductive lifespan.

In order to test how the occlusal surface changes due to progressive wear and how the enamel islets change with progressing wear, a major study was executed by one of the authors (Schubert 2019) in the DFG Research Unit 771. This study intended to clarify the basic formation and fate of enamel islets, using the great variety of rodent molars. The variability in various ontogenetic stages can be investigated and used for comparative morphological studies as well as morphofunctional analyses.

Rodents show a high diversity that is also reflected in their dentitions, and enamel islets evolved independently in several herbivorous species of the three major clades, the Ctenohystrica, mouse-related clade, and squirrel-related clade (see Thenius 1989 and Koenigswald 2016). Previous studies on these structures are based on the comparison of the occlusal surface of teeth in different wear stages (e.g., Hagmann 1907: *Cuniculus paca* and *Dasyprocta leporina*, Rusconi 1930: *Myocastor coypus*, van Weers 1990, 1993: Hystricidae, Calede & Hopkins 2012: Mylagaulidae) or on teeth that were artificially ground down (e.g., Rusconi 1934: *Cuniculus paca*, Hünermann 1966: *Castor* sp.). The former method allows an investigation of various wear stages in

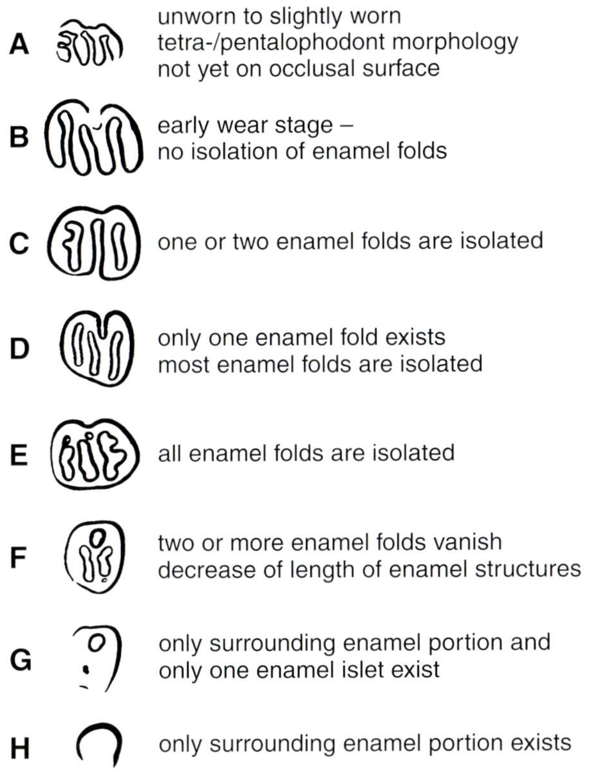

Fig. 7.4. Definition of general wear stages A–H based on rodent dentitions with enamel folds and islets. Not all wear stages (schematic sketches) can be found in each of the investigated species due to different morphologies. Modified from Schubert (2019).

one species, but does not decipher the entire ontogeny of an individual and species, respectively. The artificial grinding is a destructive method and not always applicable. Non-invasive high-resolution computed tomography (µCT) and virtual 3D reconstructions reveal the relationships of internal dental structures as shown in the Late Cretaceous gondwanatherian *Vintana sertichi* (Krause 2014). Furthermore this approach allows elucidation of future occlusal surfaces along the entire crown reflecting major parts of the life history in a single individual as already demonstrated for the fossil castorid *Steneofiber siamensis* (Suraprasit et al. 2011). Most former studies concentrated on one or very few closely related species in order to understand their variability but contributed only to a limited degree to the general understanding of tooth ontogeny and functional changes by wear.

The changes of enamel islets by progressive wear were investigated by µCT and virtual 3D reconstructions in ten extant and three fossil rodent species. The sample covers a wide range of the morphological diversity of enamel islet patterns encountered in Rodentia: extant Caviomorpha (*Cuniculus paca*, *Dasyprocta azarae*, *Dasyprocta leporina*, *Myocastor coypus*, *Erethizon dorsatum*), extant Old World Hystricognathi (*Thryonomys swinderianus*, *Hystrix cristata*, *Hystrix indica*), extant Castoridae (*Castor fiber*, *Castor canadensis*), fossil Castoridae (*Steneofiber eseri*, *Steneofiber depereti*), and fossil Aplodontidae (*Mylagaulus elassos*).

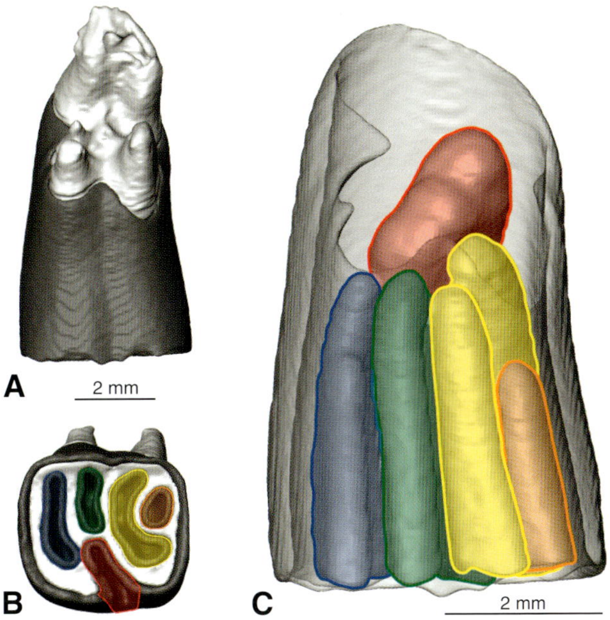

Fig. 7.5. 3D model and standardized nomenclature for enamel folds and islets as exemplified by the M2 of *Dasyprocta leporina*. **A, B,** 3D model of enamel (dark gray) and dentin (light gray); **C,** enamel portion only. Color code of enamel folds and islets: blue, enamel islet 1; green, enamel islet 2; yellow, enamel islet 3; orange, enamel islet 4; red, enamel fold V. **A** and **C** in buccal view, **B** in occlusal view.

Based on the 3D models virtual occlusal surfaces over the complete height of the enamel portion of the tooth were created. For the comparison of the generated occlusal surfaces several wear stages (A to H) were defined based on appearance and disappearance of folds and/or islets (Fig. 7.4). The traditional nomenclature for the enamel folds and islets was standardized as enamel elements are numbered from I–V (folds) and 1–5 (islets) from mesial to distal (Fig. 7.5).

Origin and splitting patterns of enamel islets

Some enamel islets occur in the occlusal surface immediately after a short period of wear, other islets are formed later. Based on hypsodont molars of the gondwanatherian *Sudamerica*, Koenigswald et al. (1999) differentiated three types of enamel islets. Type 1 comprises islets that originate from enamel invaginations from the occlusal surface. Type 2 refers to islets that derive from lateral infolds, which still exist in later wear stages. Type 3 comprises islets that are formed by complete transformation (isolation) of lateral infolds, mostly close to the base of the crown. In *Sudamerica* all three types occur in the same molar (Fig. 7.6). Enamel islets of type 1 were observed in *Mimomys*, a basal arvicoline rodent. Enamel islet type 3 occurs in many rodents (e.g., *Hystrix* and *Castor*) and in early lagomorphs such as *Palaeolagus* (Tobien 1974, 1975, Koenigswald et al. 1999, 2010). Schubert (2019) differentiated the enamel islets of type 2 in enamel islets originating from splitting of enamel folds (type 2a) and enamel islets developing from splitting of enamel islets (type 2b). Rodent molars often show a combination of enamel islets of different origins: in

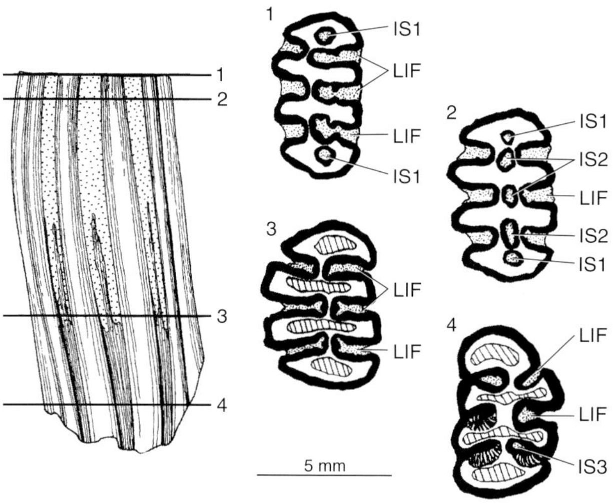

Fig. 7.6. The three types of enamel islets and their relation to lateral infolds. Cross sections (right) of a cheek tooth (left) of the Paleocene gondwanatherian *Sudamerica ameghinoi*. Enamel is black, dentin is white, the pulp cavity is hatched, and cementum is indicated by stippled areas. Abbreviations: IS1–3, type of enamel islets; LIF, lateral infold. Modified from Koenigswald et al. (1999).

the upper dentition of *Erethizon dorsatum* and *Thryonomys swinderianus* type 1 and type 3 occur together. In most investigated species enamel islets of type 2 (2a and/or 2b) and type 3 were found.

Splitting of enamel islets during further abrasion i.e., one enamel islet/fold splits into two islets, was postulated previously for e.g., *Hystrix* (van Weers 1990, 1993, 2005) and *Mylagaulus* (Black & Wood 1956, Kitts & Black 1959, Baskin 1980, Calede & Hopkins 2012), but their intra- and interspecific variability caused problems in the systematic evaluation. However, beside these variations the present study revealed also some stable patterns e.g., the sequence of isolation of the enamel folds. This sequence can be the same in all tooth positions as in the upper dentition of *Myocastor coypus* or the lower dentition of *Cuniculus paca* (Fig. 7.7). The splitting pattern of enamel islets is also comparable in all tooth positions of certain species e.g., in *Dasyprocta* or *Steneofiber* (Fig. 7.7B). In one specimen of *Myocastor coypus* under study and *Thryonomys swinderianus* no splitting of enamel elements was observed (Fig. 7.7C). In the dentition of both species of *Castor* only minor changes can be detected over most part of the crown height. Enamel folds get separated late in ontogeny in *Castor* and splitting of enamel islets occurs rarely (Fig. 7.7E), but older individuals can show crenulation of enamel structures (Hünermann 1966, Piechocki & Stiefel 1977, Mayhew 1979). This was also confirmed in later wear stages of both species by Schubert (2019).

In most investigated species the enamel elements split once or twice. Three or four splittings are very rare but occur in e.g., *Cuniculus paca* (M2 with four splittings and m1 with three splittings) or in *Mylagaulus elassos* (four splittings in one p4). If more than one splitting is present, they mostly occur from buccal to lingual in the upper and from lingual to buccal in the lower dentition. In contrast to the splitting of enamel islets, a fusion in apical direction was never observed as it is ontogenetically impossible.

A comparison of upper and lower cheek teeth reveals only slight differences in the morphology of splitting of enamel folds and islets in the specimens under study. However, more splittings of enamel elements can be found in the lower dentition (Fig. 7.7D). Since upper and lower dentitions do not differ in their function in rodents and thus their morphology often is similar (functional symmetry after Koenigswald et al. 1994), the observed differences cannot be related to function. This is also true for differences in left and right teeth, which do not differ in their function either.

In some investigated specimens (e.g., *Dasyprocta* and *Hystrix*) the enamel islets are completely closed apically but some of which are not fully mineralized in the apical part (Fig. 7.8). This may not be due to developmental constraints as not always the deepest part of the enamel islet is affected by incomplete mineralization. Instead it seems that the mineralization was stopped too early, leaving the apical part of the islet unmineralized.

Density and alignment of the enamel crests formed by enamel islets

Splitting of enamel elements is apparently a common feature among certain rodent clades and it is likely that this pattern is functionally constraint. However, there seems to be no correlation between the amount of splitting of enamel elements as well as the diet of a species or crown height of its teeth. This feature occurs in the frugivore *Cuniculus* as well as in the grazer *Mylagaulus*, in the relatively low crowned *Steneofiber* as well as in the relatively high crowned *Hystrix*.

As the above described morphological changes during ontogeny of the occlusal surface might have an impact on their functional performance, two functional parameters were tested. Functionality of the dentition can be quantified with the length of functional enamel ridges in relation to the area of the occlusal surface (functional structural density (D_e), Schmidt-Kittler 1984). A high D_e equals a high number of enamel ridges and thus a better functionality for communition as it allows a faster disintegration of food particles with fewer power strokes during mastication (Rensberger 1973). Additional enamel folds can therefore increase the functionality (Koenigswald 1980, 1982). However, too high values for D_e would mean too dense enamel ridges on the occlusal surface leading to a decrease in functionality (Schmidt-Kittler 1984). Another parameter for functionality is enamel ridge alignment (e.g., Herrmann 2002, Schmidt-Kittler 2002). The highest functionality in an evenly grinding tooth can be achieved by enamel ridges aligned mainly perpendicularly but slightly inclined to the chewing direction, leading to minimal contact area but increasing the duration of the contact of the shearing edges (Maglio 1972, Koenigswald 1980). Enamel ridges parallel to the chewing direction cannot disintegrate food particles and thus are functionless.

For the morphofunctional analyses of this study the generated occlusal surfaces were aligned in chewing direction and the functional length of the enamel ridges (all ridges not parallel to chewing direction) were measured. By this, changes in functionality by progressive wear can be detected within the investigated individuals (for details see Schubert 2019).

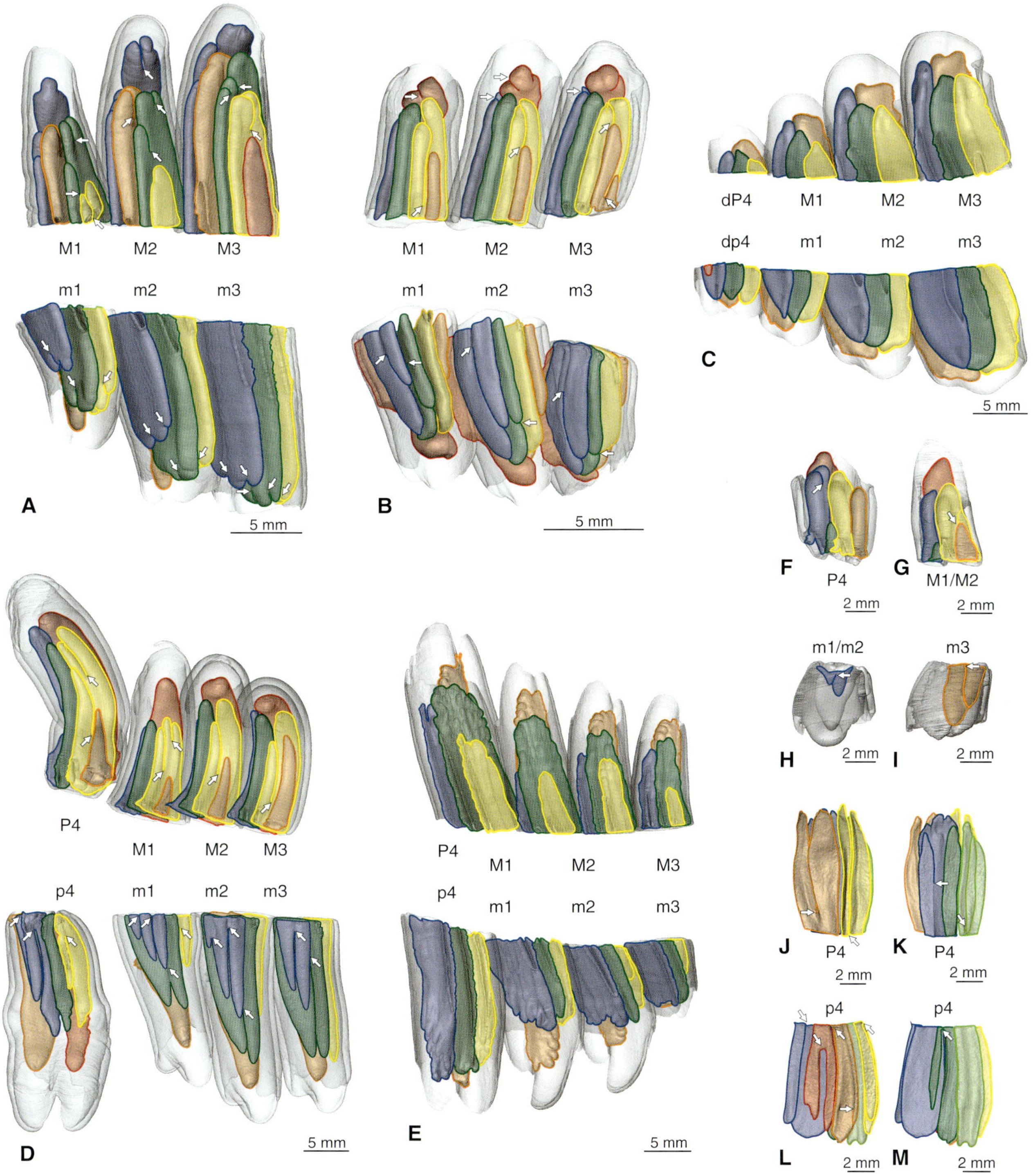

Fig. 7.7. Splitting pattern of enamel islets in upper and lower cheek teeth of selected extant and fossil rodent species. View through the transparent outer enamel portion onto colored internal enamel elements. Upper teeth in buccal view, lowers in lingual view. **A,** *Cuniculus paca*; **B,** *Dasyprocta azarae*; **C,** *Myocastor coypus*; **D,** *Hystrix indica*; **E,** *Castor fiber*; **F,** *Steneofiber eseri* (buccal view); **G,** *Steneofiber eseri* (buccal view, mirrored); **H,** *Steneofiber eseri* (mesial view); **I,** *Steneofiber eseri* (distal view); **J–K,** *Mylagaulus elassos* from lingual (**J**) and buccal (**K**, mirrored); **L–M,** *Mylagaulus elassos* from lingual (**L**) and buccal (**M**, mirrored). Color code A–I: blue, enamel fold I; green, enamel fold II; yellow, enamel fold III; orange, enamel fold IV; red, enamel fold V. Color code J–M: blue, enamel islet 1; dark green, enamel islet 2; light green, enamel islet 3; yellow, enamel islet 4; orange, enamel islet 5; red, enamel islet 6. Mesial to the left, in **H** and **I** buccal to the left. Arrows indicate splitting of enamel elements. Modified from Schubert (2019).

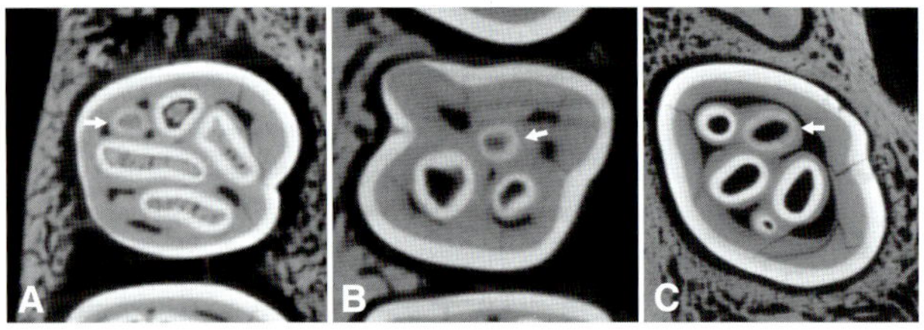

Fig. 7.8. Incomplete mineralization (arrows) deep inside the enamel islets of rodents as revealed by µCT images (cross sections). Thus, these structures show a less X-ray density and appear darker. **A,** *Cuniculus paca*, left M1. **B,** *Dasyprocta leporina*, left m1. **C,** *Hystrix cristata*, right m1. Not to scale. Modified from Schubert (2019).

In order to maintain a high functionality during the lifetime of the animal, D_e and the concentration of enamel ridge alignment should remain constant despite progressive wear. Most of the investigated species show similar values for D_e around 5–6, although they are characterized by slight differences in morphology of the occlusal surface as well as in their degree of hypsodonty. In many groups of rodents D_e decreases with increasing hypsodonty e.g., Aplodontidae, Ctenodactylidae and Heteromyidae. In euhypsodont taxa the rather fast abrasion due to the low amount of enamel ridges gets compensated with life-long growth of the dentition (Schmidt-Kittler 2002). In some of the investigated species D_e as well as the concentration of enamel ridge alignment decreases only in latest wear stages. Other species show decrease of D_e constantly with progressive wear and the enamel ridges are less concentrated (Figs. 7.9, 7.10). Though most of the investigated species show the same trend in upper and lower dentition, D_e and concentration of enamel ridge alignment are not necessarily correlated. Like in the upper dentition of *Dasyprocta azarae*, D_e can decrease with progressive wear but enamel ridge alignment stays more or less constant. In *Hystrix cristata* D_e of the upper dentition decreases constantly, whereas in the lower dentition D_e remains constant with progressive wear. Species with enamel folds and/or islets present into high age (late wear stages), as *Cuniculus paca* or *Castor*, show a more or less constant D_e with progressive wear. Only in late wear stages (E–H) enamel elements get lost and the remaining elements significantly reduce in length, leading to a decrease of D_e and thus reduction of functionality of the dentition. In species with continuous loss of enamel elements by progressive wear, as *Dasyprocta azarae* or *Hystrix indica*, D_e decreases continuously during ontogeny (Fig. 7.9).

Conclusions about functional compensation and dietary adaptations

Most of the investigated species possess a rather high functionality of their dentition during lifetime. In those species with continuously decreasing values for D_e and/or decreasing concentration of enamel ridge alignment eruption of M3/m3 and P4/p4 (if present) leads to com-

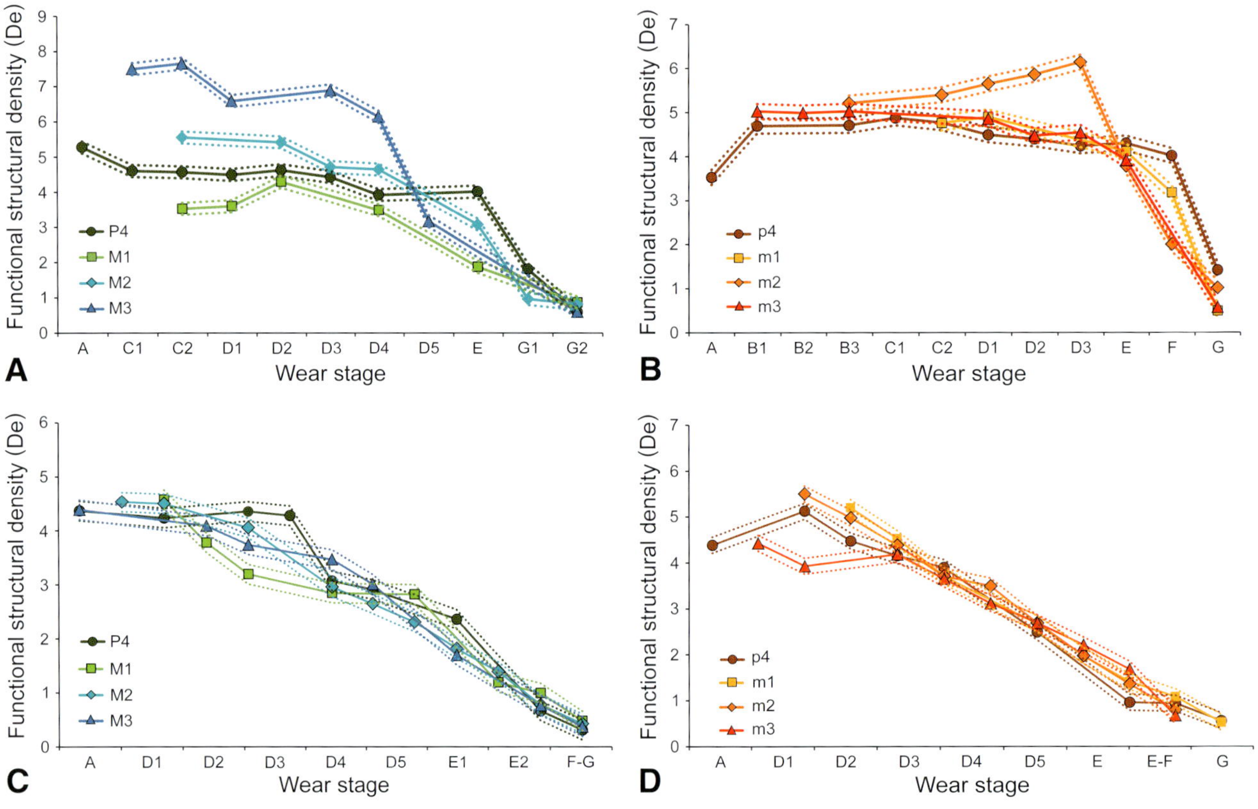

Fig. 7.9. Decrease of the functional structural density (D_e) with progressive wear in *Cuniculus paca* (**A, B**) and *Hystrix indica* (**C, D**). **A, C,** upper cheek teeth; **B, D,** lower cheek teeth. *Cuniculus paca* represents species with reduction of enamel islets only in late wear stages and *Hystrix indica* shows continuously reduction of enamel islets with progressive wear. From Schubert (2019).

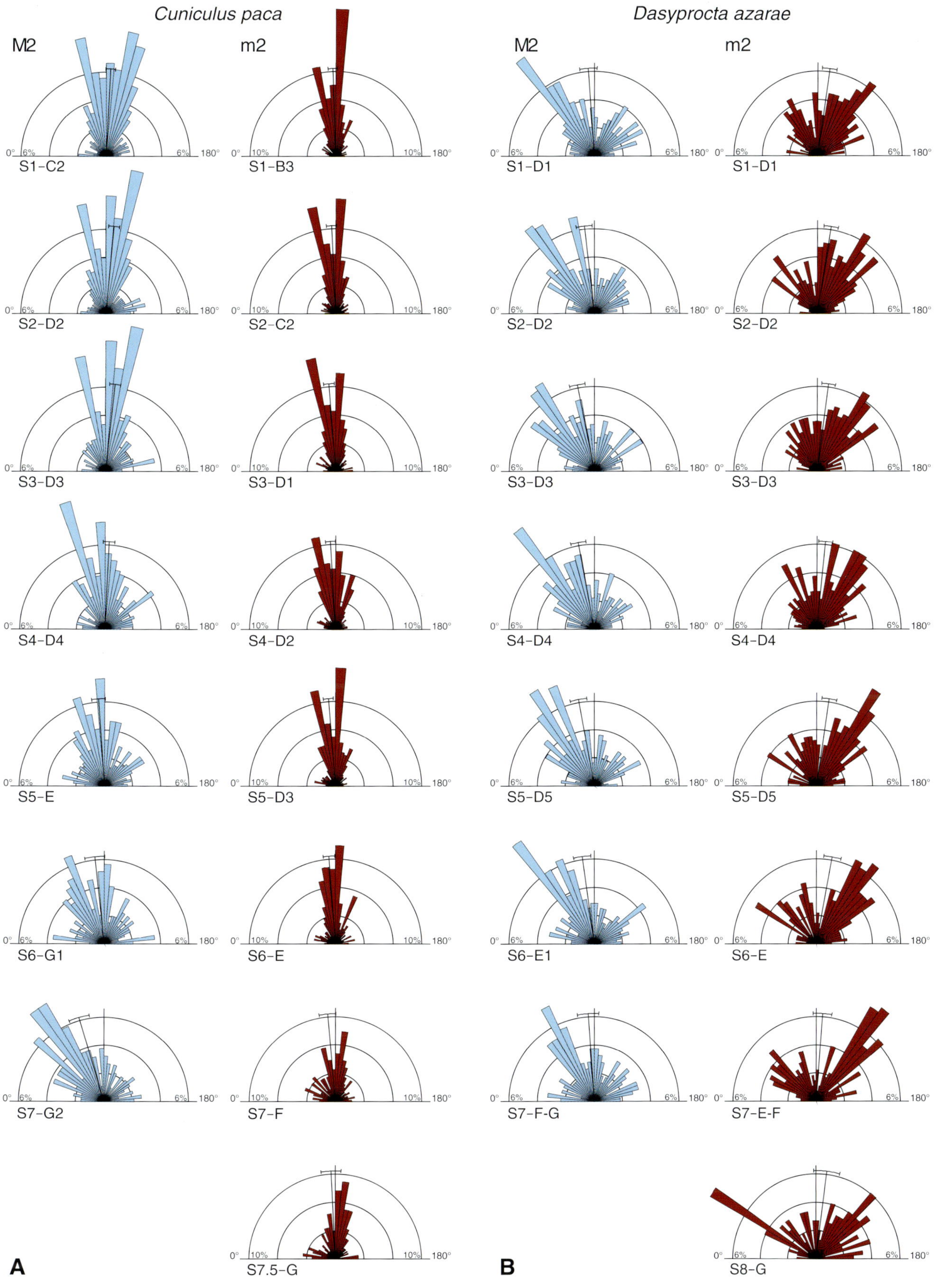

Fig. 7.10. Comparison of enamel ridge alignment with progressive wear. **A,** species with straight enamel folds (example of *Cuniculus paca*); **B,** curved enamel folds (example of *Dasyprocta azarae*). Future occlusal surfaces (S) are numbered consecutively (number next to S) defined by changing in wear. Wear stage categories (see also Fig. 7.4) are given. Values of 0° and 180° indicate ridge alignment parallel to chewing direction. Colored bars show enamel ridge alignment in steps of 5°; length of bars indicate amount of enamel ridges in percentage aligned in specific direction. Light blue: upper M2; dark red: lower m2. From Schubert (2019).

pensation of abraded material of the M1/m1 and M2/m2. In some investigated species like *Cuniculus paca*, *Dasyprocta leporina*, and *Hystrix indica* the occlusal surface and/or crown height of the P4/p4 are greater than those of the molars and thus increase functionality (Janis & Fortelius 1988). However, we should keep in mind that wild individuals seldom reach an age where the dentition gets worn down to wear stages that would show very low D_e and concentration of the angles.

The investigated rodent taxa can be divided in two groups after their pattern of enamel ridge alignment. For enamel ridge alignment the angles of an enamel ridge to the chewing direction were measured on several points along the enamel ridge; measurements were taken only on the leading edges (see Schubert 2019). The first group (taxa with parallel, straight enamel folds and islets as *Cuniculus paca* and *Castor*) possess a high concentration of enamel ridges around specific angles. The shearing edges are aligned roughly perpendicular to the chewing direction with a deviation up to 20°, and the inclination of upper and lower jaw is rather small (e.g., 20–30° in *Cuniculus paca*) (Fig. 7.10A). In the second group (e.g., *Dasyprocta*, *Hystrix*) not all elements on the occlusal surface are parallel and some enamel folds and islets are curved, leading to a higher deviation to 90°, and the enamel ridges are oriented in a wider range (e.g., 50–70° in the upper, 120–130° in the lower dentition of *Dasyprocta azarae*); the inclination of shearing edges of upper and lower jaw is 50–80° (Fig. 7.10B). It seems as if functionality is lower in the second group, but with curved enamel ridges food particles can be trapped more easily and a greater inclination of antagonistic shearing edges is possible (Rensberger 1973, Schmidt-Kittler 1984). To avoid the lateral movement of food particles that occurs in dentitions with straight enamel ridges, the inclination of the enamel ridges has to be less than 45°. Thus enamel ridge alignment gives only limited information about efficiency; other parameters as food particle transportation or the kind of food are also important (Rensberger 1973, Schmidt-Kittler 1984). Abrasive and tough food like grass needs higher values for D_e and straight, parallel shearing edges; for less abrasive food such as leaves or fruits lower values for D_e and curved enamel ridges are sufficient (Herrmann 2002). However, curved ridges may be more efficient to trap certain food items like fruits. This pattern could not be confirmed among the investigated species. Although the morphology of the occlusal surface seems to follow this tendency in most species (e.g., straight, parallel enamel ridges in tough-fiber eating taxa as *Castor*, *Myocastor* and the potential grazer *Mylagaulus* and curved elements in fruit-eating *Dasyprocta*), the values for D_e are approximately the same in most species (ca. 5–7 in first wear stages). Although *Cuniculus paca* feeds on fruits and leaves (Dubost & Henry 2006, Townsend & Croft 2008) the morphology of the dentition resembles a grazer. The relatively low-crowned teeth of *Thryonomys swinderianus* have rather low values for D_e (3–4) and enamel ridges that are not aligned perpendicular to the chewing direction. Both features do not indicate a grazer diet but *Thryonomys* strictly feeds on tough grasses. This is possible because of the highly specialized feeding procedure of *Thryonomys swinderianus*, in which the grass mostly gets disintegrated with the incisors and only afterwards with the cheek dentition (Ewer 1969). The diet of a species can differ from that of its ancestor, but as long as the inherited morphology of the dentition fulfills the required amount of functionality it does not need to be optimally adapted to the diet (Schmidt-Kittler 1984, Gailer et al. 2016).

Surprisingly, in the present case study the results for functional structural density (D_e) and enamel ridge alignment indicate no significant influence of splitting of enamel folds and islets on the masticatory function. The splitting of enamel folds and islets can increase the number of enamel elements on the occlusal surface, but as demonstrated not D_e and thus functionality, as splitting obviously only affects the number of elements but not the length of shearing blades.

Ontogeny and tooth replacement

Teeth are rarely functioning as isolated individuals but mostly within the tooth row. Therefore, the coordinated eruption of teeth into the tooth row is of selective value in placental mammals. Most eutherian mammals have a diphyodont dentition. In contrast marsupials replace only the third premolar as all other teeth are permanent (e.g., Ungar 2010). The tooth eruption sequence of the permanent dentition has been investigated in many species and can show quite different patterns among orders constrained by phylogeny as well as function and size (e.g., Shigehara 1980). For example, the eruption of all molars before permanent premolars observed in several extinct and extant clades of Artiodactyla is regarded to present a plesiomorphic condition of this group (Monson & Hlusko 2018, Gomes Rodrigues et al. 2019). Size differences among teeth may also cause sequence differences like observed in Primates, in which the enlarged canines erupt relatively late (Smith 2000). Delayed eruption and completion of the permanent postcanine dentition has been observed in Afrotheria, which is regarded to be a synapomorphic pattern of this taxon (Asher & Lehmann 2008). Another exceptional pattern of tooth eruption is linked to the horizontal tooth replacement, where the eruption of the last molars is delayed like in Elephantidae, *Trichechus* and even in Macropodidae (Domning & Hayek 1984, Sanson 1989, Gomes Rodrigues et al. 2012). Such a delay of erupting teeth allows the presence of fresh teeth even at a high individual age. The functional area of the dentition (FAD) can grow parallel to the increasing body size as observed in Elephantidae (Anders & Koenigswald 2013). Continuous tooth replacement in combination with supernumerary teeth like in the African rodent *Heliophobius argenteocinereus* is rare in mammals (Gomes Rodrigues et al. 2011).

Tooth eruption is controlled by hox gene Msx2 that targets the growth factor BMP4 (see above) as well as the osteoclast differentiation pathways in the alveolar bone (e.g., Aïoub et al. 2007). Marks & Schroeder (1996) give

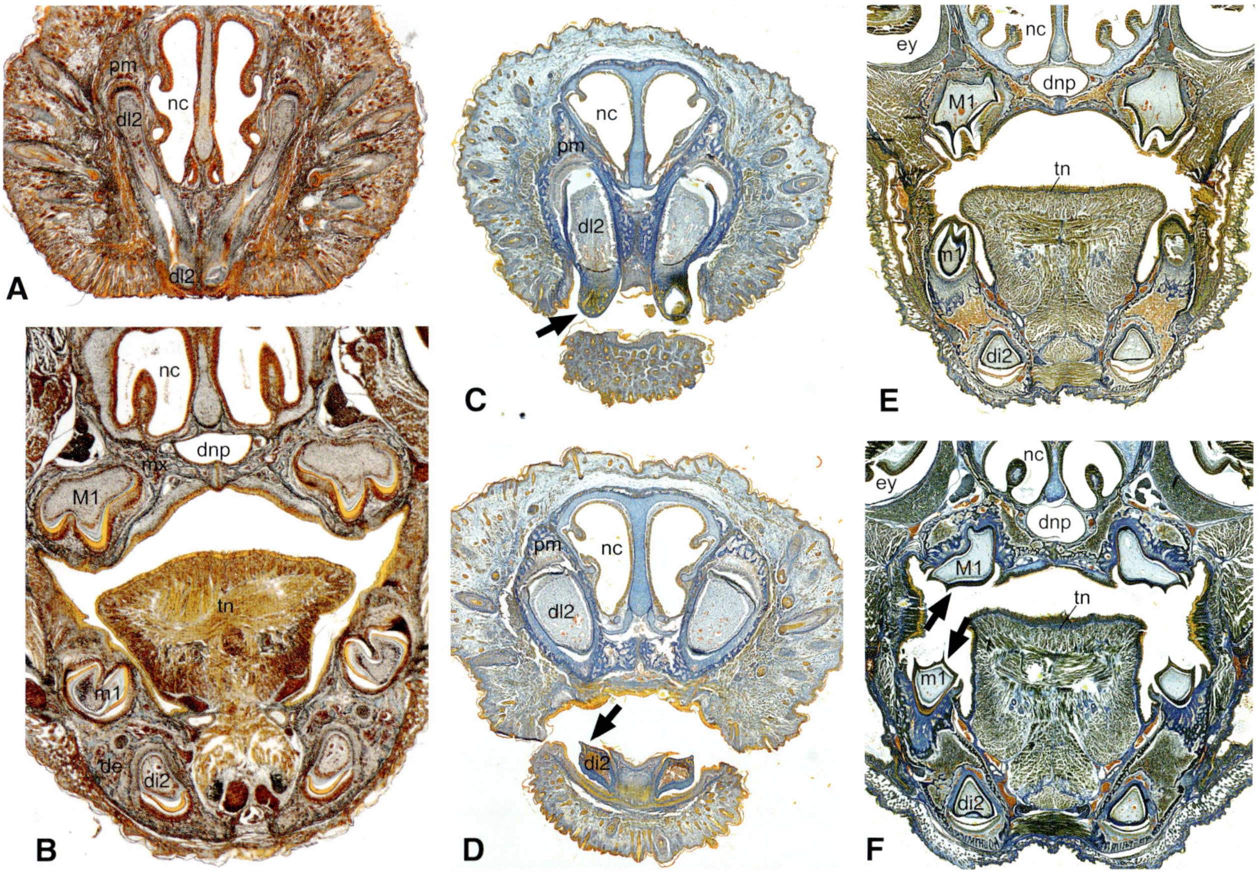

Fig. 7.11. Transversal histological serial sections through the dentition of altricial muroid rodents showing the eruption of specific tooth position in neonatal to early postnatal stages. **A, B,** *Micromys minutus*, 4 days, 14 mm head length, slices 10-5-4 (A, oral fissure) and 20-4-1 (B, oral cavity behind diastema). **C, D,** *Mesocricetus auratus*, neonate, 15 mm head length, slices 5-2-6 (C, oral fissure) and 6-2-3 (D, oral fissure slightly behind (C). **E,** *Mesocricetus auratus*, 7 days, 19 mm head length, slice 21-4-2 (oral cavity behind diastema). **F,** *Mesocricetus auratus*, 11 days, 23 mm head length, slice 25-3-2 (oral cavity behind diastema). Arrows refer to erupting teeth. Not to scale. Abbreviations: de, dentale; dI/i2, deciduous upper/lower incisor 2; dnp, ductus nasopharyngeus; ey, eye; M/m1, upper/lower molar 1; mx, maxillare; nc, nasal cavity; pm, praemaxillare; tn, tongue.

a detailed overview on the tooth eruption mechanisms and pathways in primates, dogs and rodents. During the pre-eruptive phase bone formation and root growth (brachydont and hypsodont teeth) occur at the base of the dental alveoli while the bone above and/or root of the respective deciduous tooth are resorbed until the tooth completely penetrates the gingiva. The speed of pre-eruptive movements of developing teeth and the eruption speed are stage and species specific and depend on tooth position. In *Homo sapiens* premolars the speed of eruption from gingival emergence to the complete occlusal plane can be up to 75 µm/day (Proffit et al. 1991).

While the timing of eruption of permanent teeth can easily be investigated for comparative morphofunctional and systematic purposes (e.g., Asher & Lehmann 2008, Anders et al. 2011a,b, see case study 3), the study of pre- and perinatal eruption of deciduous teeth especially in placentals is a matter of technical limitations although the development of one of the major metabolic tools, the dentition, is an important key to the understanding of life history of a species. Thus, most studies on tooth eruption focus on the permanent dentition in postnatal stages, especially in primates (e.g., Smith et al. 1994, Jekl et al. 2011).

Case study 2: Perinatal tooth eruption in rodents and lagomorphs

Growth rates differ significantly in precocial (born relative mature) and altricial (born naked with closed eyes) mammals. The former have their growth rate peak earlier in development than the latter (Gaillard et al. 1997). Perinatal eruption of teeth in placental mammals is therefore also expected to be highly constrained by the mode of postnatal development in terms of precocial versus altricial adaptations.

Rodentia and Lagomorpha are suitable case studies as both orders comprise altricial as well as precocial species (Nowak 1991). Because lagomorphs and especially rodents use their incisors not only for gnawing but also for exploration, early eruption is advantageous in order to explore their surroundings early in life. Furthermore, in altricial species tooth eruption of check teeth may be delayed well into postnatal stages compared to precocial species as the former require initially intense care and nursing. To date, only few species of rodents and lagomorphs have been investigated in terms of tooth eruption in early ontogeny

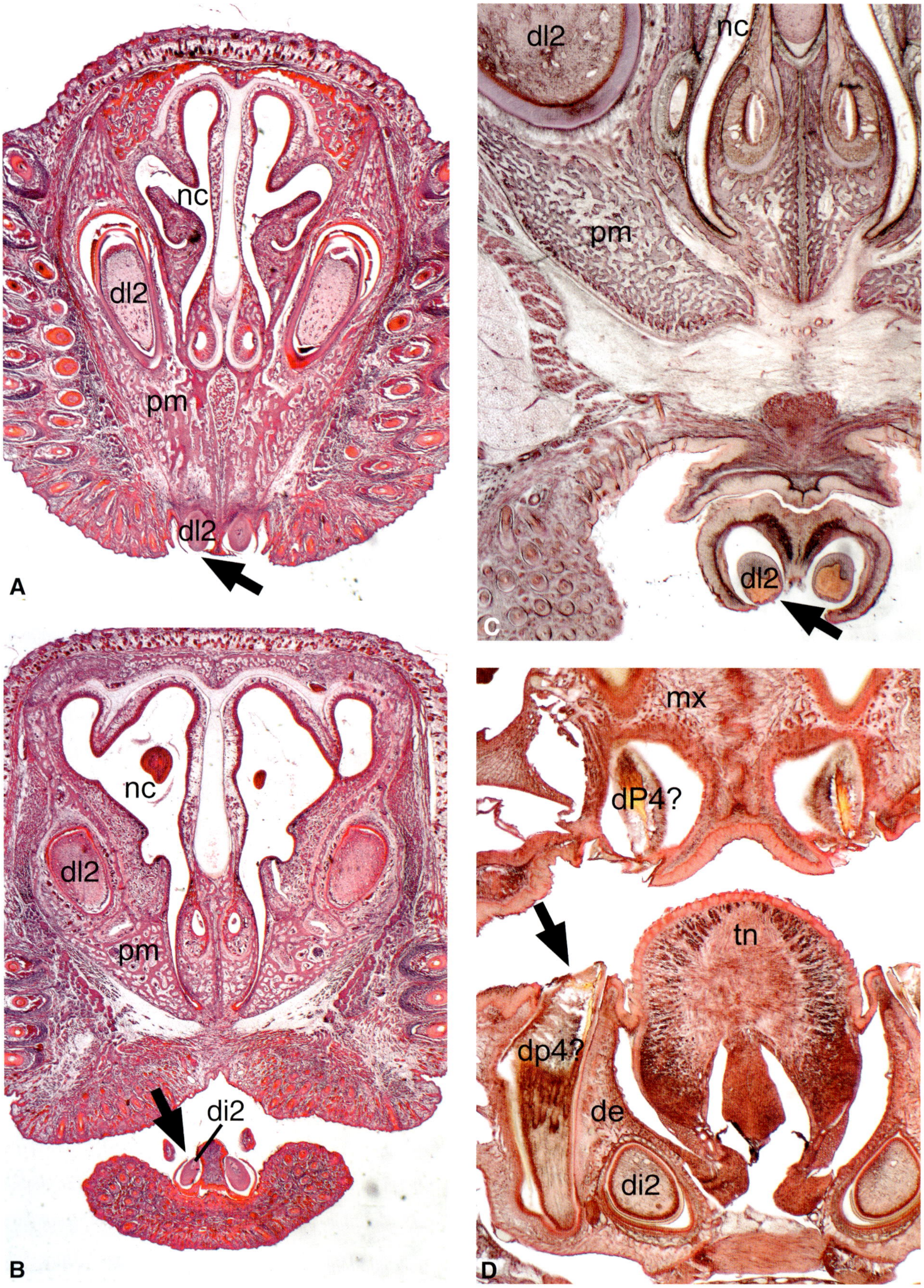

Fig. 7.12. Transversal histological serial sections through the dentition of precocial ctenohystric rodents showing the eruption of specific tooth position in fetal stages. **A, B,** *Octodon degus*, late fetal, 24 mm head length, slices 14-2-2 (A, oral fissure) and 19-2-5 (B, oral fissure slightly behind A). **C,** *Thryonomys swinderianus*, late fetal, 38 mm head length, slice 48-2-1 (anterior oral cavity). **D,** *Cavia porcellus*, late fetal, 19 mm head length, slice 67-1-1 (oral cavity behind diastema). Arrows refer to erupting teeth. Not to scale. Abbreviations: de, dentale; dI/i2, deciduous upper/lower incisor 2; nc, nasal cavity; dP/p4?, ambiguous deciduous upper/lower premolar 4; mx, maxillare; pm, praemaxillare; tn, tongue.

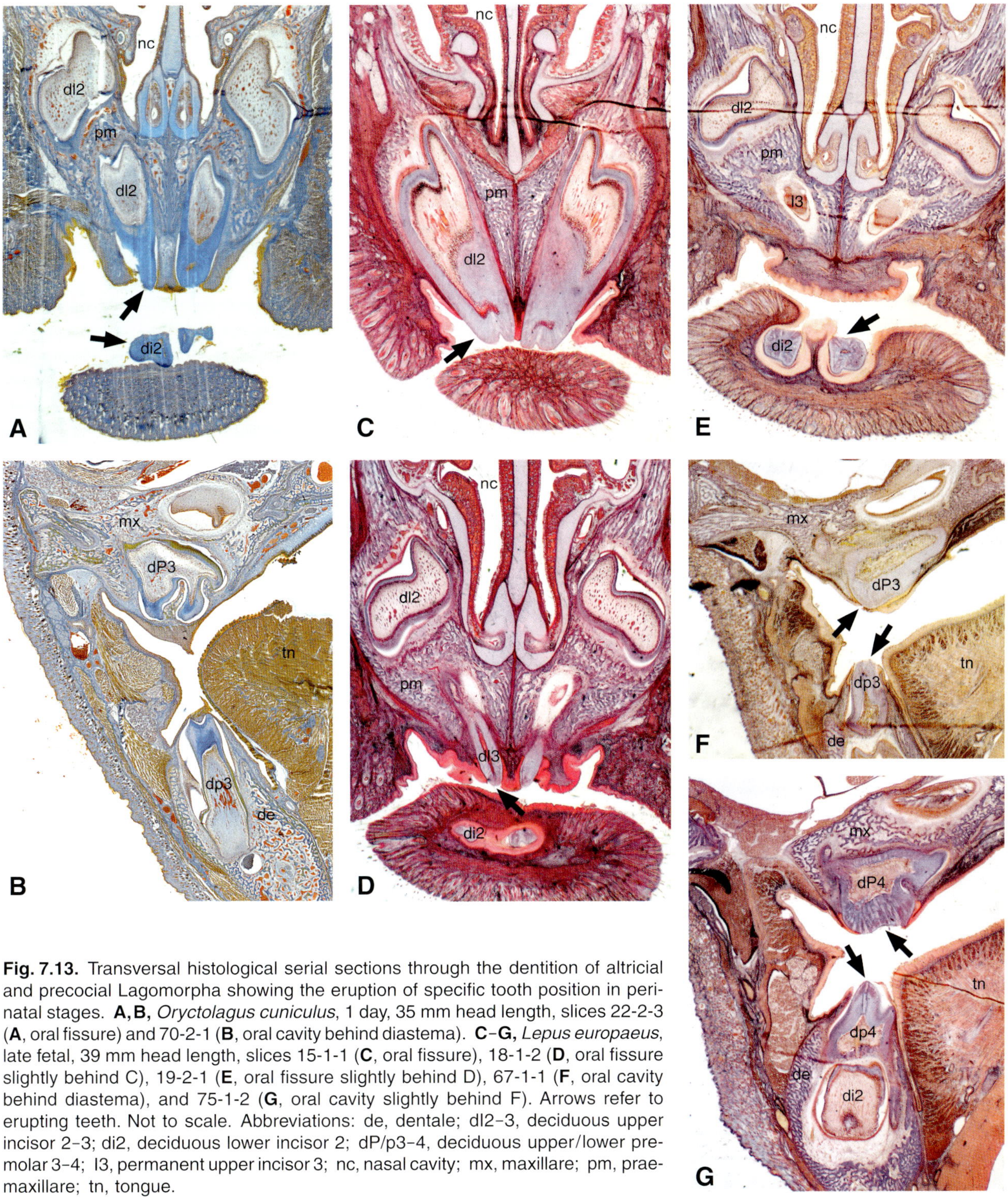

Fig. 7.13. Transversal histological serial sections through the dentition of altricial and precocial Lagomorpha showing the eruption of specific tooth position in perinatal stages. **A, B,** *Oryctolagus cuniculus*, 1 day, 35 mm head length, slices 22-2-3 (**A**, oral fissure) and 70-2-1 (**B**, oral cavity behind diastema). **C**–**G,** *Lepus europaeus*, late fetal, 39 mm head length, slices 15-1-1 (**C**, oral fissure), 18-1-2 (**D**, oral fissure slightly behind C), 19-2-1 (**E**, oral fissure slightly behind D), 67-1-1 (**F**, oral cavity behind diastema), and 75-1-2 (**G**, oral cavity slightly behind F). Arrows refer to erupting teeth. Not to scale. Abbreviations: de, dentale; dI2–3, deciduous upper incisor 2–3; di2, deciduous lower incisor 2; dP/p3–4, deciduous upper/lower premolar 3–4; I3, permanent upper incisor 3; nc, nasal cavity; mx, maxillare; pm, praemaxillare; tn, tongue.

and correlations with their reproductive biology were only rarely considered (e.g., Jekl et al. 2011).

For the present study perinatal stages of altricial and precocial Rodentia and Lagomorpha were investigated some of which are represented by several ontogenetic stages. The altricial rodent species are members of the Muroidea (*Mesocricetus auratus*, *Peromyscus maniculatus*, *Mus musculus*, and *Micromys minutus*), the precocial ones belong to the Ctenohystrica (*Petromus typicus*, *Thryonomys swinderianus*, *Octodon degus*, *Chinchilla lanigera*, *Myoprocta pratti*, and *Cavia porcellus*). Lagomorpha are represented by the altricial *Oryctolagus cuniculus* and the precocial *Lepus europaeus*. The histological sections are housed in the Institut für Evolution und Ökologie (formerly Lehrstuhl für Spezielle Zoologie), Eberhard-Karls-Universität Tübingen, Germany. Dental eruption is defined here according to previous studies (e.g., Jekl et al. 2011) in that any portion of the tooth crown is penetrating the gingiva. Therefore, information on the soft tissue structures is demanded for the precise identification of eruption of deciduous as well as permanent teeth. This can be best achieved by histological serial sections.

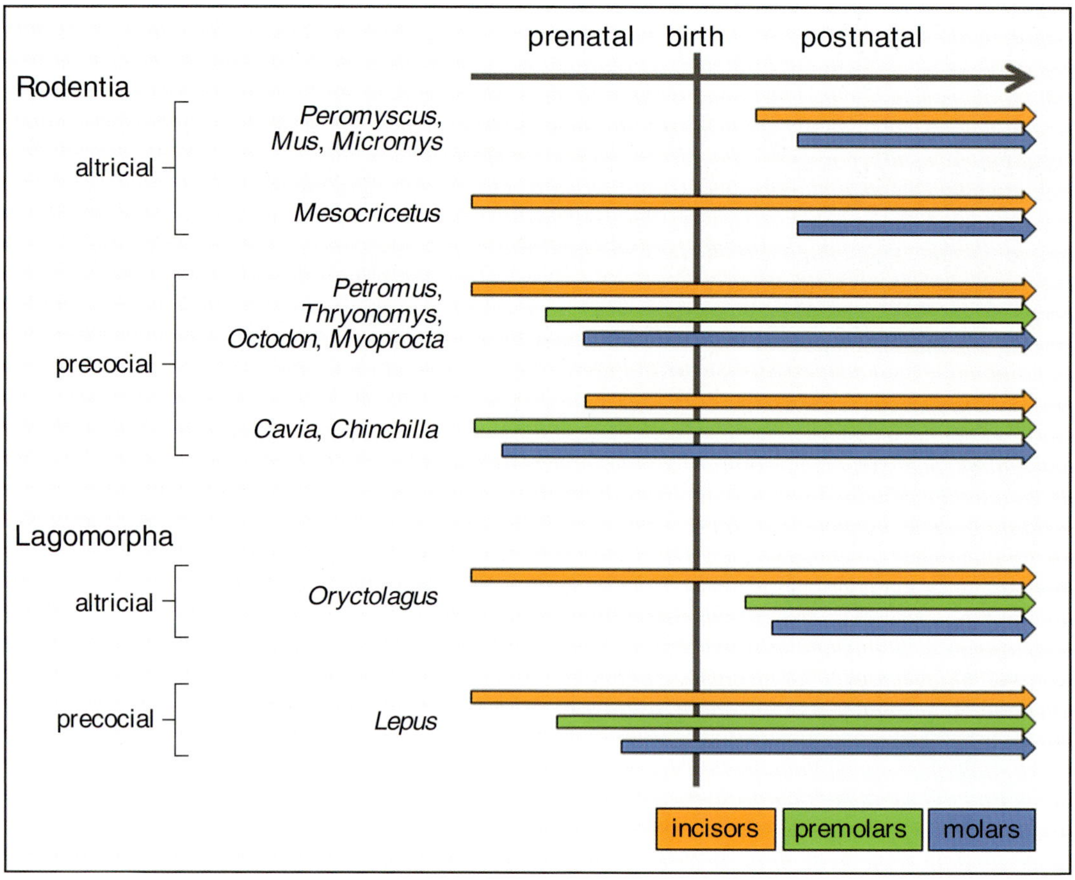

Fig. 7.14. Relative timing pattern of perinatal tooth eruption in altricial and precocial rodents and lagomorphs. Precocial species are born with all tooth types erupting whereas in the altricial species under study cheek teeth start erupting not until after birth. Interestingly among the altricial rodents only *Mesocricetus* is born with erupting incisors. In general, incisors erupt before cheek teeth except for *Cavia* and *Chinchilla* in which it is vice versa.

In all investigated altricial rodents, except for *Mesocricetus auratus*, eruption of incisors and molars is restricted to postnatal stages (Fig. 7.11). *Peromyscus maniculatus* nicely exemplifies this pattern: a neonate shows no erupted teeth whereas an early juvenile stage (25 mm head length) has a fully erupted dentition. However, a neonatal stage of *Mesocricetus auratus* shows erupted upper and lower incisors which clearly indicates prenatal eruption of the incisors (Fig. 7.11C,D). This pattern is surprising but may be explained by the species' peculiar reproduction biology. *Mesocricetus auratus* has the shortest gestation period (15–17 days) among placentals but its development and craniogenesis are accelerated and therefore mostly comparable in their timing to all other investigated muroids (Soderwall et al. 1960, Ruf 2020). This species starts nibbling food eight days after birth and after three weeks weaning is completed (Nowak 1991). However, a seven days old specimen shows no further erupted teeth; in an eleven days stage M1 and m1 have erupted. Thus, the observed eruption pattern correlates with the above described postnatal development but a fully erupted dentition is obviously not necessary after weaning.

The precocial rodents under study are generally characterized by a longer gestation period than in altricial species (Nowak 1991). Though this does not necessarily affect the timing of tooth eruption, tooth development seems to be more advanced at time of birth compared to altricial rodents. In the investigated precocial species the incisors erupt in prenatal stages before or at the same time as the premolars or first molars and this happens before birth (Fig. 7.12). As shown by a recent study in *Octodon degus* none of the premolars and molars erupts before the second day after birth. Functional occlusion of the first two cheek teeth does not start before the ninth day after birth. Eruption of lower molars is somewhat delayed compared to the upper dentition and full occlusion of all cheek teeth is accomplished on days 58–72 after birth (Jekl et al. 2011). This pattern, slower development of teeth in the lower jaw, has been also described for *Cavia* (Harman & Smith 1936, Berkovitz 1972). However, some variation can be observed in *Octodon degus* as in our sampling a neonatal specimen already shows erupting deciduous premolars. Prenatal tooth eruption has already been described in caviomorphs and the prenatal occlusion can even result in attritional tooth wear before birth as described for *Cavia* and *Myocastor* (Ainamo 1971, Sone et al. 2008).

In general the sequence of cheek tooth eruption in the investigated rodents is from anterior to posterior. However, the investigated specimens of *Cavia porcellus* and *Chinchilla lanigera* show a deviating pattern as the first molar is already erupting while the premolar is still covered by gingiva. Unfortunately, it is not clear if these premolars are deciduous or permanent as no younger or older stages of these species are at hand. At least in *Cavia* it is known that the deciduous premolar becomes resorbed in prenatal stages (Harman & Smith 1936). Surprisingly, in a late fetal *Cavia porcellus* the incisors are still covered by gingiva though the premolars, M1–2, and m1–2 are in different stages of eruption. Although intraspecific variation in the timing of tooth formation and eruption, even within the same litter, has been observed in *Cavia* (Harman & Smith 1936), a delayed eruption of the incisors has not been mentioned yet. Due to the time consuming preparation of histological serial sections and the fact that the sampling of prenatal stages is restricted in most species, analyses of intraspecific variation of perinatal tooth eruption are still pending.

The lagomorphs under study reveal that all deciduous incisors already erupt well before birth (Fig. 7.13A, C–E). However, Nagy & Daniel (1992) describe the beginning of incisor eruption for the neonate New Zealand white rabbit (*Oryctolagus cuniculus*); intraspecific variation could explain this discrepancy. In the neonatal stage of the altricial *Oryctolagus cuniculus* the cheek teeth have not pierced the gingiva yet (Fig. 7.13B). The premolars are observed to erupt after one week, the first molar after two weeks, the second after 3.5 weeks and the third molar after four weeks (Nagy & Daniel 1992). In contrast, the investigated late fetal stage of the precocial *Lepus europaeus* already shows partially erupted dP3–4 and dp3–4 (Fig. 7.13F, G). Obviously the first upper cheek tooth (dP2) shows a delayed eruption compared to all other deciduous premolars. A study on *Lepus brachyurus brachyurus* reveals that at birth all deciduous premolars and the first and second molars are erupted (Yamada et al. 1990).

Though in both lagomorph species under study the incisors erupt before birth, there is evidence for a different timing of cheek tooth eruption in the two reproduction types. This pattern is comparable to the observations in rodents, in which altricial species show postnatal eruption of cheek teeth in contrast to precocial species that are born with at least partly erupted cheek teeth (Fig. 7.14). This confirms the hypothesis that the mode of development at birth is also reflected in the development of the dentition of Rodentia although further altricial and precocial members of the mouse-related clade as well as the squirrel-related clade need to be investigated to exclude a phylogenetical effect. As many altricial muroids reach their sexual maturity within a few weeks and therefore need a functional dentition for optimal energy supply, their relative delayed tooth development is obviously compensated by highly accelerated tooth eruption.

Age differentiation and life history

Unlike bones, teeth are not remodeled during the life history. Therefore at least a limited amount of time markers in the dentition is available to reconstruct the life history of a mammal. However, classifications of age stages based on tooth wear can be used for ontogenetic and (paleo) ecological interpretations of single species as well as species communities as demonstrated in our case study below. In contrast, life history is also documented in teeth that are not affected by wear i.e., in the tusks of elephants a major part of the life history is recorded. For *Mammuthus primigenius* major events, such as severe traumata or diseases, malnutrition, giving birth in females, or rush in males, are recognizable by the changes of microstructure in the dentin (Fisher et al. 2014). Based on the carbon and nitrogen isotope analyses of a juvenile *Mammuthus primigenius* from Wrangel Island evidence of weaning has been documented (Rountrey et al. 2007). Perhaps the large dentin teeth with limited abrasion of *Monodon monoceros* or *Odobenus rosmarus* may provide similar data.

The formation of enamel during the early ontogeny of teeth contains other time markers. On the outside of teeth, perikymata (incremental growth lines) mark subsequent steps of the mineralization of the enamel. Within the enamel they are seen as Retzius lines that cross the individual prisms. The prisms themselves have a rhythm, regarded as daily, and thus the progress of tooth formation can be compared between different taxa. In dentin a comparable daily incremental pattern can be observed, the von Ebner lines (Dean et al. 1993). This topic is discussed especially in early hominins (e.g., Smith et al. 2015).

In many mammal species the degree of tooth wear is used as an indicator of individual age, especially in domesticated animals and game (Baumann 1949, Habermehl 1975, 1985). Already at medieval fares, horses were evaluated according to the wear stages of the incisors (Kroon 1929). Unfortunately, these approaches can be applied only to a limited degree on extinct species. The methods used for extant mammals aim at individual ages measured in years but are often restricted to the investigated species. For instance, the dental wear age classes that refer to the absolute age of the animals defined by Grau et al. (1970) for *Procyon lotor* (classes I–V) and by Goddard (1970) for *Diceros diceros* (classes I–XX) are not comparable or consistent. General age stages as “infantile”, “juvenile”, “mature”, “adult”, or “senile” were often used in various ways by different authors, such as Schultz (1935, 1960) and Smith (2000) for primates in general, and Knußmann & Martin (1988) for humans. Therefore the general application of these definitions to a wide range of extant and fossil taxa has been difficult to date. However, related to the aim of a study specific and detailed classifications of ontogenetic stages are still demanded as shown above by Schubert (2019).

Case study 3: Individual dental age stages

Anders et al. (2011a, b) provide an age classification for eutherian mammalian dentitions based on the tooth eruption and wear of the postcanine teeth. The “individual dental age stages” (IDAS) are independent of the occlusal morphology and the absolute age and can be applied to almost all fossil and extant eutherian species. Taxa with horizontal tooth replacement, expanded or reduced tooth formula (e.g., elephants, manatees, hyenas) can be considered by slight adaptations of the IDAS system (see Anders et al. 2011a, b).

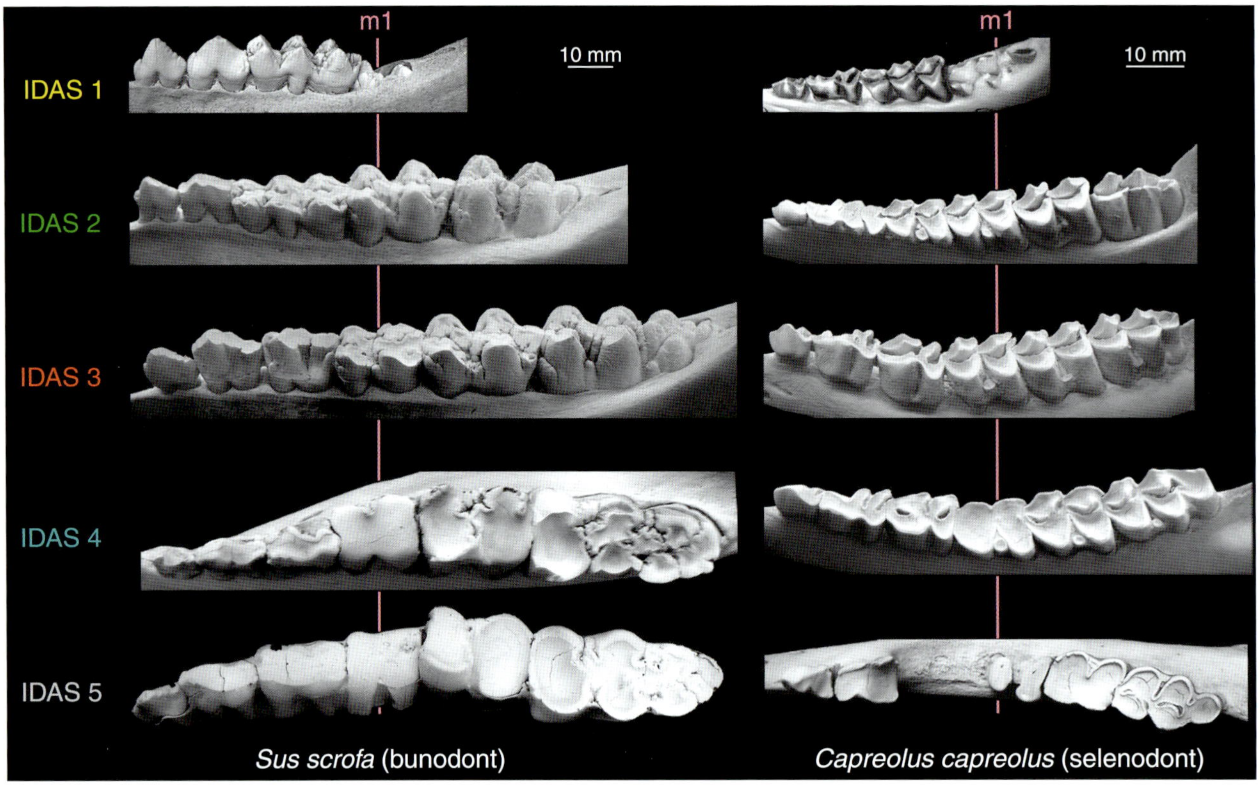

Fig. 7.15. Individual dental age stages (IDAS) 1–5 examplified in the lower dentition of two extant Artiodactyla: the bunodont *Sus scrofa* and the selenodont *Capreolus capreolus*. The vertical lines refer to the position of the lower first molar (m1). Modified from Anders et al. (2011a).

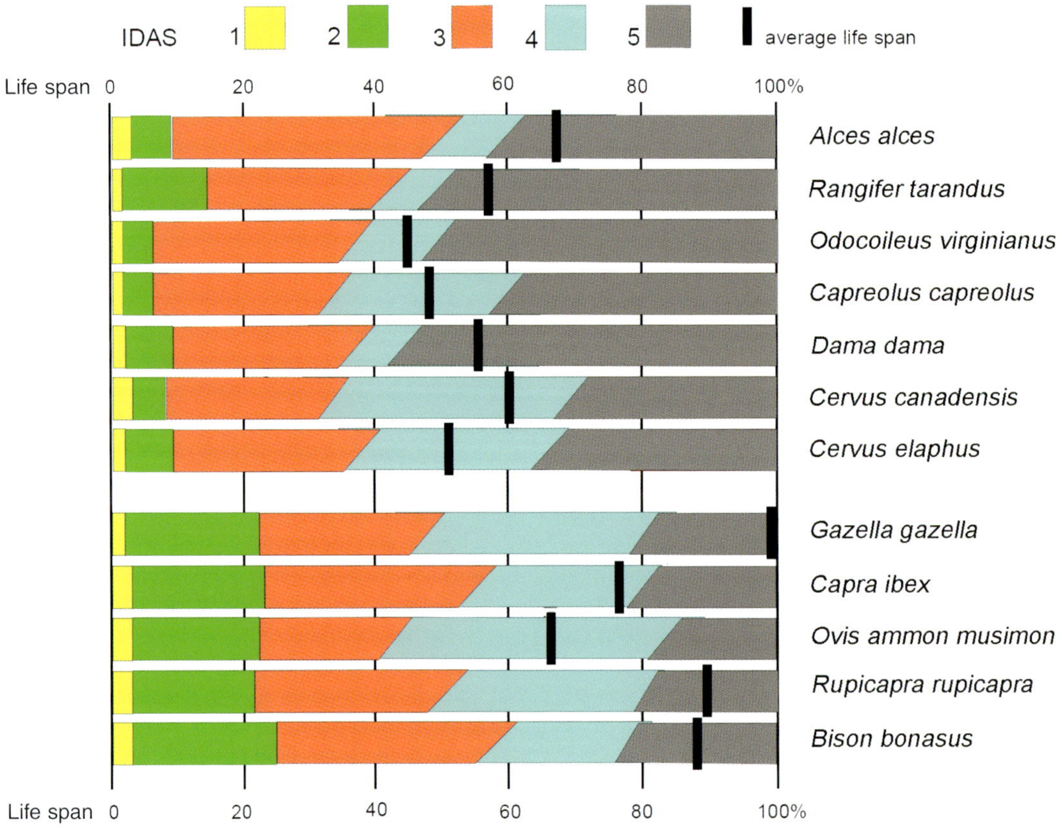

Fig. 7.16. Distribution patterns of individual dental age stages (IDAS) 1–5 in relation to the life span of selected extant cervids (top) and bovids (bottom). Compared to the maximum life span (100 %) the average life span (vertical bars) is longer in bovids (around 82 %) than in cervids (around 54 %). The elongated phases IDAS 2, 3, and 4 reflect the hypsodonty in Bovidae. Variation of the IDAS 3 and 4 is indicated by the oblique edges of the bars. Modified from Anders et al. (2011a).

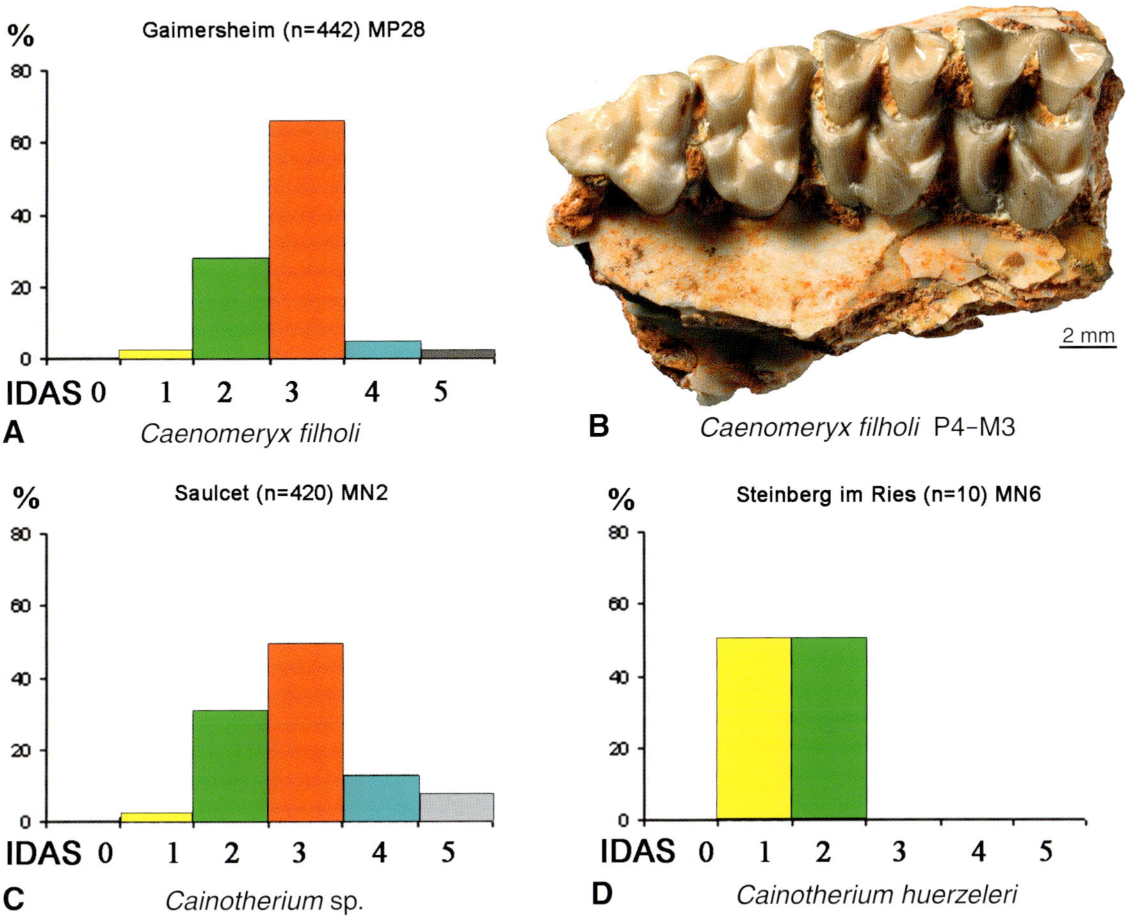

Fig. 7.17. Mortality of Cainotheriidae in three fossil communities deduced from individual dental age stages (IDAS) of the respective samples. **A,** IDAS of *Caenomeryx filholi* from the Upper Oligocene of Gaimersheim showing the predominance of IDAS 2 and 3. **B,** Left maxilla with P4-M3 (IDAS 3) of *Caenomeryx filholi* from the same locality (IGPB 7395). **C,** *Cainotherium* sp. from Saulcet (Lower Miocene) is also mostly represented by IDAS 2 and 3. **D,** *Cainotherium huerzeleri* from Steinberg im Ries (Lower Miocene) shows a different pattern in being represented by only IDAS 1 and 2. Abbreviations: IGPB, Institute of Geosciences, Section Paleontology, University of Bonn; M3, upper molar 3; MN, Mammal Neogene zone; MP, Mammal Paleogene zone; P4, upper premolar 4. Modified from Anders et al. (2011a).

The IDAS cover the entire life span from prenatal to senile stages and can be correlated to traditional terms as e.g., juvenile, adult or senile. They comprise six stages that are defined as follows (Fig. 7.15):

IDAS 0 (prenatal) – The prenatal stage of tooth development cannot be determined in the teeth directly. However, this stage has been included in order to allow comparison and characterization of prenatal specimens. In the fossil record, independent evidence for the prenatal death of the individual is needed.

IDAS 1 (infant) – This stage refers to the period from birth to the first attritional facets on the erupting first molar. The deciduous dentition dominates and the eruption of the first molar approximately marks the time of weaning in many placentals (e.g., Smith 2000, Kelley & Smith 2003).

IDAS 2 (juvenile) – This stage refers to the period from the first facets through wear on the first molar to the fully erupted permanent dentition.

IDAS 3 (adult) – This stage refers to the period from full eruption of the dentition until the loss of the inner profile of the first molar, so the occlusal surface is free of interior enamel but the outer enamel band is not broken down. IDAS 3 is supposed to be the time of maximum activity and reproduction in most Eutheria. In the fossil record this stage is common.

IDAS 4 (late adult) – This stage lasts from the loss of the inner profile of the first molar until the loss of the inner profile of the second molar.

IDAS 5 (senile) – This stage lasts from the loss of the inner profile of the second molar to the complete breakdown of the dentition by wear and/or tooth loss. The nutrition may be compromised.

The varying duration of different stages in the IDAS pattern during the life history forms an additional way to characterize mammalian species and groups by the ontogeny and wear of their dentition. The late eruption of the third upper and lower molar is discussed to be related to longevity (e.g., Smith 2000). The length of the single IDAS stages differs in the various species that were studied by Anders et al. (2011a,b). A comparison between Cervidae and Bovidae shows that the former reach IDAS 5, the senile stage, earlier in their maximal life span than the latter (Fig. 7.16). This difference might be due to the fact that bovids have hypsodont cheek teeth. The higher amount of material in these teeth can compensate the loss of

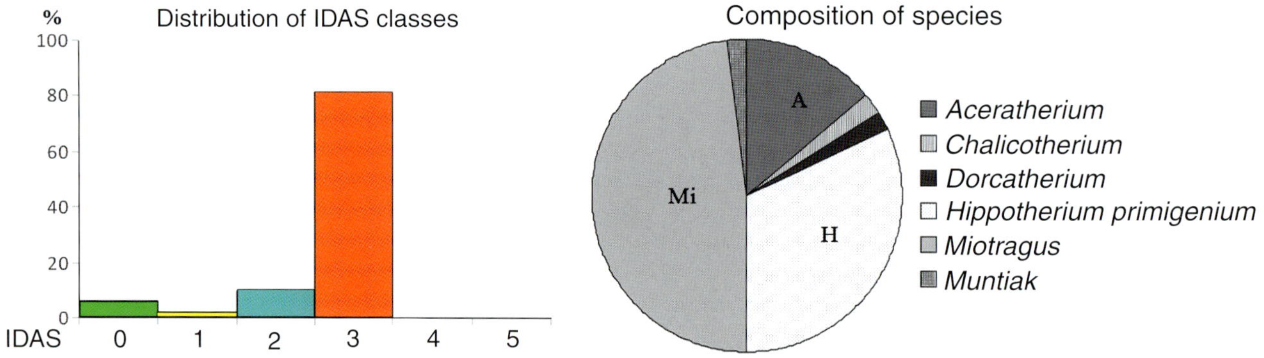

Fig. 7.18. Distribution of individual dental age stages (IDAS, left) and species composition (right) among fossil perissodactyls and artiodactyls from Höwenegg, Upper Miocene, Germany. The sampling is dominated by IDAS 3. Modified from Anders et al. (2011a).

dental material caused by chewing of highly abrasive plant matter. Therefore, these teeth function for a longer part of the average life span. This result supports the hypothesis that the functional performance of the dentition impacts the longevity of an animal (compared to similarly sized species). This is especially the case in herbivores as the functional efficiency of their cheek teeth is a key limiting factor to energy harvest, and the loss of shearing blades or blade sharpness reduces the grinding capacity of the dentition (Popowics & Fortelius 1997).

A major advantage of the IDAS is the application to fossil specimens, characterizing the mortality in monospecific populations or of entire fossil communities composed of various species. Even if no extant relatives exist from which the individual age in years could be deduced by comparison, functional age groups can be classified and used to elucidate fluctuations in population size or age structures in correlation to climatic or seasonal parameters. Anders et al. (2011a,b) provide IDAS analyses of the placental fauna of several fossil sites. One striking example refers to Cainotheriidae, extinct rabbit-sized artiodactyls from the Eocene/Miocene of Europe from three investigated localities. *Caenomeryx filholi* from Gaimersheim (Upper Oligocene fissure fillings, Germany) and *Cainotherium* sp. from Saulcet (Lower Miocene freshwater limestones, France) are represented by IDAS 1–5 (Fig. 7.17A–C). Juvenile and adult individuals are clearly dominating and IDAS 3 makes up at least 50 % of the population. This pattern resembles a stable population indicated by the dominance of individuals of reproductive age (Klein & Cruz-Uribe 1983). In contrast the fossil site Steinberg im Ries (Lower Miocene freshwater deposits, Germany) provides a quite different pattern (Fig. 7.17D). *Cainotherium huerzeleri* is mostly represented by IDAS 1 and 2. This accumulation of infant and juvenile stages might be best explained by a taphonomic phenomenon caused by birds of prey that might have nested in that area and preferred smaller prey animals (Heizmann & Fahlbusch 1983).

The lake deposits of Höwenegg (Upper Miocene, Germany) exemplify the application of IDAS to a fossil assemblage in order to elucidate age classes and mortality in a fossil mammal fauna represented by different taxa. The 46 investigated specimens comprise Perissodactyla as well as Artiodactyla (Fig. 7.18). Again the investigated sampling is dominated by IDAS 3 (about 80 %). IDAS 1 is represented by only one specimen of *Hippotherium* and IDAS 2 by one *Aceratherium*, one *Dorcatherium*, and three individuals of *Miotragocerus*. IDAS 4 and 5 are completely missing in the sample.

IDAS 0 is represented theoretically by three fetuses. A pregnant mare of *Hippotherium* shows one fetus and twins can be observed in the abdomen of a *Miotragocerus* specimen. The teeth of the fetuses are not well documented (Anders et al. 2011a,b). As both pregnant females belong to IDAS 3, these finds support the hypothesis that IDAS 3 definitely coincides with reproductive activity, although reproduction could also occur in IDAS 2 or 4. However, the observed IDAS pattern in the fossil perissodactyl and artiodactyl assemblage from Höwenegg does not reflect a natural population composition of neither of the investigated species because not all IDAS are represented. Predominance of IDAS 3 might be due to a so far unidentified selection.

Conclusion

Premolars and molars play an important role in food comminution and their reliability is correlated with longevity of the respective species. Especially in herbivorous mammals enhancement of efficiency during the power stroke is a major key to cope with abrasive materials in their diet, such as phytoliths or grit. Therefore, specialized occlusal patterns (e.g., enamel elements) and compensatory constructions (e.g., hypsodonty) evolved. These features are generally correlated to but also constrained by ontogeny. Only the deeper understanding of tooth ontogeny i.e., from pre-erupting to late adult stages are crucial to elucidate the evolution and functional performance of postcanine dentitions in herbivorous mammals.

The case studies presented in this chapter meet general research questions investigated in the DFG Research Unit 771 and clearly demonstrate how ontogeny in a single tooth as well as in a tooth row representing a functional unit can be interpreted from extant as well as fossil dentitions.

Hypsodonty and euhypsodonty guarantee the efficiency or a prolonged function of rasp facets in the dentition. This is often in concert with additional enamel crests of the occlusal surface like enamel folds and/or islets. Enamel islets can increase the length of shearing ridges on the occlusal surface. There are, however, limitations in the development of teeth with enamel islets such as the possible crown height. Especially in Rodentia enamel islets often split with progressive wear, thus an increasing amount of enamel elements functions on the occlusal surface. As shown in case study 1, these splittings are highly variable but may not directly contribute to an increase or preservation of functionality of the teeth during progressive wear. Even if the stimulus of the splitting is so far unknown, this pattern evolved several times independently.

Loss of functionality of worn teeth can be compensated to a certain degree by the normal tooth replacement of the deciduous dentition or by a horizontal tooth replacement. The latter allows a function for a longer period, the compensation of a possible increase of body size, and enhances the lifespan of an individual. The sequence and timing of tooth eruption can vary in different species. Case study 2 indicates the correlation of the tooth eruption sequence and its timing with the reproduction mode in rodents and lagomorphs. In these orders, precocial taxa show an eruption of the first cheek teeth prior to birth, whereas in altricial species cheek teeth erupt after birth. Accordingly, this heterochronic pattern allows the more mature newborn pups of precocial species to have a functional dentition and to cope with food particles relatively early after birth.

Due to the differences in tooth eruption and wear the determination of the age of an individual is only possible when the dentition can be compared to an individual of the same species with known age. This is not only difficult in many wild living species, but almost impossible for extinct mammals. An estimation of the relative age (e.g., juvenile, adult, senile) can be accomplished with the eruption pattern and wear stages of the dentition. Case study 3 provides the definition of several individual dental age stages (IDAS), which can be applied for almost all living and extinct mammal species. Furthermore, by means of IDAS a direct comparison of different species as well as the age composition of a fossil population is possible.

In systematic studies of fossil and extant mammals, unworn or only slightly worn teeth are preferred. But they gain access only to a very limited aspect of mammals. If attention to the entire ontogeny is included, as clearly demonstrated in the three case studies, many more aspects become visible that reflect adaptations to specific functions or the requirements of specific environments.

Acknowledgments

We thank Ulrike Anders and Leonie Schwermann, who provided data from their PhD theses that were executed in the DFG Research Unit 771. The case studies were part of several projects in this research group (KOE 627/36-1, RU 1496/5-1).

We are grateful to Philip D. Gingerich (University of Michigan, Ann Arbor, MI), Ernest L. Lundelius (University of Texas, Austin, TX), Kenneth D. Rose (Johns Hopkins School of Medicine, Baltimore, MD), Richard C. Hulbert Jr. (Florida Museum of Natural History, Gainesville, FL), Thomas Mörs and Daniela Kalthoff (Naturhistoriska Riksmuseet Stockholm), Thomas Kaiser (Universität Hamburg), Wolfgang Maier (Universität Tübingen) as well as the citizen scientists Frank Menger (Groß-Rohrheim), Walter and Christian Keller (Ginsheim) who provided access to fossil and extant specimens for our studies. Adrian Tröscher (Universität Tübingen, Senckenberg-HEP) assisted with histological data, Georg Oleschinski (Universität Bonn) took the photographs. Last but not least we thank the reviewers Vincent Lazzari (Université de Poitiers) and Madeleine Geiger (Universität Zürich) who helped to improve the manuscript.

References

* indicates publications that originated from the DFG Research Unit 771.

Aberg, T., Wozney, J. & Thesleff, I. (1997): Expression patterns of bone morphogenetic proteins (Bmps) in the developing mouse tooth suggest roles in morphogenesis and cell differentiation. Developmental Dynamics 210: 383–396.

Ainamo, J. (1971): Prenatal occlusal wear in guinea pig molars. European Journal of Oral Sciences 79: 69–71.

Aïoub, M., Lézot, F., Molla, M., Castaneda, B., Robert, B., Goubin, G., Néfussi, J. R. & Berdal, A. (2007): Msx2 -/- transgenic mice develop compound amelogenesis imperfecta, dentinogenesis imperfecta and periodontal osteopetrosis. Bone 41: 851–859.

Alt, K. W. & Türp, J. C. (1997): Die Evolution der Zähne, Phylogenie, Ontogenie, Variation. Quintessenz, Berlin.

*Anders, U. & Koenigswald, W. v. (2013): Increasing and decreasing functional area of the dentition (FAD) during the life history of *Mammuthus primigenius* and other elephants. Paläontologische Zeitschrift 87: 515–527.

*Anders, U., Koenigswald, W. v., Ruf, I. & Smith, B. H. (2011a): Generalized individual dental age stages for fossil and extant placental mammals. Paläontologische Zeitschrift 85: 321–339.

*Anders, U., Koenigswald, W. v., Ruf, I. & Smith, B. H. (2011b): Erratum. Paläontologische Zeitschrift 85: 341.

Asher, R. J. & Lehmann, T. (2008): Dental eruption in afrotherian mammals. BMC Biology 6: 14.

Baumann, D. F. (1949): Die freilebenden Säugetiere der Schweiz. Hans Huber, Bern.

Baskin, J. A. (1980): Evolutionary reversal in *Mylagaulus* (Mammalia, Rodentia) from the late Miocene of Florida. The American Midland Naturalist 104: 155–162.

Bei, M., Stowell, S. & Maas, R. (2004): Msx2 controls ameloblast terminal differentiation. Developmental Dynamics 231: 758–765.

Berkovitz, B. K. B. (1972): Ontogeny of tooth replacement in the guinea pig (*Cavia cobaya*). Archives of Oral Biology 17: 711–718.

Black, C. C. & Wood, A. E. (1956): Variation and tooth-replacement in a Miocene mylagaulid rodent. Journal of Paleontology 30: 672–684.

Brandt, J. F. (1864): Observationes de Elasmotherii reliquiis. Mémoires de l'Académie Impériale des Sciences de St.-Pétersbourg, 7e série 8: 1–34.

Cai, J., Cho, S. W., Kim, J. Y., Lee, M. J., Cha, Y. G. & Jung, H. S. (2007): Patterning the size and number of tooth and its cusps. Developmental Biology 304: 499–507.

Calede, J. J. M. & Hopkins, S. S. B. (2012): Intraspecific versus interspecific variation in Miocene Great Basin mylagaulids: implications for systematics and evolutionary history. Zoological Journal of the Linnean Society 164: 427–450.

Chen D., Zhao, M. & Mundy, G. R. (2004): Bone morphogenetic proteins. Growth Factors 22: 233–241.

Dean, M. C., Beynon, A. D., Reid, D. J. & Whittaker, D. K. (1993): A longitudinal study of tooth growth in a single individual based on long- and short-period incremental markings in dentine and enamel. International Journal of Osteoarchaeology 3: 249–264.
Domning, D. P. & Hayek, L.-A. C. (1984): Horizontal tooth replacement in the Amazonian manatee (*Trichechus inunguis*). Mammalia 48: 105–128.
Dubost, G. & Henry, O. (2006): Comparison of diets of the acouchy, agouti and paca, the three largest terrestrial rodents of French Guianan forests. Journal of Tropical Ecology 22: 641–651.
Ewer, R. F. (1969): Form and function in the grass cutter, *Thryonomys swinderianus* Temm. (Rodentia, Thryonomyidae). Ghana Journal of Science 9: 131–140.
Fisher, D. C., Cherney, M. D., Newton, C. & Rountrey, A. N. (2014): Taxonomic overview and tusk growth analyses of Ziegler Reservoir proboscideans. Quaternary Research 82: 518–532.
Fortelius, M. (1985): Ungulate cheek teeth: development, functional, and evolutionary interrelations. Acta Zoologica Fennica 180: 1–76.
*Gailer, J. P., Calandra, I., Schulz-Kornas, E. & Kaiser, T. M. (2016): Morphology is not destiny: discrepancy between form, function and dietary adaptation in bovid cheek teeth. Journal of Mammalian Evolution 23: 369–383.
Gaillard, J.-M., Pontier, D., Allaine, D., Loison, A., Herve, J. C. & Heizmann, A. (1997): Variation in growth form and precocity at birth in eutherian mammals. Proceedings of the Royal Society B 264: 859–868.
Goddard, J. (1970): Age criteria and vital statistics of a black rhinoceros population. East African Wildlife Journal 8: 105–121.
Gomes Rodrigues, H., Lihoreau, F., Orliac, M., Thewissen, J. G. M. & Boisserie, J.-R. (2019): Unexpected evolutionary patterns of dental ontogenetic traits in cetartiodactyl mammals. Proceedings of the Royal Society B 286: 20182417.
Gomes Rodrigues, H., Marangoni, P., Šumbera, R., Tafforeau, P., Wendelen, W. & Viriot, L. (2011): Continuous dental replacement in a hyper-chisel tooth digging rodent. Proceedings of the National Academy of Sciences of the United States of America 108: 17355–17359.
Gomes Rodrigues, H., Solé, F., Charles, C., Tafforeau, P., Vianey-Liaud, M. & Viriot, L. (2012): Evolutionary and biological implications of dental mesial drift in rodents: the case of the Ctenodactylidae (Rodentia, Mammalia). PLoS One 7: e50197.
Graf, D. & Mitsiadis, T. A. (2013): Den Geheimnissen der Zahnentwicklung auf der Spur. Swiss Medical Forum 13: 168–170.
Grau, G. A., Sanderson, G. C. & Rogers, J. P. (1970): Age determination of raccoons. Journal of Wildlife Management 34: 364–372.
Greaves, W. S. (1973): The inference of jaw motion from tooth wear facets. Journal of Paleontology 47: 1000–1001.
Habermehl, D. K.-H. (1975): Die Altersbestimmung bei Haus- und Labortieren. Paul Parey, Berlin.
Habermehl, D. K.-H. (1985): Altersbestimmung bei Wild- und Pelztieren. Paul Parey, Berlin.
Haeckel, E. (1866). Generelle Morphologie der Organismen. Allgemeine Grundzüge der organischen Formen-Wissenschaft, mechanisch begründet durch die von Charles Darwin reformirte Descendenz-Theorie (Vol. 2). Georg Reimer, Berlin.
Hagmann, G. (1907): Über das Gebiß von *Coelogenys* und *Dasyprocta* in seinen verschiedenen Stadien der Abkauung. Zeitschrift für Morphologie und Anthropologie 10: 464–480.
Harada, H. & Ohshima, H. (2004): New perspectives on tooth development and the dental stem cell niche. Achives of Histology and Cytology 67: 1–11.
Harjunmaa, E., Seidel, K., Häkkinen, T., Renvoisé, E., Corfe, I. J., Kallonen, A., Zhang, Z.-Q., Evans, A. R., Mikkola, M. L., Salazar-Ciudad, I., Klein, O. D., Jernvall, J. (2014): Replaying evolutionary transitions from the dental fossil record. Nature 512: 44–48.
Harman, M. T. & Smith, A. (1936): Some observations on the development of the teeth of *Cavia cobaya*. The Anatomical Record 66: 97–111.
Heizmann, E. & Fahlbusch, V. (1983): Die mittelmiozäne Wirbeltierfauna vom Steinberg (Nördlinger Ries). Eine Übersicht. Mitteilungen der Bayerischen Staatssammlung für Paläontologie und historische Geologie 23: 83–93.
Herrmann, N. (2002): Food-specialization and structural parameters of dental patterns of Arvicolinae (Rodentia, Mammalia). Senckenbergiana lethaea 82: 153–165.
Hünermann, K. A. (1966): Der Bau des Biber-Prämolaren und seine Verwendbarkeit für die Systematik der Castoridae. Neues Jahrbuch für Geologie und Paläontologie, Abhandlungen 125: 227–234.
*Hummel, J., Findeisen, E., Südekum, K.-H., Ruf, I., Kaiser, T. M., Bucher, M., Clauss, M. & Codron, D. (2011): Another one bites the dust – faecal silica levels in large herbivores correlate with high-crowned teeth. Proceedings of the Royal Society B 278: 1742–1747.
Janis, C. M. & Fortelius, M. (1988): On the means whereby mammals achieve increased functional durability of their dentitions, with special reference to limiting factors. Biological Reviews 63: 197–230.
Jekl, V., Hauptman, K., Jeklova, E. & Knotek, Z. (2011): Dental eruption chronology in degus (*Octodon degus*). Journal of Veterinary Dentistry 28: 16–20.
Jernvall, J. (2000): Linking development with generation of novelty in mammalian teeth. Proceedings of the National Academy of Sciences of the United States of America 97: 2641–2645.
Jernvall, J. & Thesleff, I. (2000): Reiterative signaling and patterning during mammalian tooth morphogenesis. Mechanisms of Development 92: 19–29.
Jernvall, J., Keränen, S. V. E. & Thesleff, I. (2000): Evolutionary modification of development in mammalian teeth: quantifying gene expression patterns and topography. Proceedings of the National Academy of Sciences of the United States of America 97: 14444–14448.
Kavanagh, K. D., Evans, A. R. & Jernvall, J. (2007): Predicting evolutionary patterns of mammalian teeth from development. Nature 449: 427–432.
Kay, R. F. (1975): The functional adaptations of primate molar teeth. American Journal of Physical Anthropology 43: 195–216.
Kelley, J. & Smith T. M. (2003): Age at first molar emergence in early Miocene *Afropithecus turkanensis* and life history evolution in the Hominoidea. Journal of Human Evolution 44: 307–329.
Kitts, D. B. & Black, C. C. (1959): A Pliocene vertebrate local fauna from Roger Mills County, Oklahoma. Oklahoma Geological Survey 48: 27–47.
Klein, R. & Cruz-Uribe, K. (1983): The computation of ungulate age (mortality) profiles from dental crown heights. Paleobiology 9: 70–78.
Knußman, R. R. & Martin, R. (1988): Anthropologie: Handbuch der vergleichenden Biologie des Menschen, Bd. I/1. Gustav Fischer, Stuttgart.
Koenigswald, W. v. (1980): Schmelzstruktur und Morphologie in den Molaren der Arvicolidae (Rodentia). Abhandlungen der senckenbergischen naturforschenden Gesellschaft 539: 1–129.
Koenigswald, W. v. (1982): Zum Verständnis der Morphologie der Wühlmausmolaren (Arvicolidae, Rodentia, Mammalia). Zeitschrift für Geologische Wissenschaften 10: 951–962.
*Koenigswald, W. v. (2011): Diversity of hypsodont teeth in mammalian dentitions – construction and classification. Palaeontographica A 294: 63–94.
*Koenigswald, W. v. (2016): Specialized wear facets and late ontogeny in mammalian dentition. Historical Biology 30: 7–29.
*Koenigswald, W. v. (2020): Construction and wear of mammalian teeth in terms of heterochrony. In: Martin, T. & Koenigswald, W. v. (eds.). Mammalian Teeth – Form and Function. Pfeil, Munich: 171–186.
Koenigswald, W. v. & Kolfschoten, T. v. (1996): The *Mimomys-Arvicola* boundary and the enamel thickness quotient (SDQ) of *Arvicola* as stratigraphic markers in the Middle Pleistocene. In: Turner, C. (ed.). The Early Middle Pleistocene in Europe. Balkema, Rotterdam: 211–226.
Koenigswald, W. v., Sander, P. M., Leite, M. B., Mörs, T. & Santel, W. (1994): Functional symmetries in the schmelzmuster and morphology of rootless rodent molars. Zoological Journal of the Linnean Society 110: 141–179.
Koenigswald, W. v., Goin, F. & Pascual, R. (1999): Hypsodonty and enamel microstructure in the Paleocene gondwanatherian mammal *Sudamerica ameghinoi*. Acta Palaeontologica Polonica 44: 263–300.

*Koenigswald, W. v., Anders, U., Engels, S., Schultz, J. & Ruf, I. (2010): Tooth morphology in fossil and extant Lagomorpha (Mammalia) reflects different mastication patterns. Journal of Mammalian Evolution 17: 275–299.

Krause, D. W. (2014): Dental morphology of *Vintana sertichi* (Mammalia, Gondwanatheria) from the Late Cretaceous of Madagascar. Journal of Vertebrate Paleontology 34, supplement to number 6: 137–165.

Kroon, H. M. (1929): Die Lehre der Altersbestimmung der Haustiere, 3. Auflage. M. & H. Schaper, Hannover.

Maglio, V. J. (1972): Evolution of mastication in the Elephantidae. Evolution 26: 638–658.

Marks, Jr., S. C. & Schroeder, H. E. (1996): Tooth eruption: theories and facts. The Anatomical Record 245: 374–393.

Martin, L. F., Winkler, D., Tütken, T., Codron D., De Cuyper, A., Hatt, J.-M. & Clauss, M. (2019): The way wear goes: phytolith-based wear on the dentine-enamel system in guinea pigs (*Cavia porcellus*). Proceedings of the Royal Society B 286: 20191921

Mayhew, D. F. (1979): Evolution of a dental character in the beaver *Castor fiber* L. (Mammalia: Rodentia). Zoological Journal of the Linnean Society 65: 177–184.

Merceron, G., Ramdarshan, A., Blondel, C., Boisserie, J.-R., Brunetiere, N., Francisco, A., Gautier, D., Milhet, X., Novello, A. & Pret, D. (2016): Untangling the environmental from the dietary: dust does not matter. Proceedings of the Royal Society B 283: 20161032.

Mendoza, M. & Palmqvist, P. (2008): Hypsodonty in ungulates: an adaptation for grass consumption or for foraging in open habitat? Journal of Zoology 274: 134–142.

Monson, T. A. & Hlusko, L. J. (2018): The evolution of dental eruption sequence in artiodactyls. Journal of Mammalian Evolution 25: 15–26

Nagy, N. B. & Daniel, J. C. (1992): Development of the rabbit craniomandibular joint in association with tooth eruption. Archives of Oral Biology 37: 271–280.

Nowak, R. M. (1991): Walker's Mammals of the World, 5th ed., Volume 1. Johns Hopkins, Baltimore and London.

Osborn, J. W. (2008): A model of growth restraints to explain the development and evolution of tooth shapes in mammals. Journal of Theoretical Biology 225: 338–343.

Piechocki, R. & Stiefel, A. (1977): Zahndurchbruch und Zahnwechsel beim Elbebiber, *Castor fiber albicus*. Anatomischer Anzeiger 142: 374–384.

Popowics, T. E. & Fortelius, M. (1997): On the cutting edge: tooth blade sharpness in herbivorous and faunivorous mammals. Annales Zoologici Fennici 34: 73–88.

Proffit, W. R., Prewitt, J. R., Baik, H. S. & Lee, C. F. (1991): Video microscope observations of human premolar eruption. Journal of Dental Research 70: 15–18.

Rensberger, J. M. (1973): An occlusion model for mastication and dental wear in herbivorous mammals. Journal of Paleontology 47: 515–528.

Renvoisé, E., Kavanagh, K. D., Lazzari, V., Häkkinen, T. J., Rice, R., Pantalacci, S., Salazar-Ciudad, I. & Jernvall, J. (2017): Mechanical constraint from growing jaw facilitates mammalian dental diversity. Proceedings of the National Academy of Sciences of the United States of America 114: 9403–9408.

Rinderknecht, A., Bostelmann, T. E. & Ubilla, M. (2011): A new genus of giant Dinomyinae (Rodentia, Hystricognathi: Caviomorpha) from the late Miocene of Uruguay. Journal of Mammalogy 92: 169–178.

Rountrey, A. N., Fisher, D. C., Vartanyan, S. & Fox, D. L. (2007): Carbon and nitrogen isotope analyses of a juvenile woolly mammoth tusk: evidence of weaning. Quaternary International 169/170: 166–173.

Ruf, I. (2020). Ontogenetic transformations of the ethmoidal region in Muroidea (Rodentia, Mammalia): new insights from perinatal stages. Vertebrate Zoology 70: 383–415.

Rusconi, C. (1930): Evolución cráneodental de la nutria (*Myocastor coypus bonariensis*) a través de su desarrollo postembrionario. Anales de la Sociedad Cientifica Argentina 110: 5–31.

Rusconi, C. (1934): Apuntes sobre la evolución ontogenética de los molares del género *Agouti* (Rodentia). Revista Odontológica 22: 167–177.

Sanson, G. D. (1989): Morphological adaptations of teeth to diets and feeding in the Macropodidea. In: Grigg, G., Jarman, P. & Hume, I. (eds.). Kangaroos, Wallabies and Rat-Kangaroos. Surrey Batty & Sons, Chipping Norton: 151–168.

Schmidt-Kittler, N. (1984): Pattern analysis of occlusal surfaces in hypsodont herbivores and its bearing on morpho-functional studies. Proceedings of the Koninklijke Nederlandse Akademie van Wetenschappen 87: 453–480.

Schmidt-Kittler, N. (2002): Feeding specializations in rodents. Senckenbergiana lethaea 82: 141–152.

*Schubert, A. M. (2019): Konstruktionsmorphologie hypsodonter Backenzähne bei rezenten und fossilen Rodentia (Mammalia). Dissertation, Rheinische Friedrich-Wilhelms-Universität Bonn. http://hss.ulb.uni-bonn.de/2019/5485/5485.htm

Schultz, A. H. (1935): Eruption and decay of the permanent teeth in primates. American Journal of Physical Anthropology 14: 489–581.

Schultz, A. H. (1960): Age changes in primates and their modification in man. In: Tanner, J. M. (ed.). Human Growth. Pergamon, Oxford: 1–20.

Shigehara, N. (1980): Epiphyseal union, tooth eruption, and sexual maturation in the common tree shrew, with reference to its systematic problem. Primates 21: 1–19.

Smith, B. H. (2000): 'Schultz's rule' and the evolution of tooth emergence and replacement patterns in primates and ungulates. In: Teaford, M., Smith, M. & Ferguson, M. (eds.). Development, Function and Evolution of Teeth. Cambridge University Press, Cambridge: 212–227.

Smith, B. H., Crummett, T. L. & Brandt, K. L. (1994): Ages of eruption of primate teeth: a compendium for aging individuals and comparing life histories. Yearbook of Physical Anthropology 37: 177–231.

Smith, T. M., Tafforeau, P., Le Cabec, A., Bonnin, A., Houssaye, A., Pouech, J., Moggi-Cecchi, J., Manthi, F., Ward, C., Makaremi, M. & Menter, C. G. (2015): Dental ontogeny in Pliocene and early Pleistocene hominins. PLoS One 10: e0118118.

Soderwall, A. L., Kent, Jr., H. A., Turbyfill, C. L. & Britenbaker, A. L. (1960): Variation in gestation length and litter size of the golden hamster *Mesocricetus auratus*. Journal of Gerontology 15: 246–248.

Sone, K., Koyasu, K., Kobayashi, S. & Oda, S.-i. (2008): Fetal growth and development of the coypu (*Myocastor coypus*): prenatal growth, tooth eruption, and cranial ossification. Mammalian Biology 73: 350–357.

Strömberg, C. A. E. (2006): Evolution of hypsodonty in equids: testing a hypothesis of adaptation. Paleobiology 32: 236–258.

Suraprasit, K., Chaimanee, Y., Martin, T. & Jaeger, J.-J. (2011): First castorid (Mammalia, Rodentia) from the Middle Miocene of Southeast Asia. Naturwissenschaften 98: 315–328.

Teaford, M. F., Smith, M. M. & Ferguson, M. W. J. (2009): Development, Function and Evolution of Teeth. Cambridge University Press, New York.

Thenius, E. (1989). Zähne und Gebiß der Säugetiere. De Gruyter, Berlin.

Tobien, H. (1974): Zur Gebißstruktur, Systematik und Evolution der Genera *Amphilagus* und *Titanomys* (Lagomorpha, Mammalia) aus einigen Vorkommen im jüngeren Tertiär Mittel- und Westeuropas. Mainzer geowissenschaftliche Mitteilungen 3: 95–214.

Tobien, H. (1975): Zur Gebißstruktur, Systematik und Evolution der Genera *Piezodus*, *Prolagus* und *Ptychoprolagus* (Lagomorpha, Mammalia) aus einigen Vorkommen des jüngeren Tertiär Mittel- und Westeuropas. Notizblatt des Hessischen Landesamtes für Bodenforschung zu Wiesbaden 103: 103–186.

Townsend, K. E. B. & Croft, D. A. (2008): Enamel microwear in caviomorph rodents. Journal of Mammalogy 89: 730–743.

Tummers, M. & Thesleff, I. (2003): Root or crown: a developmental choice orchestrated by the differential regulation of the epithelial stem cell niche in the tooth of two rodent species. Development 130: 1049–1057.

Ubilla, M. & Rinderknecht, A. (2016): *Lagostomus maximus* (Desmarest) (Rodentia, Chinchillidae), the extant plains vizcacha in the Late Pleistocene of Uruguay. Alcheringa 40: 354–365.

Ungar, P. S. (2010): Mammal Teeth: Origin, Evolution, and Diversity. The Johns Hopkins University Press, Baltimore.

van Weers, D. J. (1990): Dimensions and occlusal patterns in molars of *Hystrix brachyura* Linnaeus, 1758 (Mammalia, Rodentia) in a system of wear categories. Bijdragen tot de Dierkunde 60: 121–134.

van Weers, D. J. (1993): On the tooth morphology of the long-tailed porcupine *Trichys fasciculata* (Hystricidae: Rodentia), with notes on the genera *Atherurus* and *Hystrix*. Raffles Bulletin of Zoology 41: 251–261.

van Weers, D. J. (2005): A taxonomic revision of the Pleistocene *Hystrix* (Hystricidae, Rodentia) from Eurasia with notes on the evolution of the family. Contributions to Zoology 74: 301–312.

Williams, S. H. & Kay, R. F. (2001): A comparative test of adaptive explanations for hypsodonty in ungulates and rodents. Journal of Mammalian Evolution 8: 207–229.

*Winkler, D. A. & Kaiser, T. M. (2015): Uneven distribution of enamel in the tooth crown of a Plains Zebra (*Equus quagga*). PeerJ 3: e1002.

Yamada, F., Shiraishi, S., Taniguchi, A. & Uchida, T. A. (1990): Growth, development and age determination of the Japanese hare, *Lepus brachyurus brachyurus*. Journal of the Mammalogical Society of Japan 14: 65–77.

Zouvelou, V., Luder, H. U., Mitsiadis, T. A. & Graf, D. (2009): Deletion of BMP7 affects the development of bones, teeth, and other ectodermal appendages of the orofacial complex. Journal of Experimental Zoology, Part B: Molecular and Developmental Evolution 312B: 361–374.

Enamel ridge alignments in ungulates: a cut above

Daniela E. Winkler, Jordi Marcé-Nogué, and Thomas M. Kaiser

Introduction

Efficient mastication in ungulate herbivores is related to enamel ridge morphology as well as tooth crown height, both of which are found to be adapted to more browse or grass-dominated diets. During ontogeny, the dentition undergoes a change in morphology, as teeth are subsequently worn down. Consequently, a loss in functionality is expected to occur in later life stages, while the dentition should maintain a functional optimum during an individual's growth and reproductive periods.

Comminution of herbivorous diets requires post-canine morphology characterized by shear-cutting ridges combined with compression basins. During mastication, opposing enamel ridges act against each other to disintegrate larger food components into smaller pieces, which are subsequently compressed in dentin basins to extract cell components. All extant ungulate herbivores with either selenodont or selenolophodont cheek teeth show this distinct morphology. The number, complexity and width of ridges as well as tooth crown height are regarded as adaptations to the special demands of grazing and browsing diets (Archer & Sanson 2002, Heywood 2009, Kaiser et al. 2010, Famoso et al. 2013).

Despite the diversity in molar morphologies and mechanical demands of diets, adaptive trends exist in dental form, function, and biomechanical constraints in ungulates. Firstly, newly erupted post-canine teeth are not immediately fully functional, they need a certain degree of wear in order to expose enamel ridges and initiate occlusion with their antagonists. Kurtén (1983) suggested more rapid wear of the tip of the tooth crown in antelopes. For *Equus quagga*, Winkler and Kaiser (2015a) have shown that the tip of premolar and molar tooth crowns is composed of proportionally more dentin and cementum as compared to the rest of the crown. This might be useful to facilitate rapid wear of the crown apex and thus faster occlusion, while the base of the tooth is adapted to resist wear and high stress loads due to a larger enamel content. Secondly, the post-canine dentition should maintain high chewing efficiency for a prolonged time, but at least until the end of reproductive age. Rapid wear of the dentition is suggested to lead to earlier senescence, as seen in captive browsing ungulates that received more abrasive diets than they would have found in the wild, resulting in more abrasion-dominated mesowear patterns (Kaiser et al. 2009). In adult individuals of several cetartiodactyl and perissodactyl species, the upper third molar (M3) shows significantly higher relative proportions of enamel compared to the first upper molar (M1) (Winkler & Kaiser 2015b). This is in accordance with observations in museum collections, where ruminant jaws are usually found with the third molar only just fully erupted, and yet the first molar severely worn (Damuth & Janis 2011). Winkler & Kaiser (2015b) postulated the "functional compensation hypothesis", proposing that for both terrestrial Cetartiodactyla and Perissodactyla, the M3 maintains high functionality and is morphologically "enhanced" to compensate for the functional loss associated with advanced wear in the M1. For hypsodont taxa only, this had been already suggested by Solounias et al. (1994), who noted that the second and third molars are more hypsodont than the first molar and can hence retain high functionality after the first molar has been worn down. Further support for the functional compensation hypothesis comes from Kaiser (2002), who showed for the extinct three-toed horse *Cormohipparion* sp., that in all cheek tooth positions with the exception of the M3, the occlusal planes successively exposed by wear tend to lose some functional "optimization" in terms of shearing blade orientation in advanced individual age.

It can be expected that a functional compensation of M3 for loss of masticatory efficiency in more anterior tooth positions is a general trait in the evolution of herbivorous ungulates, because it is not only evident in a larger enamel proportion in M3s, but also in the orientation of enamel ridges. Rensberger (1973) suggested that enamel ridges should be aligned almost perpendicular towards the chewing direction to reach the highest shearing efficiency. If such a condition were favourable, a notion that is challenged here, one would expect enamel ridges of the M3 to be aligned at angles close to 90° with the main direction of occlusal movements. To test for functional compensation, the following hypotheses are posed:

1. The M3 compensates for functional losses in more anterior tooth positions and has hence a larger proportion of enamel ridges aligned to higher angles than M1 and M2.
2. Enamel ridge alignments of younger ontogenetic stages have significantly higher enamel ridge alignments compared to senile stages of the same individual. A "prime age" should be identifiable as representing most functionally favourable traits along the tooth row.
3. M3 is built more resistant to withstand higher forces. Enamel ridges are more strictly oriented at high angles to optimize shearing contacts during the displacement-limited power-stroke. M1 has less strictly oriented enamel ridges, which ensure shearing contacts under a higher degree of freedom in lateral displacement.

By comparing enamel ridge alignments of ruminant cetartiodactyls in the upper second molar (M2), Kaiser et al. (2010) found that species with more grass in their natural diet showed a higher proportion of enamel ridges aligned at low angles to the direction of the chewing stroke. They

T. Martin & W. v. Koenigswald (eds.): Mammalian Teeth – Form and Function. Pp. 125–170, 30 figs., 30 tabs.

DOI: http://doi.org/10.23788/mammteeth.08 Published 22 December 2020

used functional residual (FR) images of enamel ridges on the occlusal surface employing the FR algorithm (first described in Kaiser 2002). The current study employs the FR algorithm, but with a smaller Gaussian blur setting (8 pixel instead of 35 pixel in Kaiser et al. 2010). It is anticipated that this will better preserve the original morphology in order to identify subtle differences between tooth positions and ontogenetic stages. It is important to note, that the results of this chapter will thus not immediately be comparable to those of Kaiser et al. (2010).

With the exception of equids, the molar tooth row spans approximately over 50 % of the postcanine dentitions' anterio-posterior dimension (Janis 1990a, b, Mendoza et al. 2002). Hence, M1 and P4 constitute the masticatory centre of the tooth row (except for camelids, where the centre is expected to shift towards M1 and M2 as premolars are reduced (Janis 1990b)). The M3 is always closest to the temporomandibular joint. Based on their position in the jaw, teeth are always under the influence of distinct constraints imposed on them. While the M3 is expected to experience the highest bite forces (Greaves 2012), the orthal displacement (opening gape) as well as the lateral displacement (chewing stroke) at this position are smallest/shortest (Greaves 2012). The contrary applies for M1. Lower bite forces (due to the longer lever of the lower jaw corpus) and more extensive lateral displacement will immediately translate into less precise guidance of the occlusal movements as compared to more posterior tooth positions (Taylor et al. 2013). It therefore can be assumed that enamel ridge alignments, if an evolutionary response to local functional thresholds in the M1 occurs, are signified by traits indicative of a force-limited but displacement-optimized system, while the M3 is displacement-limited (due to its position in the jaw), but force-optimized.

A dataset comprised of 22 cetartiodactyl and perissodactyl taxa with diverse dental morphologies, ranging from brachydont browsing species to hypsodont grazers, was employed to investigate common dental morphological adaptations towards maintaining high efficiency of the masticatory apparatus. Hereby specifically the functional traits of enamel ridge morphology towards specific demands of life stage and tooth position in the jaw are examined.

CT scanning techniques are applied to virtually wear down upper and lower molars of one side, in order to create hypothetical future occlusal surfaces of an individual. Those hypothetical surfaces are then treated as ontogenetic wear stages. While extant Perissodactyla possess an enlarged and elongated premolar tooth row (P2–P4), ruminants often have shortened premolar tooth rows (Mendoza et al. 2002). These specific dental traits are most pronounced between specialized perissodactyl grazers like extant equids, in which the premolar tooth row exceeds the molar tooth row (M1–M3) in length, and grazing ruminant Cetartiodactyla and camelids which have elongated the third molar and reduced the premolar tooth row (Janis 1988, 1990a, b, Solounias & Dawson-Saunders 1988). The role of premolars and molars in the masticatory process is hence less comparable between Perissodactyla and Cetartiodactyla and likely such morphologies are also a result of dietary trait evolution. The molar tooth row, however, is never reduced in size and hence can be expected to be a substantial component in efficient food comminution in ruminants, camelids and perissodactyls.

Material and Methods

The material used in this chapter is curated at the Zoological Museum Hamburg (ZMH) or Museum für Naturkunde Berlin (ZMB) (compare Tab. 8.1) and consists of one individual per species. Specimens were selected according to individual age and dental wear stage: All permanent teeth should be fully mineralized and the M3 unworn or in early wear to cover the longest functional time span of the permanent dentition. If undamaged, preferably right dentitions were selected in order to create a consistent sample.

CT data of upper and lower occluding dentitions were obtained on the v|tome|x s (GE phoenix|x-ray) at Section Paleontology, Institute of Geosciences (Rheinische Friedrich-Wilhelms-Universität Bonn, Germany) and for the larger specimens (Rhinocerotidae) on a medical CT at the Universitätsklinikum Hamburg Eppendorf (UKE). Exact voxel sizes of each specimen are given in the supplementary content of Winkler and Kaiser (2015a, b). CT data were reconstructed and segmented in VGStudio MAX 2.2 (Volume Graphics Heidelberg) following the procedure described in Winkler and Kaiser (2015b) to obtain volume models composed of enamel only. These models were cut with clipping-planes to create hypothetical future ontogenetic stages (Fig. 8.1).

The four subsequent stages represent the occlusal surfaces with close to 100 % (subadult = stage 1), 75 % (adult young = stage 2), 50 % (adult old = stage 3) and 25 % (senile = stage 4) of the observed crown height without clipping. It is important to note that the hypothetical wear stages therefore take the different eruption times of the molars into account. During stage 3, for example, the M3 is worn down to 50 % of the *un*worn crown-height, because individuals were selected with unworn M3 or very early wear. For M1, on the contrary, stage 3 represents 50 % of the *initial* crown height in the specimen, which has already experienced a substantial amount of wear and hence crown height loss.

2D images in occlusal view were obtained for each stage and exported to Adobe Photoshop CS5 (Adobe, Dublin). All images were equally scaled to 4400 × 3400 pixels. Fissures in the enamel which resulted in gaps within an otherwise continuous ridge were manually filled. Then an FR algorithm with 8 pixel Gaussian blur was applied and the resulting grey-scaled matrix converted into a one-bit BMP using the 50 % dichotomisation algorithm in Adobe Photoshop CS5 to smooth enamel ridges, which often appear noisy and pixelated after extraction from VGStudio MAX. Kaiser et al. (2010) applied an FR algorithm with a 35 pixel Gaussian blur, resulting in fusion or elimination of thin enamel ridges, which do not function as shearing edges (Fig. 8.1E). Only a slight smoothing was applied in order to preserve also smaller morphological features, as these should be compared between ontogenetic stages of

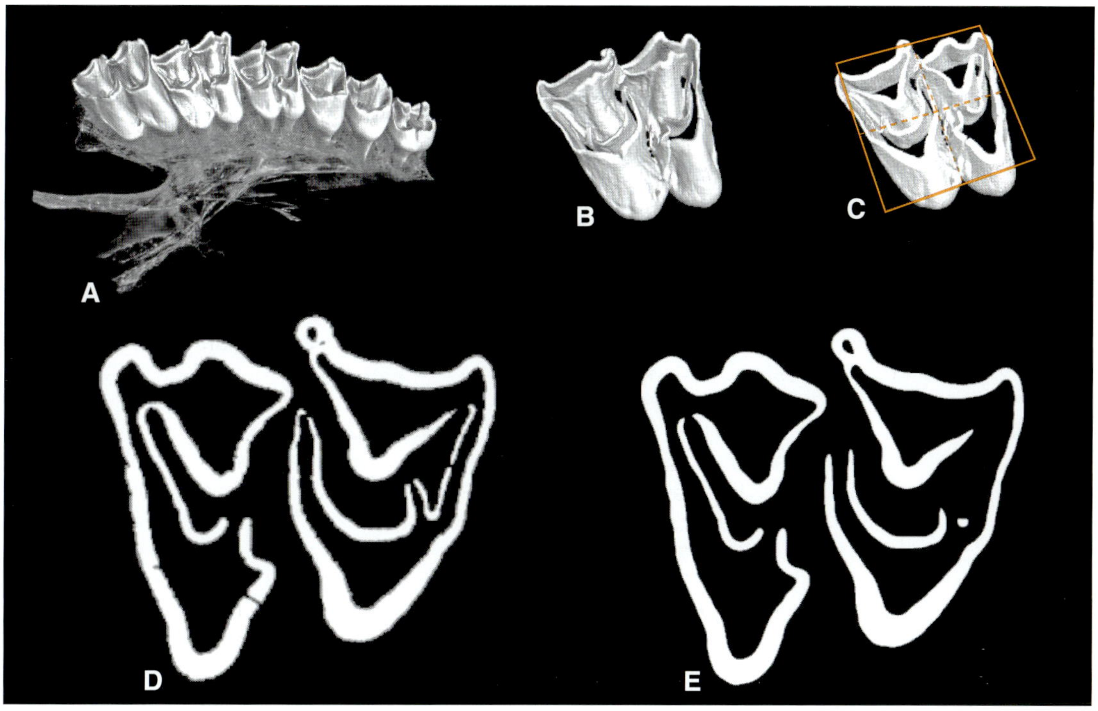

Fig. 8.1. Extraction of enamel ridge images from CT data. **A,** Original volume data of *Ozotoceros bezoarticus* (upper right dentition); **B,** extracted enamel volume of the second molar; **C,** same as in B with clipping plane at stage 2 (approx. 75 % of the initial crown height); **D,** extracted 2D image of enamel ridges at stage 2; **E,** resulting final image after FR calculation.

an individual, which a stronger Gaussian blur might have obliterated. A complete FR example image sequence is given in the Appendix (Fig. 8. A1) for *Elaphurus davidianus*.

Each image was oriented in anatomical position towards the chewing direction, which was set vertically (Fig. 8.2A). The computation of enamel ridge alignments was conducted in ArcView GIS 3.2 (esri, New York) following a script written by Andreas Musolff (Helmholtz-Zentrum für Umweltfoschung, Leipzig, Germany). First, each image was vectorized and converted to polylines on which a mark was placed every 100 pixels (Fig. 8.2B). Measurement of angles was achieved by fitting a tangent through two adjacent marks (Fig. 8.2C). Only the buccal (leading) edges were considered, as trailing edges do not significantly contribute to food disintegration in the initial chewing stroke (Fig. 8.2D).

Angle counts were collapsed to 0°–90°, with 91° equalling 89° etc. This is applicable under the assumption that

Table 8.1. Sample composition. Each species is represented by one specimen. Diet classification after Mendoza et al. (2002): HBR, high-level browser; GBR, general browser; FGR, fresh-grass grazer; GGR, general grazer; OMF, open-habitat mixed-feeder.

Species	Specimen ID	Order	Family	Diet	Side	Crown height (M3)
Antilocapra americana	ZMH-10162	Cetartiodactyla	Antilocapridae	GBR	left	hypsodont
Alcelaphus buselaphus	ZMH-7487	Cetartiodactyla	Bovidae	GGR	left	hypsodont
Antilope cervicapra	ZMH-9349	Cetartiodactyla	Bovidae	GGR	right	hypsodont
Antidorcas marsupialis	ZMH-8268	Cetartiodactyla	Bovidae	OMF	left	hypsodont
Aepyceros melampus	ZMH-5809	Cetartiodactyla	Bovidae	OMF	right	hypsodont
Camelus bactrianus	ZMH-1870	Cetartiodactyla	Camelidae	OMF	right	hypsodont
Capra ibex	ZMH-6983	Cetartiodactyla	Bovidae	OMF	right	hypsodont
Ceratotherium simum	ZMH-2552	Perissodactyla	Rhinocerotidae	GGR	left	hypsodont
Connochaetes taurinus	ZMH-6777	Cetartiodactyla	Bovidae	GGR	right	hypsodont
Diceros bicornis	ZMH-9379Z	Perissodactyla	Rhinocerotidae	GBR	right	brachydont
Damaliscus pygargus	ZMB_Mam_55243	Cetartiodactyla	Bovidae	GGR	right	hypsodont
Elaphurus davidianus	ZMB_Mam_75425	Cetartiodactyla	Cervidae	FGR	right	mesodont
Equus quagga	ZMB_Mam_70335	Perissodactyla	Equidae	GGR	right	hypsodont
Hemitragus jemlahicus	ZMH-7747	Cetartiodactyla	Bovidae	OMF	left	hypsodont
Kobus ellipsiprymnus	ZMH-447	Cetartiodactyla	Bovidae	FGR	right	hypsodont
Lama glama	ZMH-8254	Cetartiodactyla	Camelidae	OMF	left	hypsodont
Lama huanachus	ZMH-286	Cetartiodactyla	Camelidae	OMF	left	hypsodont
Litocranius walleri	ZMB_Mam_39663	Cetartiodactyla	Bovidae	HBR	right	brachydont
Naemorhedus goral	ZMH-478	Cetartiodactyla	Bovidae	OMF	left	hypsodont
Ozotoceros bezoarticus	ZMH-7800	Cetartiodactyla	Cervidae	OMF	right	brachydont
Redunca fulvorufula	ZMH-7632	Cetartiodactyla	Bovidae	FGR	right	hypsodont
Rangifer tarandus	ZMH-7104	Cetartiodactyla	Cervidae	GBR	right	brachydont

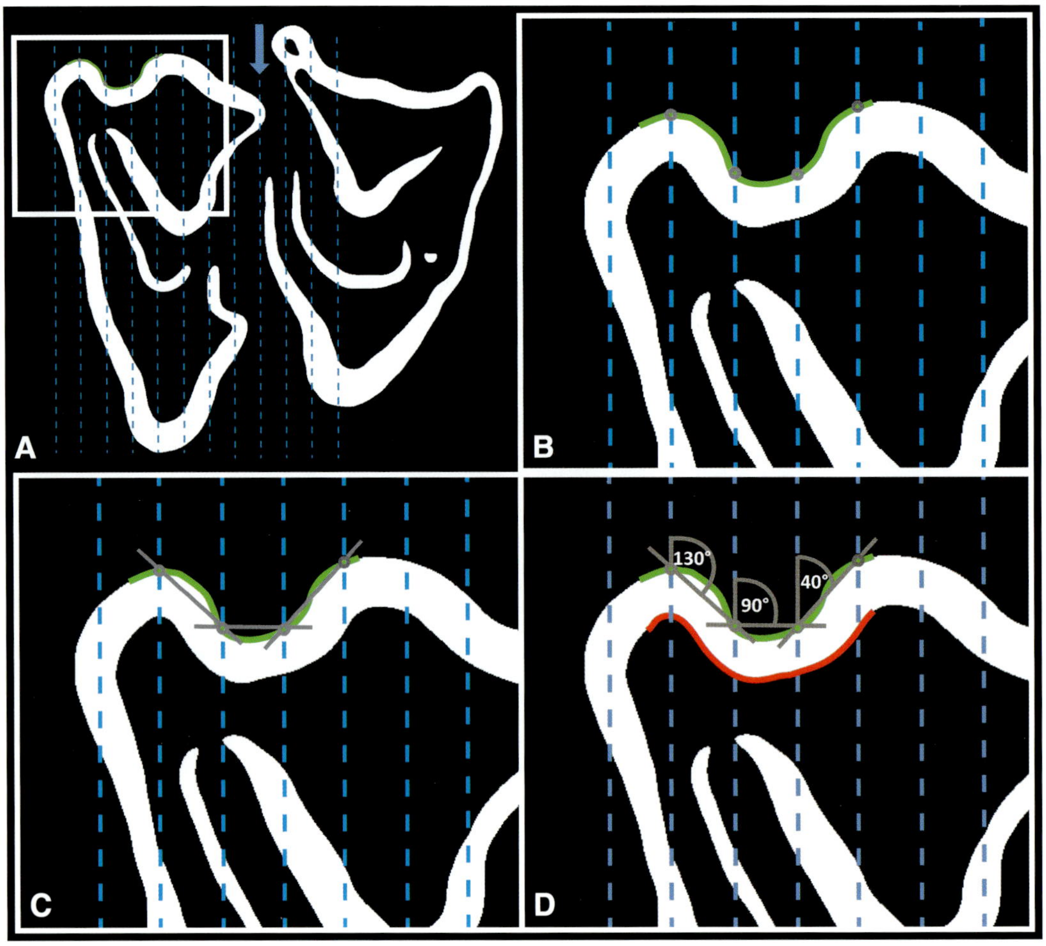

Fig. 8.2. Computation of enamel ridge alignments (ERA) for a sample part of an enamel ridge. ERAs are always computed for the leading edges of an occlusal surface. **A,** The blue arrow indicates the chewing direction; **B,** marks are placed along the leading edge (green); **C,** tangents are fitted through adjacent marks resulting in **D,** with the ERAs shown. The trailing edge (red) is not considered.

functional significance for tilts of the same angle clockwise and counter-clockwise to the chewing direction is equivalent (Kaiser et al. 2010).

Computational models for *A. buselaphus*, *A. cervicapra*, *A. melampus*, *C. simum*, *C. taurinus*, *D. bicornis*, *E. quagga*, *K. ellipsiprymnus*, and *L. walleri* were created and provided by Dr. Jordi Marcé-Nogué based on pictures sent by courtesy of Dr. Soledad de Esteban-Trivigno. Some of these models were CT-scanned and used later in Zhou et al. (2019). Bite forces were calculated in ANSYS (version 16.1) for central positions on lower m1, m2 and m3 in each species according to the equilibrium laws of static mechanics where only force values and distances from the applied points are necessary. The combined input muscle force for masseter, temporal and pterygoid muscles was set to 1 N in all models. This way allometric scaling of muscle forces with body mass is avoided and only the relative magnitude of bite forces along the jaw and between tooth positions is displayed. This procedure is favourable as the enamel ridge alignments are shown to be phylogenetically independent and do not scale with body mass.

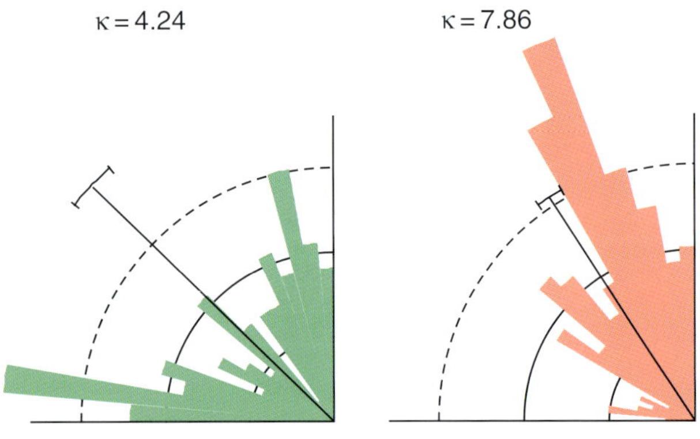

Fig. 8.3. Example ERA distributions with low (left) and high (right) concentrations.

Statistics

Linear statistical analyses were conducted in R version 2.15.1 (R.D.C. Team 2009) and circular statistics in Oriana 4 (Kovach Computing), respectively. Enamel ridge alignments (ERA) were expressed as frequencies and the circular statistical parameters circular median (θ), concentration (κ) and mean vector (μ) computed per ontogenetic stage and tooth position. Concentration is a dimensionless value that describes the dispersion of a circular dataset. The lower the κ-value (with $\kappa = 0$ corresponding to uniformity), the higher the dispersion (Fig. 8.3). Data of all ontogenetic

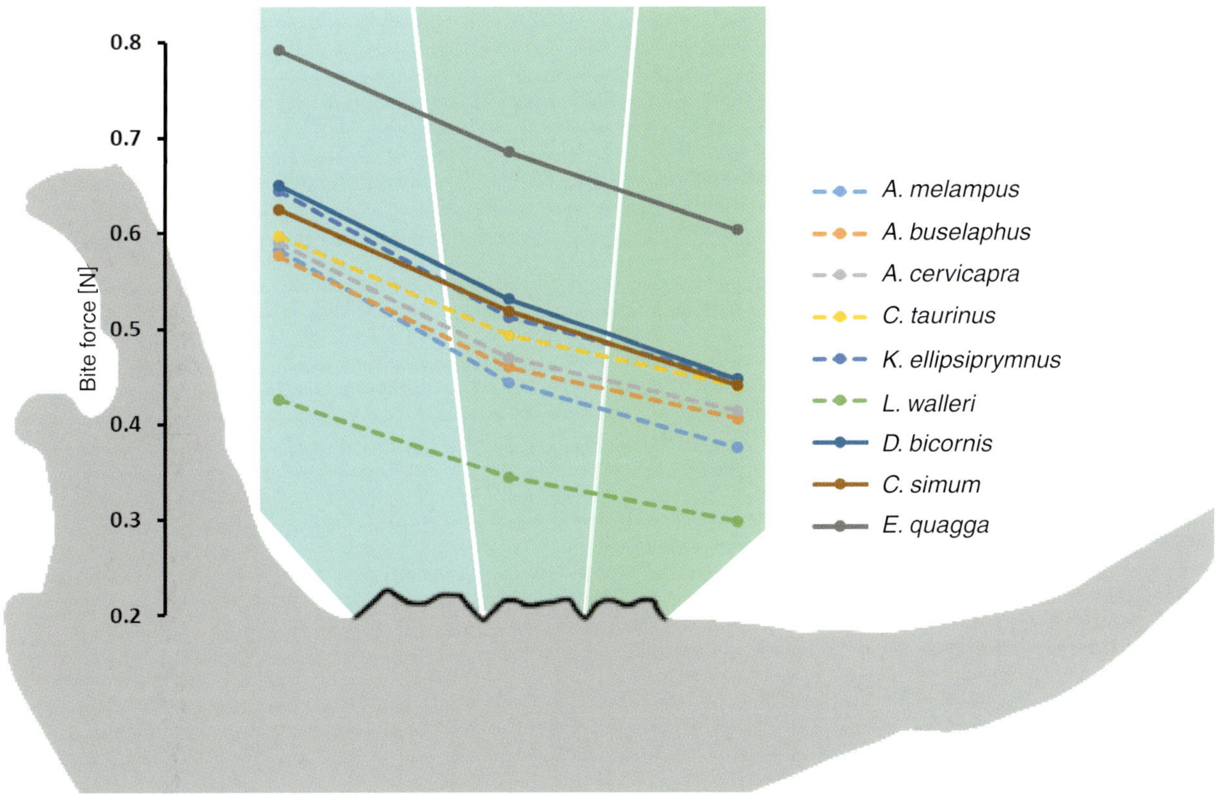

Fig. 8.4. Relative bite force per tooth position under an input muscle force of 1N. Continuous lines represent Perissodactyla, dashed lines Artiodactyla.

stages per tooth were pooled to compute an overall value for the tooth (total θ, κ and μ). In Oriana, the Watson U2 and Mardia-Whatson-Wheeler test were employed to detect significant differences between ontogenetic stages and tooth positions. Both tests are sensitive to differences in mean and variance. Results were only treated as significant if they were below a critical p-value of 0.05 in both tests.

Enamel ridge alignments were clustered in two (0°–44°, 45°–90°) and three clusters (0°–29°, 30°–59°, 60°–90°) and expressed as frequencies. Percentage data were then arcsine transformed in order to normalise them. To test for normality and heteroscedasticity of the data, the Shapiro-Wilk normality test and a robust Brown-Forsythe Levene-type test were applied. If data violated the assumption of normality, a log-transformation was applied and the tests on the transformed data were repeated. This had to be done with the circular parameter concentration (κ). If the data then met the requirements of normality and homoscedasticity, a one-way ANOVAs with post-hoc Tukey test to detect significant differences between means was performed. The packages used were 'stats' version 3.3.0 (R.D.C. Team 2009) and 'car' version 2.0.25 (Fox and Weisberg 2011). For non-normally distributed, heteroscedastic data, a robust Welch-Yuen test with 15 % trimming and a heteroscedastic pair-wise comparison test (analogous to Dunnett's T3 test; Lincon test) following the procedure of Calandra et al. (2012) and Schulz et al. (2013) were applied.

Phylogenetically informed MANOVAs using the package "geiger" version 2.0.3 (Harmon et al. 2015) were conducted, which simulates new sets of dependent variables on the phylogenetic tree under a Brownian-motion model to ensure phylogenetic independence of variables. The phylogeny is based on the supertree of Bininda-Emonds et al. (2007). Species were used as the independent variable to test the circular statistical parameters circular median (θ), concentration (κ) and mean vector (μ) as well as ERA for upper and lower molars as dependent variables.

Results

Phylogenetic independence

Phylogenetic MANOVAs were conducted for the mean circular statistical variables for each tooth position and ERA frequencies clustered in three clusters, independently for upper and lower teeth. No dependant variable showed a significant relation with phylogeny (Tab. 8.2). Hence the results can be treated independent of phylogeny and thus are not discussed further in a phylogenetic context.

Relative bite forces along the jaw

Relative bite forces are found to decrease with increasing distance from the temporomandibular joint. The resulting bite force curves are displayed for each species in Figure 8.4.

Table 8.2. Test statistics for phylogenetic MANOVAs. Species was used as the independent variable. No significant influence of phylogeny was detected. *df*, degree of freedom; *F*, test value; *p*, significance level.

Parameter	Wilks' lambda	Approx. *F*	Numerator *df*	Denominator *df*	*p*(>F)	*p*(phy)
Ccircular median (θ)	0.609	1.6023	6	15	0.214	0.811
Concentration (κ)	0.701	1.0677	6	15	0.424	0.580
Mean vector (μ)	0.627	1.4902	6	15	0.247	0.751
ERA for upper jaw	0.507	1.294	9	12	0.332	0.168
ERA for lower jaw	0.519	1.235	9	12	0.359	0.155

Table 8.3. Total number of significant variability between enamel ridge alignments between wear stages of individual tooth position. The source of significant differences is given in the Appendix Table A1.

			n Significant differences						
			Upper jaw			Lower jaw			
Species	Family	Diet	M1	M2	M3	m1	m2	m3	Total
A. americana	Antilocapridae	GBR	3	3	2	0	0	1	9
A. buselaphus	Bovidae	GGR	1	3	3	0	0	0	7
A. cervicapra	Bovidae	GGR	0	2	0	3	1	0	6
A. marsupialis	Bovidae	OMF	0	2	4	0	1	0	7
A. melampus	Bovidae	OMF	0	1	3	0	1	0	5
C. bactrianus	Camelidae	OMF	0	0	3	0	0	0	3
C. ibex	Bovidae	OMF	3	0	1	0	0	3	7
C. simum	Rhinocerotidae	GGR	1	2	2	0	1	1	7
C. taurinus	Bovidae	GGR	0	0	0	0	0	0	0
D. bicornis	Rhinocerotidae	GBR	2	0	0	0	2	3	7
D. pygargus	Bovidae	GGR	1	1	3	0	0	0	5
E. davidianus	Cervidae	FGR	3	3	2	0	0	0	8
E. quagga	Equidae	GGR	0	0	0	0	0	0	0
H. jemlahicus	Bovidae	OMF	0	3	0	2	0	0	5
K. ellipsiprymnus	Bovidae	FGR	0	0	2	0	0	0	2
L. glama	Camelidae	OMF	3	0	0	0	0	0	3
L. huanachus	Camelidae	OMF	0	0	0	0	0	2	2
L. walleri	Bovidae	HBR	2	3	3	0	0	0	8
N. goral	Bovidae	OMF	0	2	1	0	0	0	3
O. bezoarticus	Cervidae	OMF	4	1	5	4	0	1	15
R. fulvorufula	Bovidae	FGR	0	3	3	0	2	3	11
R. tarandus	Cervidae	GBR	3	4	5	0	0	0	12
total			26	33	42	9	8	14	

Table 8.4. Total number of significant differences per ontogenetic stage position between tooth positions. Note that the total number of differences is lowest for stage 2 in the upper jaw. When Cervidae are excluded, this relation is even stronger.

			n Significant differences								
			Upper jaw				Lower jaw				
Species	Family	Diet	Stage 1	Stage 2	Stage 3	Stage 4	Stage 1	Stage 2	Stage 3	Stage 4	Total
A. americana	Antilocapridae	GBR	2	1	0	0	2	1	2	1	9
A. buselaphus	Bovidae	GGR	0	0	1	1	0	0	0	1	3
A. cervicapra	Bovidae	GGR	0	0	1	0	1	0	1	0	3
A. marsupialis	Bovidae	OMF	0	0	0	1	0	0	0	0	1
A. melampus	Bovidae	OMF	2	1	2	0	1	1	1	2	10
C. bactrianus	Camelidae	OMF	0	0	0	0	0	0	1	0	1
C. ibex	Bovidae	OMF	0	0	1	2	2	1	1	1	8
C. simum	Rhinocerotidae	GGR	0	0	3	2	1	0	0	1	7
C. taurinus	Bovidae	GGR	0	0	0	0	0	0	0	0	0
D. bicornis	Rhinocerotidae	GBR	0	1	0	2	0	2	2	0	7
D. pygargus	Bovidae	GGR	2	0	1	1	0	0	0	0	4
E. davidianus	Cervidae	FGR	0	1	1	2	2	0	1	2	9
E. quagga	Equidae	GGR	0	0	0	0	0	0	0	0	0
H. jemlahicus	Bovidae	OMF	3	0	0	0	0	0	0	2	5
K. ellipsiprymnus	Bovidae	FGR	2	0	0	0	0	0	0	0	2
L. glama	Camelidae	OMF	1	0	1	2	0	0	0	0	4
L. huanachus	Camelidae	OMF	0	0	1	0	1	1	0	0	3
L. walleri	Bovidae	GBR	1	0	1	1	0	0	0	0	3
N. goral	Bovidae	OMF	1	0	1	2	1	0	0	0	5
O. bezoarticus	Cervidae	OMF	2	2	0	2	2	3	3	3	17
R. fulvorufula	Bovidae	FGR	2	1	0	1	1	0	0	0	5
R. tarandus	Cervidae	GBR	3	2	2	2	0	0	0	0	9
Total			21	9	16	21	14	9	12	13	
Without Cervidae			16	4	13	15	10	8	9	10	

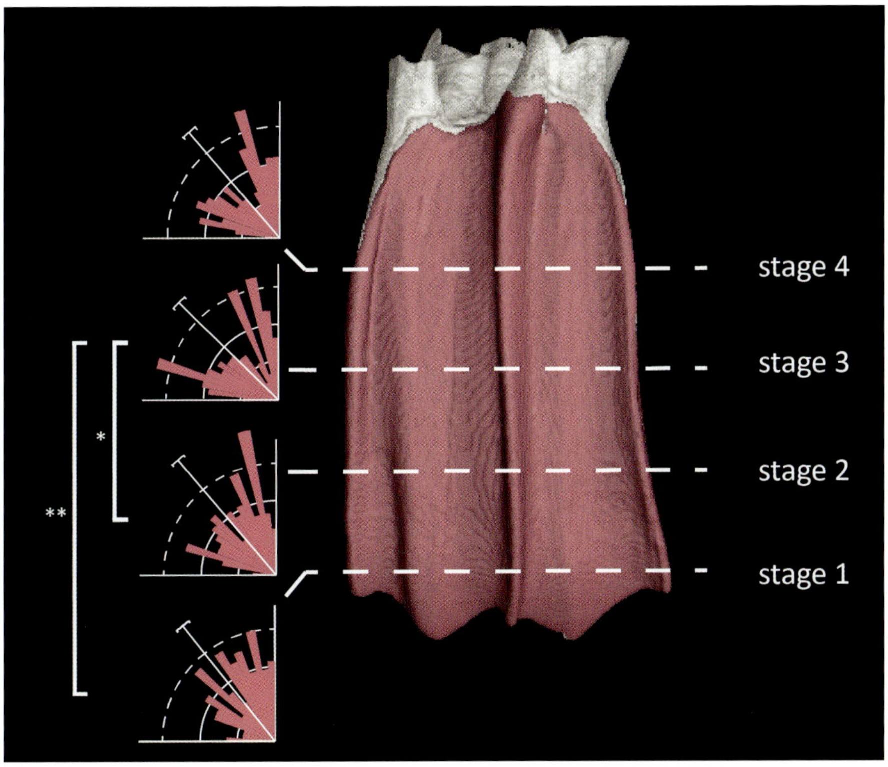

Fig. 8.5. Rose diagrams of enamel ridge alignments per ontogenetic stage in an upper M2 of *Antidorcas marsupialis*. The angle distribution pattern of stage 3 is significantly different from stage 1 and stage 2. Rose diagrams show the mean vector with 95 % confidence intervals and angle frequencies from 0°–90°. Inner circle = 3 %, middle circle = 6 %, outer dashed circle = 9 %.

ERA between wear stages of individual tooth positions

Enamel ridge alignments of successive ontogenetic stages of individual tooth positions are very variable between taxa and feeding types. The source of the observed difference can be both mean angle distribution and/or variance. However, generally two trends can be observed: 1) ERAs of upper teeth are more variable between stages of individual tooth positions than ERAs of lower teeth. 2) For both, upper and lower teeth, either the first and/or the last stage are significantly different from the wear stages in between (see Appendix Tab. 8.1).

Enamel ridge alignments of successive ontogenetic stages for upper third molars are most variable, followed by upper second molars and upper first molars. An example for significantly different enamel ridge alignments is given in Figure 8.5. When analysed according to diet, the number of differences between stages is not normally distributed and cannot be normalized by transformation. Therefore, the procedure for non-normally distributed, heteroscedastic data was followed to further analyse them.

The highest total variability was found for *O. bezoarticus* (Tab. 8.3), the lowest for *C. taurinus* and *E. quagga*. All cervids in the sample have a high variability between ontogenetic stages, while camelids, on the contrary, show few significant differences. Due to the uneven representation of families in the sample, no statistical analyses were performed regarding number of differences per family.

ERA at corresponding wear stages between teeth

When comparing ERA at corresponding stages between molar teeth, the three cervids exhibit the highest variability besides *Aepyceros melampus*. In all species, ERAs of molars are most variable during wear stages 1 and 4, though this trend is more pronounced in the upper jaw. Moreover, for the upper and lower jaw, stage 2 has the most similar ERA patterns between molars. Cervids have lower crowned teeth than the other taxa in the sample, which might be the source for the observed high variability of ERA between wear stages. If cervids are excluded from the dataset, variability is only found between tooth positions in *A. americana*, *A. melampus*, *D. bicornis* and *R. fulvorufula* for stage 2 in the upper jaw (Tab. 8.4).

Total ERA per tooth

For each tooth position, angle distributions from stages 1-4 were summed up to produce an estimate of the overall ERA count for the respective tooth. This estimate is referred to as total ERA. Total ERAs are found to be significantly different between molars in most taxa (Tab. 8.5). Upper and lower M3s are distinctly different in their ERAs from M1s and M2s, while in both, upper and lower jaw, M1s and M2 are more similar in ERAs. To detect the source of the differences, total ERAs per tooth were pooled for all species (Tab. 8.6) and described with the circular parameters θ, κ and μ.

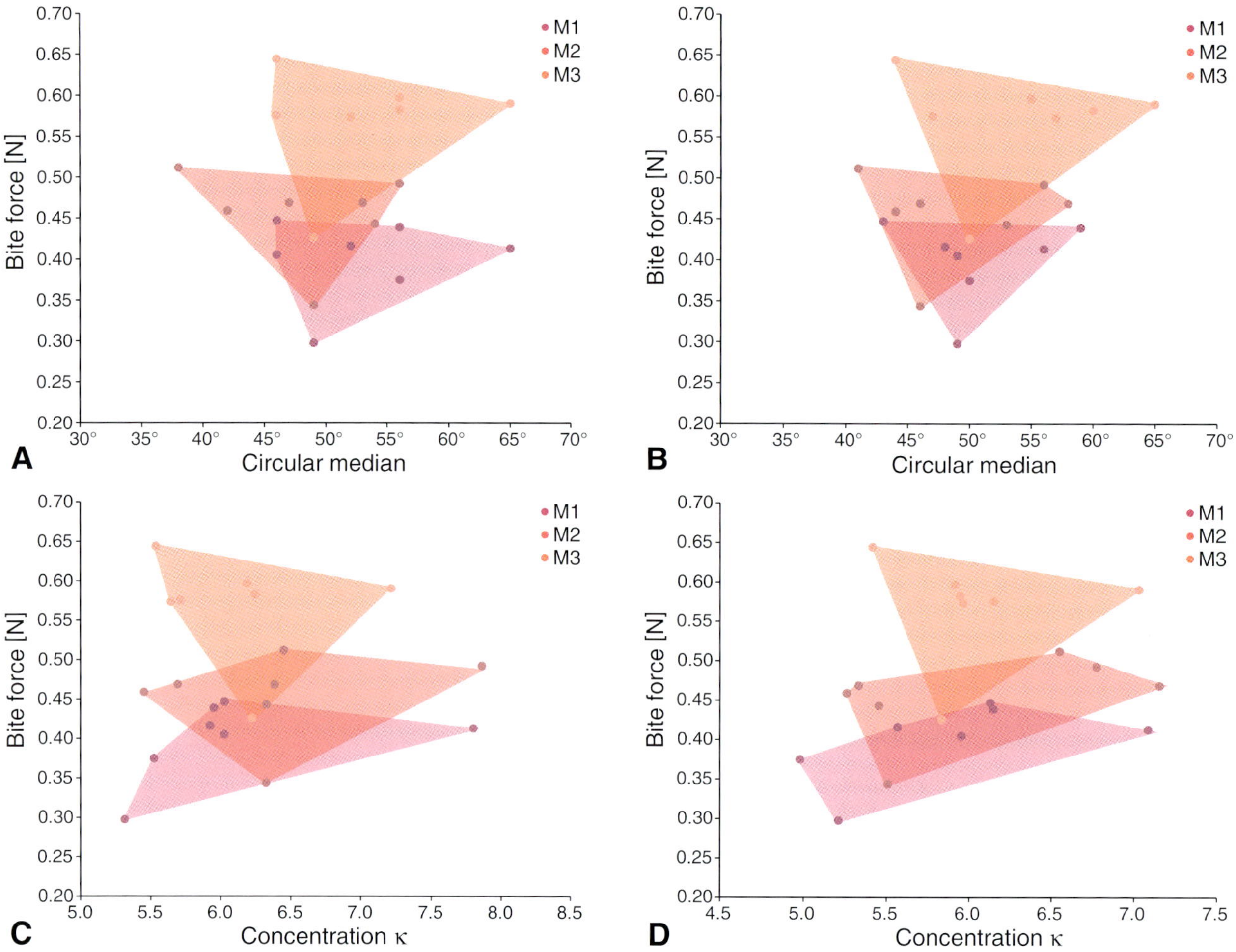

Fig. 8.6. Relative bite forces per tooth position related to the circular statistical parameters circular median (θ) and concentration (κ). **A,** circular median for stage 2; **B,** total circular median; **C,** concentration for stage 2; **D,** total concentration.

Table 8.5. Comparisons between total ERA per tooth position. Significances are only accepted if they were produced by both Watsons U^2 and Mardia-Whatson-Wheeler test. ***, $p=0.001$; **, $p=0.01$; *, $p=0.05$.

	Group					
Species	m1 vs m2	m1 vs m3	m2 vs m3	M1 vs M2	M1 vs M3	M2 vs M3
A. americana		***	***		***	
A. buselaphus	*		*	**		***
A. cervicapra		**	**		**	*
A. marsupialis		***	***		***	***
A. melampus					***	*
C. bactrianus	*	***		*		
C. ibex	**	***	***		***	
C. simum		***	**		**	*
C. taurinus		**				*
D. bicornis			*	*	***	**
D. pygargus			**		***	***
E. davidianus	***	***	**	**	**	*
E. quagga				*	**	
H. jemlahicus	***	**			***	***
K. ellipsiprymnus					*	**
L. glama	**					**
L. huanachus		**			**	
L. walleri					*	*
N. goral	*	**	*	*	***	***
O. bezoarticus	***	***	***	***	***	*
R. fulvorufula					**	***
R. tarandus			*	**	***	***

Circular median (θ)

In both upper and lower jaws, θ shows the same pattern with M3 > M1 > M2 (Tab. 8.6). This pattern is not significant for lower or upper molars. In the bovid subset, bite forces do not correlate with θ (Fig. 8.6A, B). A post-hoc Tukey test (not displayed here) showed that lower molars have significantly higher θ only than upper molars.

Table 8.6. Descriptive statistics of circular statistical parameters per tooth position pooled for all species. SD, standard deviation.

	Circular median (θ)		Concentration (κ)		Mean vector (μ)	
Tooth	Mean	SD	Mean	SD	Mean	SD
m1	55.136°	4.622°	5.634	0.612	51.750°	2.813°
m2	53.227°	5.944°	5.773	0.624	50.357°	4.026°
m3	55.955°	6.622°	6.771	1.265	53.993°	5.120°
M1	49.000°	5.782°	5.537	0.559	47.902°	4.011°
M2	47.818°	6.162°	5.896	0.542	47.135°	4.423°
M3	52.545°	6.906°	6.167	0.523	50.522°	4.918°

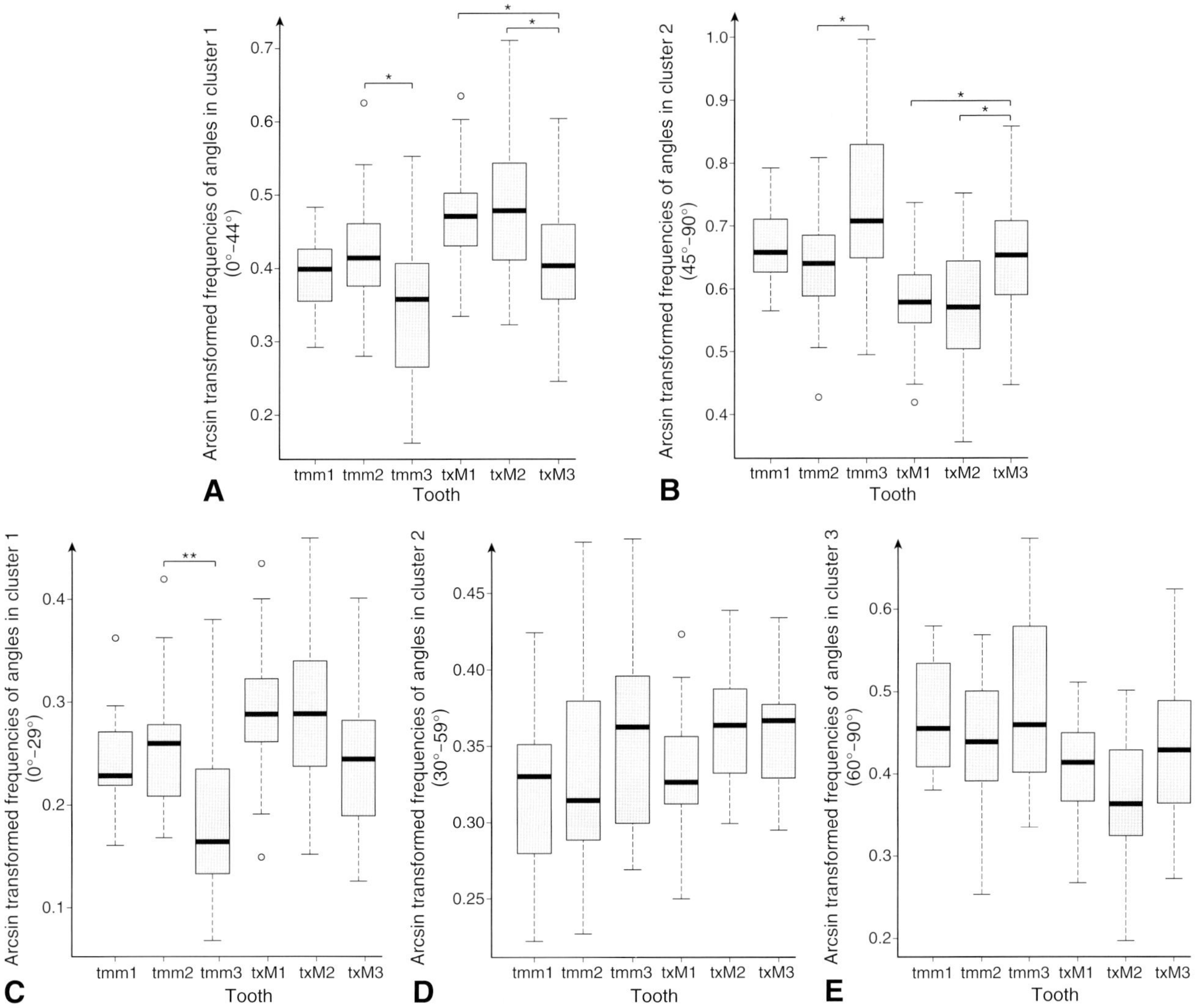

Fig. 8.7. Boxplots of significant differences between tooth positions for total ERA per cluster. **A, B,** ERA distributed in two clusters, test statistic from Lincon. **C–E,** ERA in three clusters, test statistic from Tukey. ****p*=0.001; ***p*=0.01; **p*=0.05.

Concentration (κ)

κ is increasing from M1 to M2 to M3 in upper and lower molars, with values for lower m1s and m2s being very similar. Lower m3s have significantly higher κ than m2s (Tukey: p=0.001) and m1s (Tukey: p=0.000). For upper molars, significant differences are found between M1s and M3s (Tukey: P=0.028). Test statistics are given in Table 8.7. The bite force calculations for the bovid subset illustrate that both bite forces and concentration κ follow the same gradient (M3>M2>M1, Fig. 8.7C, D).

Mean vector (μ)

As expected, μ is found to exhibit the same pattern as θ with M3>M1>M2. The significances are also very similar to those for mean θ (compare Tab. 8.7).

Clustered total ERAs per tooth

The frequency of observed ERAs is normally distributed in all clusters. When assigned to two clusters (0–44°, 45°–90°), ERAs show unequal variances (Levene: p=0.000).

Table 8.7. Results of post-hoc Tukey test for circular median, concentration and mean vector per tooth position. Significances are printed in bold. lwr, lower end point of the interval; upr, upper end point of the interval; *p*, significance level.

		Circular median (θ)				Concentration (κ)				Mean vector (μ)			
		diff. means	lwr	upr	*p*	diff. means	lwr	upr	*p*	diff. means	lwr	upr	*p*
m1	m2	−0.033	−0.125	0.059	0.901	0.011	−0.034	0.055	0.982	−0.024	−0.090	0.041	0.889
m1	m3	0.014	−0.078	0.106	0.998	0.075	0.031	0.120	**0.000**	0.039	−0.026	0.104	0.510
m2	m3	0.048	−0.045	0.140	0.668	0.065	0.020	0.109	**0.001**	0.063	−0.002	0.129	0.062
M1	M2	−0.021	−0.113	0.072	0.987	0.028	−0.017	0.072	0.466	−0.013	−0.079	0.052	0.991
M1	M3	0.062	−0.030	0.154	0.381	0.047	0.003	0.092	**0.028**	0.046	−0.020	0.111	0.333
M2	M3	0.083	−0.010	0.175	0.107	0.020	−0.024	0.064	0.787	0.059	−0.006	0.124	0.100

Therefore, the distribution into two clusters is analysed using the procedure for non-normally distributed, heteroscedastic data.

In all tooth positions, the majority of angles fall into the 50°–90° and 45°–90° clusters (Fig. 8.7A–E). The most pronounced frequency differences between clusters are observed in upper and lower M3s for distribution in two clusters. Lower m3s have significantly more ERAs in the 45–90° cluster than lower m2s (Lincon: $p=0.021$), while upper M3s have significantly more ERAs in this cluster than upper M1s (Lincon: $p=0.040$) and M2s (Lincon: $p=0.028$).

When assigned to three clusters, the only significant difference is detected in the 0–29° cluster, where lower m3s have significantly lower ERA frequencies than lower m2s (Tukey: $p=0.005$).

Discussion

No influence of phylogeny on enamel ridge alignments (ERAs) was found, thus all results obtained for ungulates here are interpreted as following a general adaptive pattern. ERAs towards the chewing direction are tooth position specific and show a consistent pattern for all tested species.

Hypothesis 1 (H1)

Total ERAs are not found to have significantly higher circular median (θ) or mean vector (μ) in M3s than in M1s and M2s. However, a trend for M3s to have higher ERAs than the anterior molars can be consistently observed in the sample. When clustered in lower (cluster 1: 0°–44°) and higher (cluster 2: 45°–90°) alignments, upper and lower M3s show significantly higher angle frequencies in the 45°–90° cluster. Moreover, upper and lower M3s show the highest concentration (κ) values. κ is a measurement of angle dispersion. Low κ-values are characteristic of a dispersed angle pattern which is created by more complexly folded enamel ridges or differently aligned enamel ridges between ontogenetic stages. M3 having the highest κ-values over the whole tooth crown shows that its enamel ridges are either less complexly folded or very precisely aligned for each ontogenetic stage. The complexity of enamel ridges can be expected to be similar between teeth, as it is a strong indicator of diet (Archer and Sanson 2002; Heywood 2009), and therefore the second explanation is considered as more likely. M3 is morphologically conservative over the tooth crown and hence has more precisely aligned angles towards the chewing direction. This conserved morphology can also be considered as indicative for a maintained functionality once anterior tooth positions wear out and lose chewing efficiency. Hypothesis 1 (H1) can hence be partly accepted and is reformulated to:

The M3 compensates for functional loss in more anterior tooth positions by having a larger proportion of enamel ridges aligned to higher angles than M1 and M2 and by maintaining precise alignments over the whole tooth crown.

The results of this chapter also show that enamel ridge alignments close to 90° do not occur as frequently as hypothesized by Rensberger (1973), at least not when alignments are analysed in a 2D projection. One has to keep in mind that all data in this study are reducing the measured angle to two dimensions, thus the real, natural curvature of enamel ridges is lost. It is therefore suggested that either adequate chewing efficiency can also be achieved by lower aligned enamel ridges, or the natural curvature would actually result in higher aligned enamel ridges, which cannot be detected here. Nevertheless, it seems that the degrees of freedom of the masticatory movement during the power stroke are higher than anticipated. Consequently, lateral chewing movements would not be as strictly aligned as expected in a rather simplistic model and a more anterio-posterior (proal) component to the chewing stroke would also make higher ERAs efficient shearing blades as suggested in this study.

Hypothesis 2 (H2)

Ontogenetic differences in ERAs are mostly related to the youngest and/or the oldest (senile) stage of a tooth position (see Appendix Tab. 8.1). This does not fully support the hypothesis that younger ontogenetic stages should be more optimized in enamel ridge alignments. Significant differences in ERAs of wear stage 2 and 3 compared to stage 4 would be expected if that was the case. However, the "prime age", which is expected to be represented by stages 2 and 3, are only for *C. simum*, *E. davidianus*, *R. tarandus* and *O. bezoarticus* in more than one tooth position distinctly different from other stages. It is striking that three Cervidae are amongst these taxa. Cervidae are, with the Rhinocerotidae and *L. walleri*, the most brachydont taxa in the sample. In brachydont taxa and GBRs, the variability of ERAs between stages is significantly higher than in all other families and dietary categories. Such obvious variability in crown morphology during wear has also been noted by Fortelius (1985b) in a Pleistocene rhinoceros and interpreted as compensation and reciprocal change during wear in brachydont taxa. It can therefore be concluded that crown hypsodonty does not only represent a proportional increase in crown height, but is also accompanied by a harmonisation of ERAs.

Stage 2 seems to be the most homogenous stage in ERAs in the upper and lower jaw, which indicates that orientation of enamel ridges is most similar in all molar teeth during the prime age. Though enamel ridges are not found to be significantly higher aligned during stage 2, they seem to undergo an optimization in terms of functional similarity. Chewing efficiency is thus least different between molars during the prime age. This optimization might be necessary in order to fulfil high energetic demands during peak reproduction on basis of a highly abrasive diet. If indeed chewing efficiency is increased through functional similarity, it would enable these species to incorporate higher amounts of abrasive forage during the prime age and process it with fewer chewing strokes, hereby not significantly increasing tooth wear. Without such an optimization, teeth would be exposed to higher wear which could only be countered by even greater crown height. As expected, a limitation to crown height exists in order to not distort ossification of the mandible and maxilla, thus an additional increase in crown height is unlikely.

With older age, differences between tooth positions become more pronounced, especially when M1 wears out (Damuth and Janis 2011). The M3 (in both upper and lower jaw) meanwhile maintains higher angulated and more precisely aligned enamel ridges than the M1 and M2, which supports functional compensation of the M3 as proposed in H1 and by Winkler et al. (2015b). In other words, during the prime age, the first and second molars are not distinguishable in ERAs from M3 and optimized chewing efficiency during this stage can be inferred, which is more pronounced in hypsodont than in brachydont species.

Hypothesis 2 (H2) is only partly supported and should be rephrased to: Enamel ridge alignments are optimized during the prime age in terms of homogenous function, but not in terms of general higher alignments.

Hypothesis 3 (H3)

The lower overall concentration values for M1s and M2s compared to M3s indicate that enamel ridges are less strictly oriented in the first two molars. Upper and lower M1s have the lowest concentration values. This is likely to be related to a higher number of thin ridges aligned at lower angles and to a changing enamel ridge orientation over the tooth crown in M1s and M2s and more consistent enamel ridge morphology in M3s.

The temporomandibular joint allows for both lateral and anterio-posterior movements, though the chewing stroke for most ungulates is described as unilateral, transverse chewing characterized by pronounced medio-lateral jaw movements (Fortelius 1985a, Williams et al. 2007, Kaiser et al. 2010). During the opening of the mouth, the condylar head of the mandible rotates within the mandibular fossa up to a certain point, when the condylar head and articular disc translate to allow for further jaw opening. It is intuitively comprehensible that the opening is smaller in close proximity to the temporomandibular joint. Therefore, the lateral displacement is also smallest close to the joint. Mastication on the tooth positions closest to the jaw joint hence has to be adapted to disintegrating food particles efficiently under displacement-limited conditions. A precise guidance and orientation of enamel ridges is thus favourable in this setting. Moreover, the data showed that largest bite forces can be generated closest to the jaw joint. The highest relative bite force was always computed for the third molar. The biomechanical constraints of the jaw morphology (shorter lateral displacement in combination with higher forces close to the temporomandibular joint) are combined with more precisely aligned enamel ridges at the third molar.

Disintegration of plant matter requires crack propagation through anisotropic, tough tissues. The dental shear-cutting system of extant herbivorous ungulates achieves this by tooth position-specific adaptations: On the third molar, cracks are driven through the tissue by shear-cutting with a higher force over enamel ridges which have predominant directions. On the first molar, cracks are propagated through a longer chewing stroke over dispersed enamel ridges, which ensure multiple contacts. The more anterior tooth positions might also display more degrees of freedom during translation. The presence of less precisely aligned enamel ridges at more anterior tooth positions could hence be an advantage, as it would maximize contacts during the longer, less strictly guided chewing stroke. Lower guidance and more anterio-posterior components would bring lower aligned enamel ridges in more favourable positions for shearing.

The hypothesis of a more robust build in general for M3s could not be proven by this study; however, data on enamel volume by Winkler and Kaiser (2015b) show that M3s have higher enamel proportions which either indicates wider enamel ridges or smaller dentin basins. The assumption that enamel ridges of the M3s are wider is likely, because wider ridges would better withstand the higher bite forces. The lower variability in enamel ridge orientation seen in M3s could also be a consequence from the higher enamel content: as wider ridges take a larger space on the occlusal surface, there would be less space for changes in ridge orientation (e.g., plications). Higher concentration of ERAs in M3s needs to be accounted for and might not be solely adaptive to occlusal function, but result from space constraints in the occlusal surface and also related to the narrow space close to the jaw's angle, where bending forces are large and thus lateral alveolar bone sheets would have to be inflated in order to transmit forces. The latter could be a biomechanical constraint not yet considered in FEA models accordingly. Future FEA studies on stress distribution along the jaw using artificially low enamel thicknesses for third molars might contribute to the understanding of this phenomenon and decide whether it is adaptive or a morphological constraint.

Conclusion

Virtual 3D reconstructions of enamel ridge morphologies based on computer tomographic data have great value in measuring chewing direction of subsequent ontogenetic stages using the GIS-based approach described in this chapter. Enamel ridge alignments of third molars are found to be higher angulated and significantly more concentrated than those of first and second molars. For subsequent ontogenetic stages, the second wear stage, which is considered to resemble the prime reproductive age of an individual, shows the fewest differences between molar tooth positions on enamel ridge orientation. These findings might represent an adaptation towards more homogenous tooth function during peak reproduction. Tooth-position specific bite forces along the jaw decline along the molar tooth row with increasing distance from the temporomandibular joint. This indicates that enamel ridge alignments are an adaptation towards more load-optimized and displacement-limited conditions close to the jaw joint, while more anterior tooth positions are adapted towards higher degrees of lateral displacement. The higher dispersion of enamel ridge angles in first molars hereby suggests a benefit under a less guided and less powerful chewing stroke, as it ensures maximum contacts between enamel ridges. The shorter displacement of the posterior lower jaw is compensated by higher bite forces at the M3.

Molar teeth of Perissodactyla and Cetartiodactyla do not form a homogenous functional unit during the whole ontogeny. Their functionality is optimized during the prime age of an individual, but afterwards M1 and M2 functionality declines. The M3 compensates for functional loss of M1 and M2. This study has important implications for future studies on tooth function and dietary reconstruction: Wear stage and tooth position need to be carefully considered in order to accommodate morphological variability within the tooth crown, but also general biomechanical constraints that universally apply, independent of diet and taxonomy.

Acknowledgments

We thank the Museum für Naturkunde Berlin for specimen loan, Jan-Hendrik Buhk at Universitätsklinikum Eppendorf for CT scanning of large specimens, and all members of the DFG research unit 771 for fruitful discussions during and after the project duration. We highly acknowledge the helpful comments and suggestions from Gina Semprebon and Indrė Žliobaitė who reviewed this chapter.

References

* indicates publications that originated from the DFG Research Unit 771.

Archer, D. & Sanson, G. (2002): Form and function of the selenodont molar in southern African ruminants in relation to their feeding habits. Journal of Zoology 257: 13-26.

Bininda-Emonds, O. R. P., Cardillo, M., Jones, K. E., MacPhee, R. D. E., Beck, R. M. D., Grenyer, R., Price, S. A., Vos, R. A., Gittleman, J. L. & Purvis, A. (2007): The delayed rise of present-day mammals. Nature 446: 507-512.

*Calandra, I., Schulz, E., Pinnow, M., Krohn, S. & Kaiser, T. M. (2012): Teasing apart the contributions of hard items on 3D dental microtextures in primates. Journal of Human Evolution 63: 85-98.

Damuth, J. & Janis, C. M. (2011): On the relationship between hypsodonty and feeding ecology in ungulate mammals, and its utility in palaeoecology. Biological Reviews 86: 733-758.

Famoso, N. A., Feranec, R. S. & Davis, E. B. (2013): Occlusal enamel complexity and its implications for lophodonty, hypsodonty, body mass and diet in extinct and extant ungulates. Palaeogeography, Palaeoclimatology, Palaeoecology 387: 211-216.

Fortelius, M. (1985a): Ungulate cheek teeth: developmental, functional, and evolutionary interrelations. Acta Zoologica Fennica 180: 1-76.

Fortelius, M. (1985b). The functional significance of wear-induced change in the occlusal morphology of herbivore cheek teeth, exemplified by *Dicerorhinus etruscus* upper molars. Acta Zoologica Fennica 170: 157-158.

Fox, J. & Weisberg, S. (2011): An {R} Companion to Applied Regression, Second Edition. Sage Publishing, Thousand Oaks.

Greaves, W. S. (2012): The Mammalian Jaw – A Mechanical Analysis. Cambridge University Press, Cambridge.

Harmon, L., Weir, J., Brock, C., Glor, R., Challenger, W., Hunt, G. et al. (2015): Package 'geiger'. In: R package version 2.0.3.

Heywood, J. J. N. (2009): Functional anatomy of bovid upper molar occlusal surfaces with respect to diet. Journal of Zoology 281: 1-11.

Janis, C. M. (1988): An estimation of tooth volume and hypsodonty indices in ungulate mammals, and the correlation of these factors with dietary preference. In: Russel, D. E., Santoro, J. P. & Signogneau-Russel, D. (eds.). Teeth Revisited: Proceedings of the VII International Symposium on Dental Morphology. Mémoires du Muséum national d'Histoire naturelle C 53: 367-387.

Janis, C. M. (1990a): Correlation of cranial and dental variables with dietary preferences: a comparison of macropodoid and ungulate mammals. Memoirs of the Queensland Museum 28: 349-366.

Janis, C. M. (1990b): Correlation of cranial and dental variables with body size in ungulates and macropodoids. In: Damuth J. & MacFadden, B. J. (eds.). Body Size in Mammalian Paleobiology. Cambridge University Press, Cambridge: 255-299.

Kaiser, T. M. (2002): Functional significance of ontogenetic gradients in the enamel ridge pattern of the upper cheek dentition of the Miocene hipparionine horse *Cormohipparion occidentale* (Equidae, Perissodactyla): Senckenbergiana lethaea 82: 167-180.

Kaiser, T. M., Brasch, J., Castell, J. C., Schulz, E. & Clauss, M. (2009): Tooth wear in captive wild ruminant species differs from that of free-ranging conspecifics. Mammalian Biology 74: 425-437.

Kaiser, T. M., Fickel, J., Streich, W. J., Hummel, J. & Clauss, M. (2010): Enamel ridge alignment in upper molars of ruminants in relation to their natural diet. Journal of Zoology 281: 12–25.

Kurten, B. (1983): Variation and dynamics of a fossil antelope population. Paleobiology 9: 62-69.

Mendoza, M., Janis, C. M. & Palmqvist, P. (2002): Characterizing complex craniodental patterns related to feeding behaviour in ungulates: a multivariate approach. Journal of Zoology 258: 223-246.

R. D. C. Team (2009): R: A language and environment for statistical computing. In: R Foundation for Statistical Computing, Vienna.

Rensberger, J. M. (1973): An occlusion model for mastication and dental wear in herbivorous mammals. Journal of Paleontology 47: 515-528.

*Schulz, E., Calandra, I. & Kaiser, T. M. (2013): Feeding ecology and chewing mechanics in hoofed mammals: 3D tribology of enamel wear. Wear 300: 169-179.

Solounias, N. & Dawson-Saunders, B. (1988): Dietary adaptations and palaeoecology of the late Miocene ruminants from Pikermi and Samos in Greece. Palaeogeography, Palaeoclimatology, Palaeoecology 65: 149-172.

Solounias, N., Fortelius, M. & Freeman, P. (1994): Molar wear rates in ruminants: A new approach. Annales Zoologici Fennici 31: 219-227.

*Taylor, L. A., Kaiser, T. M., Schwitzer, C., Müller, D. W. H., Codron, D., Clauss, M. & Schulz, E. (2013): Detecting inter-cusp andilnter-tooth wear patterns in rhinocerotids. PLoS One 8: e80921.

Williams, S. H., Vinyard, C. J., Wall, C. E. & Hylander, W. L. (2007): Masticatory motor patterns in ungulates: a quantitative assessment of jaw-muscle coordination in goats, alpacas and horses. Journal of Experimental Zoology Part A: Comparative Experimental Biology 307A: 226-240.

*Winkler, D. E. & Kaiser, T. M. (2015a): Uneven distribution of enamel in the tooth crown of a Plains Zebra (*Equus quagga*). PeerJ 3: e1002.

*Winkler, D. E. & Kaiser, T. M. (2015b): Structural morphology of molars in large mammalian herbivores: Enamel content varies between tooth positions. PLoS One 10 (8): e0135716.

*Zhou, Z., Winkler, D. E., Fortuny, J., Kaiser, T. M. & Marcé-Nogue, J. (2019). Why ruminants chew sloppily: Biomechanics discern a phylogenetic pattern. PLoS ONE 14 (4): e0214510.

Appendix

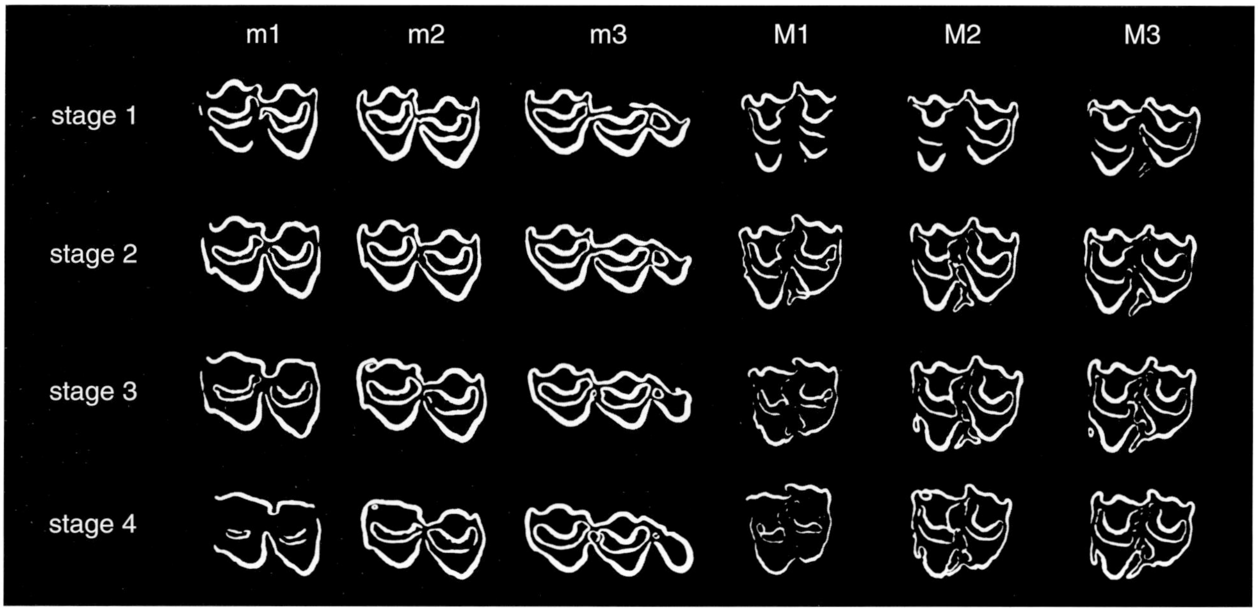

Fig. 8.A1. FR image sequence of the lower and upper molar tooth row of *Elaphurus davidianus*. The individual images are not true to scale.

Table 8.A1. Source of detected significant differences between ontogenetic stages. ns, no significant difference found.

	Difference between stages					
Species	M1	M2	M3	m1	m2	m3
A. americana	1 vs. 2/3/4	1 vs. 2/3/4	1 vs. 3/4	ns	ns	1 vs. 4
A. buselaphus	1 vs. 4	1 vs. 2/3/4	1 vs. 2/3/4	ns	ns	ns
A. cervicapra	ns	1 vs. 2/3	ns	4 vs. 1/2/3	2 vs. 4	ns
A. marsupialis	ns	3 vs. 1/2	1 vs. 2/3/4 2 vs. 4	ns	1 vs. 4	ns
A. melampus	ns	1 vs. 4	1 vs. 2/3/4	ns	1 vs. 4	ns
C. bactrianus	ns	ns	1 vs. 2/3/4	ns	ns	ns
C. ibex	1 vs. 3/4 2 vs. 4	ns	1 vs. 3	ns	ns	1 vs. 2/3/4
C. simum	2 vs. 3	5 vs. 1/2	1 vs. 2/3	ns	2 vs. 4	3 vs. 4
C. taurinus	ns	ns	ns	ns	ns	ns
D. bicornis	1 vs. 2/3	ns	ns	ns	3 vs. 1/2	4 vs. 1/2/3
D. pygargus	1 vs. 3	1 vs. 4	1 vs. 2/3/4	ns	ns	ns
E. davidianus	1 vs. 2/4 2 vs. 4	1 vs. 3/4 2 vs. 4	1 vs. 3 2 vs. 4	ns	ns	ns
E. quagga	ns	ns	ns	ns	ns	ns
H. jemlahicus	ns	1 vs. 2/3/4	ns	1 vs. 3/4	ns	ns
K. ellipsiprymnus	ns	ns	1 vs. 3/4	ns	ns	ns
L. glama	1 vs. 3/4 2 vs. 4	ns	ns	ns	ns	ns
L. huanachus	ns	ns	ns	ns	ns	4 vs. 1/2
L. walleri	1 vs. 3/4	4 vs. 1/2/3	1 vs. 2/3/4	ns	ns	ns
N. goral	ns	1 vs. 2/3	1 vs. 4	ns	ns	ns
O. bezoarticus	1 vs. 3/4 4 vs. 2/3	1 vs. 4	1 vs. 2/3/4 2 vs. 3/4	1 vs. 2/3/4 2 vs. 4	ns	3 vs. 4
R. fulvorufula	ns	1 vs. 2/3/4	1 vs. 2/3/4	ns	1 vs. 3/4	1 vs. 2/3/4
R. tarandus	1 vs. 3/4 2 vs. 4	1 vs. 2/3/4 2 vs. 4	1 vs. 2/3/4 2 vs. 3/4	ns	ns	ns

Table 8.A2. Descriptive circular statistics for *A. americana*. n, number of observed angles; SD, circular standard deviation; SE, standard error of mean.

	Stage	n	μ Mean vector	θ Circular median	κ Concentration	Circular variance	SD	SE	95 % Confidence interval (±) for μ	
m1	1	170	52.145°	58°	5.785	0.091	25.020°	1.916°	48.388°	55.901°
	2	154	52.847°	55°	5.594	0.094	25.493°	2.051°	48.826°	56.868°
	3	133	50.016°	51°	5.096	0.104	26.865°	2.325°	45.458°	54.575°
	4	118	51.358°	55°	4.722	0.113	28.059°	2.577°	46.305°	56.411°
m2	1	153	53.609°	58°	6.880	0.076	22.743°	1.837°	50.008°	57.210°
	2	132	53.501°	58°	6.087	0.086	24.324°	2.114°	49.356°	57.646°
	3	129	47.268°	50°	5.608	0.094	25.458°	2.238°	42.880°	51.655°
	4	130	47.836°	50°	5.310	0.100	26.248°	2.298°	43.330°	52.341°
m3	1	95	65.604°	67°	13.238	0.039	16.064°	1.648°	62.374°	68.835°
	2	130	59.369°	64°	6.083	0.086	24.333°	2.131°	55.191°	63.548°
	3	135	57.613°	62°	6.986	0.075	22.555°	1.939°	53.811°	61.415°
	4	139	57.718°	61°	7.043	0.074	22.455°	1.903°	53.987°	61.448°
M1	1	180	44.084°	41°	6.596	0.079	23.274°	1.733°	40.687°	47.481°
	2	179	41.917°	40°	5.250	0.101	26.416°	1.971°	38.053°	45.781°
	3	175	38.028°	38°	5.505	0.096	25.721°	1.941°	34.223°	41.834°
	4	159	38.555°	38°	5.205	0.102	26.546°	2.102°	34.435°	42.675°
M2	1	169	48.093°	49°	6.689	0.078	23.095°	1.775°	44.614°	51.572°
	2	180	43.626°	42°	5.592	0.094	25.499°	1.898°	39.906°	47.346°
	3	183	42.388°	41°	4.987	0.107	27.197°	2.007°	38.454°	46.322°
	4	170	44.424°	43.5°	5.159	0.103	26.679°	2.042°	40.419°	48.428°
M3	1	99	58.052°	63°	6.732	0.078	23.015°	2.311°	53.522°	62.582°
	2	127	51.533°	50°	7.001	0.074	22.530°	1.997°	47.617°	55.448°
	3	151	45.826°	47°	5.922	0.089	24.697°	2.007°	41.891°	49.761°
	4	191	42.614°	42°	5.811	0.091	24.957°	1.803°	39.079°	46.149°
Total	m1	575	51.685°	55°	5.314	0.100	26.238°	1.092°	49.544°	53.827°
	m2	544	50.962°	56°	5.879	0.089	24.797°	1.062°	48.881°	53.043°
	m3	499	59.664°	64°	7.236	0.072	22.128°	0.990°	57.724°	61.604°
	M1	693	40.744°	39°	5.550	0.095	25.605°	0.971°	38.840°	42.648°
	M2	702	44.595°	44°	5.485	0.096	25.775°	0.971°	42.691°	46.499°
	M3	568	48.183°	49°	5.933	0.089	24.672°	1.034°	46.157°	50.210°

Table 8.A3. Descriptive circular statistics for *A. buselaphus*. n, number of observed angles; SD, circular standard deviation; SE, standard error of mean.

	Stage	n	μ Mean vector	θ Circular median	κ Concentration	Circular variance	SD	SE	95 % Confidence interval (±) for μ	
m1	1	216	47.856°	49°	5.936	0.088	24.665°	1.676°	44.570°	51.142°
	2	184	50.429°	52°	6.121	0.086	24.250°	1.785°	46.928°	53.929°
	3	121	51.413°	51°	6.104	0.086	24.287°	2.205°	47.090°	55.736°
	4	84	54.381°	58°	6.400	0.082	23.662°	2.579°	49.326°	59.437°
m2	1	223	50.302°	50°	5.598	0.094	25.481°	1.704°	46.962°	53.642°
	2	226	49.661°	50°	5.076	0.105	26.925°	1.788°	46.156°	53.165°
	3	214	47.162°	47°	5.579	0.095	25.531°	1.743°	43.745°	50.578°
	4	168	47.828°	49°	5.305	0.100	26.261°	2.023°	43.863°	51.793°
m3	1	162	52.562°	54°	6.164	0.085	24.156°	1.896°	48.846°	56.278°
	2	170	53.704°	52°	6.242	0.084	23.989°	1.838°	50.102°	57.307°
	3	178	50.137°	51°	5.950	0.088	24.631°	1.844°	46.522°	53.751°
	4	178	54.240°	58°	6.247	0.084	23.978°	1.795°	50.721°	57.759°
M1	1	159	50.956°	51°	6.784	0.077	22.919°	1.816°	47.396°	54.516°
	2	188	47.004°	48°	6.024	0.087	24.465°	1.782°	43.510°	50.497°
	3	157	46.274°	46°	5.704	0.092	25.218°	2.010°	42.334°	50.214°
	4	143	47.229°	48°	5.513	0.096	25.701°	2.146°	43.022°	51.436°
M2	1	191	52.025°	52°	6.249	0.084	23.975°	1.733°	48.628°	55.421°
	2	239	43.445°	42°	5.451	0.097	25.864°	1.670°	40.170°	46.719°
	3	235	42.063°	41°	5.156	0.103	26.687°	1.738°	38.656°	45.47°
	4	210	40.781°	38.5°	4.940	0.108	27.344°	1.883°	37.089°	44.473°
M3	1	216	51.185°	53°	6.676	0.078	23.120°	1.571°	48.105°	54.266°
	2	266	44.582°	46°	5.711	0.092	25.200°	1.543°	41.558°	47.607°
	3	257	44.020°	44°	6.537	0.080	23.387°	1.457°	41.163°	46.877°
	4	261	42.715°	41°	6.242	0.084	23.989°	1.483°	39.808°	45.623°
Total	m1	605	50.261°	51°	6.043	0.087	24.422°	0.992°	48.317°	52.205°
	m2	831	48.817°	49°	5.370	0.098	26.083°	0.903°	47.047°	50.588°
	m3	688	52.654°	53°	6.123	0.086	24.246°	0.923°	50.844°	54.464°
	M1	647	47.859°	49°	5.953	0.088	24.626°	0.967°	45.964°	49.755°
	M2	875	44.341°	44°	5.260	0.101	26.388°	0.891°	42.595°	46.087°
	M3	1000	45.383°	47°	6.151	0.085	24.184°	0.764°	43.886°	46.881°

Table 8.A4. Descriptive circular statistics for *A. cervicapra*. n, number of observed angles; SD, circular standard deviation; SE, standard error of mean.

	Stage	n	μ Mean vector	θ Circular median	κ Concentration	Circular variance	SD	SE	95 % Confidence interval (±) for μ	
m1	1	79	56.334°	55°	8.496	0.061	20.297°	2.282°	51.860°	60.808°
	2	81	55.325°	59°	6.294	0.083	23.881°	2.600°	50.130°	60.521°
	3	71	54.378°	55°	6.161	0.085	24.162°	2.864°	48.764°	59.993°
	4	57	63.265°	70°	5.617	0.094	25.434°	3.364°	56.671°	69.859°
m2	1	117	61.090°	67°	8.061	0.064	20.875°	1.929°	57.309°	64.871°
	2	111	55.376°	59°	7.225	0.072	22.147°	2.100°	51.259°	59.494°
	3	95	56.549°	59°	7.038	0.074	22.465°	2.303°	52.035°	61.063°
	4	132	54.027°	62°	5.114	0.104	26.812°	2.329°	49.460°	58.593°
m3	1	75	65.002°	68°	10.059	0.051	18.553°	2.141°	60.804°	69.200°
	2	86	63.715°	67°	10.413	0.049	18.218°	1.964°	59.866°	67.565°
	3	98	59.913°	59.5°	9.038	0.057	19.637°	1.983°	56.026°	63.799°
	4	95	60.862°	64°	8.114	0.064	20.803°	2.133°	56.681°	65.043°
M1	1	61	58.014°	61°	8.760	0.059	19.967°	2.555°	53.005°	63.023°
	2	97	55.618°	55°	7.804	0.066	21.243°	2.155°	51.392°	59.843°
	3	98	53.889°	54°	6.634	0.079	23.201°	2.341°	49.300°	58.479°
	4	61	56.022°	55°	5.892	0.089	24.766°	3.167°	49.814°	62.229°
M2	1	71	63.327°	67°	12.819	0.040	16.336°	1.938°	59.527°	67.126°
	2	153	50.439°	53°	6.383	0.082	23.697°	1.914°	46.688°	54.190°
	3	94	54.981°	56°	7.691	0.067	21.410°	2.207°	50.655°	59.307°
	4	73	59.380°	62°	7.063	0.074	22.421°	2.622°	54.240°	64.520°
M3	1	36	61.333°	66°	7.612	0.068	21.531°	3.586°	54.303°	68.362°
	2	85	62.377°	65°	7.217	0.072	22.160°	2.401°	57.669°	67.085°
	3	82	57.702°	65°	6.127	0.086	24.235°	2.673°	52.462°	62.942°
	4	93	59.777°	64°	7.800	0.066	21.248°	2.202°	55.461°	64.094°
Total	m1	288	56.920°	61°	6.451	0.081	23.559°	1.387°	54.202°	59.639°
	m2	455	56.734°	61°	6.481	0.081	23.498°	1.100°	54.577°	58.892°
	m3	354	62.179°	64°	9.153	0.056	19.506°	1.036°	60.147°	64.210°
	M1	317	55.631°	56°	7.087	0.073	22.380°	1.256°	53.169°	58.093°
	M2	391	55.607°	58°	7.156	0.073	22.263°	1.125°	53.402°	57.812°
	M3	296	60.148°	65°	7.033	0.074	22.472°	1.305°	57.590°	62.706°

Table 8.A5. Descriptive circular statistics for *A. marsupialis*. n, number of observed angles; SD, circular standard deviation; SE, standard error of mean.

	Stage	n	μ Mean vector	θ Circular median	κ Concentration	Circular variance	SD	SE	95 % Confidence interval (±) for μ	
m1	1	160	52.125°	58°	5.315	0.100	26.236°	2.071°	48.066°	56.184°
	2	132	53.451°	62°	4.946	0.107	27.325°	2.374°	48.798°	58.104°
	3	109	50.162°	60°	4.215	0.128	29.968°	2.862°	44.551°	55.774°
	4	80	55.114°	64°	4.549	0.118	28.668°	3.198°	48.846°	61.383°
m2	1	154	55.136°	61°	6.564	0.080	23.336°	1.878°	51.454°	58.819°
	2	156	51.915°	59°	5.103	0.104	26.844°	2.145°	47.709°	56.120°
	3	160	51.613°	61°	5.136	0.103	26.746°	2.111°	47.476°	55.751°
	4	132	49.200°	50°	4.994	0.106	27.175°	2.361°	44.572°	53.828°
m3	1	94	64.460°	65.5°	11.618	0.044	17.198°	1.773°	60.984°	67.936°
	2	113	60.292°	62°	8.724	0.059	20.010°	1.881°	56.604°	63.980°
	3	112	57.007°	60°	7.878	0.066	21.135°	1.996°	53.095°	60.919°
	4	128	53.450°	57°	7.255	0.072	22.098°	1.951°	49.625°	57.276°
M1	1	134	52.443°	55°	6.077	0.086	24.347°	2.101°	48.326°	56.561°
	2	125	51.362°	57°	5.201	0.102	26.556°	2.371°	46.714°	56.010°
	3	112	48.765°	49.5°	5.081	0.104	26.910°	2.538°	43.789°	53.740°
	4	124	45.716°	49°	4.990	0.106	27.187°	2.437°	40.939°	50.493°
M2	1	155	51.399°	55°	6.084	0.086	24.331°	1.952°	47.573°	55.226°
	2	160	49.576°	50°	6.224	0.084	24.027°	1.897°	45.856°	53.295°
	3	178	44.936°	42°	5.039	0.105	27.037°	2.023°	40.971°	48.901°
	4	168	49.617°	50°	5.289	0.100	26.308°	2.026°	45.644°	53.589°
M3	1	108	61.817°	64°	9.542	0.054	19.079°	1.835°	58.219°	65.414°
	2	159	53.065°	54°	7.147	0.073	22.277°	1.765°	49.605°	56.526°
	3	166	52.501°	52°	6.966	0.075	22.591°	1.752°	49.067°	55.935°
	4	168	51.313°	57°	6.145	0.085	24.197°	1.864°	47.658°	54.968°
Total	m1	495	49.672°	53°	5.275	0.100	26.347°	1.182°	47.354°	51.990°
	m2	661	48.784°	50°	5.541	0.095	25.630°	0.995°	46.832°	50.735°
	m3	601	54.028°	56°	6.908	0.075	22.693°	0.925°	52.215°	55.841°
	M1	481	52.547°	60°	4.780	0.112	27.860°	1.268°	50.062°	55.032°
	M2	602	52.086°	58°	5.359	0.099	26.114°	1.063°	50.003°	54.169°
	M3	447	58.427°	61°	8.145	0.064	20.761°	0.981°	56.504°	60.351°

Table 8.A6. Descriptive circular statistics for *A. melampus*. n, number of observed angles; SD, circular standard deviation; SE, standard error of mean.

	Stage	n	μ Mean vector	θ Circular median	κ Concentration	Circular variance	SD	SE	95 % Confidence interval (±) for μ	
m1	1	86	57.980°	65.5°	6.366	0.082	23.731°	2.556°	52.969°	62.991°
	2	92	55.605°	60°	6.439	0.081	23.583°	2.456°	50.790°	60.420°
	3	68	48.175°	50°	5.155	0.103	26.689°	3.231°	41.842°	54.509°
	4	62	53.247°	61°	5.775	0.091	25.044°	3.176°	47.020°	59.473°
m2	1	80	62.355°	69°	6.970	0.075	22.584°	2.523°	57.410°	67.300°
	2	96	52.614°	58°	5.712	0.092	25.199°	2.568°	47.580°	57.649°
	3	81	51.681°	57°	5.846	0.090	24.875°	2.760°	46.270°	57.092°
	4	76	45.570°	51.5°	5.316	0.099	26.231°	3.004°	39.681°	51.458°
m3	1	73	55.959°	63°	5.402	0.098	25.996°	3.038°	50.004°	61.914°
	2	93	61.858°	67°	7.002	0.074	22.527°	2.334°	57.283°	66.433°
	3	93	61.143°	66°	9.256	0.056	19.390°	2.010°	57.203°	65.082°
	4	107	55.689°	61°	6.671	0.078	23.130°	2.234°	51.310°	60.068°
M1	1	109	49.312°	56°	5.013	0.106	27.115°	2.592°	44.230°	54.393°
	2	105	49.240°	49°	5.522	0.096	25.678°	2.502°	44.335°	54.145°
	3	92	46.107°	44.5°	4.714	0.113	28.086°	2.922°	40.379°	51.834°
	4	59	44.447°	42°	4.662	0.115	28.266°	3.672°	37.250°	51.645°
M2	1	110	53.795°	55°	5.962	0.088	24.604°	2.343°	49.202°	58.388°
	2	111	52.136°	54°	6.324	0.083	23.818°	2.258°	47.710°	56.563°
	3	120	47.886°	49°	5.065	0.105	26.957°	2.456°	43.071°	52.701°
	4	122	48.503°	48°	4.986	0.107	27.199°	2.458°	43.685°	53.321°
M3	1	68	68.417°	69°	16.478	0.031	14.340°	1.739°	65.008°	71.825°
	2	113	52.265°	56°	6.244	0.084	23.986°	2.254°	47.847°	56.683°
	3	127	51.795°	54°	5.224	0.101	26.491°	2.347°	47.195°	56.395°
	4	122	53.009°	58°	5.654	0.093	25.342°	2.291°	48.518°	57.500°
Total	m1	308	54.187°	60°	5.841	0.090	24.885°	1.416°	51.411°	56.963°
	m2	333	53.170°	58°	5.615	0.094	25.438°	1.392°	50.442°	55.899°
	m3	366	58.727°	64°	6.810	0.077	22.871°	1.194°	56.385°	61.068°
	M1	365	47.711°	50°	4.980	0.107	27.217°	1.422°	44.923°	50.498°
	M2	463	50.497°	53°	5.453	0.097	25.860°	1.200°	48.144°	52.849°
	M3	430	55.019°	60°	5.947	0.088	24.640°	1.187°	52.692°	57.345°

Table 8.A7. Descriptive circular statistics for *C. bactrianus*. n, number of observed angles; SD, circular standard deviation; SE, standard error of mean.

	Stage	n	μ Mean vector	θ Circular median	κ Concentration	Circular variance	SD	SE	95 % Confidence interval (±) for μ	
m1	1	142	56.169°	60°	6.390	0.082	23.682°	1.985°	52.277°	60.060°
	2	148	50.291°	55°	5.765	0.091	25.069°	2.058°	46.257°	54.325°
	3	102	48.311°	50°	5.366	0.099	26.093°	2.579°	43.254°	53.367°
	4	76	52.125°	52.5°	6.815	0.077	22.862°	2.620°	46.989°	57.261°
m2	1	155	60.994°	64°	7.331	0.071	21.972°	1.763°	57.537°	64.451°
	2	166	54.534°	58.5°	5.876	0.089	24.804°	1.923°	50.766°	58.303°
	3	178	54.894°	62°	5.352	0.099	26.131°	1.955°	51.060°	58.727°
	4	160	51.578°	57°	5.749	0.092	25.106°	1.982°	47.692°	55.463°
m3	1	112	67.813°	71.5°	10.425	0.049	18.207°	1.720°	64.442°	71.184°
	2	140	60.146°	64°	6.717	0.078	23.043°	1.945°	56.332°	63.960°
	3	143	57.764°	62°	6.067	0.086	24.369°	2.035°	53.775°	61.754°
	4	162	52.064°	56.5°	5.566	0.095	25.564°	2.005°	48.132°	55.995°
M1	1	130	52.069°	55°	5.895	0.089	24.760°	2.169°	47.818°	56.321°
	2	144	48.736°	50°	4.834	0.110	27.683°	2.302°	44.223°	53.249°
	3	138	47.695°	49°	4.815	0.111	27.745°	2.357°	43.074°	52.315°
	4	107	48.737°	55°	4.487	0.119	28.896°	2.787°	43.274°	54.200°
M2	1	170	54.414°	56°	6.534	0.080	23.394°	1.792°	50.900°	57.927°
	2	173	52.225°	54°	6.058	0.087	24.390°	1.852°	48.595°	55.856°
	3	191	50.593°	52°	5.364	0.099	26.098°	1.885°	46.897°	54.289°
	4	185	50.209°	53°	5.185	0.102	26.604°	1.952°	46.382°	54.037°
M3	1	117	57.323°	60°	8.450	0.061	20.355°	1.881°	53.636°	61.009°
	2	136	53.369°	55.5°	5.770	0.091	25.055°	2.145°	49.164°	57.575°
	3	146	52.296°	55°	5.741	0.092	25.127°	2.077°	48.225°	56.367°
	4	170	50.182°	53°	5.311	0.100	26.244°	2.009°	46.243°	54.122°
Total	m1	468	51.960°	56°	5.912	0.089	24.720°	1.141°	49.723°	54.197°
	m2	659	55.462°	61°	5.862	0.090	24.837°	0.966°	53.568°	57.356°
	m3	557	58.817°	64°	6.274	0.083	23.921°	1.012°	56.832°	60.802°
	M1	519	49.312°	52°	4.954	0.107	27.300°	1.196°	46.967°	51.656°
	M2	719	51.806°	54°	5.687	0.093	25.258°	0.941°	49.962°	53.650°
	M3	569	52.992°	57°	5.933	0.089	24.673°	1.033°	50.967°	55.017°

Table 8.A8. Descriptive circular statistics for *C. ibex*. n, number of observed angles; SD, circular standard deviation; SE, standard error of mean.

	Stage	n	μ Mean vector	θ Circular median	κ Concentration	Circular variance	SD	SE	95 % Confidence interval (±) for μ	
m1	1	171	49.221°	51°	5.306	0.100	26.258°	2.005°	45.291°	53.150°
	2	146	50.626°	50°	5.359	0.099	26.112°	2.157°	46.396°	54.855°
	3	122	49.260°	47°	5.033	0.105	27.055°	2.445°	44.467°	54.052°
	4	90	55.572°	60°	6.244	0.084	23.985°	2.525°	50.621°	60.522°
m2	1	175	49.457°	55°	5.167	0.103	26.656°	2.011°	45.514°	53.400°
	2	166	48.909°	55°	5.663	0.093	25.319°	1.962°	45.062°	52.756°
	3	168	48.800°	55°	5.719	0.092	25.182°	1.940°	44.996°	52.603°
	4	151	49.349°	57°	5.337	0.099	26.173°	2.126°	45.181°	53.518°
m3	1	169	56.131°	59°	8.838	0.058	19.873°	1.528°	53.136°	59.126°
	2	191	50.434°	52°	6.936	0.075	22.644°	1.637°	47.225°	53.643°
	3	192	48.513°	49°	7.092	0.073	22.371°	1.613°	45.351°	51.675°
	4	201	47.359°	48°	6.722	0.078	23.034°	1.623°	44.177°	50.541°
M1	1	183	52.384°	54°	6.782	0.077	22.923°	1.693°	49.066°	55.703°
	2	190	48.532°	50°	5.138	0.103	26.740°	1.936°	44.736°	52.328°
	3	173	49.061°	50°	4.840	0.110	27.664°	2.099°	44.946°	53.175°
	4	115	53.715°	62°	4.619	0.116	28.415°	2.644°	48.532°	58.897°
M2	1	167	51.373°	55°	6.067	0.086	24.370°	1.883°	47.681°	55.065°
	2	173	48.234°	51°	5.647	0.093	25.358°	1.925°	44.460°	52.008°
	3	161	48.403°	51°	5.645	0.093	25.365°	1.996°	44.490°	52.316°
	4	167	46.858°	49°	5.492	0.096	25.757°	1.990°	42.957°	50.760°
M3	1	124	56.660°	62°	6.327	0.083	23.811°	2.136°	52.473°	60.847°
	2	164	53.130°	56°	6.426	0.081	23.609°	1.841°	49.520°	56.740°
	3	173	53.272°	52°	7.694	0.067	21.406°	1.626°	50.084°	56.460°
	4	138	55.317°	54.5°	7.384	0.070	21.887°	1.862°	51.668°	58.967°
Total	m1	529	50.714°	51°	5.352	0.099	26.131°	1.134°	48.490°	52.937°
	m2	660	49.126°	55°	5.460	0.097	25.841°	1.004°	47.157°	51.095°
	m3	753	50.425°	53°	7.117	0.073	22.329°	0.813°	48.831°	52.018°
	M1	661	50.648°	52°	5.267	0.100	26.368°	1.024°	48.641°	52.655°
	M2	668	48.722°	51°	5.683	0.093	25.270°	0.976°	46.808°	50.636°
	M3	599	54.402°	56°	6.913	0.075	22.684°	0.926°	52.587°	56.218°

Table 8.A9. Descriptive circular statistics for *C. simum*. n, number of observed angles; SD, circular standard deviation; SE, standard error of mean.

	Stage	n	μ Mean vector	θ Circular median	κ Concentration	Circular variance	SD	SE	95 % Confidence interval (±) for μ	
m1	1	99	52.119°	56°	5.210	0.102	26.532°	2.662°	46.901°	57.338°
	2	90	49.877°	52°	4.503	0.119	28.834°	3.032°	43.933°	55.821°
	3	60	53.334°	59°	4.488	0.119	28.893°	3.721°	46.040°	60.629°
	4	59	55.311°	61°	5.298	0.100	26.281°	3.416°	48.615°	62.007°
m2	1	106	54.399°	61°	5.471	0.097	25.813°	2.503°	49.492°	59.306°
	2	121	48.943°	52°	5.626	0.094	25.411°	2.307°	44.421°	53.465°
	3	125	48.708°	51°	4.956	0.107	27.292°	2.436°	43.932°	53.484°
	4	127	49.508°	57°	4.248	0.127	29.831°	2.640°	44.333°	54.683°
m3	1	67	46.758°	47°	6.662	0.078	23.147°	2.825°	41.220°	52.295°
	2	101	50.438°	51°	5.550	0.095	25.605°	2.544°	45.451°	55.425°
	3	115	47.615°	48°	5.036	0.105	27.046°	2.517°	42.680°	52.550°
	4	121	55.506°	58°	7.105	0.073	22.348°	2.030°	51.527°	59.485°
M1	1	188	51.648°	56°	5.640	0.093	25.377°	1.848°	48.025°	55.271°
	2	204	51.892°	58°	5.298	0.100	26.283°	1.837°	48.291°	55.494°
	3	187	48.228°	50°	4.867	0.109	27.578°	2.013°	44.283°	52.173°
	4	137	47.454°	53°	4.598	0.116	28.492°	2.429°	42.693°	52.215°
M2	1	102	57.899°	62°	6.730	0.078	23.018°	2.277°	53.435°	62.362°
	2	144	55.677°	58.5°	5.963	0.088	24.603°	2.048°	51.663°	59.691°
	3	172	51.374°	55°	5.825	0.090	24.923°	1.898°	47.653°	55.094°
	4	194	48.849°	54°	5.241	0.101	26.441°	1.895°	45.134°	52.565°
M3	1	151	40.594°	42°	4.762	0.112	27.923°	2.267°	36.149°	45.039°
	2	107	54.106°	55°	6.027	0.087	24.458°	2.361°	49.477°	58.736°
	3	126	56.933°	63°	6.163	0.085	24.159°	2.150°	52.719°	61.147°
	4	160	56.395°	63°	5.475	0.096	25.801°	2.036°	52.403°	60.388°
Total	m1	308	52.323°	58°	5.086	0.104	26.895°	1.003°	48.072°	52.006°
	m2	479	50.248°	54°	5.309	0.100	26.252°	0.949°	48.495°	52.215°
	m3	404	50.574°	51°	5.873	0.090	24.812°	0.994°	54.097°	57.995°
	M1	716	50.039°	55°	4.828	0.110	27.703°	1.575°	49.235°	55.411°
	M2	763	50.355°	55°	4.954	0.107	27.299°	1.245°	47.808°	52.689°
	M3	621	56.046°	62°	5.817	0.090	24.943°	1.239°	48.145°	53.004°

Table 8.A10. Descriptive circular statistics for *C. taurinus*. n, number of observed angles; SD, circular standard deviation; SE, standard error of mean.

	Stage	n	μ Mean vector	θ Circular median	κ Concentration	Circular variance	SD	SE	95 % Confidence interval (±) for μ	
m1	1	80	60.382°	58.5°	7.183	0.072	22.217°	2.482°	55.517°	65.247°
	2	104	51.477°	52.5°	6.116	0.086	24.262°	2.376°	46.819°	56.135°
	3	78	56.017°	61.5°	6.021	0.087	24.472°	2.767°	50.592°	61.442°
	4	63	51.518°	58°	5.158	0.103	26.682°	3.356°	44.940°	58.096°
m2	1	130	53.017°	54°	6.928	0.075	22.657°	1.985°	49.125°	56.909°
	2	127	50.422°	50°	6.733	0.077	23.014°	2.040°	46.423°	54.421°
	3	90	53.450°	56°	7.031	0.074	22.477°	2.367°	48.810°	58.091°
	4	141	49.149°	48°	5.525	0.095	25.671°	2.158°	44.917°	53.380°
m3	1	156	53.125°	52°	6.958	0.075	22.605°	1.808°	49.580°	56.670°
	2	156	54.132°	56°	7.887	0.066	21.122°	1.690°	50.819°	57.445°
	3	149	55.834°	54°	8.995	0.057	19.687°	1.612°	52.674°	58.994°
	4	145	53.965°	55°	6.887	0.076	22.731°	1.886°	50.268°	57.662°
M1	1	90	57.164°	61°	6.729	0.078	23.022°	2.424°	52.411°	61.916°
	2	108	54.612°	58.5°	5.951	0.088	24.631°	2.367°	49.971°	59.252°
	3	99	54.483°	59°	6.182	0.085	24.117°	2.421°	49.737°	59.229°
	4	77	51.746°	55°	5.930	0.089	24.680°	2.809°	46.239°	57.252°
M2	1	143	53.226°	57°	6.503	0.080	23.455°	1.959°	49.385°	57.067°
	2	122	54.455°	56°	7.862	0.066	21.158°	1.914°	50.703°	58.208°
	3	112	52.711°	54°	7.010	0.074	22.513°	2.125°	48.545°	56.878°
	4	93	54.906°	58°	5.895	0.089	24.759°	2.564°	49.880°	59.932°
M3	1	187	49.927°	49°	6.126	0.086	24.239°	1.770°	46.457°	53.398°
	2	163	53.151°	56°	6.187	0.085	24.107°	1.886°	49.454°	56.848°
	3	133	55.718°	58°	6.118	0.086	24.256°	2.101°	51.600°	59.836°
	4	170	53.084°	56°	5.473	0.096	25.807°	1.976°	49.210°	56.958°
Total	m1	325	54.797°	58°	5.966	0.088	24.595°	1.363°	52.126°	57.468°
	m2	488	51.319°	52°	6.388	0.082	23.686°	1.071°	49.219°	53.418°
	m3	606	54.258°	54°	7.574	0.069	21.589°	0.876°	52.540°	55.976°
	M1	374	54.609°	59°	6.147	0.085	24.193°	1.249°	52.160°	57.058°
	M2	470	53.754°	56°	6.774	0.077	22.937°	1.057°	51.682°	55.826°
	M3	653	52.732°	55°	5.916	0.089	24.710°	0.966°	50.839°	54.625°

Table 8.A11. Descriptive circular statistics for *D. bicornis*. n, number of observed angles; SD, circular standard deviation; SE, standard error of mean.

	Stage	n	μ Mean vector	θ Circular median	κ Concentration	Circular variance	SD	SE	95 % Confidence interval (±) for μ	
m1	1	92	50.476°	55°	4.903	0.108	27.459°	2.857°	44.876°	56.077°
	2	70	60.147°	68.5°	4.968	0.107	27.255°	3.251°	53.773°	66.520°
	3	48	54.269°	59°	5.438	0.097	25.900°	3.732°	46.952°	61.586°
	4	42	61.873°	69°	7.225	0.072	22.146°	3.414°	55.180°	68.566°
m2	1	71	43.934°	49°	4.429	0.121	29.114°	3.447°	37.178°	50.691°
	2	110	44.162°	44°	4.237	0.127	29.877°	2.841°	38.593°	49.730°
	3	78	59.826°	67°	5.737	0.092	25.137°	2.842°	54.254°	65.397°
	4	52	52.850°	61°	4.833	0.110	27.688°	3.832°	45.338°	60.361°
m3	1	74	39.551°	40°	4.623	0.116	28.403°	3.294°	33.094°	46.009°
	2	88	48.909°	49°	5.116	0.104	26.807°	2.852°	43.317°	54.501°
	3	95	46.010°	49°	5.021	0.106	27.092°	2.774°	40.571°	51.449°
	4	86	58.567°	65°	5.412	0.098	25.968°	2.796°	53.086°	64.047°
M1	1	78	57.057°	62°	5.998	0.088	24.522°	2.773°	51.621°	62.493°
	2	142	40.785°	38.5°	4.514	0.119	28.795°	2.411°	36.060°	45.511°
	3	130	42.434°	40°	4.951	0.107	27.309°	2.390°	37.748°	47.121°
	4	111	48.499°	54°	4.677	0.114	28.213°	2.672°	43.261°	53.737°
M2	1	126	50.718°	54.5°	5.703	0.092	25.221°	2.244°	46.320°	55.117°
	2	154	46.779°	45.5°	5.269	0.100	26.364°	2.121°	42.622°	50.937°
	3	187	48.908°	50°	5.700	0.092	25.226°	1.842°	45.297°	52.519°
	4	195	49.766°	55°	5.216	0.102	26.512°	1.895°	46.051°	53.481°
M3	1	78	54.041°	61°	6.859	0.076	22.781°	2.577°	48.990°	59.093°
	2	123	48.463°	51°	6.879	0.076	22.746°	2.049°	44.447°	52.480°
	3	139	46.673°	49°	6.132	0.086	24.225°	2.052°	42.649°	50.696°
	4	154	45.197°	48°	6.590	0.079	23.286°	1.874°	41.523°	48.872°
Total	m1	252	55.824°	62°	5.153	0.103	26.696°	1.679°	52.533°	59.115°
	m2	311	49.587°	54°	4.457	0.120	29.007°	1.641°	46.371°	52.804°
	m3	343	48.548°	50°	4.786	0.111	27.841°	1.500°	45.608°	51.489°
	M1	461	45.908°	47°	4.691	0.114	28.164°	1.309°	43.342°	48.474°
	M2	662	49.013°	51°	5.434	0.097	25.911°	1.005°	47.042°	50.984°
	M3	494	47.827°	50°	6.466	0.081	23.529°	1.057°	45.754°	49.900°

Table 8.A12. Descriptive circular statistics for *D. pygargus*. n, number of observed angles; SD, circular standard deviation; SE, standard error of mean.

	Stage	n	μ Mean vector	θ Circular median	κ Concentration	Circular variance	SD	SE	95 % Confidence interval (±) for μ	
m1	1	222	52.127°	56°	5.870	0.090	24.817°	1.663°	48.866°	55.387°
	2	207	50.110°	55°	5.335	0.099	26.179°	1.817°	46.549°	53.671°
	3	159	54.582°	59°	5.626	0.094	25.412°	2.012°	50.637°	58.527°
	4	155	52.236°	57°	5.484	0.096	25.779°	2.067°	48.183°	56.289°
m2	1	195	49.129°	51°	5.476	0.096	25.799°	1.845°	45.513°	52.745°
	2	177	51.423°	55°	5.591	0.094	25.501°	1.914°	47.671°	55.174°
	3	175	53.072°	59°	5.784	0.091	25.022°	1.889°	49.369°	56.775°
	4	213	48.708°	50°	5.216	0.102	26.515°	1.814°	45.153°	52.264°
m3	1	144	59.872°	63°	7.206	0.072	22.179°	1.847°	56.252°	63.492°
	2	157	55.125°	56°	6.980	0.075	22.566°	1.799°	51.598°	58.652°
	3	177	55.384°	57°	6.995	0.074	22.539°	1.693°	52.066°	58.702°
	4	183	54.002°	57°	6.460	0.081	23.541°	1.738°	50.594°	57.410°
M1	1	239	45.579°	43°	5.839	0.090	24.892°	1.608°	42.427°	48.732°
	2	235	47.857°	47°	5.923	0.089	24.696°	1.609°	44.703°	51.011°
	3	218	48.343°	53°	5.346	0.099	26.148°	1.768°	44.877°	51.809°
	4	185	50.434°	52°	5.213	0.102	26.522°	1.946°	46.618°	54.250°
M2	1	232	48.602°	49°	5.948	0.088	24.636°	1.615°	45.435°	51.768°
	2	293	45.872°	47°	5.692	0.093	25.247°	1.473°	42.985°	48.759°
	3	301	45.297°	44°	5.166	0.103	26.660°	1.534°	42.290°	48.304°
	4	314	42.441°	42°	4.928	0.108	27.382°	1.542°	39.417°	45.464°
M3	1	117	61.007°	63°	9.282	0.055	19.361°	1.789°	57.500°	64.514°
	2	187	52.206°	52°	5.644	0.093	25.365°	1.852°	48.575°	55.837°
	3	205	52.328°	56°	5.886	0.089	24.780°	1.728°	48.939°	55.716°
	4	192	52.722°	56°	5.497	0.096	25.744°	1.855°	49.085°	56.358°
Total	m1	743	52.116°	57°	5.561	0.095	25.576°	0.937°	50.280°	53.953°
	m2	760	50.460°	55°	5.469	0.097	25.816°	0.935°	48.627°	52.293°
	m3	661	55.923°	58°	6.821	0.076	22.851°	0.888°	54.182°	57.664°
	M1	877	47.893°	48°	5.567	0.095	25.561°	0.862°	46.203°	49.582°
	M2	1140	45.347°	46°	5.332	0.099	26.189°	0.774°	43.829°	46.865°
	M3	701	53.897°	57°	5.965	0.088	24.599°	0.928°	52.078°	55.716°

Table 8.A13. Descriptive circular statistics for *E. davidianus*. n, number of observed angles; SD, circular standard deviation; SE, standard error of mean.

	Stage	n	μ Mean vector	θ Circular median	κ Concentration	Circular variance	SD	SE	95 % Confidence interval (±) for μ	
m1	1	218	50.818°	54°	6.052	0.087	24.402°	1.651°	47.583°	54.054°
	2	235	48.285°	51°	5.794	0.091	24.999°	1.628°	45.093°	51.478°
	3	233	47.996°	51°	5.323	0.099	26.212°	1.714°	44.635°	51.356°
	4	162	52.857°	57°	5.338	0.099	26.170°	2.053°	48.833°	56.881°
m2	1	257	47.094°	46°	5.854	0.090	24.856°	1.548°	44.059°	50.130°
	2	238	47.356°	47°	6.026	0.087	24.460°	1.583°	44.252°	50.460°
	3	255	47.751°	48°	5.674	0.093	25.292°	1.582°	44.650°	50.851°
	4	242	47.837°	49°	5.808	0.091	24.964°	1.602°	44.696°	50.979°
m3	1	233	51.815°	54°	6.513	0.080	23.435°	1.534°	48.808°	54.821°
	2	236	50.845°	50°	6.910	0.075	22.690°	1.476°	47.952°	53.737°
	3	243	46.759°	49°	6.654	0.078	23.162°	1.484°	43.850°	49.669°
	4	253	47.260°	48°	6.792	0.077	22.905°	1.439°	44.440°	50.080°
M1	1	156	48.489°	49°	6.617	0.079	23.232°	1.858°	44.847°	52.132°
	2	239	42.571°	41°	5.537	0.095	25.639°	1.656°	39.325°	45.817°
	3	172	47.564°	47°	5.578	0.095	25.533°	1.944°	43.753°	51.375°
	4	159	47.935°	54°	4.723	0.113	28.054°	2.220°	43.583°	52.287°
M2	1	175	45.819°	44°	6.478	0.081	23.506°	1.775°	42.340°	49.299°
	2	247	41.996°	38°	6.711	0.078	23.054°	1.465°	39.124°	44.869°
	3	243	39.103°	36°	6.197	0.085	24.084°	1.543°	36.078°	42.128°
	4	248	39.913°	38°	5.367	0.098	26.092°	1.654°	36.670°	43.156°
M3	1	200	48.309°	49°	7.291	0.071	22.038°	1.557°	45.256°	51.361°
	2	230	43.134°	42°	6.537	0.080	23.388°	1.540°	40.114°	46.154°
	3	259	42.939°	42°	6.171	0.085	24.141°	1.498°	40.002°	45.876°
	4	267	38.686°	39°	5.634	0.094	25.390°	1.552°	35.644°	41.728°
Total	m1	848	49.730°	51°	5.598	0.094	25.482°	0.874°	48.017°	51.443°
	m2	992	47.507°	47°	5.834	0.090	24.903°	0.790°	45.959°	49.054°
	m3	965	49.110°	49°	6.660	0.078	23.150°	0.744°	47.650°	50.569°
	M1	726	46.203°	46°	5.478	0.096	25.793°	0.956°	44.329°	48.076°
	M2	913	41.403°	39°	6.060	0.087	24.384°	0.806°	39.823°	42.983°
	M3	956	42.949°	42°	6.178	0.085	24.125°	0.779°	41.421°	44.477°

Table 8.A14. Descriptive circular statistics for *E. quagga*. n, number of observed angles; SD, circular standard deviation; SE, standard error of mean.

	Stage	n	μ Mean vector	θ Circular median	κ Concentration	Circular variance	SD	SE	95 % Confidence interval (±) for μ	
m1	1	207	51.003°	56°	5.099	0.104	26.855°	1.863°	47.351°	54.655°
	2	196	51.104°	57°	5.230	0.101	26.474°	1.888°	47.404°	54.805°
	3	198	48.411°	50°	5.482	0.096	25.782°	1.829°	44.825°	51.997°
	4	155	46.518°	49°	5.040	0.105	27.035°	2.167°	42.269°	50.767°
m2	1	196	48.225°	51°	4.820	0.111	27.730°	1.977°	44.350°	52.100°
	2	209	48.636°	53°	5.020	0.106	27.093°	1.87°	44.970°	52.303°
	3	199	48.579°	49°	5.164	0.103	26.666°	1.887°	44.880°	52.278°
	4	167	47.640°	49°	5.369	0.098	26.084°	2.015°	43.690°	51.591°
m3	1	175	52.674°	58°	5.064	0.105	26.960°	2.034°	48.686°	56.662°
	2	195	51.926°	57°	5.403	0.098	25.992°	1.858°	48.283°	55.569°
	3	190	52.600°	57°	5.840	0.090	24.888°	1.803°	49.065°	56.134°
	4	167	53.647°	59°	5.435	0.097	25.908°	2.002°	49.723°	57.571°
M1	1	210	50.445°	55°	5.257	0.101	26.397°	1.818°	46.880°	54.009°
	2	210	48.731°	51°	5.541	0.095	25.629°	1.766°	45.269°	52.193°
	3	230	48.187°	51°	5.760	0.091	25.079°	1.651°	44.950°	51.425°
	4	198	48.624°	52°	5.461	0.097	25.838°	1.833°	45.030°	52.218°
M2	1	275	47.944°	49°	6.018	0.087	24.477°	1.474°	45.054°	50.834°
	2	253	46.037°	46°	5.731	0.092	25.151°	1.579°	42.941°	49.132°
	3	265	45.159°	46°	5.493	0.096	25.754°	1.58°	42.063°	48.256°
	4	217	47.307°	49°	5.569	0.095	25.556°	1.732°	43.912°	50.703°
M3	1	210	52.113°	54°	6.418	0.081	23.625°	1.628°	48.920°	55.305°
	2	241	51.924°	55°	5.676	0.093	25.286°	1.626°	48.735°	55.112°
	3	223	52.198°	57°	5.612	0.094	25.448°	1.702°	48.863°	55.534°
	4	225	51.319°	57°	5.308	0.100	26.254°	1.747°	47.894°	54.745°
Total	m1	756	49.431°	53°	5.193	0.102	26.580°	0.965°	47.539°	51.323°
	m2	771	48.300°	51°	5.073	0.105	26.934°	0.968°	46.402°	50.198°
	m3	727	52.677°	57°	5.425	0.097	25.933°	0.96°	50.795°	54.560°
	M1	848	48.980°	52°	5.499	0.096	25.737°	0.882°	47.250°	50.710°
	M2	1010	46.602°	48°	5.694	0.092	25.241°	0.793°	45.047°	48.157°
	M3	899	51.886°	56°	5.712	0.092	25.197°	0.839°	50.241°	53.531°

Table 8.A15. Descriptive circular statistics for *H. jemlahicus*. n, number of observed angles; SD, circular standard deviation; SE, standard error of mean.

	Stage	n	μ Mean vector	θ Circular median	κ Concentration	Circular variance	SD	SE	95 % Confidence interval (±) for μ	
m1	1	106	57.856°	63°	7.345	0.071	21.950°	2.130°	53.680°	62.032°
	2	105	57.225°	64°	6.515	0.080	23.431°	2.284°	52.747°	61.703°
	3	113	50.936°	55°	4.953	0.107	27.303°	2.563°	45.911°	55.961°
	4	81	50.892°	54°	5.124	0.103	26.780°	2.970°	45.069°	56.715°
m2	1	182	58.087°	64°	6.922	0.075	22.669°	1.679°	54.796°	61.378°
	2	195	57.433°	61°	6.714	0.078	23.050°	1.649°	54.201°	60.666°
	3	210	57.056°	61°	6.633	0.079	23.202°	1.599°	53.921°	60.192°
	4	176	56.486°	62°	5.690	0.093	25.252°	1.901°	52.760°	60.212°
m3	1	155	61.108°	64°	7.906	0.066	21.095°	1.693°	57.789°	64.427°
	2	150	62.660°	65°	10.276	0.050	18.345°	1.497°	59.725°	65.595°
	3	108	65.149°	68°	11.736	0.044	17.107°	1.646°	61.923°	68.375°
	4	141	64.422°	68°	10.850	0.047	17.827°	1.501°	61.480°	67.364°
M1	1	154	50.134°	51°	5.564	0.095	25.570°	2.057°	46.100°	54.167°
	2	145	51.208°	54°	5.653	0.093	25.344°	2.102°	47.088°	55.328°
	3	158	50.486°	51.5°	5.410	0.098	25.975°	2.063°	46.442°	54.531°
	4	130	53.182°	58°	5.463	0.097	25.833°	2.262°	48.748°	57.617°
M2	1	149	53.997°	55°	9.296	0.055	19.345°	1.584°	50.892°	57.102°
	2	168	49.635°	51°	6.494	0.080	23.473°	1.809°	46.088°	53.181°
	3	190	47.302°	49°	5.768	0.091	25.062°	1.816°	43.743°	50.861°
	4	192	49.098°	52°	5.811	0.091	24.959°	1.799°	45.572°	52.625°
M3	1	93	60.452°	65°	8.868	0.058	19.837°	2.056°	56.422°	64.482°
	2	126	53.787°	54°	7.036	0.074	22.467°	2.000°	49.867°	57.708°
	3	136	52.221°	55°	6.730	0.078	23.019°	1.972°	48.356°	56.087°
	4	145	53.580°	56°	6.511	0.080	23.440°	1.944°	49.768°	57.392°
Total	m1	405	54.425°	61°	5.753	0.091	25.098°	1.245°	51.984°	56.866°
	m2	763	57.270°	62°	6.464	0.081	23.532°	0.851°	55.602°	58.939°
	m3	554	63.172°	66°	9.752	0.053	18.860°	0.801°	61.602°	64.743°
	M1	587	51.169°	53°	5.511	0.096	25.708°	1.059°	49.092°	53.246°
	M2	699	49.817°	52°	6.409	0.082	23.642°	0.893°	48.066°	51.568°
	M3	500	54.562°	58°	6.944	0.075	22.630°	1.011°	52.580°	56.544°

Table 8.A16. Descriptive circular statistics for *K. ellipsiprymnus*. n, number of observed angles; SD, circular standard deviation; SE, standard error of mean.

Stage		n	μ Mean vector	θ Circular median	κ Concentration	Circular variance	SD	SE	95 % Confidence interval (±) for μ	
m1	1	152	47.020°	49°	5.721	0.092	25.175°	2.039°	43.023°	51.017°
	2	141	47.546°	48°	5.692	0.093	25.247°	2.123°	43.384°	51.708°
	3	155	46.964°	49°	5.776	0.091	25.041°	2.009°	43.026°	50.901°
	4	123	48.845°	51°	6.169	0.085	24.145°	2.174°	44.582°	53.108°
m2	1	153	46.977°	50°	6.377	0.082	23.708°	1.915°	43.223°	50.730°
	2	143	48.388°	49°	6.460	0.081	23.542°	1.966°	44.533°	52.243°
	3	175	46.094°	45°	6.367	0.082	23.728°	1.792°	42.582°	49.606°
	4	177	47.834°	48°	5.957	0.088	24.616°	1.848°	44.212°	51.456°
m3	1	174	47.196°	48°	5.628	0.094	25.406°	1.923°	43.426°	50.966°
	2	195	45.430°	47°	5.560	0.095	25.578°	1.829°	41.845°	49.015°
	3	197	44.629°	43°	5.748	0.092	25.110°	1.786°	41.126°	48.131°
	4	170	44.451°	43°	5.806	0.091	24.969°	1.912°	40.702°	48.200°
M1	1	138	44.241°	43°	6.289	0.083	23.891°	2.031°	40.259°	48.223°
	2	162	40.877°	41°	6.026	0.087	24.461°	1.919°	37.114°	44.640°
	3	170	45.236°	44°	6.164	0.085	24.156°	1.850°	41.609°	48.864°
	4	146	44.633°	48°	6.179	0.085	24.124°	1.994°	40.724°	48.542°
M2	1	151	39.988°	39°	7.101	0.073	22.355°	1.818°	36.425°	43.551°
	2	152	40.313°	38°	6.449	0.081	23.563°	1.909°	36.570°	44.055°
	3	172	43.211°	42°	6.292	0.083	23.883°	1.819°	39.646°	46.777°
	4	177	45.538°	47°	6.684	0.078	23.106°	1.735°	42.136°	48.939°
M3	1	112	48.191°	54°	4.790	0.111	27.827°	2.624°	43.047°	53.334°
	2	129	46.851°	46°	5.535	0.095	25.644°	2.254°	42.431°	51.270°
	3	133	43.308°	39°	6.044	0.087	24.420°	2.115°	39.162°	47.454°
	4	151	43.844°	41°	5.450	0.097	25.866°	2.102°	39.724°	47.964°
Total	m1	571	47.530°	49°	5.815	0.090	24.948°	1.043°	45.486°	49.574°
	m2	648	47.283°	48°	6.264	0.084	23.944°	0.939°	45.442°	49.125°
	m3	736	45.406°	45°	5.673	0.093	25.295°	0.931°	43.581°	47.231°
	M1	616	43.726°	43°	6.129	0.086	24.232°	0.975°	41.814°	45.638°
	M2	652	42.418°	41°	6.551	0.080	23.361°	0.914°	40.626°	44.210°
	M3	525	45.360°	44°	5.415	0.098	25.961°	1.131°	43.142°	47.577°

Table 8.A17. Descriptive circular statistics for *L. glama*. n, number of observed angles; SD, circular standard deviation; SE, standard error of mean.

Stage		n	μ Mean vector	θ Circular median	κ Concentration	Circular variance	SD	SE	95 % Confidence interval (±) for μ	
m1	1	56	44.716°	44°	4.176	0.129	30.131°	4.015°	36.845°	52.587°
	2	81	48.432°	47°	5.146	0.103	26.717°	2.963°	42.624°	54.241°
	3	96	47.878°	48°	4.887	0.109	27.512°	2.802°	42.385°	53.372°
	4	65	49.880°	51°	5.094	0.104	26.870°	3.327°	43.359°	56.402°
m2	1	72	40.458°	37°	5.387	0.098	26.036°	3.063°	34.453°	46.463°
	2	96	38.497°	37°	5.432	0.097	25.915°	2.641°	33.321°	43.674°
	3	106	40.109°	41°	5.654	0.093	25.341°	2.458°	35.291°	44.927°
	4	141	38.604°	36°	4.808	0.111	27.769°	2.334°	34.029°	43.178°
m3	1	63	43.195°	34°	4.212	0.128	29.979°	3.766°	35.812°	50.579°
	2	107	45.472°	44°	4.909	0.108	27.440°	2.647°	40.282°	50.662°
	3	110	46.802°	48°	5.125	0.103	26.777°	2.548°	41.806°	51.798°
	4	135	39.141°	37°	5.645	0.093	25.364°	2.180°	34.868°	43.414°
M1	1	91	43.499°	40°	6.665	0.078	23.141°	2.423°	38.748°	48.250°
	2	157	44.910°	45°	5.611	0.094	25.449°	2.028°	40.934°	48.886°
	3	127	44.707°	45°	4.690	0.114	28.167°	2.494°	39.818°	49.596°
	4	78	34.355°	27°	5.174	0.102	26.634°	3.010°	28.454°	40.257°
M2	1	66	40.646°	35°	6.108	0.086	24.277°	2.985°	34.795°	46.496°
	2	155	46.926°	46°	6.247	0.084	23.978°	1.924°	43.155°	50.697°
	3	136	46.116°	47°	6.666	0.078	23.139°	1.982°	42.231°	50.002°
	4	126	45.336°	44.5°	5.390	0.098	26.028°	2.315°	40.798°	49.874°
M3	1	55	38.302°	27°	4.656	0.115	28.286°	3.805°	30.842°	45.762°
	2	73	42.904°	39°	5.073	0.105	26.934°	3.147°	36.736°	49.073°
	3	119	43.883°	43°	5.003	0.106	27.148°	2.484°	39.014°	48.752°
	4	117	42.217°	42°	5.534	0.095	25.646°	2.367°	37.576°	46.858°
Total	m1	298	47.888°	48°	4.823	0.110	27.719°	1.602°	44.747°	51.030°
	m2	415	39.290°	38°	5.239	0.101	26.448°	1.296°	36.749°	41.830°
	m3	415	43.402°	41°	4.988	0.107	27.195°	1.332°	40.790°	46.014°
	M1	453	42.761°	41°	5.297	0.100	26.284°	1.233°	40.344°	45.178°
	M2	483	45.429°	45°	6.044	0.087	24.419°	1.110°	43.253°	47.604°
	M3	364	42.314°	40°	5.095	0.104	26.867°	1.406°	39.558°	45.069°

Table 8.A18. Descriptive circular statistics for *L. huanachus*. n, number of observed angles; SD, circular standard deviation; SE, standard error of mean.

	Stage	n	μ Mean vector	θ Circular median	κ Concentration	Circular variance	SD	SE	95 % Confidence interval (±) for μ	
m1	1	168	51.717°	53.5°	7.422	0.070	21.826°	1.683°	48.418°	55.015°
	2	172	48.579°	49°	6.266	0.084	23.939°	1.823°	45.004°	52.153°
	3	145	51.425°	52°	5.872	0.090	24.814°	2.058°	47.391°	55.460°
	4	91	50.414°	51°	6.266	0.084	23.938°	2.506°	45.500°	55.328°
m2	1	182	55.767°	58°	7.752	0.067	21.319°	1.579°	52.672°	58.863°
	2	188	50.690°	53°	6.354	0.082	23.756°	1.731°	47.297°	54.083°
	3	158	50.236°	52°	5.693	0.092	25.243°	2.005°	46.305°	54.168°
	4	139	50.597°	52°	6.315	0.083	23.837°	2.019°	46.639°	54.556°
m3	1	163	62.544°	65°	8.132	0.064	20.777°	1.626°	59.356°	65.732°
	2	210	55.190°	59°	6.281	0.083	23.907°	1.648°	51.959°	58.420°
	3	209	53.475°	54°	6.251	0.084	23.970°	1.656°	50.229°	56.722°
	4	201	53.127°	54°	7.545	0.069	21.633°	1.525°	50.138°	56.116°
M1	1	127	51.741°	52°	6.335	0.083	23.795°	2.109°	47.607°	55.876°
	2	135	49.592°	53°	6.514	0.080	23.434°	2.015°	45.643°	53.542°
	3	102	55.205°	59°	7.405	0.070	21.853°	2.162°	50.967°	59.444°
	4	73	57.641°	59°	7.045	0.074	22.453°	2.625°	52.494°	62.788°
M2	1	141	51.818°	53°	6.722	0.078	23.033°	1.938°	48.019°	55.617°
	2	149	47.612°	47°	6.092	0.086	24.312°	1.989°	43.712°	51.512°
	3	183	48.304°	48°	6.525	0.080	23.412°	1.729°	44.915°	51.693°
	4	178	48.942°	50°	6.418	0.082	23.626°	1.769°	45.475°	52.410°
M3	1	143	49.252°	50°	5.993	0.088	24.533°	2.049°	45.235°	53.269°
	2	152	50.239°	53.5°	5.511	0.096	25.706°	2.082°	46.157°	54.320°
	3	178	49.880°	52°	5.848	0.090	24.870°	1.862°	46.231°	53.529°
	4	183	49.571°	48°	6.245	0.084	23.982°	1.771°	46.099°	53.042°
Total	m1	576	50.504°	52°	6.429	0.081	23.604°	0.982°	48.578°	52.430°
	m2	667	51.970°	54°	6.423	0.081	23.615°	0.913°	50.180°	53.761°
	m3	783	55.751°	57°	6.736	0.077	23.007°	0.821°	54.140°	57.361°
	M1	437	52.883°	55°	6.631	0.079	23.207°	1.109°	50.709°	55.057°
	M2	651	49.085°	49°	6.408	0.082	23.646°	0.926°	47.270°	50.900°
	M3	656	49.739°	51°	5.899	0.089	24.752°	0.965°	47.847°	51.631°

Table 8.A19. Descriptive circular statistics for *L. walleri*. n, number of observed angles; SD, circular standard deviation; SE, standard error of mean.

	Stage	n	μ Mean vector	θ Circular median	κ Concentration	Circular variance	SD	SE	95 % Confidence interval (±) for μ	
m1	1	140	52.558°	58°	5.927	0.089	24.685°	2.083°	48.473°	56.642°
	2	143	51.525°	55°	5.882	0.089	24.789°	2.070°	47.466°	55.583°
	3	131	46.455°	50°	5.349	0.099	26.140°	2.280°	41.986°	50.925°
	4	90	54.717°	62°	5.536	0.095	25.642°	2.699°	49.426°	60.007°
m2	1	104	55.335°	62.5°	6.495	0.080	23.472°	2.299°	50.828°	59.843°
	2	154	49.647°	55°	5.729	0.092	25.155°	2.024°	45.679°	53.615°
	3	161	49.988°	56°	5.468	0.097	25.819°	2.032°	46.005°	53.971°
	4	139	47.504°	50°	5.159	0.103	26.680°	2.259°	43.076°	51.932°
m3	1	85	56.841°	62°	7.984	0.065	20.984°	2.274°	52.382°	61.300°
	2	157	51.484°	57°	5.598	0.094	25.484°	2.031°	47.503°	55.465°
	3	171	50.075°	53°	5.455	0.097	25.853°	1.974°	46.205°	53.944°
	4	169	46.998°	49°	5.276	0.100	26.344°	2.023°	43.033°	50.964°
M1	1	117	52.933°	57°	6.904	0.075	22.701°	2.097°	48.823°	57.043°
	2	186	45.605°	49°	5.313	0.100	26.241°	1.921°	41.839°	49.370°
	3	199	43.081°	47°	4.763	0.112	27.920°	1.975°	39.210°	46.953°
	4	169	42.947°	46°	5.156	0.103	26.688°	2.049°	38.930°	46.965°
M2	1	141	51.257°	51°	7.214	0.072	22.164°	1.865°	47.601°	54.913°
	2	203	47.737°	49°	6.323	0.083	23.820°	1.670°	44.463°	51.011°
	3	204	44.088°	42°	5.669	0.093	25.303°	1.769°	40.620°	47.555°
	4	222	40.672°	40°	4.520	0.118	28.774°	1.926°	36.895°	44.449°
M3	1	118	58.77°	60°	8.254	0.063	20.613°	1.896°	55.052°	62.487°
	2	182	48.766°	49°	6.222	0.084	24.031°	1.779°	45.278°	52.254°
	3	192	45.548°	47°	5.775	0.091	25.044°	1.805°	42.010°	49.086°
	4	197	45.538°	46°	5.305	0.100	26.263°	1.868°	41.876°	49.200°
Total	m1	504	51.073°	56°	5.615	0.094	25.438°	1.131°	48.855°	53.291°
	m2	558	50.291°	55°	5.565	0.095	25.566°	1.081°	48.173°	52.410°
	m3	582	50.586°	54°	5.618	0.094	25.431°	1.053°	48.522°	52.649°
	M1	671	45.508°	49°	5.210	0.102	26.532°	1.022°	43.504°	47.513°
	M2	770	45.447°	46°	5.507	0.096	25.716°	0.925°	43.633°	47.261°
	M3	689	48.714°	50°	5.835	0.090	24.902°	0.947°	46.857°	50.571°

Table 8.A20. Descriptive circular statistics for *N. goral*. n, number of observed angles; SD, circular standard deviation; SE, standard error of mean.

	Stage	n	μ Mean vector	θ Circular median	κ Concentration	Circular variance	SD	SE	95 % Confidence interval (±) for μ	
m1	1	127	52.072°	55°	6.609	0.079	23.248°	2.061°	48.033°	56.112°
	2	125	47.669°	46°	6.891	0.076	22.725°	2.031°	43.688°	51.650°
	3	108	48.553°	52°	6.377	0.082	23.708°	2.279°	44.086°	53.021°
	4	98	53.342°	57°	6.091	0.086	24.315°	2.453°	48.533°	58.151°
m2	1	235	48.145°	49°	7.258	0.072	22.092°	1.440°	45.323°	50.968°
	2	233	47.976°	49°	7.403	0.070	21.857°	1.431°	45.171°	50.780°
	3	219	47.978°	49°	7.254	0.072	22.099°	1.492°	45.053°	50.903°
	4	201	49.845°	50°	6.933	0.075	22.648°	1.596°	46.716°	52.973°
m3	1	202	54.808°	54°	8.877	0.058	19.826°	1.394°	52.075°	57.541°
	2	214	53.352°	53°	8.628	0.060	20.130°	1.375°	50.656°	56.048°
	3	210	52.695°	53°	8.571	0.060	20.201°	1.393°	49.964°	55.426°
	4	220	49.564°	49°	8.088	0.064	20.839°	1.404°	46.811°	52.316°
M1	1	154	47.813°	52°	5.458	0.097	25.846°	2.079°	43.736°	51.889°
	2	169	45.723°	46°	5.296	0.100	26.286°	2.019°	41.766°	49.680°
	3	131	48.317°	52°	5.361	0.099	26.107°	2.277°	43.853°	52.781°
	4	120	43.274°	46°	4.562	0.117	28.621°	2.607°	38.164°	48.384°
M2	1	188	48.367°	50.5°	6.442	0.081	23.576°	1.718°	45.000°	51.734°
	2	231	41.924°	42°	5.576	0.095	25.540°	1.678°	38.635°	45.213°
	3	226	40.995°	41°	5.302	0.100	26.270°	1.745°	37.575°	44.415°
	4	225	40.850°	42°	5.493	0.096	25.753°	1.714°	37.489°	44.210°
M3	1	169	54.711°	56°	8.427	0.061	20.385°	1.567°	51.639°	57.784°
	2	189	49.038°	51°	6.183	0.085	24.114°	1.752°	45.603°	52.472°
	3	201	49.769°	50°	6.835	0.076	22.826°	1.608°	46.616°	52.922°
	4	237	45.536°	48°	5.764	0.091	25.070°	1.626°	42.348°	48.724°
Total	m1	458	50.306°	52°	6.448	0.081	23.564°	1.100°	48.150°	52.462°
	m2	888	48.443°	49°	7.209	0.072	22.172°	0.743°	46.985°	49.900°
	m3	846	52.556°	52°	8.451	0.061	20.353°	0.699°	51.185°	53.927°
	M1	574	46.379°	48°	5.153	0.103	26.697°	1.112°	44.198°	48.559°
	M2	870	42.817°	44°	5.572	0.095	25.549°	0.865°	41.121°	44.512°
	M3	796	49.422°	51°	6.450	0.081	23.560°	0.834°	47.786°	51.057°

Table 8.A21. Descriptive circular statistics for *O. bezoarticus*. n, number of observed angles; SD, circular standard deviation; SE, standard error of mean.

	Stage	n	μ Mean vector	θ Circular median	κ Concentration	Circular variance	SD	SE	95 % Confidence interval (±) for μ	
m1	1	203	49.096°	51°	5.254	0.101	26.404°	1.850°	45.470°	52.723°
	2	151	40.631°	34°	4.604	0.116	28.468°	2.311°	36.100°	45.162°
	3	121	46.406°	48°	4.002	0.135	30.901°	2.800°	40.916°	51.895°
	4	95	55.936°	61°	4.045	0.134	30.705°	3.141°	49.780°	62.093°
m2	1	247	46.184°	46°	5.248	0.101	26.422°	1.678°	42.894°	49.474°
	2	266	45.156°	46°	5.700	0.092	25.228°	1.545°	42.129°	48.184°
	3	260	45.531°	43.5°	5.705	0.092	25.214°	1.561°	42.470°	48.592°
	4	174	40.302°	35.5°	5.365	0.099	26.097°	1.975°	36.430°	44.174°
m3	1	146	57.769°	61°	8.669	0.060	20.078°	1.661°	54.514°	61.025°
	2	216	52.289°	53°	6.814	0.077	22.864°	1.554°	49.243°	55.336°
	3	228	51.867°	56°	6.047	0.087	24.412°	1.615°	48.702°	55.032°
	4	216	50.773°	57°	5.364	0.099	26.098°	1.773°	47.298°	54.249°
M1	1	168	46.538°	46°	6.036	0.087	24.438°	1.883°	42.846°	50.229°
	2	223	38.761°	34°	6.182	0.085	24.116°	1.613°	35.599°	41.923°
	3	238	36.900°	32°	5.613	0.094	25.444°	1.647°	33.671°	40.128°
	4	219	38.317°	34°	4.450	0.120	29.035°	1.957°	34.480°	42.154°
M2	1	233	41.618°	40°	6.860	0.076	22.780°	1.491°	38.695°	44.541°
	2	258	36.084°	33°	6.872	0.076	22.758°	1.415°	33.309°	38.859°
	3	257	34.738°	31°	6.935	0.075	22.645°	1.411°	31.971°	37.504°
	4	278	35.098°	33°	6.217	0.084	24.041°	1.440°	32.274°	37.921°
M3	1	207	51.437°	55°	7.339	0.071	21.959°	1.525°	48.448°	54.427°
	2	224	43.147°	43°	6.620	0.079	23.228°	1.550°	40.108°	46.187°
	3	243	38.340°	34°	6.529	0.080	23.403°	1.500°	35.400°	41.280°
	4	252	35.854°	31°	6.287	0.083	23.895°	1.503°	32.907°	38.801°
Total	m1	570	47.408°	50°	4.431	0.121	29.108°	1.216°	45.024°	49.792°
	m2	947	44.638°	43°	5.481	0.096	25.784°	0.837°	42.998°	46.278°
	m3	806	52.788°	56°	6.301	0.083	23.865°	0.840°	51.142°	54.433°
	M1	848	39.680°	35°	5.357	0.099	26.118°	0.895°	37.925°	41.435°
	M2	1026	36.739°	33°	6.611	0.079	23.245°	0.725°	35.317°	38.160°
	M3	926	41.775°	40°	6.282	0.083	23.905°	0.785°	40.237°	43.313°

Table 8.A22. Descriptive circular statistics for *R. fulvorufula*. n, number of observed angles; SD, circular standard deviation; SE, standard error of mean.

	Stage	n	μ Mean vector	θ Circular median	κ Concentration	Circular variance	SD	SE	95 % Confidence interval (±) for μ	
m1	1	111	52.878°	52°	8.694	0.059	20.047°	1.902°	49.150°	56.606°
	2	168	45.871°	48°	5.733	0.092	25.146°	1.937°	42.073°	49.668°
	3	145	51.269°	52°	7.497	0.069	21.708°	1.801°	47.738°	54.801°
	4	120	50.873°	50.5°	6.750	0.077	22.982°	2.096°	46.764°	54.982°
m2	1	129	57.248°	61°	8.027	0.065	20.923°	1.841°	53.640°	60.857°
	2	180	48.247°	52°	6.519	0.080	23.424°	1.744°	44.828°	51.666°
	3	176	48.074°	49°	5.710	0.092	25.202°	1.897°	44.356°	51.793°
	4	171	46.364°	48°	5.869	0.090	24.820°	1.895°	42.648°	50.080°
m3	1	107	62.100°	64°	11.174	0.046	17.554°	1.696°	58.774°	65.425°
	2	159	53.838°	55°	6.777	0.077	22.932°	1.817°	50.276°	57.399°
	3	159	51.421°	51°	6.187	0.085	24.106°	1.909°	47.678°	55.164°
	4	160	54.058°	55°	6.878	0.076	22.747°	1.797°	50.536°	57.580°
M1	1	81	51.199°	48°	6.083	0.086	24.334°	2.700°	45.905°	56.493°
	2	126	46.329°	45°	5.185	0.102	26.605°	2.366°	41.691°	50.967°
	3	119	45.871°	44°	5.526	0.095	25.667°	2.349°	41.266°	50.476°
	4	82	47.803°	48°	5.499	0.096	25.739°	2.838°	42.239°	53.366°
M2	1	85	59.460°	60°	10.811	0.047	17.861°	1.937°	55.663°	63.256°
	2	159	40.515°	36°	5.866	0.090	24.827°	1.966°	36.661°	44.370°
	3	157	41.117°	37°	5.453	0.097	25.859°	2.060°	37.077°	45.156°
	4	186	40.652°	38°	5.433	0.097	25.913°	1.897°	36.934°	44.371°
M3	1	57	65.461°	67°	13.703	0.037	15.778°	2.089°	61.365°	69.557°
	2	128	47.991°	49°	6.282	0.083	23.906°	2.110°	43.854°	52.128°
	3	124	45.201°	49°	5.676	0.093	25.287°	2.268°	40.756°	49.646°
	4	176	46.564°	50°	5.714	0.092	25.193°	1.896°	42.847°	50.281°
Total	m1	544	49.878°	50°	6.772	0.077	22.941°	0.983°	47.952°	51.804°
	m2	656	49.515°	52°	6.171	0.085	24.141°	0.941°	47.670°	51.361°
	m3	585	54.801°	56°	6.950	0.075	22.618°	0.934°	52.970°	56.633°
	M1	408	47.468°	46°	5.478	0.096	25.793°	1.275°	44.969°	49.968°
	M2	587	43.553°	41°	5.604	0.094	25.467°	1.050°	41.495°	45.610°
	M3	485	48.901°	52°	5.903	0.089	24.741°	1.122°	46.702°	51.101°

Table 8.A23. Descriptive circular statistics for *R. tarandus*. n, number of observed angles; SD, circular standard deviation; SE, standard error of mean.

	Stage	n	μ Mean vector	θ Circular median	κ Concentration	Circular variance	SD	SE	95 % Confidence interval (±) for μ	
m1	1	136	60.706°	66°	7.267	0.072	22.077°	1.891°	56.999°	64.414°
	2	179	56.943°	62°	6.015	0.087	24.486°	1.828°	53.359°	60.526°
	3	170	53.884°	61°	5.067	0.105	26.953°	2.063°	49.839°	57.929°
	4	112	57.067°	61°	5.992	0.088	24.536°	2.315°	52.528°	61.606°
m2	1	215	57.787°	60°	6.208	0.084	24.061°	1.639°	54.574°	61.000°
	2	242	53.998°	57°	6.000	0.087	24.518°	1.574°	50.912°	57.084°
	3	227	54.272°	59°	5.866	0.090	24.828°	1.646°	51.046°	57.498°
	4	217	55.698°	63°	5.994	0.088	24.532°	1.663°	52.437°	58.958°
m3	1	163	60.318°	65°	6.463	0.081	23.534°	1.841°	56.709°	63.928°
	2	179	58.338°	64°	6.735	0.077	23.009°	1.718°	54.970°	61.706°
	3	199	55.101°	58°	6.208	0.084	24.062°	1.704°	51.762°	58.441°
	4	188	56.438°	62°	7.006	0.074	22.521°	1.641°	53.221°	59.655°
M1	1	165	50.019°	50°	6.293	0.083	23.883°	1.857°	46.378°	53.659°
	2	240	46.720°	46°	5.969	0.088	24.588°	1.585°	43.613°	49.828°
	3	229	49.379°	49°	5.398	0.098	26.008°	1.716°	46.015°	52.742°
	4	170	51.104°	56°	4.667	0.114	28.248°	2.162°	46.866°	55.341°
M2	1	192	56.344°	57°	9.267	0.056	19.377°	1.398°	53.604°	59.084°
	2	250	48.957°	49°	6.545	0.080	23.373°	1.477°	46.063°	51.852°
	3	254	48.383°	47°	5.566	0.095	25.565°	1.602°	45.243°	51.523°
	4	221	50.875°	52°	5.189	0.102	26.591°	1.785°	47.375°	54.376°
M3	1	103	65.053°	70°	8.450	0.061	20.356°	2.004°	61.123°	68.982°
	2	196	56.962°	61°	7.863	0.066	21.157°	1.510°	54.002°	59.922°
	3	224	53.164°	56°	7.181	0.072	22.220°	1.483°	50.256°	56.072°
	4	237	51.304°	54°	6.350	0.082	23.764°	1.542°	48.282°	54.327°
Total	m1	597	56.984°	62°	5.870	0.090	24.818°	1.014°	54.995°	58.972°
	m2	901	55.383°	59°	5.992	0.088	24.537°	0.816°	53.783°	56.984°
	m3	729	57.409°	62°	6.544	0.080	23.375°	0.865°	55.713°	59.104°
	M1	804	49.069°	50°	5.509	0.096	25.713°	0.905°	47.294°	50.844°
	M2	917	50.847°	52°	6.130	0.086	24.229°	0.799°	49.280°	52.413°
	M3	760	55.202°	58°	6.932	0.075	22.650°	0.821°	53.592°	56.811°

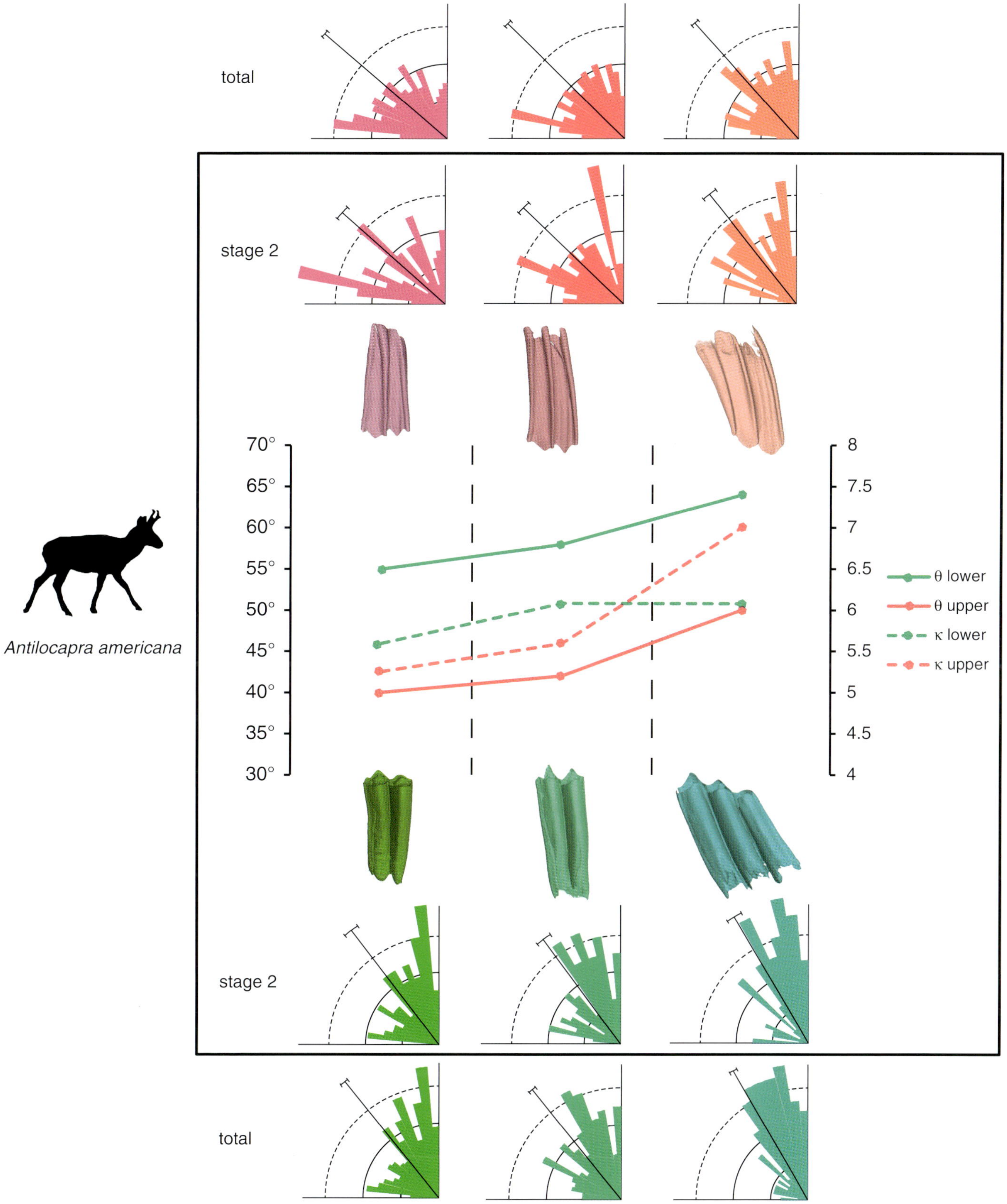

Fig. 8.A2. Summary of the most important statistical parameters for *Antilocapra americana*. ERAs are displayed as rose diagrams per tooth position for wear stage 2 and in total. The upper row shows upper M1–M3 (from left to right); the antagonistic lower teeth are in the lower row. The graphs show circular median (θ) and concentration (κ) values per tooth position for stage 2.

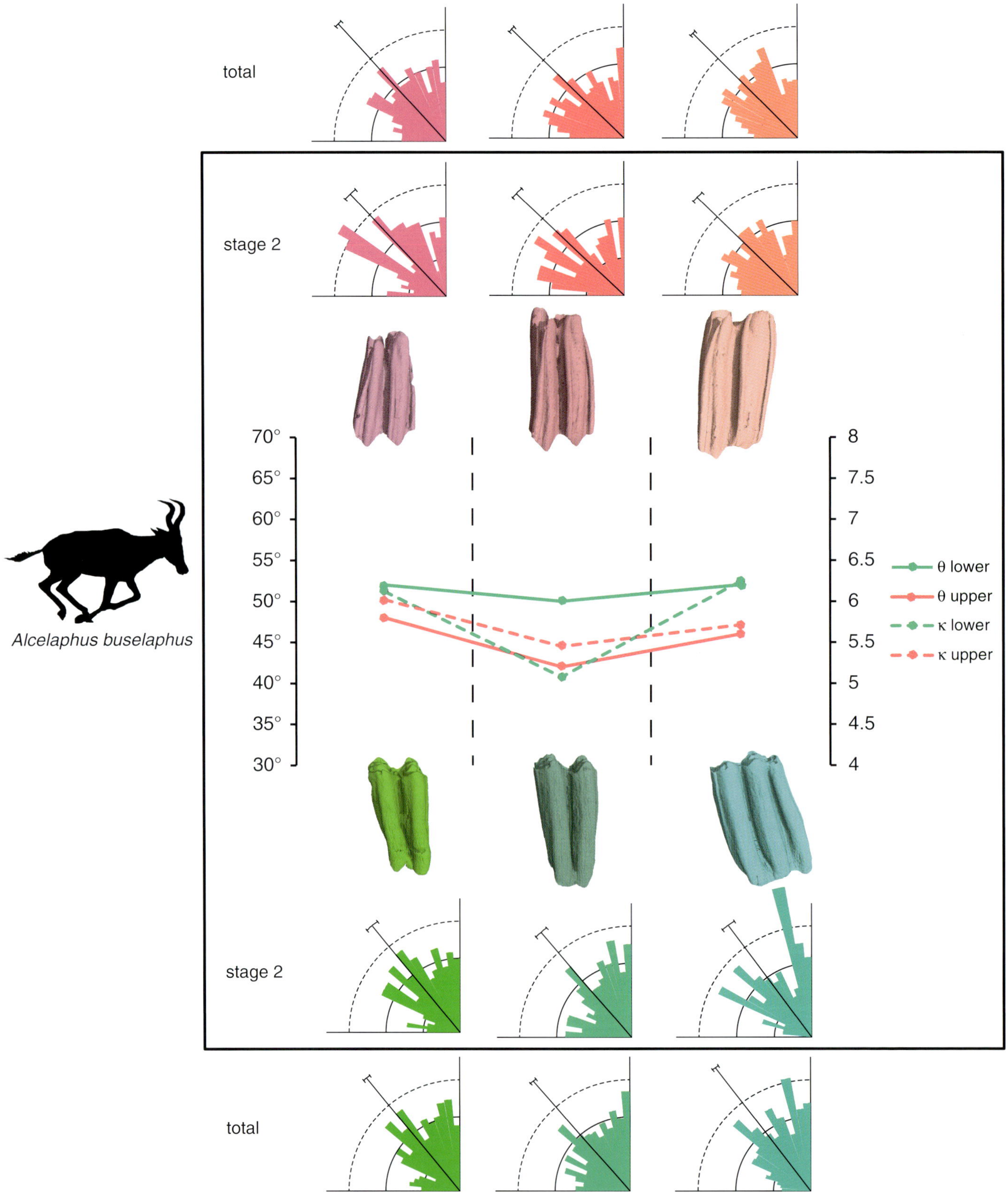

Fig. 8.A3. Summary of the most important statistical parameters for *Alcelaphus buselaphus*. ERAs are displayed as rose diagrams per tooth position for wear stage 2 and in total. The upper row shows upper M1–M3 (from left to right); the antagonistic lower teeth are in the lower row. The graphs show circular median (θ) and concentration (κ) values per tooth position for stage 2.

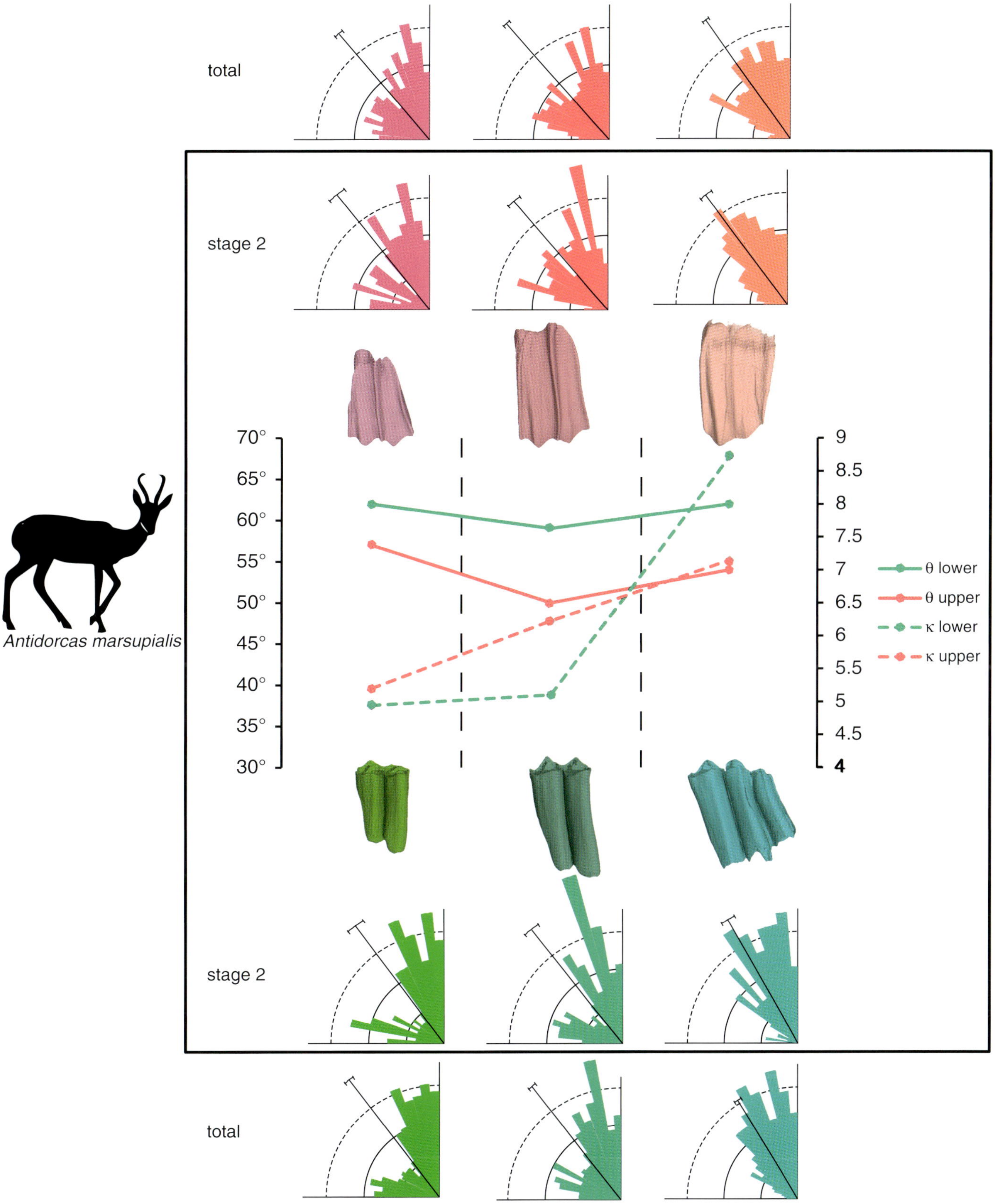

Fig. 8.A4. Summary of the most important statistical parameters for *Antidorcas marsupialis*. ERAs are displayed as rose diagrams per tooth position for wear stage 2 and in total. The upper row shows upper M1–M3 (from left to right); the antagonistic lower teeth are in the lower row. The graphs show circular median (θ) and concentration (κ) values per tooth position for stage 2.

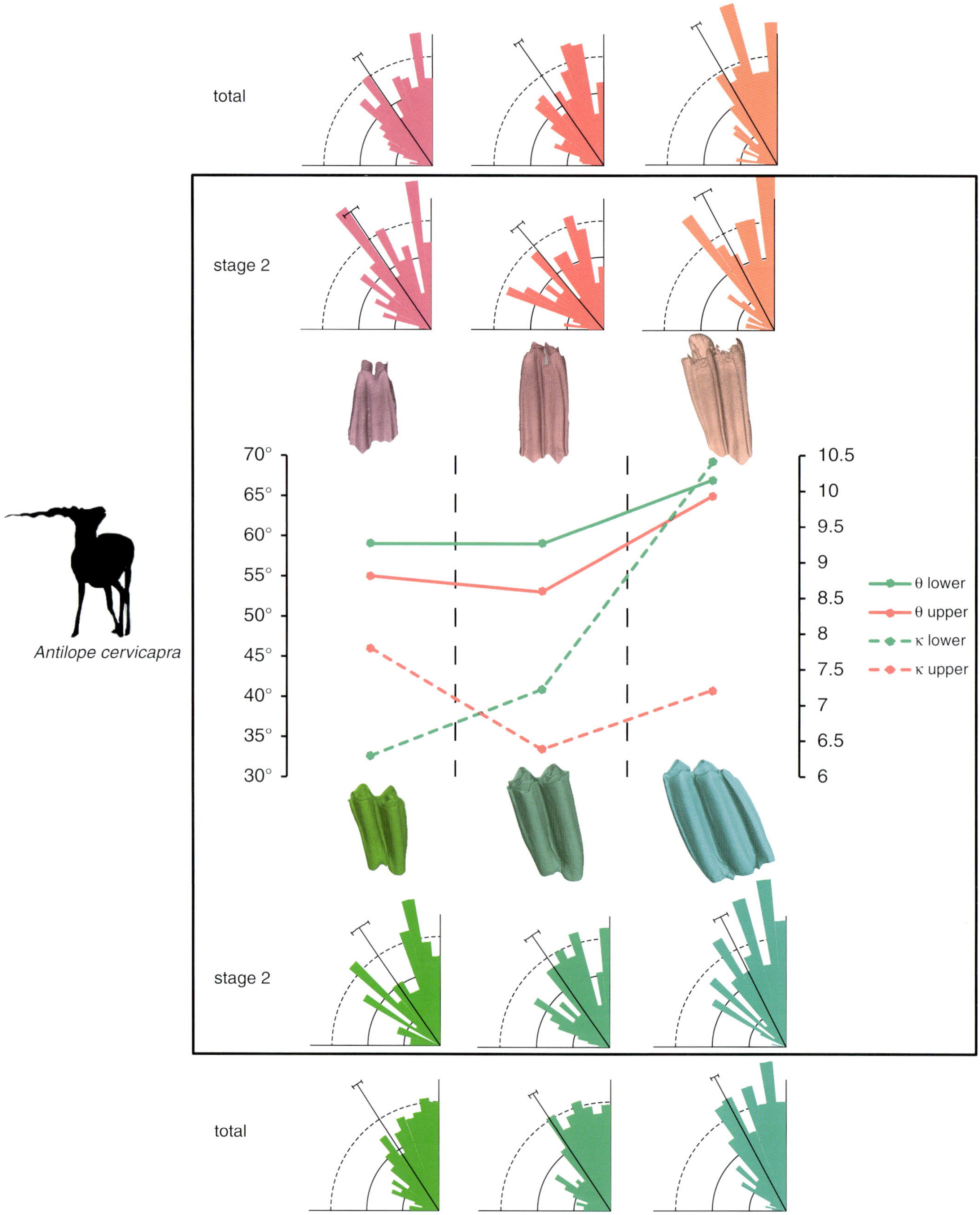

Fig. 8.A5. Summary of the most important statistical parameters for *Antilope cervicapra*. ERAs are displayed as rose diagrams per tooth position for wear stage 2 and in total. The upper row shows upper M1–M3 (from left to right); the antagonistic lower teeth are in the lower row. The graphs show circular median (θ) and concentration (κ) values per tooth position for stage 2.

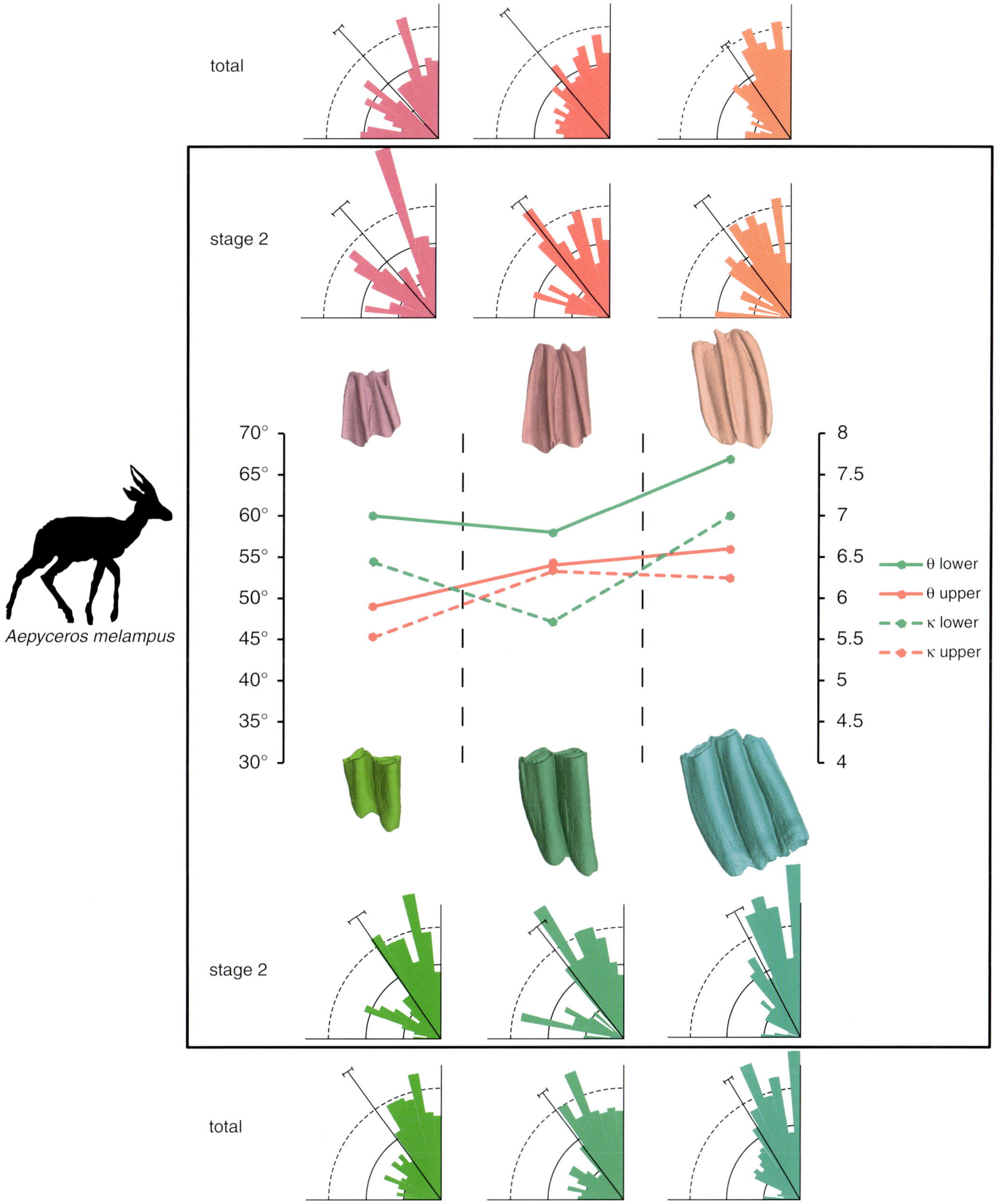

Fig. 8.A6. Summary of the most important statistical parameters for *Aepyceros melampus*. ERAs are displayed as rose diagrams per tooth position for wear stage 2 and in total. The upper row shows upper M1–M3 (from left to right); the antagonistic lower teeth are in the lower row. The graphs show circular median (θ) and concentration (κ) values per tooth position for stage 2.

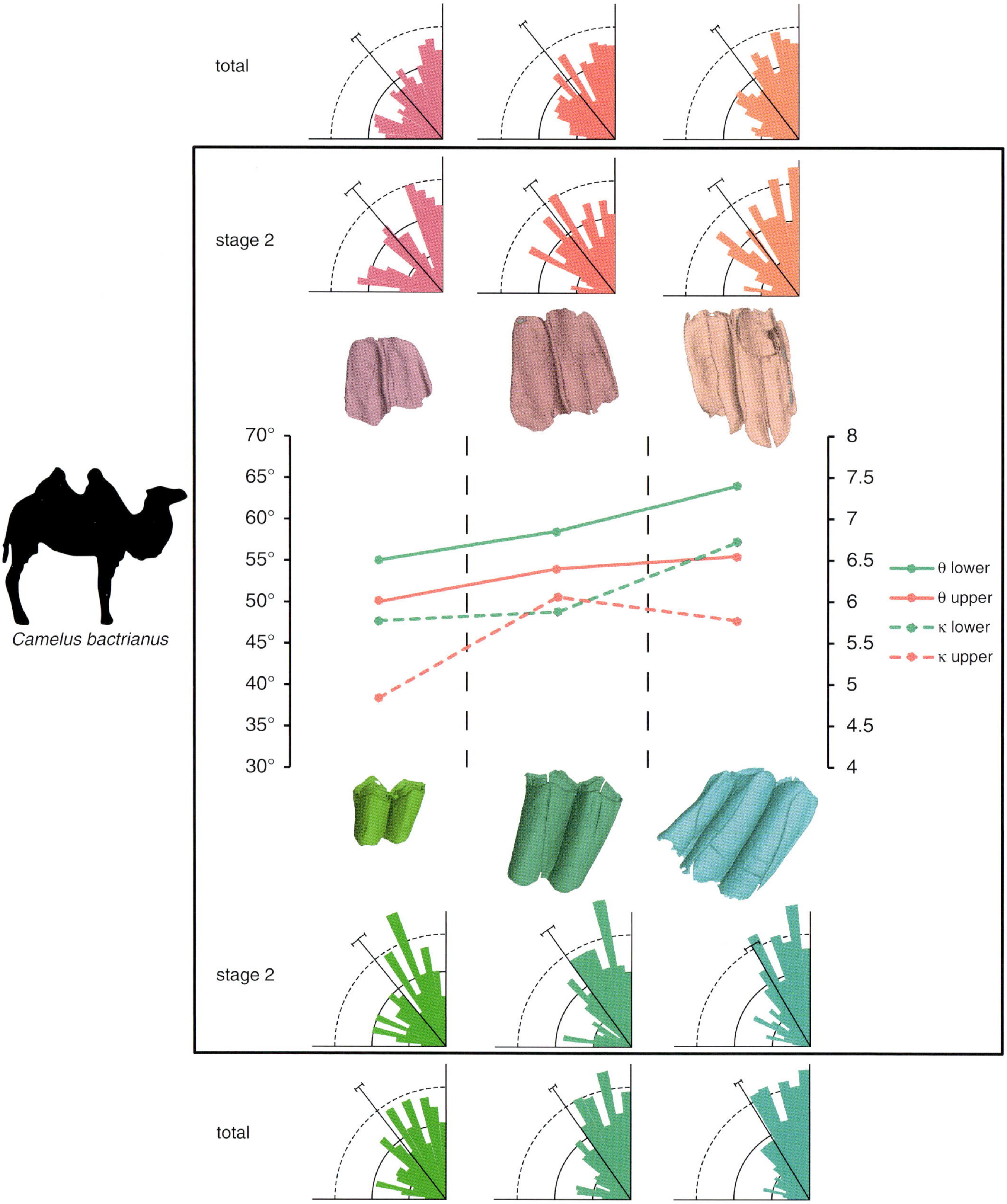

Fig. 8.A7. Summary of the most important statistical parameters for *Camelus bactrianus*. ERAs are displayed as rose diagrams per tooth position for wear stage 2 and in total. The upper row shows upper M1–M3 (from left to right); the antagonistic lower teeth are in the lower row. The graphs show circular median (θ) and concentration (κ) values per tooth position for stage 2.

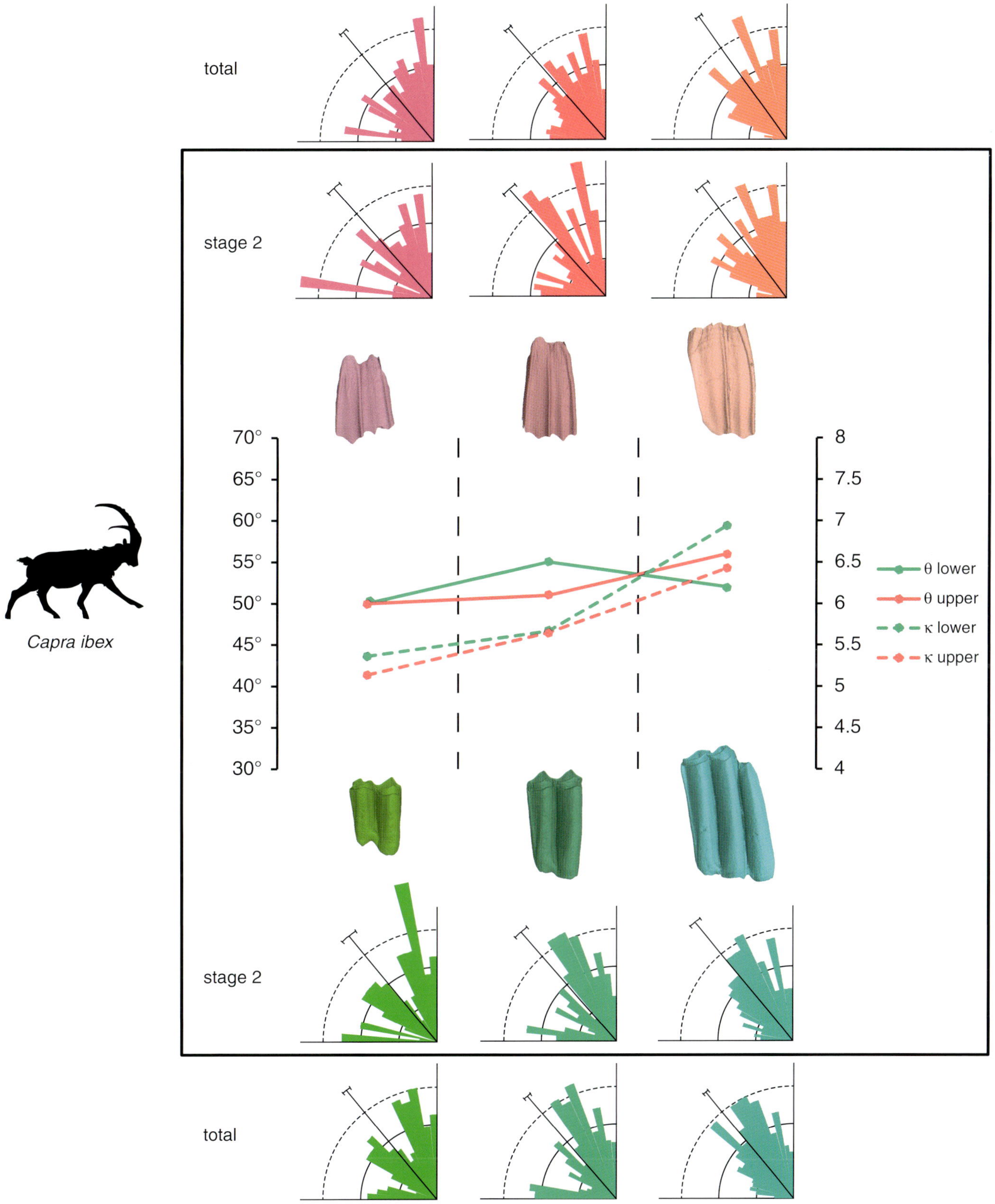

Fig. 8.A8. Summary of the most important statistical parameters for *Capra ibex*. ERAs are displayed as rose diagrams per tooth position for wear stage 2 and in total. The upper row shows upper M1–M3 (from left to right); the antagonistic lower teeth are in the lower row. The graphs show circular median (θ) and concentration (κ) values per tooth position for stage 2.

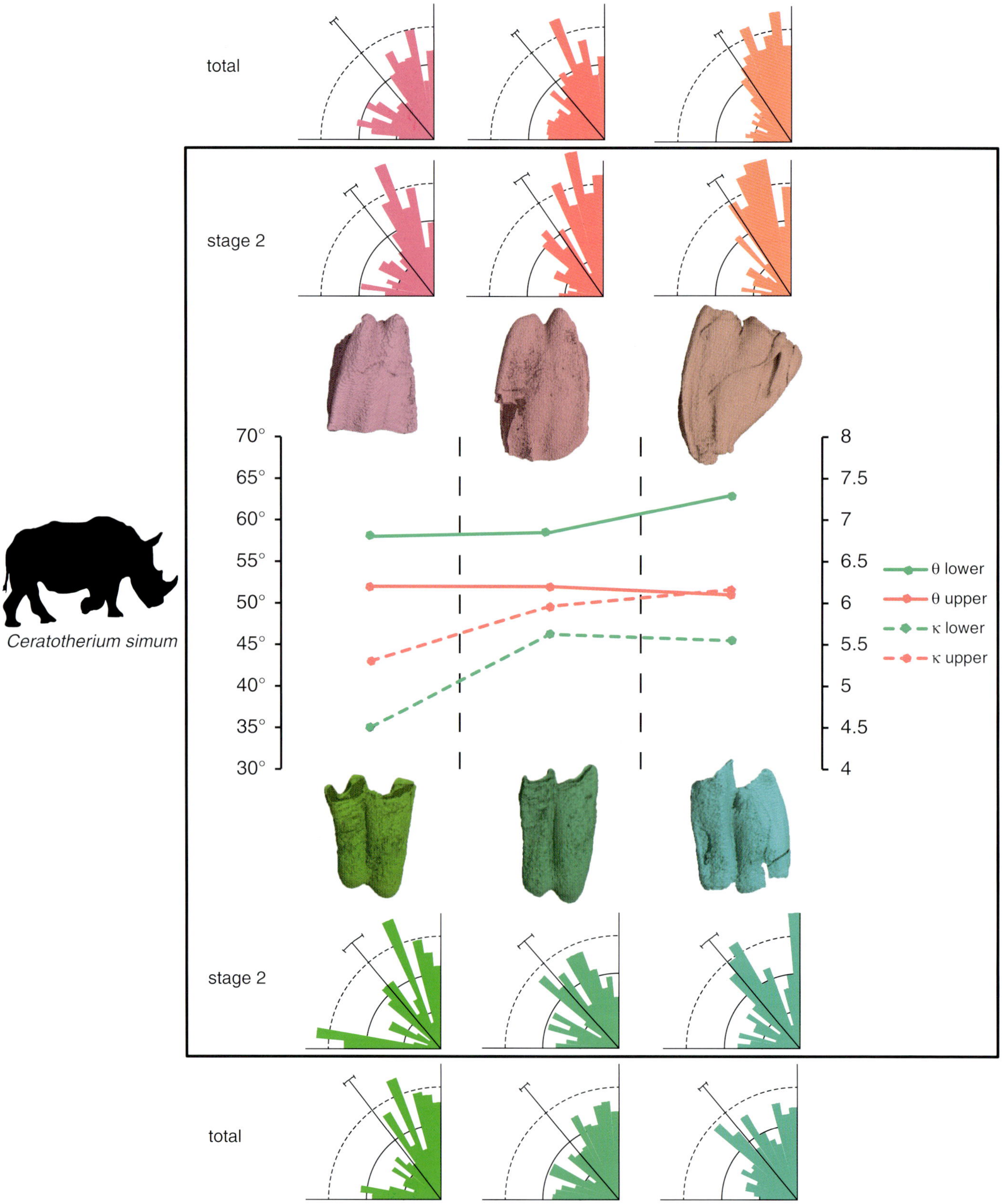

Fig. 8.A9. Summary of the most important statistical parameters for *Ceratotherium simum*. ERAs are displayed as rose diagrams per tooth position for wear stage 2 and in total. The upper row shows upper M1–M3 (from left to right); the antagonistic lower teeth are in the lower row. The graphs show circular median (θ) and concentration (κ) values per tooth position for stage 2.

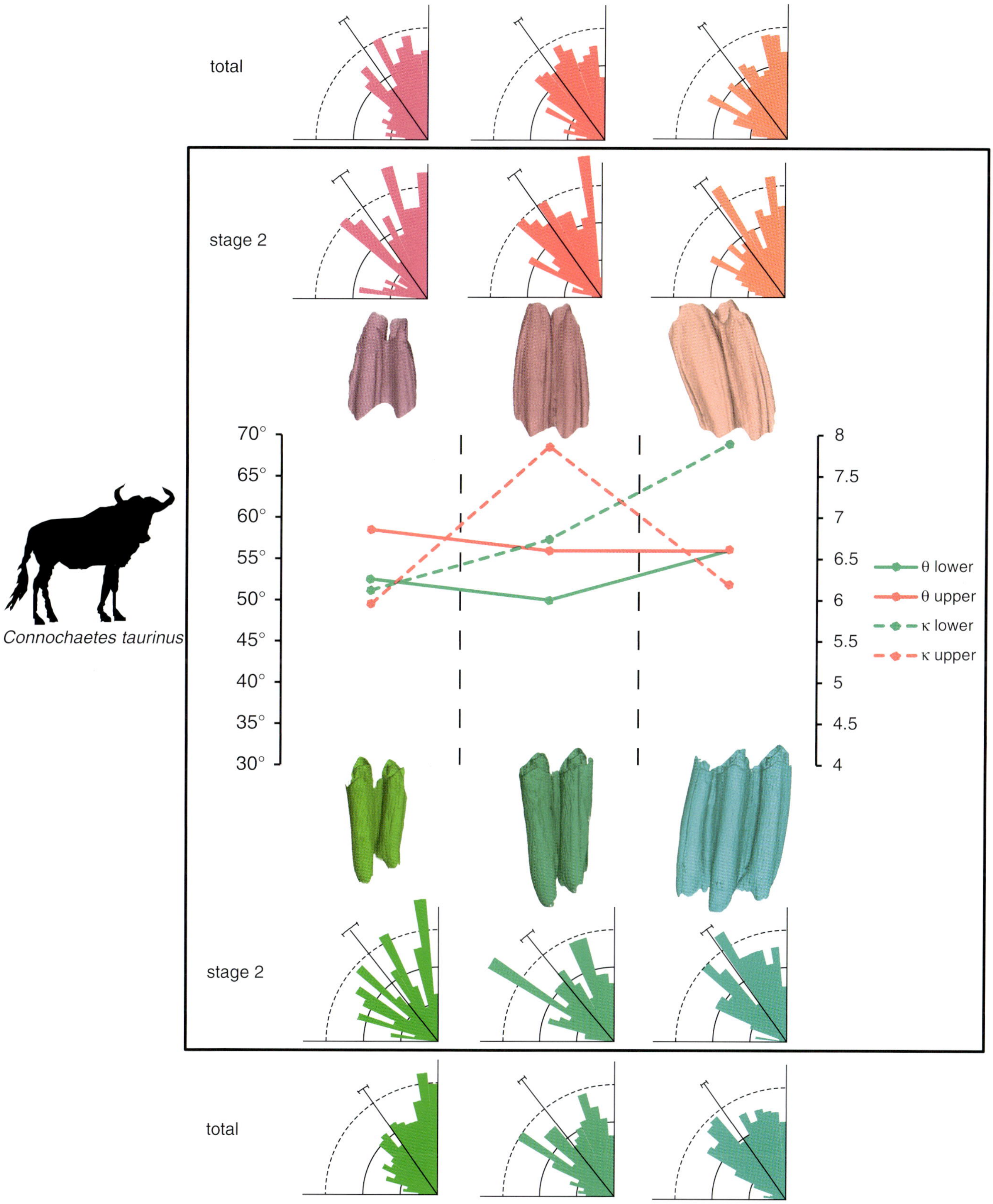

Fig. 8.A10. Summary of the most important statistical parameters for *Connochaetes taurinus*. ERAs are displayed as rose diagrams per tooth position for wear stage 2 and in total. The upper row shows upper M1–M3 (from left to right); the antagonistic lower teeth are in the lower row. The graphs show circular median (θ) and concentration (κ) values per tooth position for stage 2.

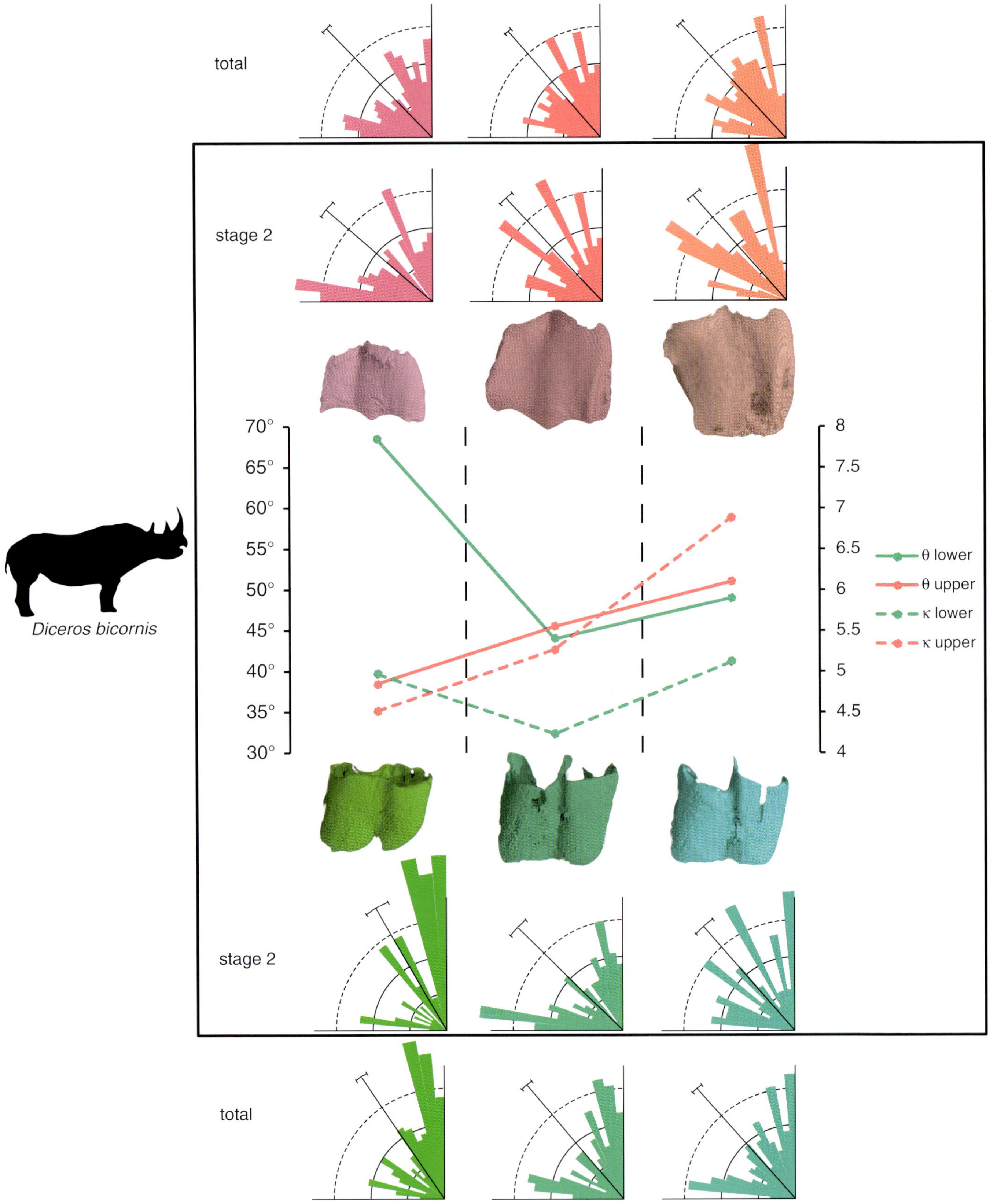

Fig. 8.A11. Summary of the most important statistical parameters for *Diceros bicornis*. ERAs are displayed as rose diagrams per tooth position for wear stage 2 and in total. The upper row shows upper M1–M3 (from left to right); the antagonistic lower teeth are in the lower row. The graphs show circular median (θ) and concentration (κ) values per tooth position for stage 2.

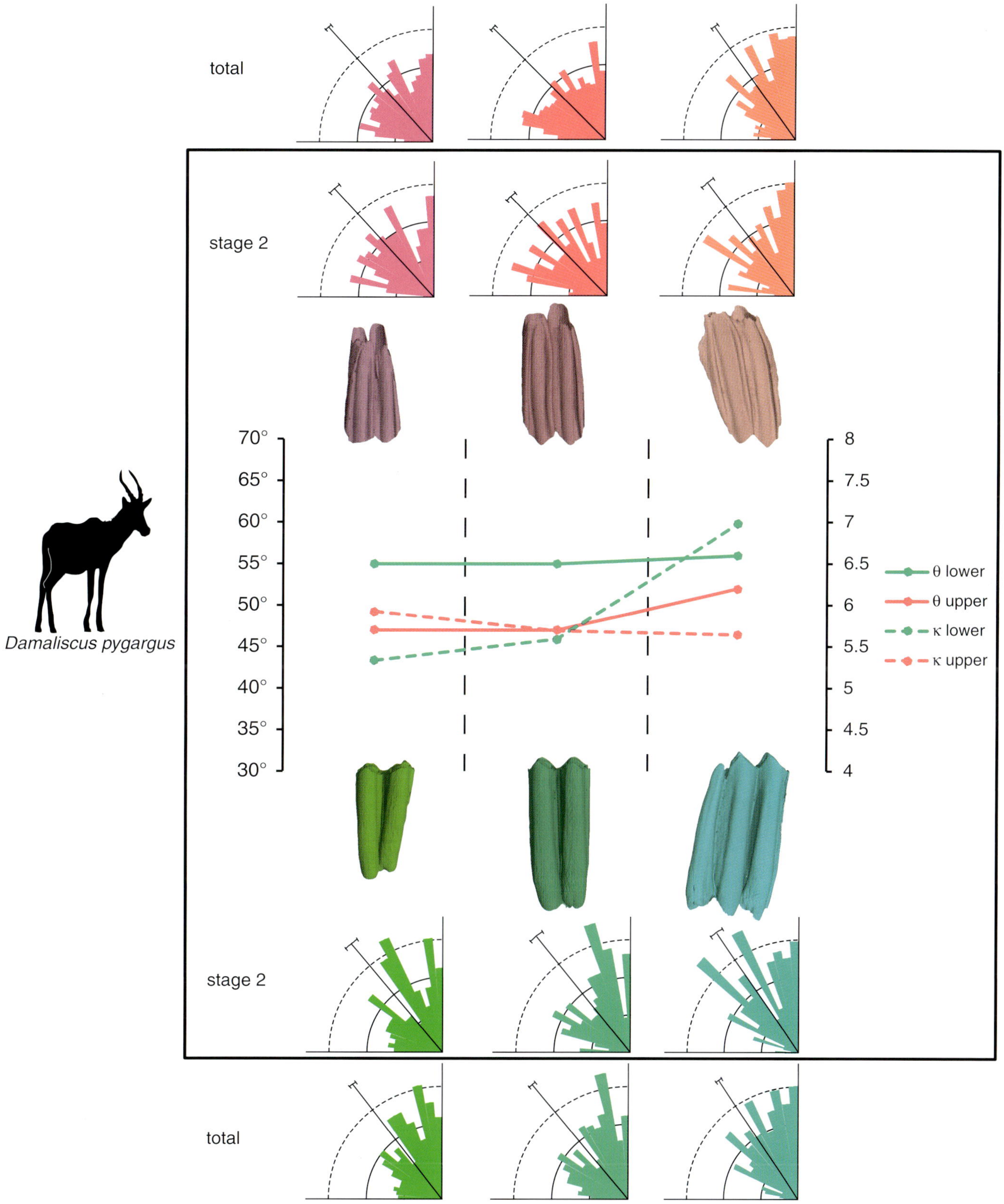

Fig. 8.A12. Summary of the most important statistical parameters for *Damaliscus pygargus*. ERAs are displayed as rose diagrams per tooth position for wear stage 2 and in total. The upper row shows upper M1–M3 (from left to right); the antagonistic lower teeth are in the lower row. The graphs show circular median (θ) and concentration (κ) values per tooth position for stage 2.

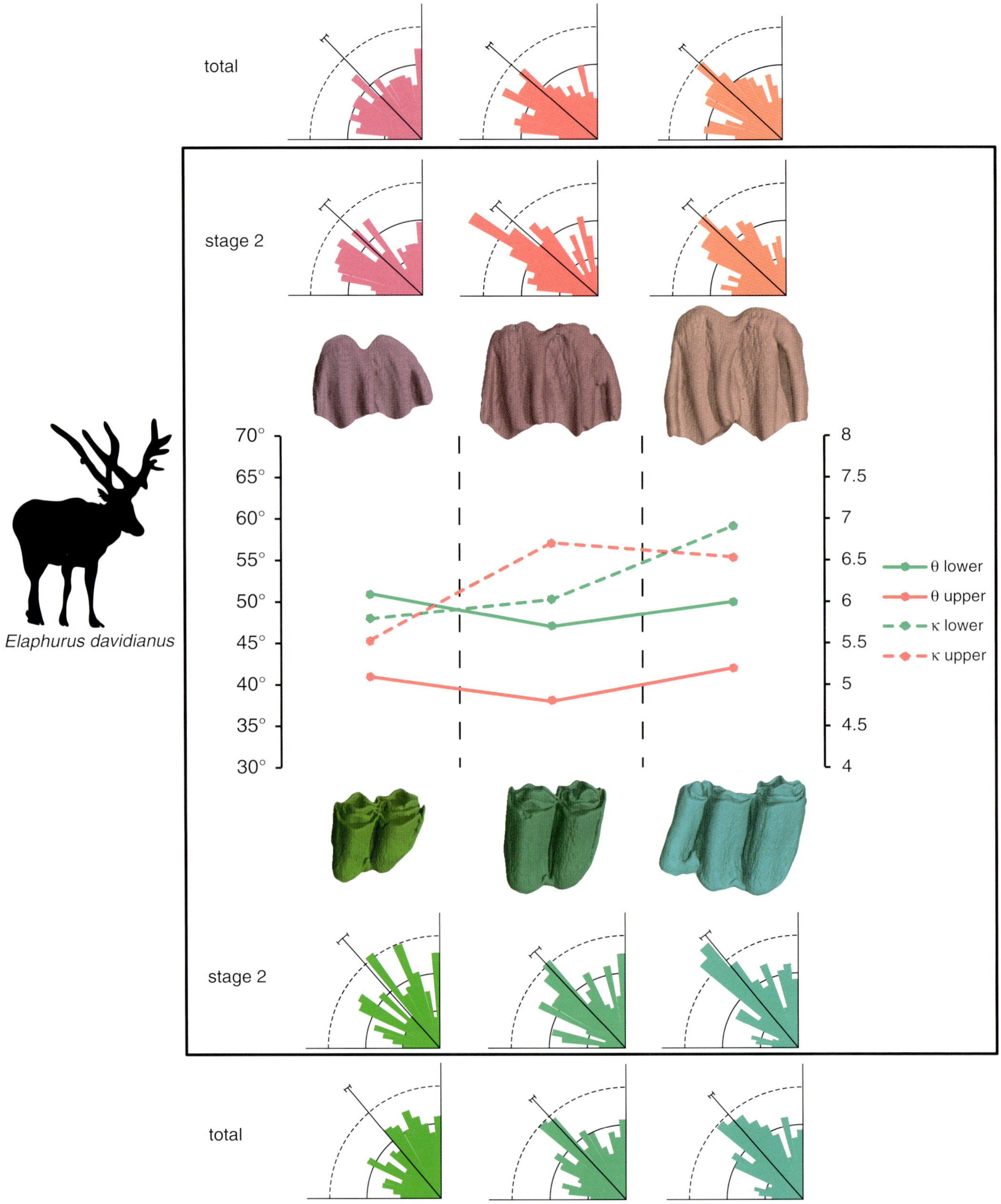

Fig. 8.A13. Summary of the most important statistical parameters for *Elaphurus davidianus*. ERAs are displayed as rose diagrams per tooth position for wear stage 2 and in total. The upper row shows upper M1–M3 (from left to right); the antagonistic lower teeth are in the lower row. The graphs show circular median (θ) and concentration (κ) values per tooth position for stage 2.

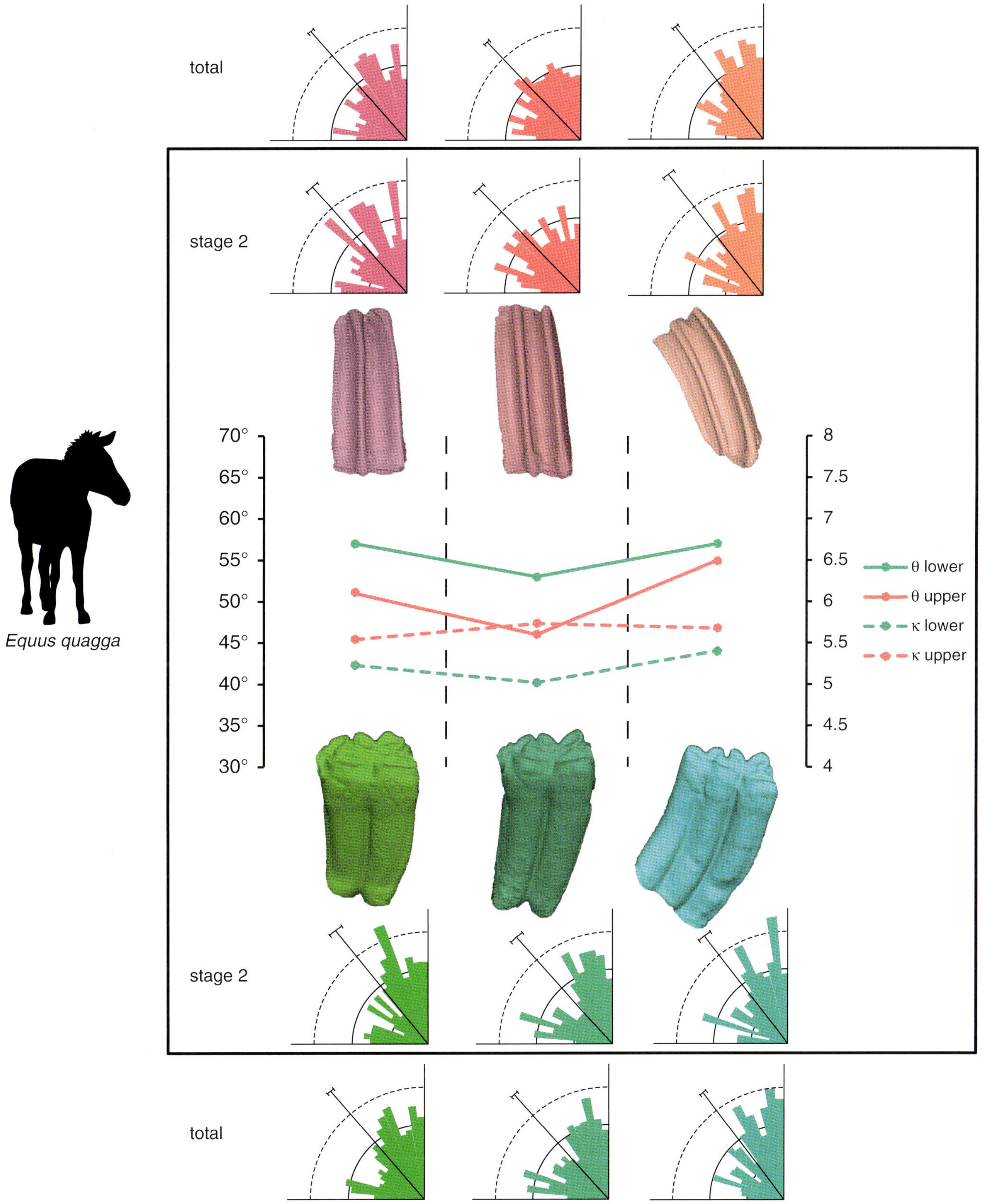

Fig. 8.A14. Summary of the most important statistical parameters for *Equus quagga*. ERAs are displayed as rose diagrams per tooth position for wear stage 2 and in total. The upper row shows upper M1–M3 (from left to right); the antagonistic lower teeth are in the lower row. The graphs show circular median (θ) and concentration (κ) values per tooth position for stage 2.

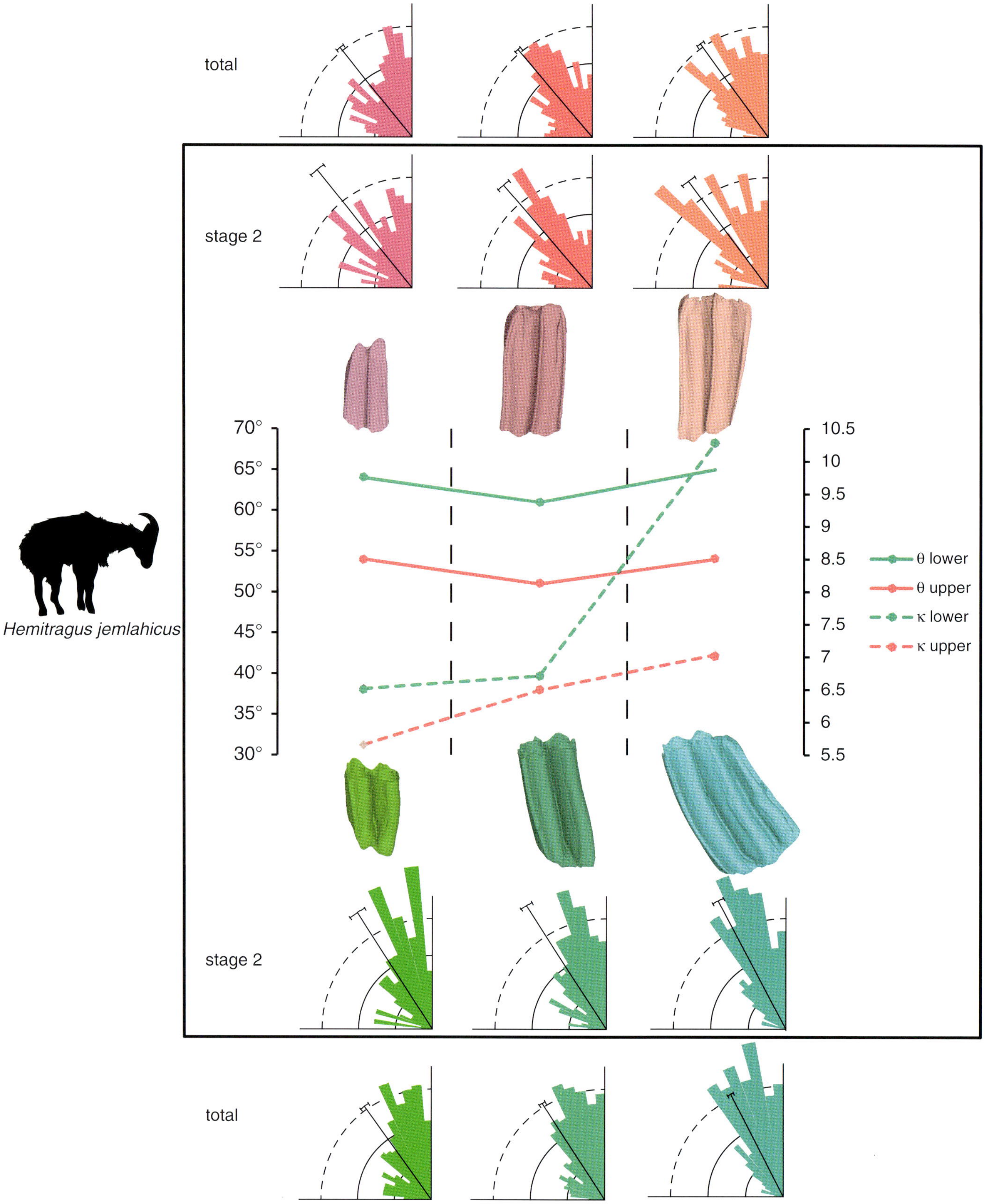

Fig. 8.A15. Summary of the most important statistical parameters for *Hemitragus jemlahicus*. ERAs are displayed as rose diagrams per tooth position for wear stage 2 and in total. The upper row shows upper M1–M3 (from left to right); the antagonistic lower teeth are in the lower row. The graphs show circular median (θ) and concentration (κ) values per tooth position for stage 2.

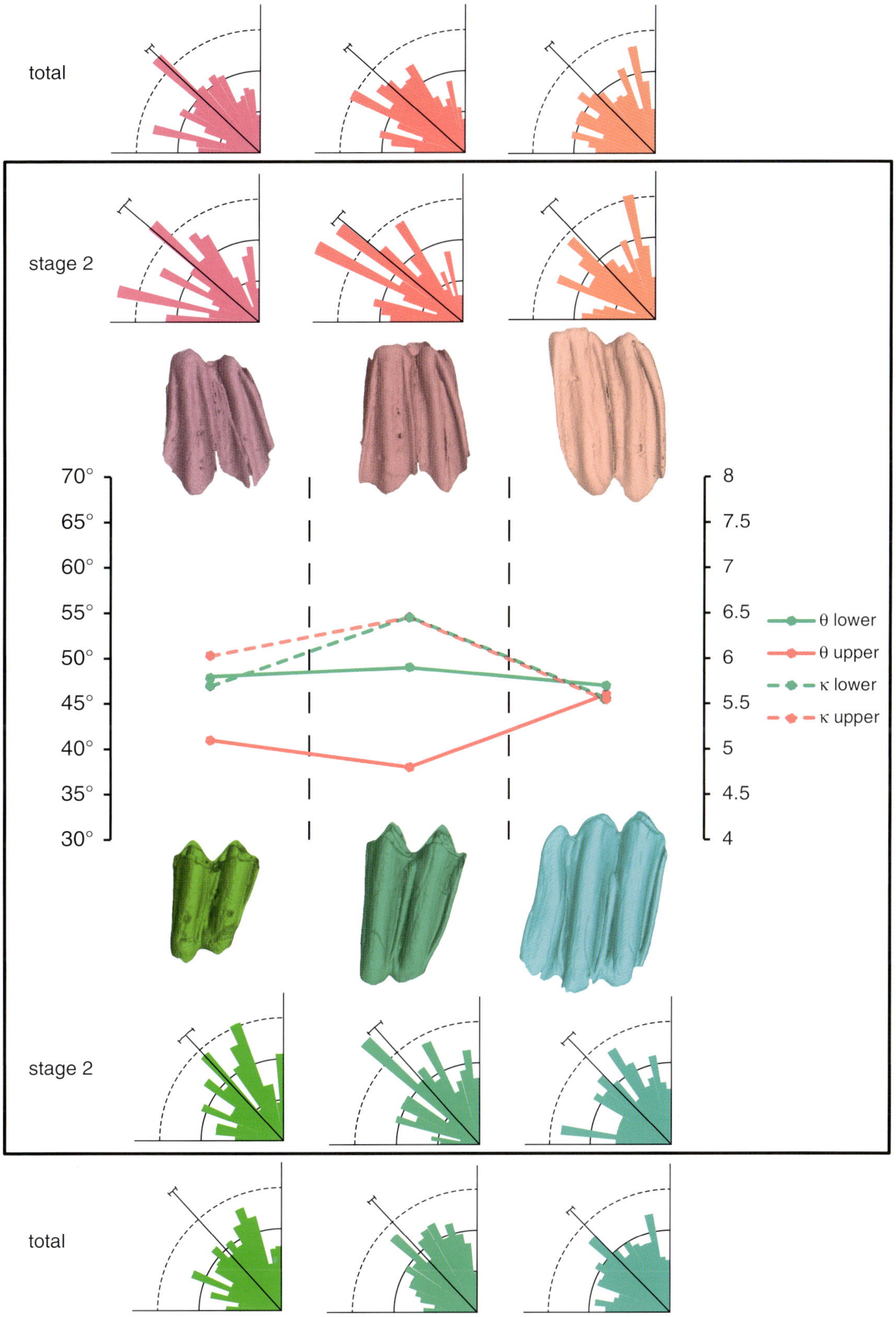

Fig. 8.A16. Summary of the most important statistical parameters for *Kobus ellipsiprymnus*. ERAs are displayed as rose diagrams per tooth position for wear stage 2 and in total. The upper row shows upper M1–M3 (from left to right); the antagonistic lower teeth are in the lower row. The graphs show circular median (θ) and concentration (κ) values per tooth position for stage 2.

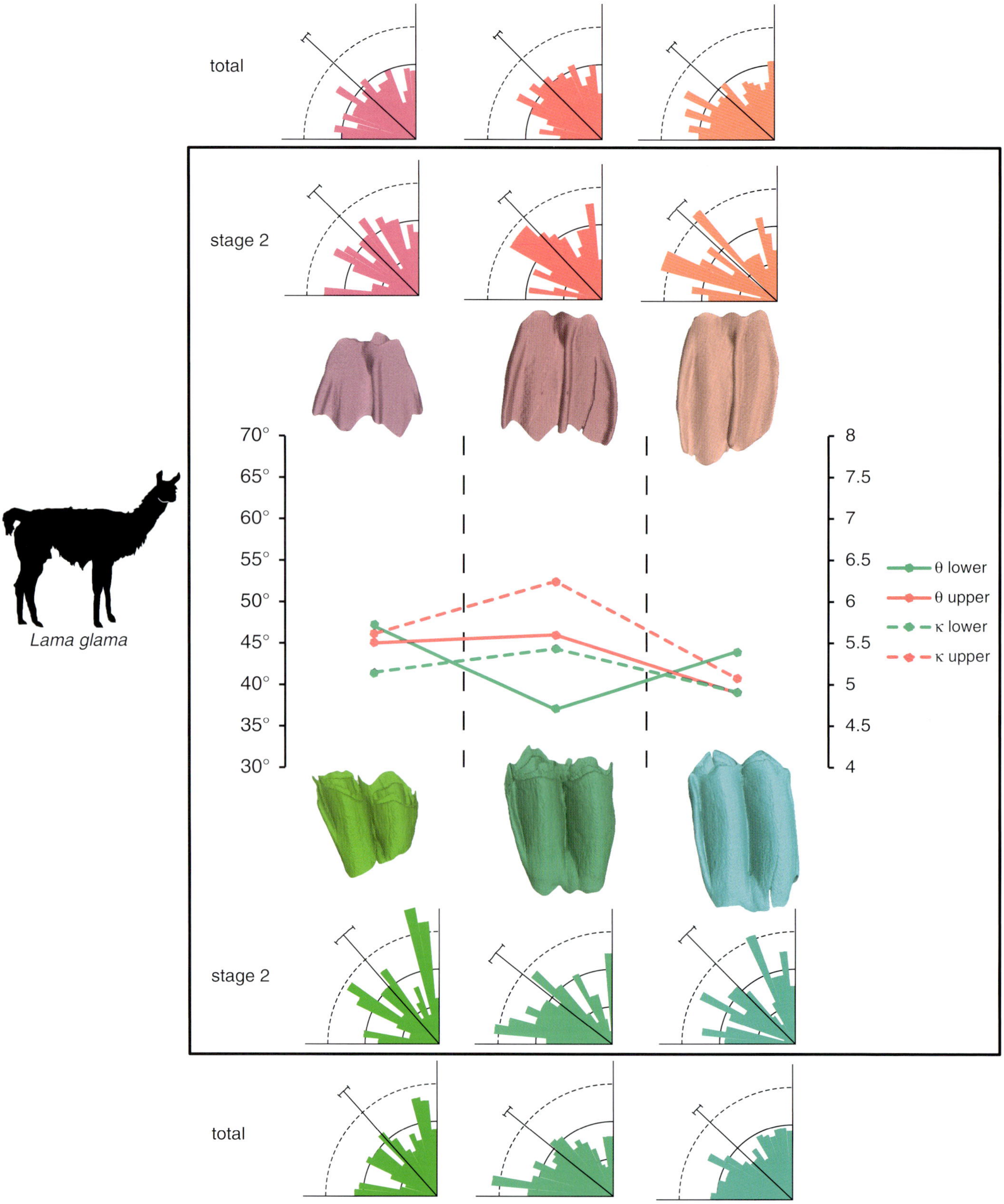

Fig. 8.A17. Summary of the most important statistical parameters for *Lama glama*. ERAs are displayed as rose diagrams per tooth position for wear stage 2 and in total. The upper row shows upper M1–M3 (from left to right); the antagonistic lower teeth are in the lower row. The graphs show circular median (θ) and concentration (κ) values per tooth position for stage 2.

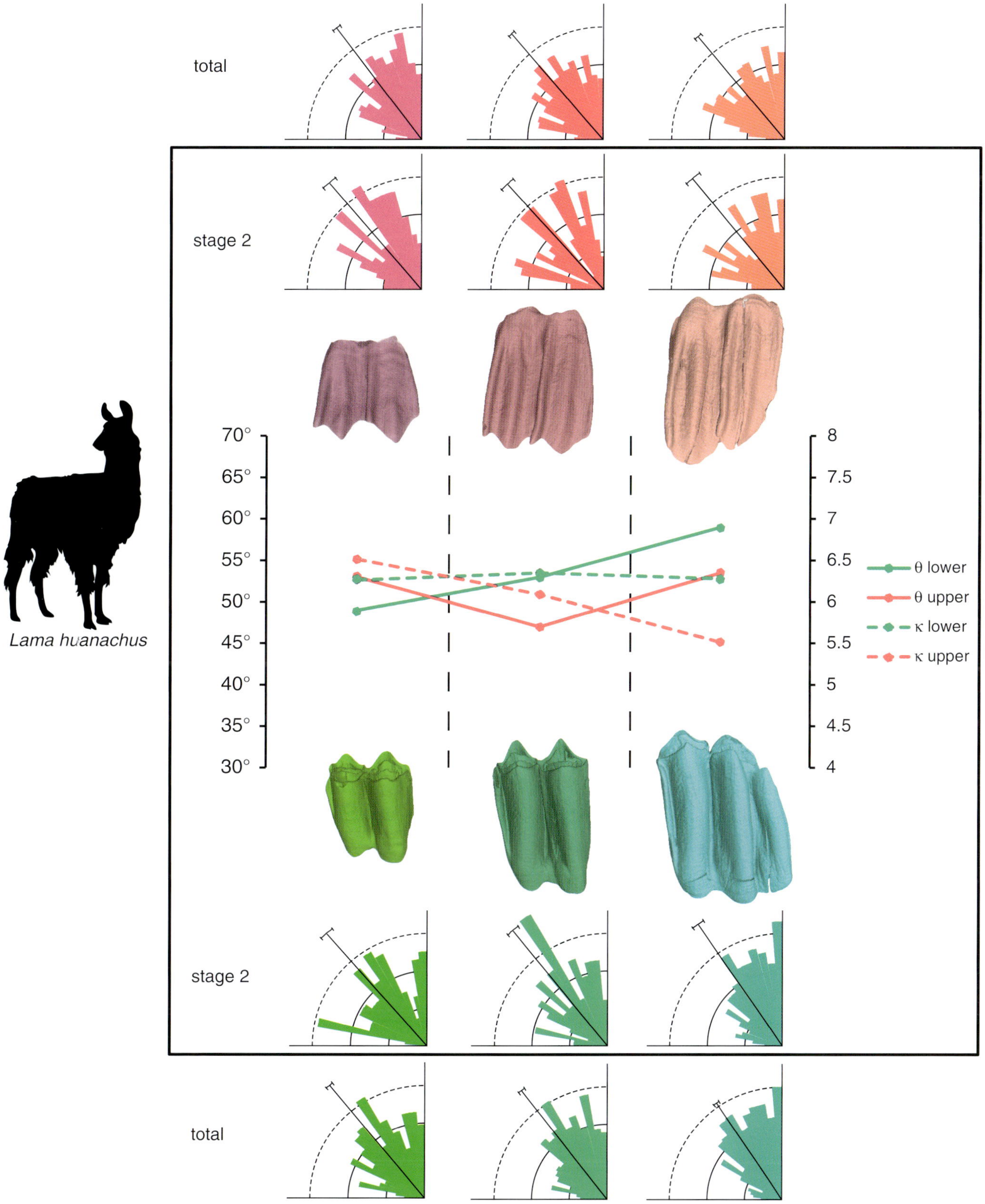

Fig. 8.A18. Summary of the most important statistical parameters for *Lama huanachus*. ERAs are displayed as rose diagrams per tooth position for wear stage 2 and in total. The upper row shows upper M1–M3 (from left to right); the antagonistic lower teeth are in the lower row. The graphs show circular median (θ) and concentration (κ) values per tooth position for stage 2.

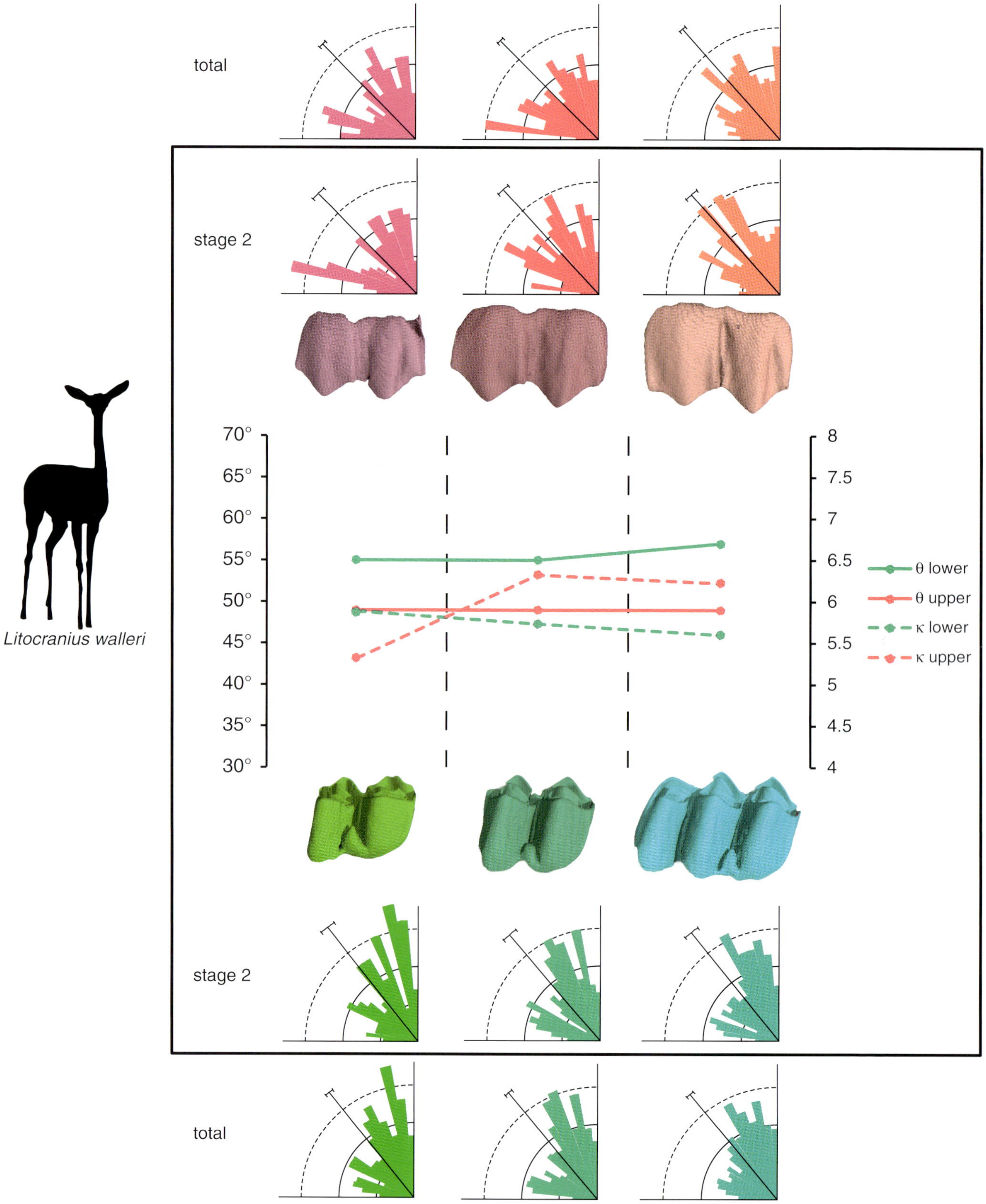

Fig. 8.A19. Summary of the most important statistical parameters for *Litocranius walleri*. ERAs are displayed as rose diagrams per tooth position for wear stage 2 and in total. The upper row shows upper M1–M3 (from left to right); the antagonistic lower teeth are in the lower row. The graphs show circular median (θ) and concentration (κ) values per tooth position for stage 2.

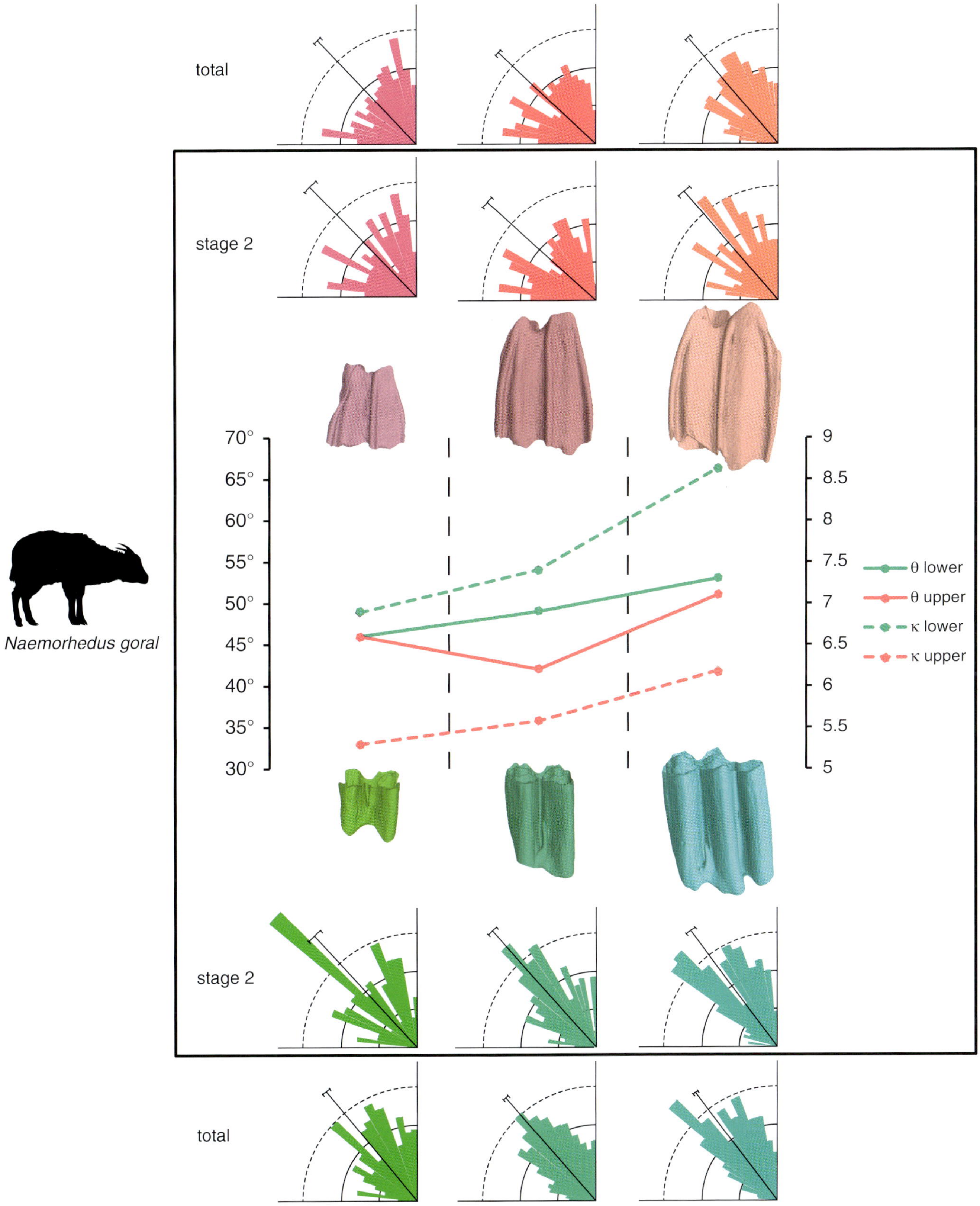

Fig. 8.A20. Summary of the most important statistical parameters for *Naemorhedus goral*. ERAs are displayed as rose diagrams per tooth position for wear stage 2 and in total. The upper row shows upper M1–M3 (from left to right); the antagonistic lower teeth are in the lower row. The graphs show circular median (θ) and concentration (κ) values per tooth position for stage 2.

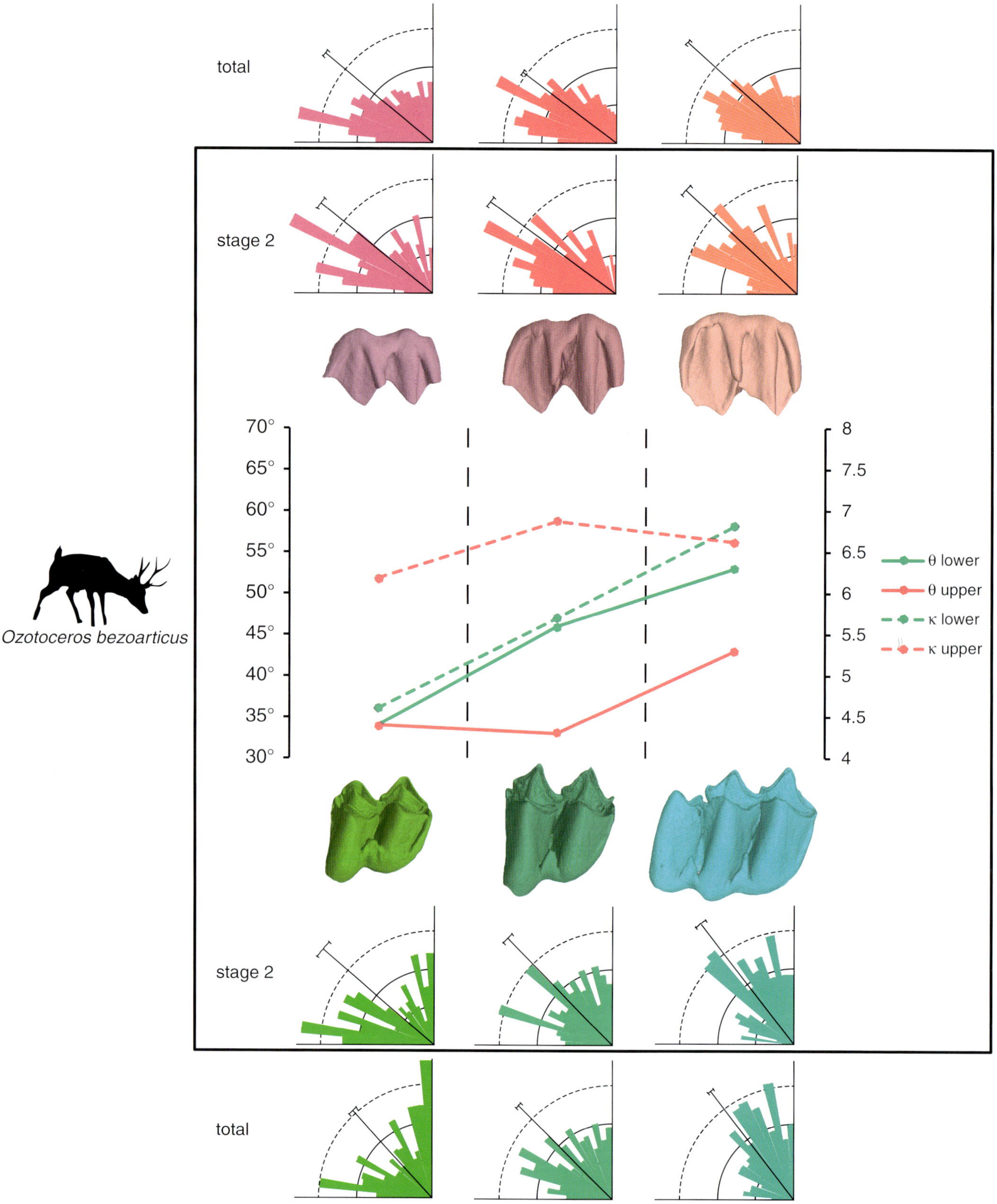

Fig. 8.A21. Summary of the most important statistical parameters for *Ozotoceros bezoarticus*. ERAs are displayed as rose diagrams per tooth position for wear stage 2 and in total. The upper row shows upper M1–M3 (from left to right); the antagonistic lower teeth are in the lower row. The graphs show circular median (θ) and concentration (κ) values per tooth position for stage 2.

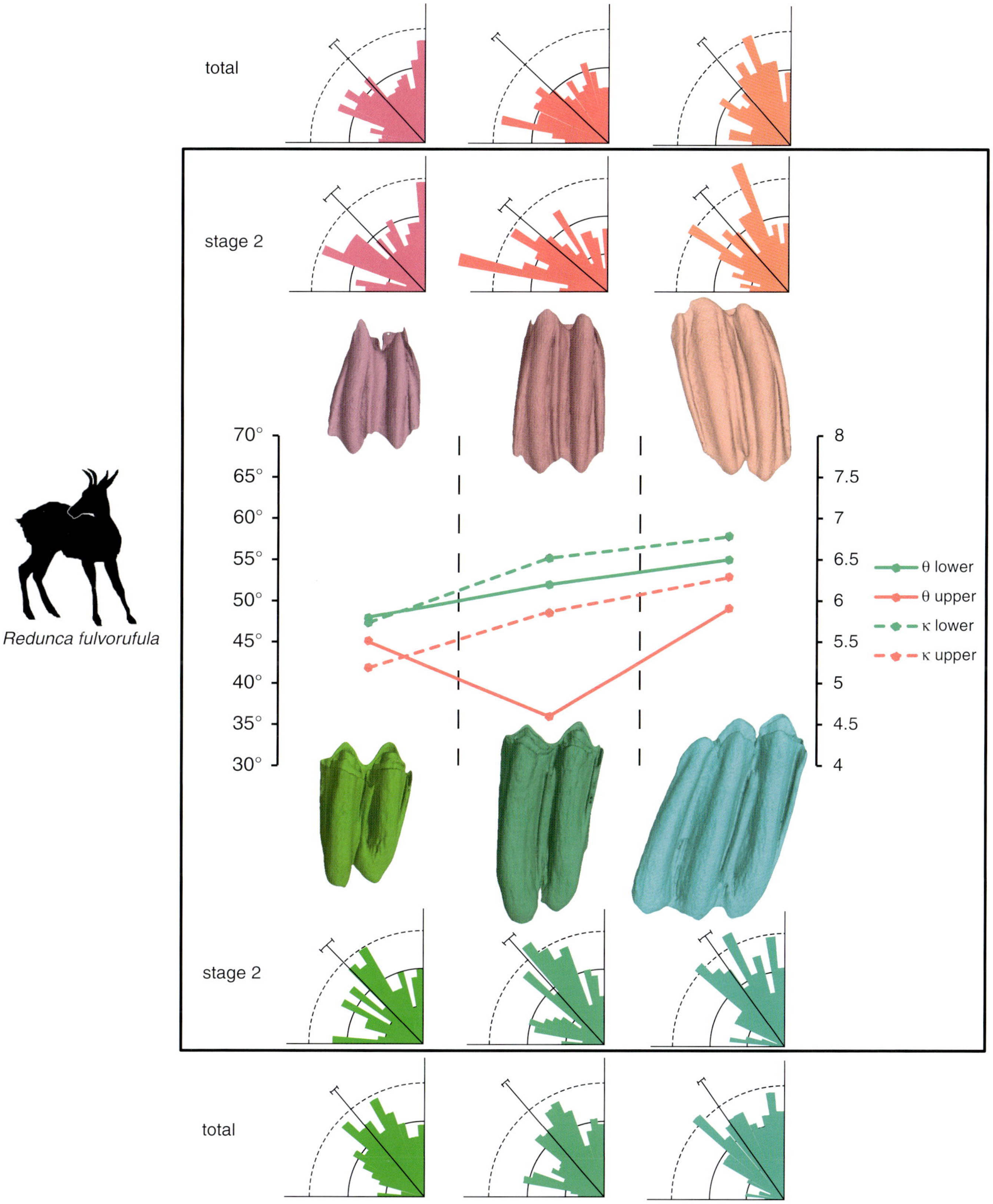

Fig. 8.A22. Summary of the most important statistical parameters for *Redunca fulvorufula*. ERAs are displayed as rose diagrams per tooth position for wear stage 2 and in total. The upper row shows upper M1–M3 (from left to right); the antagonistic lower teeth are in the lower row. The graphs show circular median (θ) and concentration (κ) values per tooth position for stage 2.

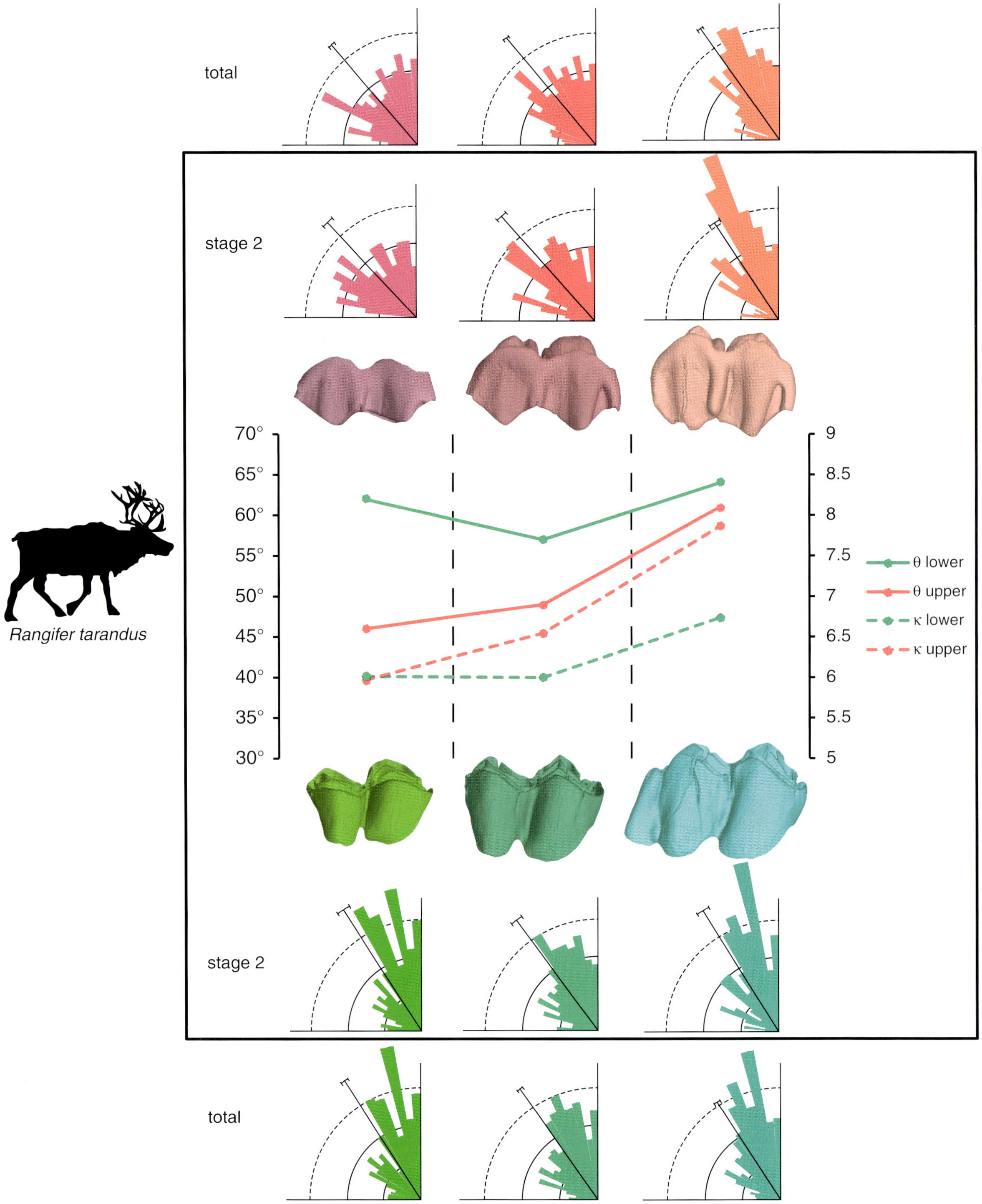

Fig. 8.A23. Summary of the most important statistical parameters for *Rangifer tarandus*. ERAs are displayed as rose diagrams per tooth position for wear stage 2 and in total. The upper row shows upper M1–M3 (from left to right); the antagonistic lower teeth are in the lower row. The graphs show circular median (θ) and concentration (κ) values per tooth position for stage 2.

Construction and wear of mammalian teeth in terms of heterochrony

Wighart v. Koenigswald

Introduction

Heterochrony describes the elongation of a specific phase during ontogeny at the cost of others. Ontogeny includes the entire life history of an animal or its organs (Haeckel 1866). The ontogeny of a tooth includes two parts, early ontogeny and late ontogeny. During early ontogeny the tooth is formed in the crypt. During late ontogeny it is functioning and exposed to wear. Heterochrony during early ontogeny affects the final shape of the tooth and forms the basis for its function during late ontogeny that is characterized by wear. The result of tooth wear is not uniform with time. Only in a few tooth types the primary enamel surface function for a major part of the life span. In most other tooth types some initial wear is required to reach full functionality, when dentin is exposed and enamel ridges are carved out. When this functional surface is reached, further changes caused by continuing wear should be limited, allowing similar functionality to continue over a long time period. Some types of facets, in various mammalian groups, fulfill this requirement, and they stimulated the search for general patterns in the process of tooth wear. Examples are discussed here as "specialized wear facets". Their function is compared with common tools, such as scissors, guillotine cutters, rasps, or end-cutting pliers. This allows us to bypass the various inconsistent definitions of "cutting", "shearing", and "grinding".

The great variability of mammalian teeth and their different functions prohibit sophisticated and very detailed comparisons, but some more general thoughts are provided in this paper. Observing the life history of teeth offers a better understanding of the relationship between the wear-related changes in tooth morphology and function.

Teeth are formed almost completely in the crypt of the jaws, and after eruption they are exposed to wear. Wear, as a destructive and unrenewable process, causes changes in morphology and function.

The general life history of teeth is divided into several phases that allow some comparisons between various tooth types and wear patterns. The estimated length of these phases can be interpreted in terms of heterochrony (McNamara 1990, Smith 2003), because some phases are extended at the cost of others. The evolution of the different kinds of hypsodonty and various specialized wear facets can be used as examples of how heterochrony affects tooth morphology and the maintenance of function during wear.

This is a summary paper based on two earlier papers where additional discussion and citations are provided (Koenigswald 2011 and 2017).

Abbreviations of collections for figured specimens

BSPG	Bayerische Staatssammlung für Paläontologie und Geologie, München
CM	Carnegie Museum, Pittsburgh
GPIT	Institute of Geosciences, Paleontology Department and Museum, Universität Tübingen
HLMD	Hessisches Landesmuseum, Darmstadt
IGPB	Institute of Geosciences, Section Paleontology, Rheinische Friedrich-Wilhelms-Universität Bonn
LACM	Los Angeles County Museum
MNHN	Muséum national d'Histoire naturelle, Paris
NHMW	Naturhistorisches Museum, Wien
SLGM	Collection Frank Menger, Groß-Rohrheim
SMF	Senckenberg Forschungsinstitut und Naturmuseum Frankfurt
UF	University of Florida, Gainesville
UMZC	University Museum of Zoology, Cambridge
UW	Universität Wien
ZFMK	Zoologische Forschungsmuseum Alexander Koenig, Bonn

The life history of teeth

Various different heterochronies occur during early and late ontogeny. A most impressive example of heterochrony during early ontogeny is provided by the development of hypsodont teeth. In other words, the different kinds of hypsodonty can be explained by the heterochronic elongation of different phases during tooth formation. Similarly, specific stages of wear exist for a long time, while other phases are passed through rapidly. Such specializations for specific long lasting phases of wear can thus be interpreted as examples of heterochrony as well. For comparison the various ontogenetic phases have to be defined.

Tooth mineralization starts from the top of the cusps and proceeds towards the base of the crown (see Ruf et al. 2019). This direction of growth is significant for the sequence of the four phases of early ontogeny. In brachydont teeth these phases are sequential but in euhypsodont teeth they may be compressed or even be eliminated (Fig. 9.1). The four phases are:

T. Martin & W. v. Koenigswald (eds.): Mammalian Teeth – Form and Function. Pp. 171–186, 11 figs.

DOI: http://doi.org/10.23788/mammteeth.09 Published 22 December 2020

Ontogenetic phase I – formation of the cusped surface. If a tooth has several cusps, the formation starts with each cusp separately. The individual cusps merge later into a closed surface.
Ontogenetic phase II – formation of the side walls continues until the base of the crown is reached. There, the enamel formation comes to an end.
Ontogenetic phase III – formation of dentin-covered surface at the base of the tooth
Ontogenetic phase IV – formation of differentiated roots.

The root formation in phase IV does not end at a specific point. The continuous modification of roots helps to compensate for wear and corrects the position of the tooth within the tooth row. The horizontal tooth displacement occurring in some marsupials, some rodents, proboscideans and sirenians (Gomes Rodrigues et al. 2012) requires an intensive modification of roots throughout the lifetime. In other hypsodont teeth the peripheral enamel is reduced in dentin tracts, to allow periodontic ligaments an easier anchoring of the teeth. Solounias et al. (2019) drew attention to an additional function of hypsodont teeth. As long as a major part of them sticks in the jaw, they may serve as additional roots.

For the discussion of wear facets, life history is divided into generalized phases identified as A–G. Capital letters are used to discriminate this sequence from that used for early ontogeny. Late ontogeny, during which erupted teeth are exposed to wear, are discussed as wear stages C, D, E, F, and G. In this model, phases A and B summarize the early ontogeny, included here, because in hypsodont teeth tooth formation and abrasion during the wear stages C to G overlap in time. The wear stages C–G are discriminated by the degree of exposed dentin:
Wear stage C – The enamel cap is intact, providing facets within the enamel.
Wear stage D – Dentin is partially exposed.
Wear stage E – Dentin is widely exposed and dominant.
Wear stage F – Enamel is present only on one side.
Wear stage G – All enamel is worn away, but the dentin core is functioning.

Defining ontogenetic phases and wear stages is an artificial tool, especially when such phases are deduced from the tooth morphology, not the genetic background. This point of view was derived through discussions in the DFG (Deutsche Forschungsgemeinschaft) Research Unit 771 where these studies originated. This research group dealt mainly with fossil material that is not open for genetic analyses. However, the models derived from the fossil material may inspire other researchers to study teeth with their methods. That will widen the understanding of the variability in mammalian dentitions.

Heterochrony, the key to the various types of hypsodonty

High-crowned teeth are generally described as hypsodont. The use of this term, however, varies in different papers according to their authors. Some authors regard mainly or exclusively the cheek-teeth (e.g., Fortelius 1985, Fortelius & Solounias 2000, Kaiser et al. 2003), while others include enlarged incisors or canines (e.g., Hershkovitz 1962, Martin 1993). The most useful definition for hypsodonty compares the crown-height with the bucco-lingual width (e.g., White 1959, Van Valen 1960, Janis 1988). Comparing the height of a tooth to its length is applicable only if the same tooth position is compared.

Hypsodont teeth are rooted. Several lineages evolved rootless tooth types that are classified as euhypsodont (Mones 1982). The term euhypsodont is used in favor of hypselodont, a term that was controversially used and is easily mixed up (Koenigswald 2011).

Hypsodont teeth may occur in two kinds – external and internal hypsodonty. External hypsodont teeth protrude from the mouth like tusks or canines. Thus, they are restricted to the anterior part of the dentition. They are characterized by an eruption rate that exceeds wear. Their function is multifold; besides catching food items, they may be used for example as weapons or to demonstrate the social position of an individual.

In contrast, internal hypsodonty occurs in the postcanine part of the dentition. The true height of such premolars and molars is hidden within the jaw, because eruption rate and abrasion are in an almost perfect equilibrium. In fact, both control each other inconspicuously but fairly perfectly. The function of internal hypsodont teeth is exclusively for breaking down food items.

An intermediate form between internal and external hypsodonty is seen in rodent incisors and the canines of pigs or *Hippopotamus*. Although they are characterized by an equilibrium of eruption rate and wear, these teeth in the anterior dentition have social functions as well.

Compensating intensive wear as in internal hypsodonty is an important reason for the evolution of hypsodont teeth, but not the only one. External hypsodont teeth offer other selective values.

There have been several attempts to define types of hypsodonty and to assign the great diversity of hypsodont teeth to them. However, most such classifications focus on specific mammalian groups (e.g., Hershkovitz 1962, 1967, Schmidt-Kittler 2002), and often only on specific tooth positions. A discussion of the various attempts and a comparison of the different terminologies was given in Koenigswald (2011).

The different types of hypsodont teeth (presented in Fig. 9.1) thus depend on which one of the various ontogenetic phases is extended to form the critical height of the tooth, the cusps, the sidewalls, or the dentin. Each of them can then be present to varying degrees in internal or external hypsodonty. These aspects allow us to describe the evolution from hypsodont to euhypsodont teeth.

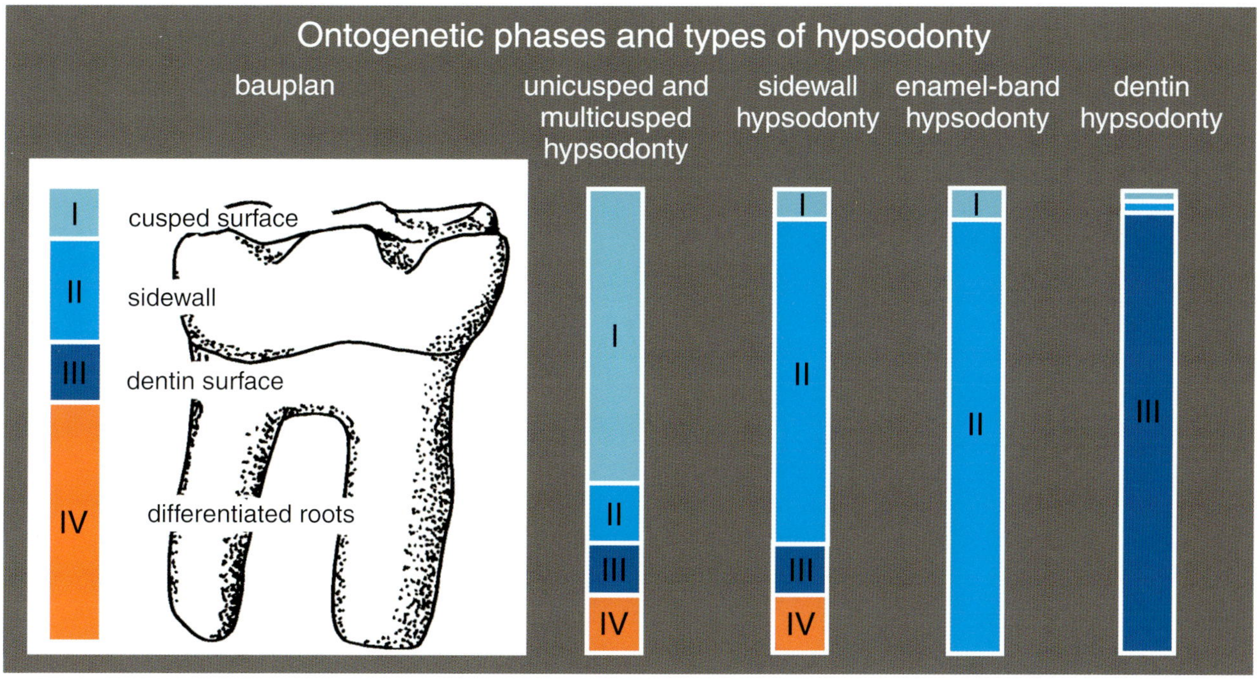

Fig. 9.1. Phases of the early ontogeny and types of hypsodonty: Idealized model of the ontogenetic phases I–IV during the early ontogeny, demonstrated in a lower murid molar. A heterochronic elongation of one of these ontogenetic phases leads to four different types of hypsodonty.

Elongation of ontogenetic phase I – "unicusped hypsodonty"

(Fig. 9.2)

Ontogenetic phase I starts with the forming of the cusp or cusps of a tooth. If this initial phase is significantly elongated, this cusp or cusps gain height and become uni- or multicusped hypsodont. These two types are differentiated, because teeth of unicusped and those of multicusped hypsodonty function differently, in addition to the differences in their morphological appearance.

Teeth such as canines with a high and pointed crown are regarded as unicusped hypsodont. The wear is generally small compared to the eruption rate. The enamel surface is of functional importance and dentin is exposed only rarely. Wear usually is limited to large teeth with an external hypsodonty. Therefore, such teeth occur in the anterior part of the dentition only. The most prominent examples of such unicusped hypsodont teeth are the enamel-covered canines of carnivores, creodonts, some artiodactyls (e.g., *Moschus*, *Deinohyus* Fig. 9.2A), most bats, and even in several primates (very prominent in *Papio*). Unicusped hypsodonty is not restricted to canines because enlarged incisors that are surrounded by enamel may be constructed in the same way (e.g. in Soricidae, lemurs, and diprotodont marsupials such as *Petaurus*). The main function of huge canines is catching prey, but the external hypsodonty facilitates a social significance too.

In the unicusped hypsodont teeth, it is difficult to separate the elongation of phase I from phase II that forms the side walls. Here, those teeth are regarded as unicusped hypsodont if their tips are of functional importance, in contrast to those teeth where the initial tips are worn away rapidly. The reason for why the ontogenetic phase I (forming cusps) is separated from phase II (forming the sidewalls) becomes apparent in multicusped teeth.

Elongation of the ontogenetic phase I – "multicusped hypsodonty"

(Fig. 9.2)

An elongation of phase I occurs in multicusped teeth as well. These teeth are characterized by an equilibrium of wear with the tooth eruption rate. They occur in the postcanine region of the dentition. Due to intensive wear, the surface of the enamel is not of functional importance, but rather the exposed cross-section of the enamel surrounding the cusps is. The molars of *Phacochoerus* (Fig. 9.2C) form an outstanding example. Each of the individual cusps forms isolated columns that merge only on the base of the crown. This is an impressive example of phase I being extended with phase II being very short in comparison. Similarly constructed are the molars of desmostylids (Fig. 9.2B). The molars of *Elephas* and *Mammuthus* provide another variant of multicusped hypsodonty (Fig. 9.2E). The transverse lophs are as high as the molars but fused only at the base of the crown. The stability of these isolated elements is provided by the cementum between the lophs. A similar construction of molar teeth is found in several rodents, e.g., the Late Miocene giant dinomyid *Phoberomys*, the caviid *Hydrochoerus* (Fig. 9.2D), and the murid *Otomys*.

The molars of Elephantidae deliver an insight into the evolutionary sequence of how various characters were generated, as the initial cusps were aligned in transverse lophs prior to the acquisition of hypsodonty.

Fig. 9.2. Uni- and multicusped hypsodonty. **A,** *Deinohyus* sp., large upper canine with strong roots, where phases III and IV are continuous. **B,** *Hydrochoerus hydrochaeris*, M3 with equally spaced enamel crests, Recent, Argentina (IGPB M 5684). **C,** *Phacochoerus aethiopicus*, m3, Recent, Uganda, (IGPB1208). **D,** *Desmostylus hesperus*, Miocene, California, (BSPG 2009 I 50). **E,** *Mammuthus primigenius*, lower m3, occlusal surface, Late Pleistocene, Rhine River, Germany (IGPB M3287). Not to scale. Photos: G. Oleschinski.

Elongation of ontogenetic phase II – "sidewall hypsodonty"

The most frequent type of hypsodonty is sidewall hypsodonty which evolved in many herbivores independently. The high sidewalls are formed during an elongation of phase II (Fig. 9.1). The initial tips of the teeth formed during phase I are of minor functional significance and worn away rapidly. The subsequent phases including root formation are delayed. Mostly this tooth type occurs in the post-canine dentition and thus represents an internal hypsodonty.

Premolars and molars of *Bos* and *Equus* (Fig. 9.3B, C) and many more herbivorous mammals are typical examples of teeth with sidewall hypsodonty. Despite the elongation of phase II, the later phases are postponed, but often present and continuous with the root formation. Some species evolve euhypsodont teeth, such as *Lepus europaeus* (Fig. 9.3D) and several rodents.

The intensive abrasion in these rasp-like grinding teeth is compensated by increasing hypsodonty. It allows faster eruption to replace abraded tooth material. Abrasion and eruption rates are in a perfect equilibrium with one another; if not, a pathological malfunction is the consequence.

Here, teeth in this category are only those whose sidewalls are predominantly covered by enamel, forming the ridges of the rasp. On the occlusal surface, the surrounding enamel forms the functional ridges of the rasp-like grinding tool. The number of ridges increases when the surrounding enamel is folded, such as in arvicolids (Fig. 9.3A), or when enamel islets are introduced.

The transition from hypsodonty to euhypsodonty was probably a relatively straight forward evolutionary step. It was observed in the phylogeny of Arvicolidae, where rooted molars show a gradual increase in hypsodonty due to an elongation of phase II (Koenigswald 1982, 1993, Chaline & Sevillia 1990). The short phase III and the root formation in phase IV are gradually delayed, allowing for a higher crown to be built during phase II. Therefore, rooted or unrooted molars are just a matter of the time when the animal dies. If individuals regularly die off before the root formation starts, then their teeth are euhypsodont. In essence, the root formation is delayed to a moment which is not outlived by the individuals. If this delay becomes part of the program, full euhypsodonty is achieved. That means that in terms of heterochrony, ontogenetic phase II continues until death in euhypsodont teeth, and the later phases III and IV are reduced or simply not present anymore (Koenigswald 1982). Therefore, the formation of euhypsodont teeth with surrounding enamel is assigned to ontogenetic phase II.

Increasing hypsodonty and the transition to euhypsodonty causes problems with the formation of enamel islets. The supply of the enamel organ forming the islets is interrupted as the rim of the islet becomes part of the occlusal surface (for details see Ruf et al. 2020, this volume). Lateral infolds of the enamel are not affected in the same way, they can be built continuously.

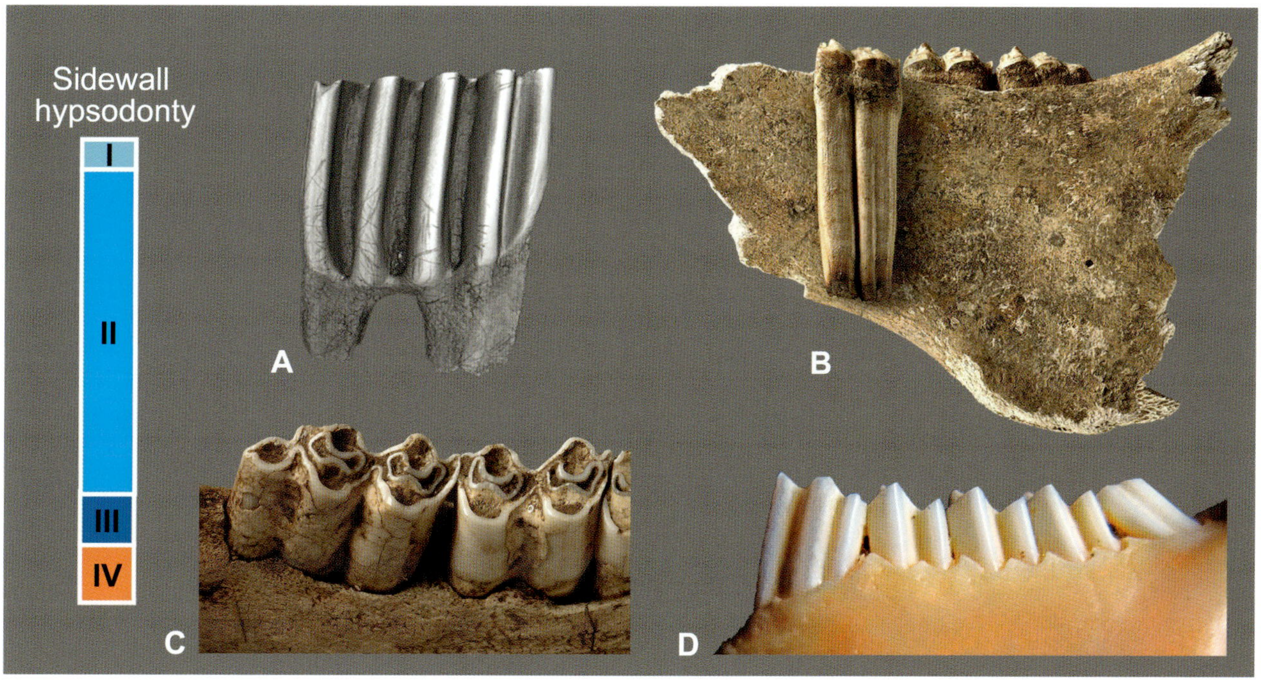

Fig. 9.3. Sidewall hypsodonty – extended ontogenetic phase II. Very common are teeth where the enamel of the elongated side walls forms the rasp-like grinding surfaces. **A,** *Mimomys reidi*, lower m1, Pliocene, Schambach, Germany (BSPG 11975 XXXI). The sidewall is deeply folded to provide proper ridges of the enamel band in the occlusal surface. **B,** *Equus caballus*, fragment of a mandible, subfossil, Germany (IGPB M589). The extracted m1 shows the hidden height of the crown and the postponed root formation. **C,** *Bos taurus*, m2 and m3 of the left mandible, subfossil, Germany. The enamel of the sidewalls and the islets form distinct ridges of the occlusal surfaces (IGBP). **D,** *Lepus europaeus*, Recent, Germany (IGPB 6241). Euhypsodont p4 and molars of the mandible. Not to scale. Photos B, C, D: G. Oleschinski.

Increasing hypsodonty retards the development of roots. The height of the tooth within the jaw may function to some degree as a substitute for a root. Dentin areas are preferred for periodontal ligament attachment, dentin tracts on hypsodont teeth are important for anchoring, their gradual extension has been observed in molars of arvicolines and other mammals (Rabeder 1981, Koenigswald et al. 2010b).

Elongation of ontogenetic phase II – "enamel-band hypsodonty"

The most prominent examples of euhypsodont teeth are the incisors of rodents and lagomorphs. They are formed during ontogenetic phase II, and they differ from the described sidewall hypsodonty by restriction of the enamel to one side or a limited part of the tooth. The presence of a preceding phase I with a tiny enamel cap can be assumed, and the later phases following are irrelevant in euhypsodont teeth. They are either totally reduced or the animal does not outlive them. The continuous formation of the enamel on one side implies an active enamel organ located on this side of the tooth. The growing tooth gradually passes along the enamel organ and gets a continuous cover of enamel. On the other side, the surface is formed by dentin.

Thus, the ontogenetic phases II and III are active at the same time, but locally separated. This modification does not invalidate the information deduced from the general scheme.

The most typical examples of the category of enamel-band hypsodonty are the large and strongly curved incisors of rodents (Fig. 9.4 A, B) and lagomorphs. In upper and lower teeth of these taxa, the enamel is restricted to the anterior and parts of the mesial sides. The occlusal surface of the enamel forms a sharp cutting edge because the dentin is abraded by the antagonist. The equilibrium between the eruption rate and wear is generally balanced, and lack of abrasion may cause severe functional difficulties or even death.

Enamel-band hypsodonty evolved in various mammalian lineages independently, e.g., in rodents, in the multituberculate *Taeniolabis*, and the marsupials *Diprotodon*, *Phascolonus*, and *Vombatus*. A similar construction of the incisors is known from hyracoids, tillodontids, pyrotheres, and the primate *Daubentonia* (Koenigswald 2011). The enamel band is generally placed on the anterior sides of the incisors, but the way in which the incisors occlude varies among species.

The canines of *Hippopotamus* (Fig. 9.4C) provide a special kind of enamel-band hypsodonty. In lower canines the enamel is situated on the anterior side, whereas in upper canines the enamel is on the posterior side. Because the occlusal surface is oblique, in both sides the cutting enamel is situated on the leading edge.

Another modification of hypsodont canines with a one-sided enamel is known from the marsupial *Thylacosmilus* and the cervid *Procervulus*. Both totally unrelated animals have enamel only on the buccal side of their large protruding upper canines which have no proper antagonist. The dentin on the lingual side is abraded by soft tissue to such a degree that the edges of the enamel are sharpened like a blade (Koenigswald & Goin 2000).

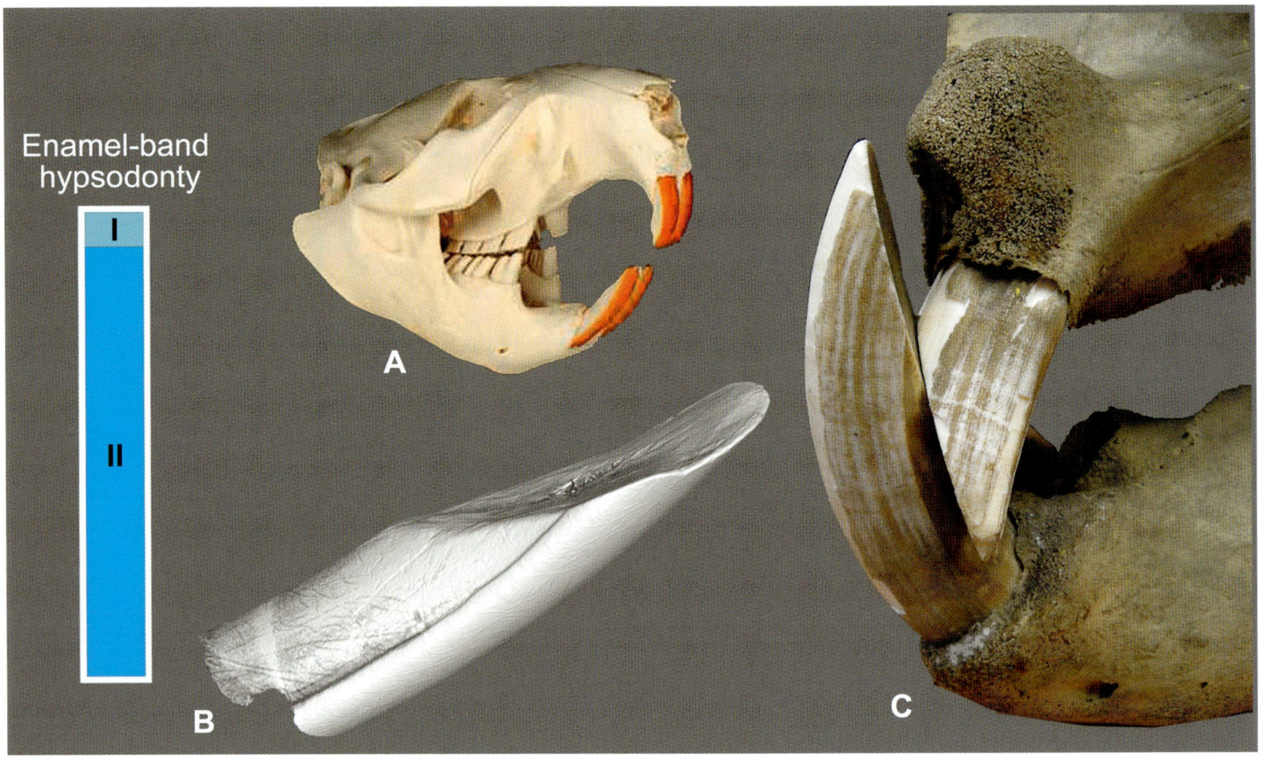

Fig. 9.4. Enamel band hypsodonty – extended ontogenetic phase III. Teeth characterized by enamel on one side are often euhypsodont and form nipper- or blade facets. **A,** *Castor canadensis*, cranium, Recent, Wyoming (IGPB MAÜ 2). Enamel on the anterior side of the teeth is pigmented red. **B,** *Dicrostonyx torquatus*, lower incisor, Pleistocene, Germany (IGPB KOE93). As typical in rodents, the thin enamel layer forms a sharp cutting edge. **C,** *Hippopotamus amphibius*, upper and lower canines in occlusion, Recent, Africa (IGPB M2365). In contrast to rodents, in *Hippopotamus* lower canine the enamel is on the anterior side and in upper on the posterior one, functioning as symmetrical blade facets. Not to scale. Photos: G. Oleschinski.

An atypical enamel-band hypsodonty without a functional context is present in some proboscidean tusks. In *Cuvieronius*, they form a spiral around the tooth, but have almost no functional significance and are a remainder of the former enamel cover.

Elongation of ontogenetic phase III – "dentin hypsodonty"

Dentin teeth that have no enamel occur in various mammalian orders and tooth positions, and include the euhypsodont tusks of elephants (Fig. 9.5B) and cheek teeth of xenarthrans (Fig. 9.5C). An assignment to phase III of the proposed ontogenetic sequence is rectified by tiny enamel caps that occur in various groups. They represent the preceding ontogenetic phases I and II, which are nearly entirely absent. Subsequent phase IV is also generally missing, for specific roots are not needed. Thus, dentin teeth represent an enormously expanded phase III.

Dentin teeth with an external hypsodonty may protrude from the mouth. Eruption rate is distinctly higher than abrasion by wear. In *Mammuthus*, erupting tusks, especially deciduous ones, show tiny caps of enamel (Fisher et al. 2014). Although they are functionally unimportant, they are still verifiable. Similarly, enamel caps were reported from the tusk of the cetacean *Monodon* and the canines of the fissiped carnivore *Odobenus* and suid *Babyrousa*.

Among whales, dolphins have enamel-covered teeth, whereas the dentin teeth of the sperm whale, *Physeter*, are well known as scrimshaws. The erupting teeth have an initial enamel cap as well, indicating reduction of the previously present ontogenetic phases I and II that were shortened during evolution.

It might be annotated that the large tusks of proboscideans that grow continuously throughout life and experience very little wear, document individual life history in their internal structure (e.g., Fisher et al. 2014).

Besides dentin teeth with external hypsodonty, dentin teeth with an internal hypsodonty occur in various xenarthrans where wear and eruption rates are in equilibrium. Most prominent are the premolars and molars of xenarthrans, e.g., *Megatherium* and *Glyptodon*. In a few xenarthrans, traces of an enamel cap have been observed (Simpson 1932, Ciancio et al. 2014, Green & Kalthoff 2015), indicating the presence of preceding ontogenetic phases I and II in this order.

Therefore, dentin teeth fit very well into the general sequence of ontogenetic phases, with the high specialization of an extremely extended phase III.

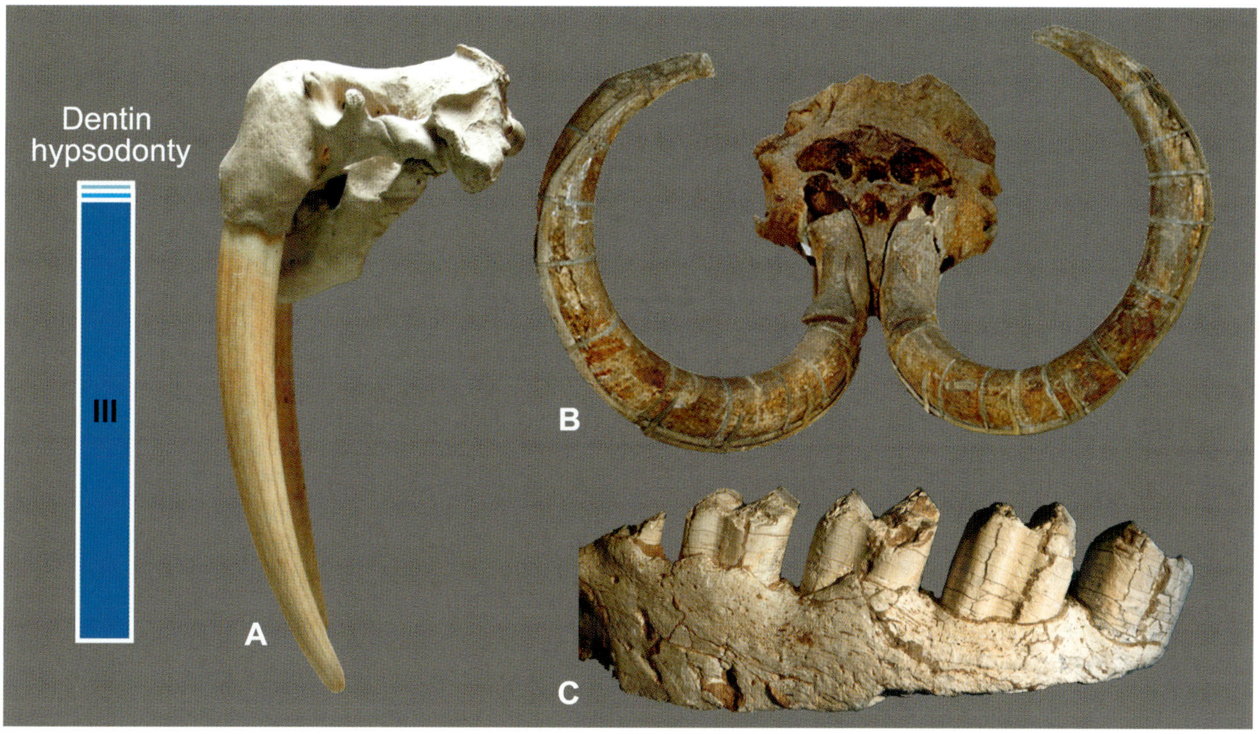

Fig. 9.5. Dentin hypsodonty – extended ontogenetic phase III. In dentin teeth, a primary enamel cap is often present but of no functional significance. **A,** *Odobenus rosmarus*, cranium with enlarged canines, Recent, Bering Sea (ZFMK94.138). **B,** *Mammuthus primigenius*, Pleistocene, Horb, Germany (GPIT/MA/2129). The strongly curved upper tusks of this male individual represent its social status. **C,** *Eremotherium* sp., part of the mandibular dentition, Late Pleistocene, South America (NHMW IX/21/3). The profile of the teeth originates of the different hardness of the dentin. The ridges of the lower teeth intercalate with those in the upper teeth. Not to scale. Photos: A and C by the author, B, I. Werneburg.

Heterochronic variability and evolution in the *bauplan* of hypsodont teeth

According to the observations above, the different types of hypsodonty can easily be interpreted as an elongation of a specific phase within a very general model of ontogenetic phases. Such an elongation of one phase occurs at the cost of the other phases. This is a typical pattern of heterochrony. In this way heterochrony modifies the *bauplan* of a normal brachydont tooth during the early ontogeny. In a similar way, the evolution from hypsodont into euhypsodont teeth can be understood.

Such an explanation illustrates the technical process during tooth formation, but it does not identify the reason why teeth become hypsodont or euhypsodont.

A chronological accordance of the expansion of grasslands and the increase of hypsodont teeth can be observed in herbivores during the Miocene. Grasses and the dust in an open habitat force a higher abrasion of teeth that is compensated by increased tooth material in hypsodont teeth. However, hypsodont teeth evolved in earlier periods as well; for example, Taeniodonta with euhypsodont teeth flourished during the Paleocene and Eocene epochs (Rose 2006).

The heterochronic expansion of phase I, II or III creates different types of hypsodonty. Functional selection may provide the explanation for why a specific phase of the ontogenetic sequence was elongated. But, in general, the achievement of hypsodonty has occurred relatively late in phylogeny, with the new character superimposed on a tooth morphology already present.

Discussing hypsodonty in terms of heterochrony is only one way of looking at this kind of teeth. Most of the other authors provided classifications concentrating on specialized tooth types or systematic groups (Koenigswald 2011). However, considering heterochrony as a useful guide allows an understanding of the evolution of specific modifications of tooth morphology.

Teeth adapted for specific wear stages, another type of heterochrony

Tooth wear is a destructive process that modifies the morphology of teeth. For estimating individual age, the wear of the entire tooth row has been considered and individual age stages (IDAS) have been described (Anders et al. 2011). Here the attempt is made to differentiate some general wear stages for individual teeth that allow comparisons of the specializations in various dentitions.

The second part of the life history of teeth starts when teeth are incorporated into the tooth row. The main purpose of teeth is breaking down food items, and this function

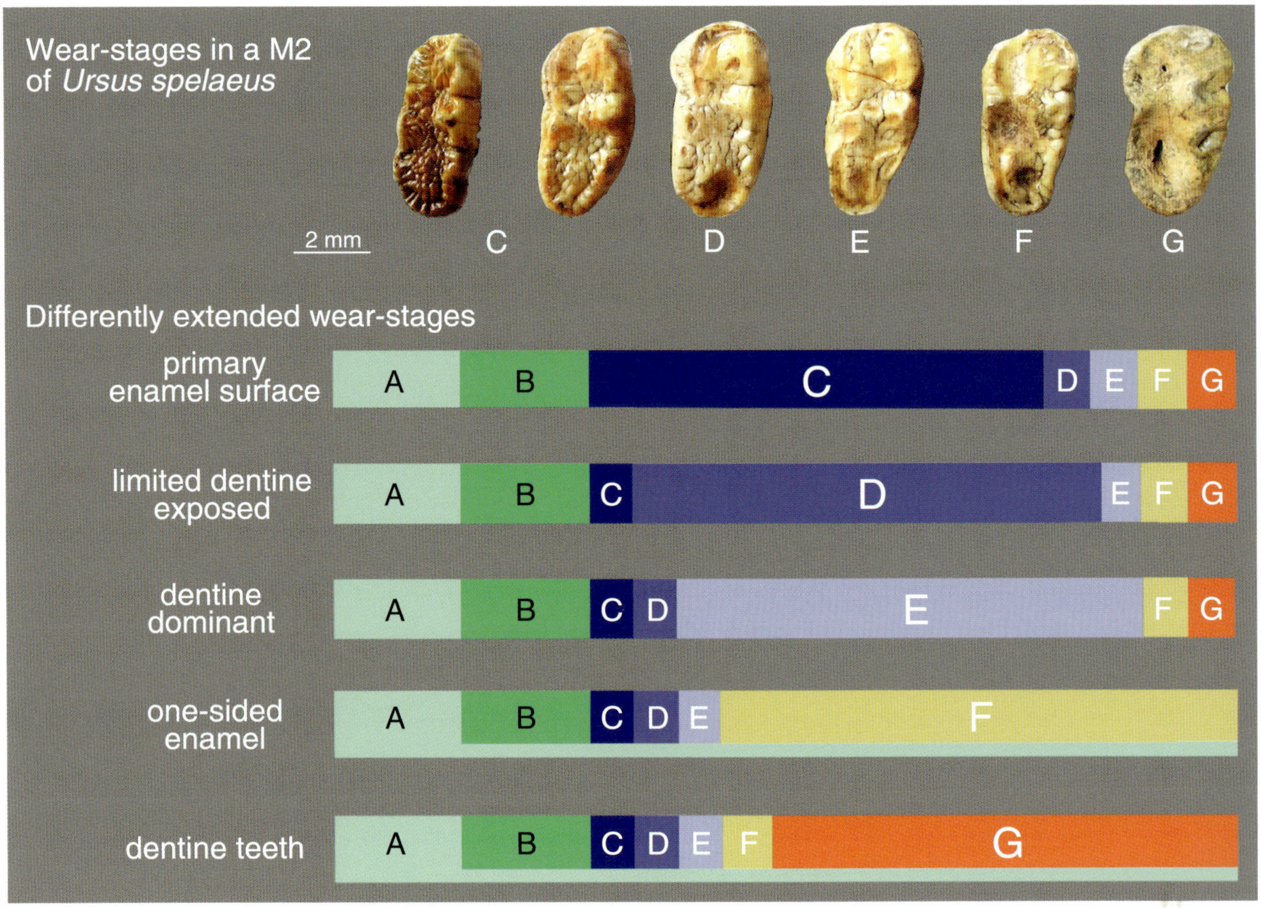

Fig. 9.6. Wear stages and heterochrony. Heterochronic differentiation of the wear stages in mammalian teeth. The stages are symbolized by capital letters C, D, E, F, and G. Wear stages A and B represent tooth formation in the crypt. During wear stage C the enamel surface is functioning. Wear stages D, E and F characterize wear stages with progressive exposure of dentin. In D, the dentin is partially opened. In E, the dentin becomes dominant. In F, only one side is covered with enamel. In wear stage G only dentin is left. Most teeth pass through these stages if the bearer lives long enough. For most teeth, the period of their optimal function can be attributed to a specific wear stage of this sequence. This phase is then often expanded, with earlier or later wear stages being reduced or missing. This allows for the characterizing of specialized wear facets that function in the same manner for a relatively long time. As an example for normal wear, various M2s of *Ursus spelaeus* (Late Pleistocene, Austria, UW collection without numbers) representing the stages C to G, are depicted in the top row.

causes destructive wear at different scales. The great variability of the morphology of mammalian teeth makes it difficult to identify comparable ontogenetic phases of wear if the scale is very detailed. In a more general aspect, when the main focus of wear is regarded, a model of wear stages can be established that seems to be generally applicable and informative. The phases of late ontogeny are indicated by capital letters C to G (Fig. 9.6), and are deduced from the amount of dentin exposed on the occlusal surface. As noted earlier, phases A and B cover early ontogeny. Introducing such wear stages allows to characterize and compare the different evolutionary trends that have occurred in order to cope with the destructive abrasion.

If teeth are functioning best with an intact enamel cover, they are attributed to wear stage C. Wear stages D, E, and F are characterized by different degrees of exposed dentin. As the exposure of dentin is affected by the different mechanical properties of enamel and dentin, these wear stages function in different ways and are described here as "specialized wear facets" (Koenigswald 2017). Their functions are comparable with those of common day tools. Teeth in which only dentin is left are added as wear stage G.

Teeth adapted to wear stage C, the primary enamel surface

(Fig. 9.7)

Using the enamel surface is common in most reptiles and several Mesozoic mammals, but it has also persisted in various mammalian groups. Therefore, not all such teeth can be regarded as specialized, but a specialization can be assumed, if the enamel is thickened in order to withstand abrasion for a longer time period.

The orientation of the enamel surfaces to one another during occlusion is of significance. In the molars of most Mesozoic mammals, the enamel surfaces of the antagonists slide along each other. A similar situation occurs in the enlarged canines of bats (Fig. 9.7A). The contact of the antagonists may cause facets within the enamel or even penetrate to the dentin. The different quality of these two materials does not affect the function in these types of movements. The molar of *Morganucodon* (Fig. 9.7B) provides an example.

The wide range of tribosphenic molars provides revealing insights. Generally, the triangles of the antagonistic

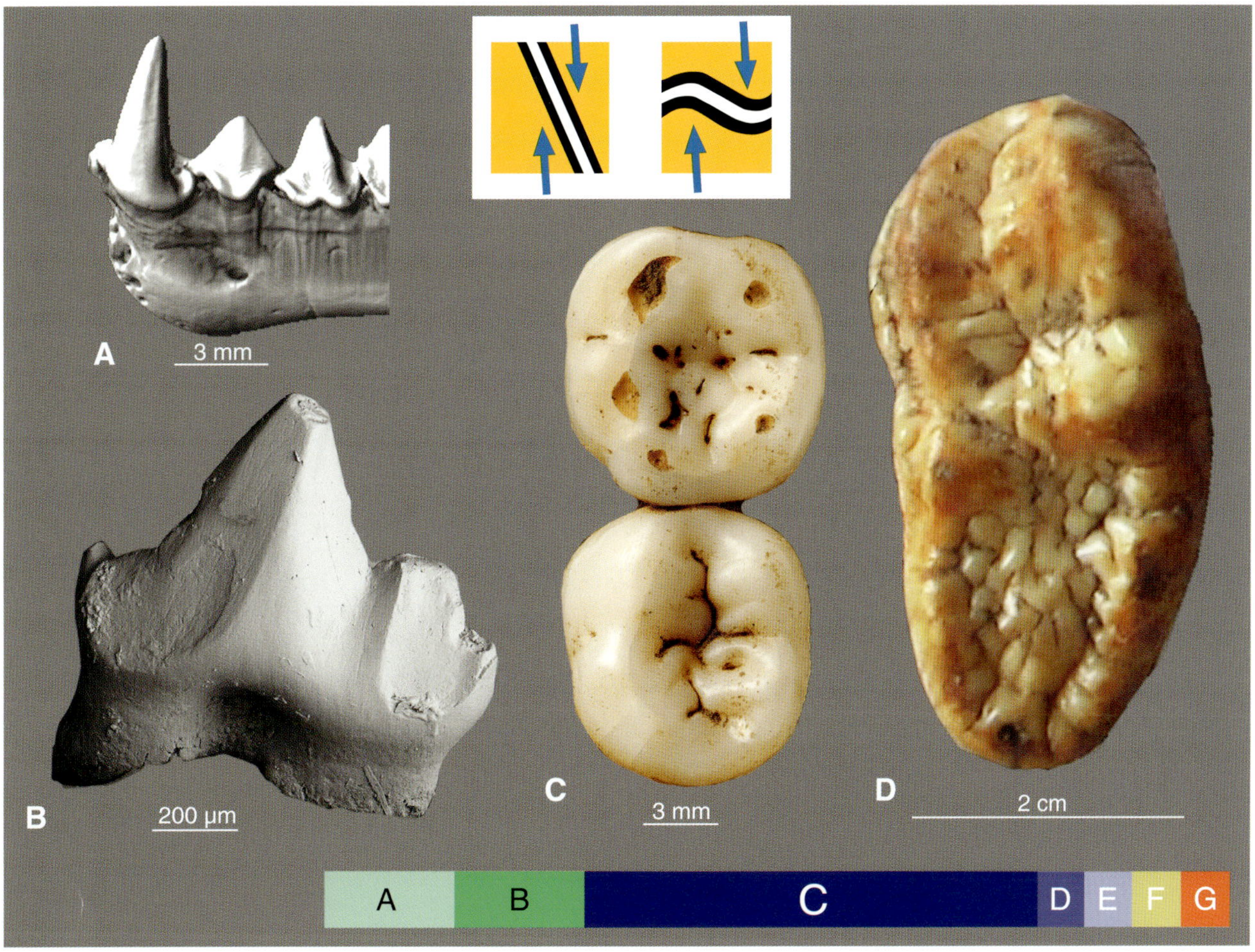

Fig. 9.7. Occlusal surfaces specialized for wear stage C. These teeth function best using their primary enamel surface. **A,** *Phyllostomus hastatus*, left dentary, Recent (ZMFK 59201). The canine and premolars function best when the enamel cover is intact (SEM micrograph). **B,** *Morganucodon watsoni*, lower m1, Late Triassic (UMZC EO.mr (Lr).32). The wear facet is parallel to the enamel surface. (SEM micrograph). **C,** *Homo sapiens*, lower m1 and m2, medieval times, Euskirchen, Germany. In the m1 wear stage D is just beginning, whereas the m2 shows stage C. The main function of these enamel covered teeth is compression. **D,** *Ursus spelaeus*, upper M2, Late Pleistocene, Austria (UW uncatalogued). Food items are compressed between the enamel covered surfaces. Photos: A, R. Hielscher; B, K. Jäger; C, G. Oleschinski.

trigons and trigonids interdigitate in a zig-zag line. As seen in pretribosphenic stem therians (Schultz & Martin 2011), the primary trigons and trigonids function like a punching die. The lateral enamel surfaces pass each other and process the food. Because these side walls are slightly curved, the contact area is reduced and the pressure raised. The enamel is very thin and does not form a functional crest when the dentin is exposed.

In the more derived molars of insectivorous mammals including Didelphidae, Soricidae, and Talpidae, the enamel functions slightly differently. Because the enamel is thicker, it forms sharp edges at the occlusal surface when it is exposed besides the dentin. It contributes to the effect of the punching die.

If the enamel-covered occlusal surface is more horizontal, food is mainly compressed or squeezed between the teeth. These teeth are often multicusped or crenulated, as in bears, humans, and other primates (Fig. 9.7C, D). Abrasion reduces the enamel, and the exposing of the dentin is a measure of the individual's age. To compensate a more intensive wear, such as the crushing of hard shells, the enamel may be thickened, e.g. in the sea otter *Enhydra* or in australopithecines. A more derived way to cope with intensive wear is evolving a multicusped hypsodonty, as in *Phacochoerus*.

Teeth adapted to wear stages D, E, and F with partially exposed dentin

(Figs. 9.8, 9.9)

A great variety of teeth function best when part of the dentin is exposed in specific areas, and the exposed cross-section of the enamel functions as a ridge, across which the food items are squeezed. The difference in hardness of enamel and dentin contributes to the function. To reach this level, wear stage C has to be passed through. Wear stages D, E, and F are characterized by an increasing amount of exposed dentin during the time of their optimal function.

Each of these phases characterizes different types of teeth and is usually functioning over a long part of the life span. Teeth that have their optimal function in D or E may pass through the later wear stages with a limited functionality until the end of life.

In teeth specialized for wear stages D and E, the cross-sections of the worn peripheral enamel are of functional significance and not the enamel surface as in teeth assigned to wear stage C. Of course, the exposure of the cross-sections of the enamel cover requires some opening of the dentin. The difference between the hard enamel and the softer dentin causes protruding enamel crests to be of significant functional importance.

Tooth morphology causes different types of facets that show a selective opening of the dentin. They are described as "specialized wear facets", and they can be related to the sequence of wear stages. "Blade facets", characterized by a single blade per facet, occur in two manners, as symmetrical or asymmetrical blade facets. But in either case, dentin is exposed in a limited area and of minor functional significance. Therefore, blade facets are assigned to wear stage D. "Rasp facets" generally have several enamel ridges, and thus more dentin is exposed. Rasp facets are assigned to wear stage E. In "nipper facets", representing wear stage F, the amount of dentin is highest, but functionally the narrow enamel band is most significant.

Symmetrical and asymmetrical blade facets

(Fig. 9.8)

"Blade facets" are facets formed by a single, well-sharpened enamel blade, forming a perfect cutting device. Due to the special morphology of the teeth, these blade facets are long-lasting and function continuously despite continuous wear. Two types of blade facets are distinguished, because they function slightly differently. The symmetrical arrangement of two blades in antagonistic teeth works like a pair of scissors. In the asymmetrical arrangement, one blade is sharp, but the antagonist is blunt and functions as an abutment. Thus, this structure is comparable to guillotine cutters. The blade facets are attributed to an expanded wear stage D.

Symmetrical blade facets

Carnassials function as typical symmetrical blade facets. The buccal side of the lower m1 slides along the lingual side of the upper P4. It requires some initial wear to expose and sharpen the antagonistic blades. The enamel blade is always on the leading edge of the facet. The edge of each enamel blade is sharpened by attrition and merges smoothly into a plane of exposed dentin. The occlusal plane is oblique, sloping on the buccal side.

The carnassials of Carnivora (Fig. 9.8B) and Hyaenodonta have typical facets forming symmetrical blades. The same tool is present in the marsupial lion *Thylacoleo carnifex*. In *Thylacoleo*, the margin of the blade is almost straight. In most Carnivora, the blade is characterized by a deep notch. This allows the cutting point to shift along the blade, which increases the applied forces and enables the dentin plane to control the exact fitting of the cutting edges. Cutting is more efficient if the subject is compressed beforehand. In carnassials, the compression is created by the general tooth morphology. The dentin is not involved specifically in this compression.

In carnivorous mammals, the symmetrical blades are mostly aligned along an anterior-posterior axis. The main function occurs during phase I of the chewing cycle. The power stroke is directed almost vertically with a slight transverse component in lingual direction. With increasing age, the angle of inclination becomes less steep, but the direction remains unchanged.

Symmetrical blade facets occur in the dentitions of herbivores as well, but the blades are oriented transversely in a bucco-lingual direction. Because they occur mostly in bilophodont molars, one may talk about 'bilophodont scissors' in contrast to the carnassials, although they function similarly. The well sharpened enamel blade is always on the leading side of the facet. During phase I of the power stroke, the food items are cut at the edges of the antagonistic enamel blades. Subsequently the food is compressed between the enamel surfaces of antagonistic lophs, when the teeth further interdigitate. The dentin platform controls this movement. Phase II of the power stroke is very limited or lacking. If present, such as in Macropodidae, it is directed in a lingual direction parallel to the lophs. But it has to be stressed that not all bilophodont teeth function with such bilophodont scissors, as, for example, the mastication pattern in *Papio* is very different.

Bilophodont scissors occur in various marsupials, such as *Diprotodon* and in almost all Macropodidae. Among extant Eutheria, only the tapirs have preserved this type of transverse blade facets. Among fossil herbivores, however, such bilophodont scissors with transverse blade facets are widespread. They occur in specific proboscideans (e.g., *Daouitherium* and *Deinotherium*), pyrotheres (*Pyrotherium*, Fig. 9.8C), xenungulates (e.g., *Carodnia*, Bergqvist & Koenigswald 2017), and of course fossil tapirs (e.g., *Lophiodon*, Koenigswald 2014) (Fig. 9.8A). Most of these teeth are brachydont, and only in *Arsinoitherium* (Embrithopoda) do bilophodont scissors occur in combination with hypsodonty.

Mastication with bilophodont scissors seems to have had limited evolutionary success; it was frequent among various mammalian groups in the Paleogene and faded gradually away, being almost extinguished in extant fauna (Hooker 2000).

Asymmetrical blade facets

A specific type of blade facets was found in the premolars and molars of *Dicerorhinus* (Fig. 9.8E–F). The ectoloph of the upper teeth forms a sharp cutting-edge without an equivalent blade in the lowers. An antagonist is required for applying any destructive forces to the food items: It is very inconspicuous and formed by the buccal edge of the lower premolars and molars. During the function of this structure, comparable with the rare guillotine cutters, the ectoloph of the upper tooth, and more precisely the cross-section of the buccal enamel, is cutting. It continues smoothly into the opened dentin. The enamel of the blade is often sharpened by attrition.

This type of blade facets was termed guillotine cutters or ectoloph-guillotine and occurs in a great number of early Perissodactyla: Brontotheriidae (e.g., *Megacerops*, Fig. 9.8G), Chalicotheriidae, and early Equoidea (e.g., *Palaeotherium*, *Anchitherium*, Fig. 98D). It was found as well in the South American Litopterna and Astrapotheria. This type of facet is rare in extant fauna, because the increasing grasses required more grinding than cutting.

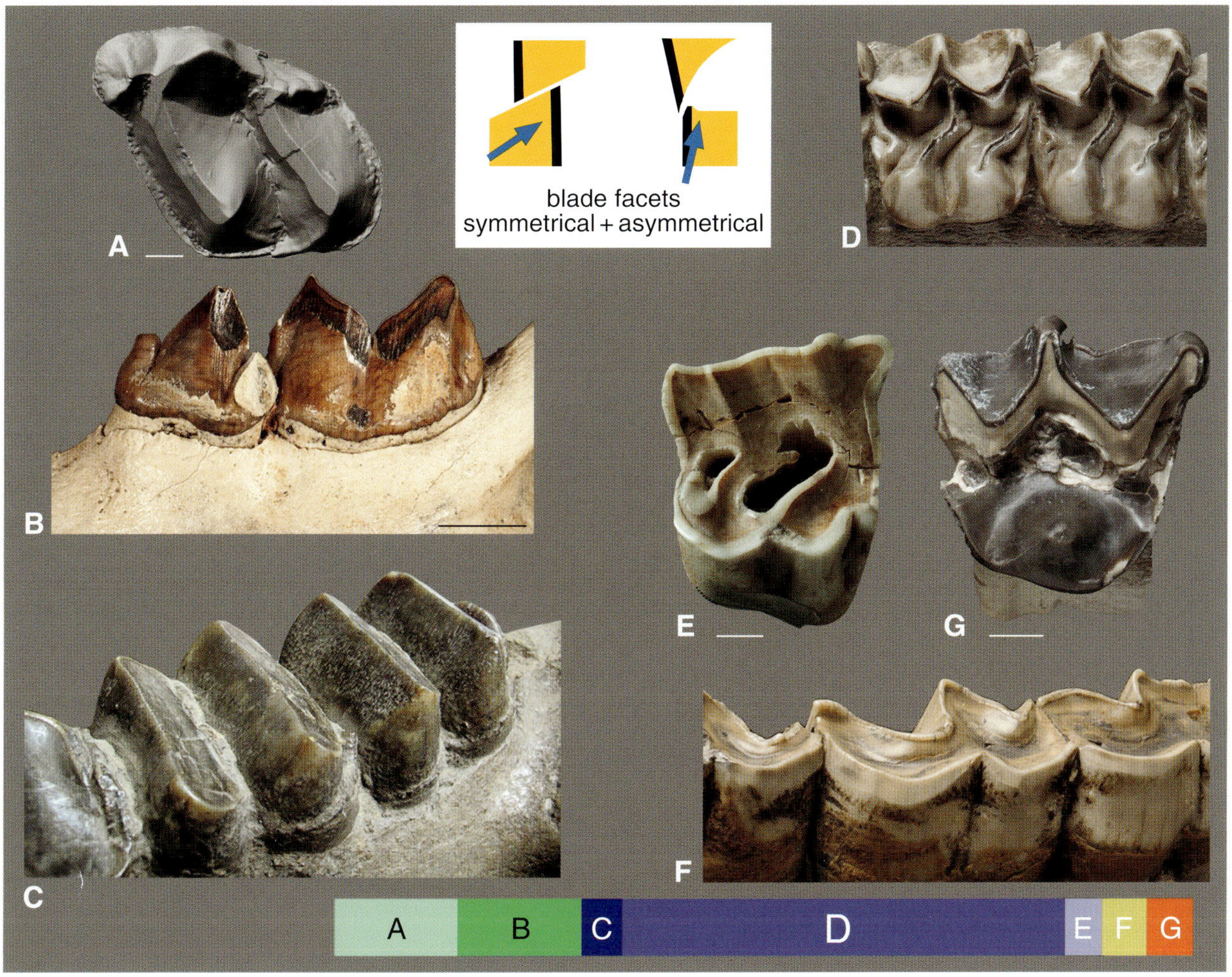

Fig. 9.8. Occlusal surfaces specialized for wear stage D. Symmetrical and asymmetrical blades are formed by the cross-section of the enamel. The dentin is partially exposed. **A,** *Lophiodon lautricense*, left upper M3, Eocene, Robiac, France (HLMD Ro 17c). The transverse lophs have symmetrical blades, that cut with the cross lophs of the lower molars. The function of the enamel blades requires an opening of the dentin that occurs after a short period of initial war. **B,** *Xenosmilus hudsonae*, left p4 and m1. Late Pleistocene, Florida (UF 60000). The dentin is opened in the blade facet with the enamel on the leading side. The functional surface of the upper molar is reciprocal. The optimal function of the symmetrical blades is provided by the exposed dentin. **C,** *Pyrotherium romeroi*, left m2 and m3, Deseadan, Rio Desado, Argentina (MNHN-F-DES 1238). The lophs have an enamel-covered anterior side, whereas the planar facet on the distal side is formed by dentin. The upper molars have a reciprocal orientation. Cutting of the food items occurs between the symmetrical blades at the enamel ridge, followed by compression against the enamel-covered surface. **D,** *Anchitherium aurelianense steinheimense*, right P4 and M1, Miocene, Steinheim am Albuch, Germany (BSPG 1897 XIII4a). The ectoloph forms the cutting facet for phase I of mastication, whereas proto- and metaloph show initial grinding facets for phase II of mastication. The lower molars carry mainly shearing facets. **E, F,** *Dicerorhinus* sp., upper M1, Pleistocene, Weimar-Ehringsdorf, Germany (IGPB M 4125) and lower molars, Pleistocene, Brüggen (IGPB 2863). The main mastication in rhinos occurs along asymmetrical facets. During phase I, the ectolophs of the upper premolars and molars form a cutting edge backed-up by dentin. The antagonistic lower premolars and molars have only a small facet as an abutment. During phase II the lower molars are in full function, grinding across the lingual part of the upper molars. The specialized facets of these teeth require a partial opening of the dentin. **G,** *Megacerops* sp., right M2, Orellan, Badlands, South Dakota (HLMD WT 735). The Brontotheriidae and Chalicotheriidae have a cutting device like that in the rhinos, but it is followed by compression against the ectoloph. Scale bars equal 10 mm. B, C, D, and F not to scale. Photos: G. Oleschinski.

The main power stroke with cutting is related to phase I of the chewing cycle and is oriented in Brontotheriidae in a strict lingual direction and stopped by the protocone. In Perissodactyla with an asymmetrical blade facet, the power stroke of phase I is usually directed in a mesiolingual direction. It is followed by a phase II with grinding that increases during further evolution.

Rasp facets

(Fig. 9.9)

Teeth that are mostly grinding often have flat occlusal surfaces with enamel ridges that are separated by dentin and/or cementum. These enamel ridges are either formed by the surrounding side walls of the teeth, lateral infolds of the side walls, or by enamel islets. Antagonists in the lower and upper dentition are usually formed very similarly. Their

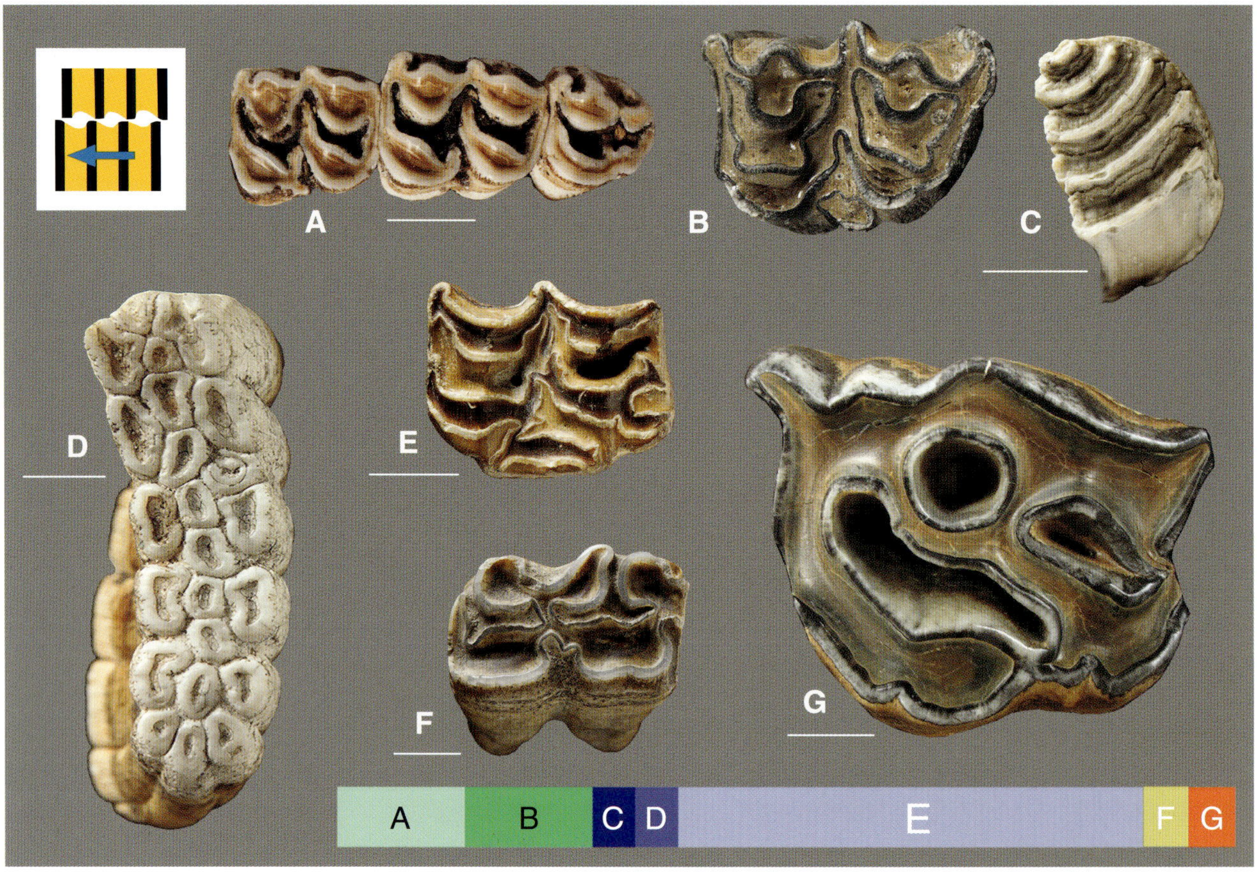

Fig. 9.9. Occlusal surfaces specialized for wear stage E. The enamel ridges between the widely exposed dentin form a rasp facet. The ridges are formed by the surrounding enamel, enamel islets and lateral infolds. Some of the teeth are hypsodont. **A,** *Rangifer tarandus*, left M1–M3, Recent, Banks Island, Canada (IGPB M 1508). The enamel islets of the low-crowned teeth are not filled with cementum. **B,** *Bison priscus*, Late Pleistocene, Groß-Rohrheim, Germany (SLGM). The enamel islets and the lateral infolds are filled with cementum. **C,** Dinomyidae indet., M3 sin., Late Miocene/Pliocene, Playa Arazatí, San José, Uruguay (IGPB KOE 4007). The enamel ridges occur on one side of the lophs only, but have exactly the same distance. **D,** *Phacochoerus aethiopicus*, Recent, Okavango, Botswana (IGPB-M 7028]. The rasp facet is formed by elevated cones that are surrounded by cementum. **E, F,** *Equus ferus*, subfossil, Adendorf near Meckenheim, Germany (IGPB-M 1850). Upper molar with enamel islets and lateral infolds and lower molar with lateral infolds only. **G,** *Coelodonta antiquitatis*, right M2 with enamel islets, Late Pleistocene, Bobenheim, Germany (SLGM). In comparison to earlier rhinos, the ectoloph of *Coelodonta* is lowered and incorporated into the flat occlusal surface. With progressive wear, the lateral infolds become when deeply worn enamel islets. Scale bars equal 10 mm, B not to scale. Photos: G. Oleschinski.

function is compared to a rasp. A great variety of teeth function in this way (Fig. 9.9), therefore the term rasp-facet is used here. In the sequence of wear stages presented here, a large amount of dentin is exposed between the enamel ridges, thus it is assigned to wear stage E.

The function of these teeth differs greatly from blade facets, because the food is compressed by the entire occlusal surface where the dentin plays a major role. The enamel ridges are thresholds where food items are squeezed across and broken. These enamel ridges mostly have a rounded surface and show striations only rarely. It is conspicuous that distances between the enamel ridges tend to be equal even though they may originate from different parts of the tooth morphology (Fig. 9.9A–G). That is especially obvious in *Mammuthus* (Fig. 9.2E) or rodents such as *Hydrochoerus* (Fig. 9.2D) and *Dinomys* (Fig. 9.9C). The occlusal surface is surrounded mainly by the side walls (e.g., *Cryptomys*, *Aplodontia*, *Thomomys*), or differentiated by enamel islets (Ruf et al. 2019). Such enamel islets, e.g. in *Mylagaulus* and *Castor* are traces of furrows of the occlusal surface as in the gondwanathere *Vintana* (Krause 2014) or basal parts of lateral infolds, e.g., in the gondwanathere *Sudamerica* and in lagomorphs (Koenigswald et al. 2000, 2010a). Lateral infolds also occur on the molars of arvicolids and the rhinocerotoid *Elasmotherium*, and on lower molars of Equidae.

Of functional significance are the enamel ridges that tower above the enclosed dentin. It is the cross-section of the worn enamel, and not the primary enamel surface that forms the enamel crests. Lateral infolds and/or enamel islets enrich the number of enamel crests (Ruf et al. 2019). Very often, these ridges are oriented closely, but not exactly perpendicularly, to the direction of mastication (Maglio 1972, Koenigswald et al. 1994). With the increasing number of enamel ridges, the efficiency of the grinding teeth grows. Rasp facets are predominantly flat in the direction of the power stroke that varies between lingual and proal. It is nearly horizontal in some perissodactyl dentitions wherein the uniform power stroke is ultimately derived from a two-phased power stroke, where the direction and inclination of both phases have become more and more identical (Ruf et

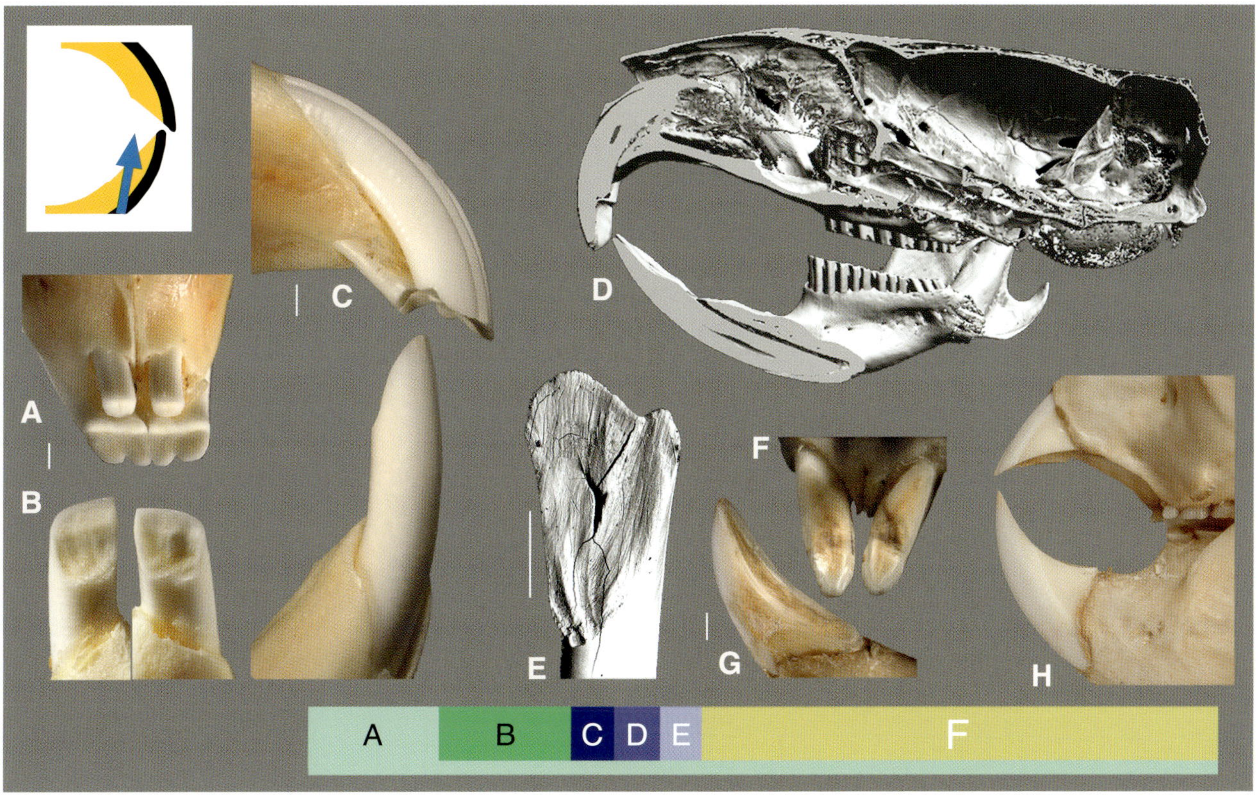

Fig. 9.10. Occlusal surfaces specialized for wear stage F. These teeth have enamel only on one side and may form nipper facets. The antagonist sharpens the blade by removing the dentin. **A–C,** *Lepus europaeus*, Recent, Germany (IGPB M 6241). **A,** wear facets on the upper incisors with a typical step caused by the lower incisor. **B,** lower incisors with marks caused by the upper ones during sharpening. Especially for sharpening the lower jaw is pushed forward so far that the upper incisors contact the posterior sides of the lower incisors. **C,** functional interaction of the two upper incisors with the lower incisor. **D,** *Microtus gregalis*, Recent, Fairbanks, Alaska (IGPB M 1005). Virtual section showing the interaction of lower and upper incisors when molars are well separated. **E,** *Otomys angoliensis*, Recent, Makapansgat, South Africa (IGPB-KOE 3356). Occlusal surface of lower incisor with scratches made by the upper incisor during sharpening. **F–H,** *Daubentonia madagascarienis*, Recent, Madacascar (SMF 1583). The nipper facets of upper (**F**) and lower incisors (**G**) are well polished and show no striations from the antagonists. They are very pointed due to the high-oval cross-sections. **F,** the upper facets grooves indicate where the tips of the lower incisors normally occlude. **H,** lateral aspect of the anterior dentition. The incisors occlude when the molars are very close. Scale bars equal 1 mm, D not to scale. Photos: G. Oleschinki.

al. 2019). Therefore, it is hard to decide whether the uniform power stroke represents phase I or phase II of the chewing cycle. Either the horizontal movement combines phases I and II with an adjusted inclination, or the horizontal grinding results from an expanded phase II found in teeth with an ectoloph-guillotine, as in most early Equoidea.

Nipper facets

(Fig. 9.10)

Nipper facets are best compared with a pair of end-cutting pliers. They are well known in rodent and lagomorph incisors. The one-sided enamel of the euhypsodont teeth is restricted more or less to the anterior side and forms a cutting edge, while the widely opened dentin is exposed on the lingual side. The two antagonistic blades cooperate like chisels, but during mastication they do not touch each other. Mostly the upper tooth is holding the item, whereas the lower one is cutting or gnawing into it. This type of blade facet is attributed to wear stage F because enamel is limited to one side and the dentin dominates intensively.

The teeth require a mutual sharpening by the antagonist. The cross-section of the incisors determines the shape of the cutting edge. It may be broad as in *Lepus* (Fig. 9.10A–C) or pointed as in *Glis*. The edge itself is formed by enamel. Because no attritional facets are recognizable on the enamel, it seems to break off. But the dentin has to be removed regularly to expose the enamel crest. During the sharpening process, the mandible with the incisor has to be shifted forward or backward in relation to the upper incisor. It has to be moved forward to such a degree that the cutting crest of the upper incisor can scrape its dentin the full length of the facet. Normally the upper incisor is positioned more anteriorly than the tip of the lower incisor. A distinct notch in the dentin of the upper incisors marking the contact with the lower incisors (Fig. 9.10A, C) occurs in some rodents and lagomorphs.

Besides lagomorph and rodent incisors, nipper facets occur in various other mammalian groups. The incisors of the primate *Daubentonia* have nipper facets mistakable for those of Glires (Fig. 9.10F–H).

Euhypsodont incisors with an enamel band occur in various mammalian orders, e.g., *Groeberia* and *Patagonia* (Gondwantheria), *Vombatus* (Marsupialia), *Tillodus* (Taeniodonta), and *Pyrotherium* (Pyrotheria), but not all form nipper facets due to the different orientation of the teeth.

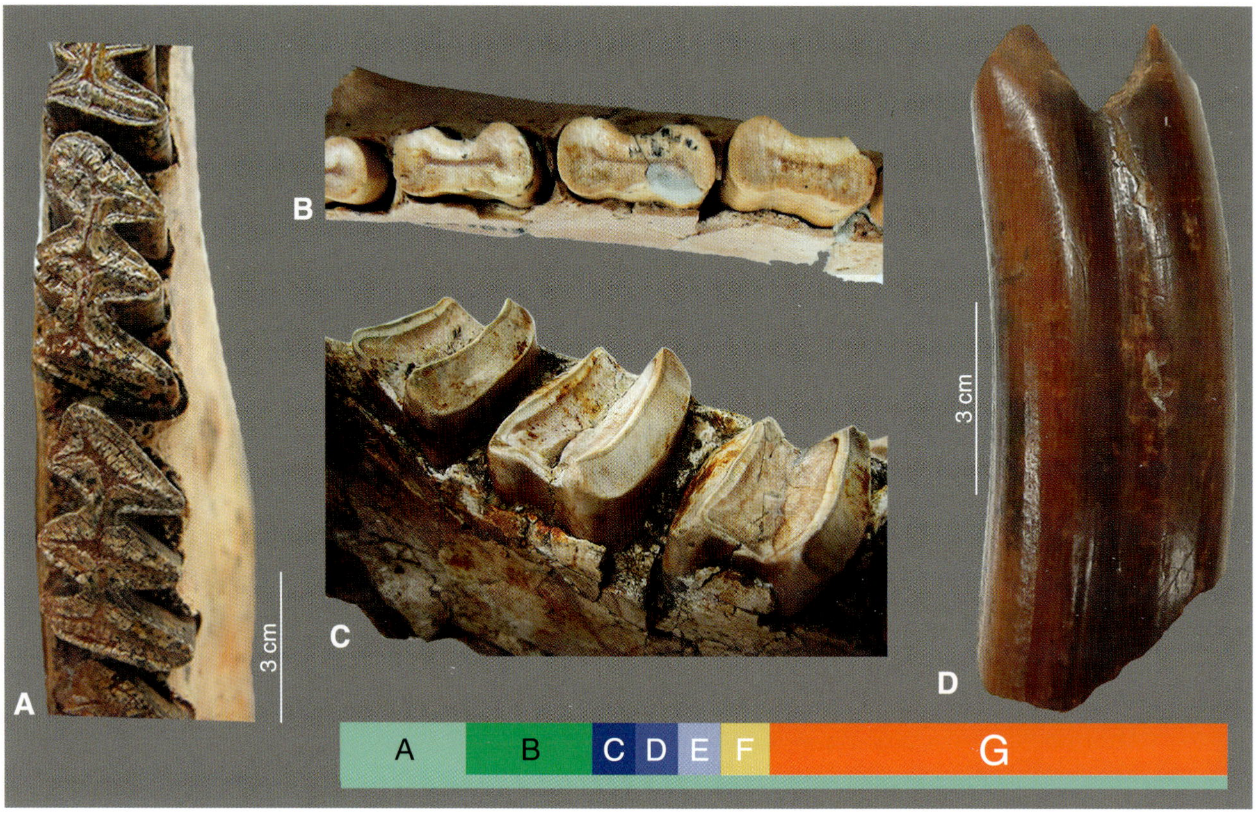

Fig. 9.11. Occlusal surfaces specialized for wear stage G. Dentin teeth may have had a primary enamel cap, but this is functionally insignificant. **A,** *Glyptodon* sp., part of the dentary, Pleistocene, Salta Prov., Argentina (LACM 64153). The cheek teeth show the differentiated dentin. **B,** *Holmesina floridanus*, cheek dentition of the mandible, Pliocene, Haile, Alachia Co., Florida (UF 223813). **C,** *Eremotherium*, part of the mandible, Pleistocene, Florida (UF). The ridges of the bilophodont occlusal surfaces are of dentin. **D,** *Eremotherium carolinense*, fragment of cheek tooth, Ecuador (LACM, no number). The entire tooth is formed by dentin, the crests are of orthodentin. B and C not to scale. Photos by the author.

Teeth adapted to function without enamel – dentin teeth

(Fig. 9.10)

In many teeth, the final wear stage (G) is reached, when all enamel is gone and only a stump of dentin may be left. Several mammals, however, evolved well-functioning dentin teeth. To complete the picture of wear stages, such dentin teeth are included here and briefly discussed. Due to a differentiation of the dentin, it may show distinct facets comparable with those of enamel.

Dentin teeth with external hypsodonty have almost always a significant function in the social environment that causes no regular wear.

In dentin teeth with an internal hypsodonty, the teeth are either abraded smoothly and form a well-rounded cusp, e.g., in *Physeter.* In Xenarthra, the dentin is not uniform but differentiated into orthodentin, osteodentin, and vasodentin. They differ somewhat in their physical properties, causing characteristic ridges on the occlusal surfaces in different patterns (Green & Kalthoff 2015). In armadillos, the outer layer is of relatively hard orthodentin surrounding a core of vasodentin. In Glyptodontidae, the trilobite cheek teeth have inner ridges where osteodentin surmounts the surrounding orthodentin (Fig. 9.11A). In contrast to the relatively flat occlusal surfaces seen in glyptodonts or *Holmesina* (Fig. 9.11B), Megatheriidae, (e.g., the giant *Megatherium* or *Eremotherium*, Figs. 9.11C,D) have bilophodont cheek teeth with a high profile. The crests are formed by orthodentin, whereas the slopes are of vasodentin or cementum. They allow a perfect self-sharpening of the euhypsodont teeth.

The correlation of the heterochrony of specialized wear facets with ontogenetic wear stages

Although limited in comparison with the vast variety of mammalian teeth, this survey shows that establishing general ontogenetic phases allows one to characterize the heterochrony of tooth formation and function, even if they are very divergent. Such a comparison illustrates the significance of heterochronies and general tendencies in the evolution of mammalian dentitions.

Tooth morphology allows to reconstruct the sequences and duration of ontogenetic phases. Although these ontogenetic phases of early life history remain theoretical, the evolution and variability of hypsodont teeth show a tight correlation between early ontogeny and the final tooth. They illustrate the manifold ways in which hypsodont teeth can be built in order to provide tooth material that compensates for intensive wear.

Another example of heterochrony in teeth occurs when tooth wear causes destructive abrasion. Only in some teeth is the primary enamel surface of functional importance. In many other teeth, for optimal function part of the enamel has to be worn away and the dentin has to be exposed in the secondary occlusal surface. The abrasion of the enamel occurs not randomly but in specific areas to prepare specialized facets. Although wear is very variable, a number of specialized wear facets can be defined. They are characterized by their long-lasting similar function, despite progressing wear. They are assigned to a series of subsequent wear stages, using the amount of exposed dentin as a guideline.

Thus, similar to the heterochrony in the ontogenetic stages, the functional duration of specific wear stages is extended at the cost of other (theoretical) wear stages.

Heterochronic shifts occur frequently in mammalian phylogeny because altering the duration of ontogenetic phases seems to be less problematic than inventing new characters. Thus, the life history of teeth is determined by different kinds of heterochrony.

Acknowledgments

This study arose from the DFG Research Unit 771 and I am very grateful for the fruitful discussions in this group. Mark Teaford and Mikael Fortelius reviewed the manuscript and gave inspiring comments; M. Teaford and Jaeylin Eberle improved the English. The illustration of this chapter was made possible by Romina Hielscher, Kai Jäger, I. Werneburg, and the excellent work of Georg Oleschinski. To all of them I want to express my cordial thanks.

References

* indicates publications that originated from the DFG Research Unit 771.

*Anders, U., Koenigswald, W. v., Ruf, I. & Smith, B. H. (2011): Generalized individual dental age stages (IDAS) for fossil and extant placental mammals. Paläontologische Zeitschrift 85: 321–341. https://doi.10.1007/s12542-011-0098-9

Bergqvist, L. P. & Koenigswald, W. v. (2017): The enamel microstructure of *Carodnia vieirae* (Mammalian, Xenungulata) and some comments to its pattern of its mastication. Palaeontologica Electronica 20.2.30A: 1–15.

Chaline, J. & Sevillia, P. (1990): Phyletic gradualism and developmental heterochronies in a European Plio-Pleistocene *Mimomys* lineage (Arvicolidae, Rodentia). In: Fejfar, O. & Heinrich, W. D. (eds.). International Symposium: Evolution, Phylogeny and Biostratigraphy of Arvicolids (Rodentia, Mammalia). Pfeil, Munich: 85–98.

Ciancio, M. R., Vieytes, E. C. & Carlini, A. A. (2014): When xenarthrans had enamel: insights on the evolution of their hypsodonty and paleontological support for independent evolution in armadillos. Naturwissenschaften 101: 715–725.

Fisher, D. C., Cherney, M. D., Newton, C., Rountrey, A. N., Calamari, Z. T., Stucky, R.K., Lucking, C. & Petrie, L. (2014): Taxonomic overview and tusk growth analyses of Ziegler Reservoir proboscideans. Quaternary Research 82: 518–532.

Fortelius, M. (1985): Ungulate cheek teeth: developmental, functional and evolutionary interrelations. Acta Zoologica Fennica 180: 1–76.

Fortelius, M. & Solounias, N. (2000): Functional characterization of ungulate molars using the abrasion-attrition wear gradient: A new method for reconstructing paleodiets. American Museum Novitates 3301: 1–36.

Gomes Rodrigues, H., Solé, F., Charles, C., Tafforeau, P., Vianey-Liaud, M. & Viriot, L. (2012): Evolutionary and biological implications of dental mesial drift in rodents: The case of the Ctenodactylidae (Rodentia, Mammalia). PLoS One 7(11): e50197. https://doi.10.1371/journal.pone.0050197

Green, J. L. & Kalthoff, D. C. (2015): Xenarthran dental microstructure and dental microwear analyses, with new data for *Megatherium americanum* (Megatheriidae). Journal of Mammalogy 96: 645–657.

Haeckel, E. (1866): Generelle Morphologie der Organismen. Allgemeine Grundzüge der organischen Formenwissenschaft. Georg Reimer, Berlin.

Hershkovitz, P. (1962): Evolution of neotropical cricetine rodents (Muridae) with special reference to the phyllotine group. Fieldiana Zoology 46: 1–524.

Hershkovitz, P. (1967): Dynamics of rodent molar evolution: A study based on New World Cricetinae, family Muridae. Journal of Dental Reseach 46: 829–842.

Hooker, J. J. (2000): Ecological response of mammals to global warming in the late Paleocene and early Eocene. Geologiska föreningens förhandlingen GFF 122: 77–79

Janis, C. M. (1988): An estimation of tooth volume and hypsodonty indices in ungulate mammals and the correlation of these factors with dietary preferences. In: Russel, D. E., Santoro, J. P. & Signogneau-Russel, D. (eds.). Teeth Revisited: Proceedings of the VII International Symposium on Dental Morphology. Mémoires du Muséum national d'Histoire naturelle C 53: 367–387

Kaiser, T. M., Bernor, R. L., Scott, R. S., Franzen, J. L. & Solounias, N. (2003): New interpretations of the systematics and palaeoecology of the Dorn-Dürkheim 1 hipparions (Late Miocene, Turolian Age [MN11]), Rheinhessen, Germany. Senckenbergiana lethaea 83: 103–133.

Koenigswald, W. v. (1982): Zum Verständnis der Morphologie der Wühlmausmolaren (Arvicolidae, Rodentia, Mammalia). Zeitschrift für geologische Wissenschaften Berlin 10: 951–962.
Koenigswald, W. v. (1993): Heterochronies in morphology and schmelzmuster of hypsodont molars in the Muroidea (Rodentia). Quaternary International 19: 57–61.
*Koenigswald, W. v. (2011): Diversity of hypsodont teeth in mammalian dentitions – Construction and classification. Palaeontographica A 294: 63–94.
*Koenigswald, W. v. (2014): Mastication and wear *Lophiodon* (Perissodactyla, Mammalia) compared to lophodont dentitions in some other mammals. Annales Zoologici Fennici 51: 162–176.
*Koenigswald, W. v. (2017): Specialized wear facets and late ontogeny in mammalian dentitions. Historical Biology 30: 7–29. https://doi.org/10.1080/08912963.2016.1256399
Koenigswald, W. v. & Goin, F. (2000): Enamel differentiation in South American marsupials and a comparison of placental and marsupial enamel. Palaeontographica A 255: 129–168.
Koenigswald, W. v., Goin, F. & Pascual, R. (2000): Hypsodonty and enamel microstructure in the Paleocene gondwanatherian mammal *Sudamerica ameghinoi*. Acta Palaeontologica Polonica 44: 263–300.
*Koenigswald, W. v., Anders, U., Engels, S., Schultz, J. A. & Ruf, I. (2010a): Tooth morphology in fossil and extant Lagomorpha (Mammalia) reflects different mastication patterns. Journal of Mammalian Evolution 17: 275–299. https://doi.10.1007/s10914-010-9140-z
*Koenigswald, W. v., Kalthoff, D. C. & Semprebon, G. (2010b): The microstructure of enamel, dentine and cementum in advanced Taeniodonta (Mammalia) with comments on their dietary adaptations. Journal of Vertebrate Paleontology 30: 1797–1804.
Koenigswald, W. v., Sander, M., Leite, M., Mörs, T. & Santel, W. (1994): Functional symmetries in the schmelzmuster and morphology in rootless rodent molars. Zoological Journal of the Linnean Society 110: 141–179.
Krause, D. W. (2014): Dental morphology of *Vintana sertichi* (Mammalia, Gondwantheria) from the Late Cretaceous of Madagascar. Journal of Vertebrate Paleontology Memoir 6: 137–165.
Maglio, V. J. (1972): Evolution of mastication in the Elephantidae. Evolution 26: 638–658.
Martin, L. D. (1993): Evolution, hypsodonty, and enamel structure in Plio-Pleistocene rodents. In: Martin, R. A. & Barnosky, A. D. (eds.). Morphological Change in Quaternary Mammals of North America. Cambridge University Press, Cambridge: 205–225.
McNamara, K. J. (1990): Heterochrony. In: Briggs, D. E. G & Crowther, P. E. (eds.). Palaeobiology. Blackwell, London: 111–119
Mones, A. (1982): An equivocal nomenclature: What means hypsodonty? Paläontologische Zeitschrift 56: 107–111.
Rabeder, G. (1981): Die Arvicoliden (Rodentia, Mammalia) aus dem Pliozän und älteren Pleistozän von Niederösterreich. Beiträge zur Paläontologie von Österreich 8: 1–373.
Rose, K. D. (2006): The Beginning of the Age of Mammals. Johns Hopkins University Press, Baltimore.
*Ruf, I., Schubert, A. M. & Koenigswald, W. v. (2020): Case studies on functional aspects and constraints in early and late tooth ontogeny. In: Martin, T. & Koenigswald, W. v. (eds.). Mammalian Teeth – Form and Function. Pfeil, Munich: 102–124.
Schmidt-Kittler, N. (2002): Feeding specializations in rodents. Senckenbergiana lethaea 82: 141–152.
*Schultz, J. A. & Martin, T. (2011): Wear pattern and functional morphology of dryolestoid molars (Mammalia, Cladotheria). Paläontologische Zeitschrift 85: 269–285.
Simpson, G. G. (1932): Enamel on the teeth of an Eocene edentate. American Museum Novitates 567: 1–4.
Smith, K. K. (2003): Time's arrow: heterochrony and evolution development. International Journal of Developmental Biology 47: 613–621.
Solounias, N., Danowitz, M., Buttlar I. & Coopee, Z. (2019): Hypsodont crowns as additional roots: A new explanation for hypsodonty. Frontiers in Ecology and Evolution 7. https://doi.10.3389/fevo.201900135
Van Valen, L. (1960): A functional index of hypsodonty. Evolution 14: 531–532.
White, T. E. (1959): The endocrine glands and evolution, no. 3: os cementum, hypsodonty, and diet. Contributions from the Museum of Paleontology, University of Michigan 13: 211–265.

Molar diversity and functional adaptations in Mesozoic mammals

Thomas Martin, Kai R. K. Jäger, Thorsten Plogschties, Achim H. Schwermann, Janka J. Brinkkötter, and Julia A. Schultz

Introduction

The vast diversity of molars seen in therian mammals is derived from the apomorphic tribosphenic respectively pretribosphenic molar pattern of the Mesozoic mammalian precursors (Fig. 10.1). The tribosphenic pattern appeared in the fossil record in two different lineages of mammals (Luo et al. 2001), the Middle Jurassic australosphenidans (e.g., *Ambondro*, Flynn et al. 1999; *Asfaltomylos*, Martin & Rauhut 2005) and Late Jurassic boreosphenidans (*Juramaia*, Luo et al. 2011). The Southern Hemisphere australosphenidans are considered stem monotremes that have completely reduced their permanent teeth in extant taxa, whereas boreosphenidans comprise all other mammals (metatherians and eutherians) with a striking diversity of dentitions (Luo 2007). Mammals with tribosphenic molars (boreosphenidans or tribosphenidans) are characterized by a two-phased masticatory power stroke with piercing and cutting function during phase I and subsequent grinding and crushing function during newly developed phase II, in contrast to the single-phased power stroke in pretribosphenidans (only phase I present) (Hiiemäe & Kay 1972, 1973, Kay & Hiiemäe 1974, Hiiemäe 1976, 1978). The two-phased power stroke with crushing and grinding function allowed the exploitation of more diverse food sources and likely fostered the great evolutionary success of therian mammals in the Cenozoic. But even before the appearance of tribosphenic molars in mammal evolution, other Mesozoic mammal lineages had already developed remarkably diverse dentitions and dietary specializations, and a second chewing phase seems to have evolved independently in several groups, including docodontans, multituberculates, and gondwanatherians. The selective pressure to enhance the ingestion and mastication of new food sources triggered a number of attempts in mammalian evolutionary history to add a crushing function to the plesiomorphic piercing and cutting action of the molariforms among early mammaliaform lineages, independently from the evolution of the tribosphenic molar. Vivid examples are the Jurassic and Early Cretaceous docodontans that evolved complex molars with crushing functions (e.g., Kermack et al. 1987, Butler 1988, Pfretzschner et al. 2005), suitable for processing diverse diet, and also with disparate molar shapes and very divergent morphologies among docodontan taxa, such as some specialized teeth for piscivory of *Castorocauda* (Ji et al. 2006). Another example is the Late Cretaceous Malagasy gondwanatherian *Vintana* with the first known euhypsodont molars in mammalian evolutionary history with complex enamel folding and enamel islets, suitable for processing tough plant materials (Krause et al. 2014, Schultz et al. 2014).

Within the Deutsche Forschungsgemeinschaft (DFG) Research Unit FOR 771 "Function and performance enhancement in the mammalian dentition – phylogenetic and ontogenetic impact on the masticatory apparatus" several projects were devoted to the functional analysis of Mesozoic mammaliaform molariforms and molars for a better understanding of the evolutionary history of mammalian mastication. 3D investigation techniques such as µCT combined with the Occlusal Fingerprint Analyser (OFA) software (Kullmer et al. 2009, 2020, this volume) that was developed within the FOR 771 made a virtual reconstruction of the masticatory cycle and a quantitative analysis of dental function and occlusal relationships possible. The following chapter provides a review of the morphological diversity of Mesozoic mammalian molariforms and molars and their various functional and dietary adaptations.

Triconodont molariforms – three cusps in a row for piercing and cutting

The most plesiomorphic mammaliaform molariform teeth are of the triconodont type with three primary cusps aligned in a mesio-distal row. This molariform type characterizes stem mammalian Morganucodonta and has been retained as a symplesiomorphic character by crown mammalian Eutriconodonta.

Plesiomorphically, central cusp A/a is the highest, whereas mesial cusp B/b and distal cusp C/c are lower (Fig. 10.2). This molariform type characterizes morganucodontans and "amphilestids" (a paraphyletic assemblage of dentally plesiomorphic eutriconodontans) among Eutriconodonta. In more derived Eutriconodonta, the height difference between the central and mesial respectively distal cusps is much smaller, or they are of equal height, as in the eutriconodontan family Triconodontidae. In derived Eutriconodonta, an elongation of the cutting edges occurs compared to the plesiomorphic condition in Morganucodonta with relatively shorter cutting edges. The

T. Martin & W. v. Koenigswald (eds.): Mammalian Teeth – Form and Function. Pp. 187–214, 22 figs.

DOI: http://doi.org/10.23788/mammteeth.10 Published 22 December 2020

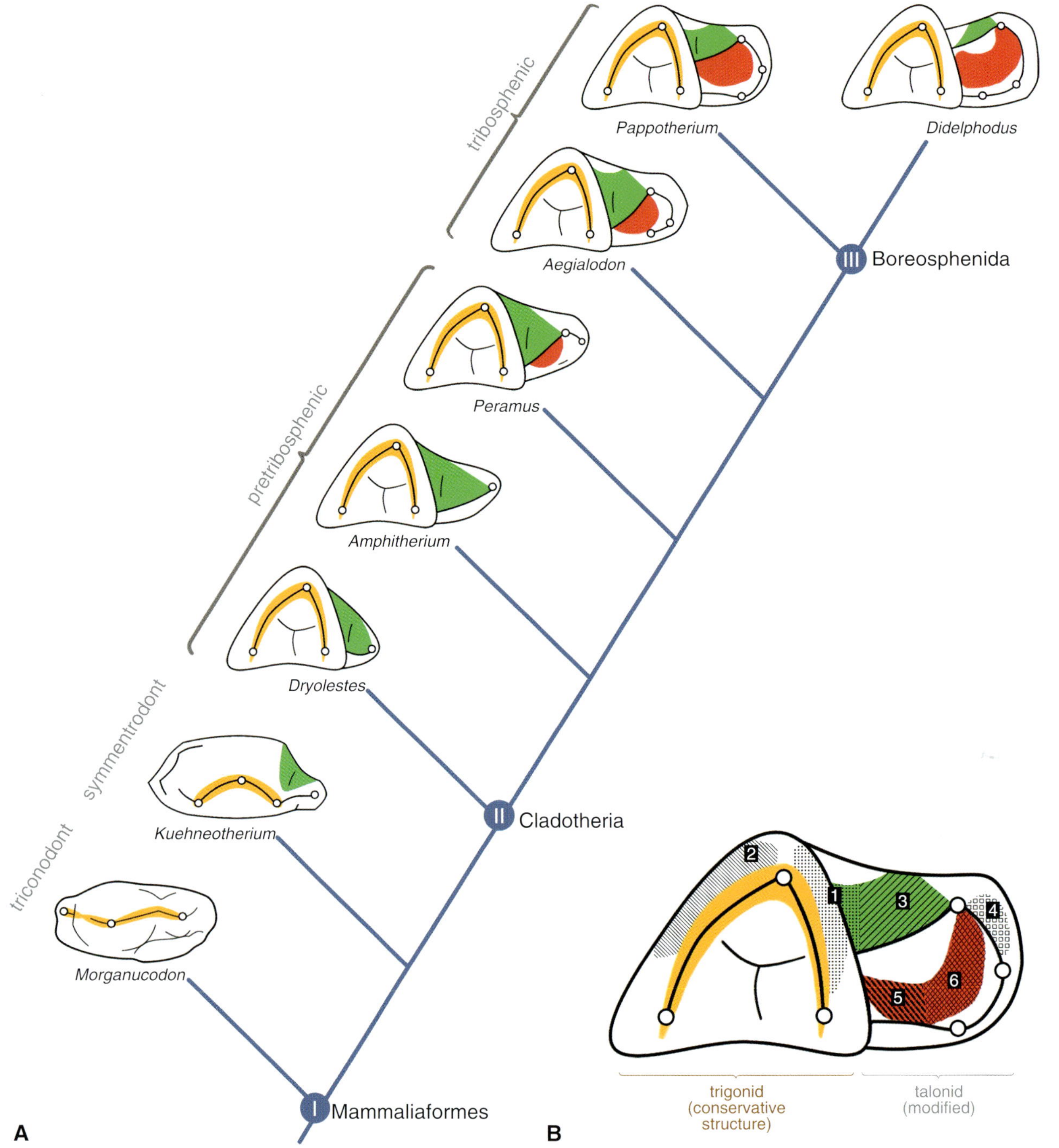

Fig. 10.1. Evolution of the tribosphenic lower molar. **A,** schematic diagram with functional transformation. Yellow, cutting edges of trigonid; green, hypoflexid groove (mainly shearing); red, crushing and grinding area within the (incipient) talonid basin. **B,** lower tribosphenic molar with wear facets (hatched areas); numbers of wear facets after Crompton (1971).

dental function of the triconodont molariform dentition is mainly piercing and cutting with a predominantly orthal jaw movement. Smaller taxa such as the morganucodontans *Morganucodon* and *Megazostrodon* were insectivorous. Based on a microtextural analysis of molariforms, Gill et al. (2014) concluded that *Morganucodon* mainly fed on hard-shelled insects such as coleopterans. Larger taxa such as the eutriconodontan *Gobiconodon* were faunivorous (Jenkins & Schaff 1988), and *Repenomamus*, the largest known Mesozoic mammal, was carnivorous, either scavenging or predatory, or both. Hu et al. (2005) reported on a *Repenomamus* specimen with remains of a juvenile psittacosaur in the stomach region. The morphological diversity of the cusps and the tooth size disparity of the triconodont molars suggest that, even with the same triconodont pattern, there occurred functional differences, leading to a dietary differentiation.

The chewing stroke of *Morganucodon* had been reconstructed as predominantly orthal with a slight movement in anterior or posterior direction (Crompton & Jenkins 1968). Jäger et al. (2019), however, found evidence for a relatively high degree of freedom during the masticatory cycle according to the varying orientations of striations on molars of *Morganucodon*. Due to the lack of guiding structures, the molars of *Morganucodon* have no auto-occlusion (*sensu* Mellet 1985, Evans & Sanson 2006, based on the teeth of

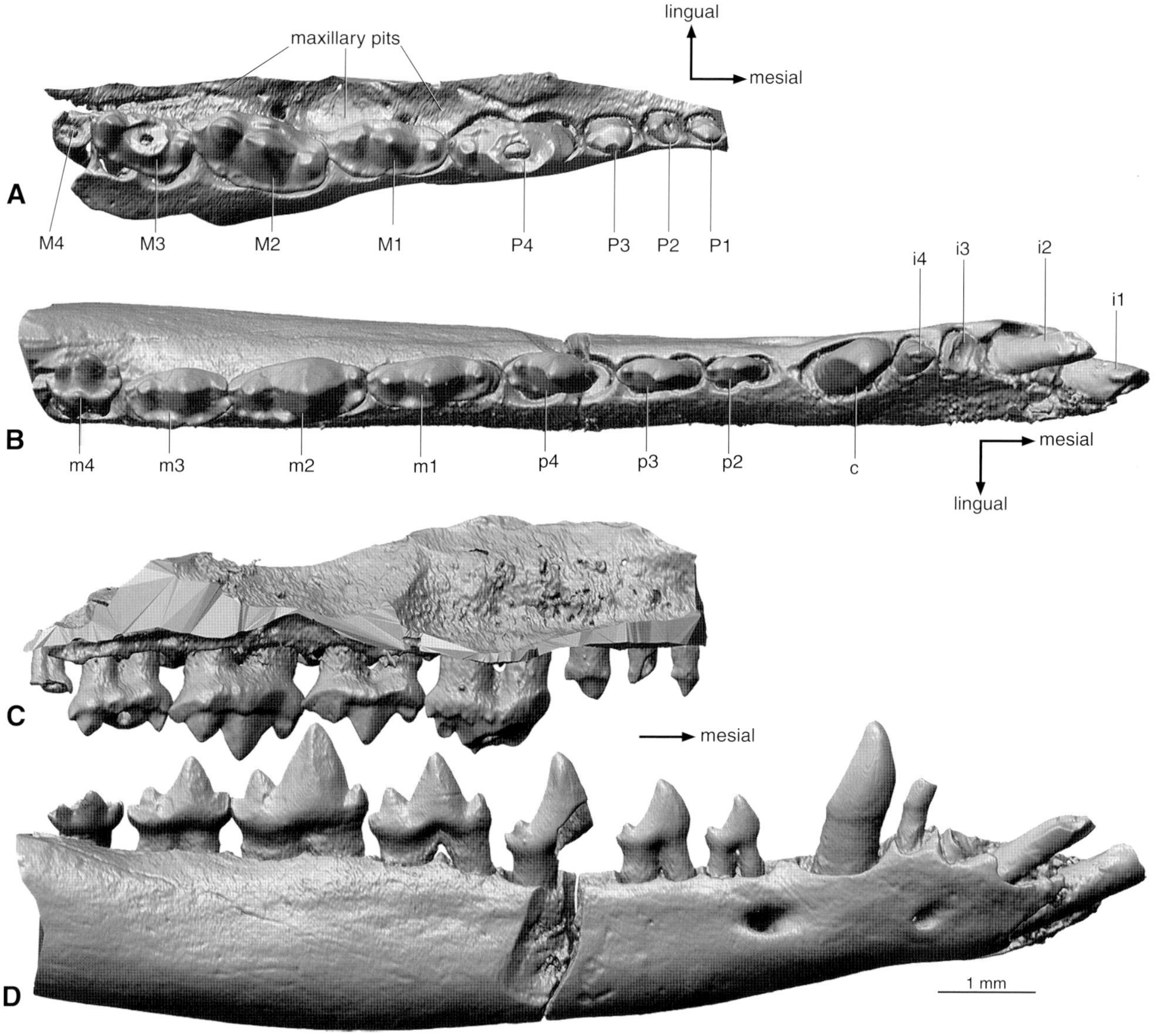

Fig. 10.2. Dentition of *Morganucodon*. **A,** left upper and **B,** left lower tooth row in occlusal view; **C,** left upper tooth row in lingual view and **D,** left lower tooth row in buccal view (mirrored). From Jäger et al. (2019).

Carnivora). Further, it is now demonstrated by OFA analysis that the occlusal movements in *Morganucodon* can have chewing strokes in orthal, distal and mesial orientations, and are more versatile than only the orthal movement as previously emphasized (Jäger et al. 2019). Most individuals apparently developed a preferred chewing direction, which suggests precise muscular control of the jaw movements which had been inferred previously based on the matching wear facets of *Morganucodon* (Crompton & Parker 1978). According to the crown morphology, distribution of wear facets, and the OFA analysis (Jäger et al. 2019), *Morganucodon* and *Megazostrodon* had a single-phased chewing stroke. This study supported the notion that piercing and shear-cutting were the main functions during the masticatory cycle of *Morganucodon* and *Megazostrodon*. According to Jäger et al. (2019), the teeth of *Morganucodon* were fully functional for piercing, shear-cutting, and (to a lesser extent) shearing upon tooth eruption. The extensive wear facets on the teeth in later ontogenetic stages, which were emphasized in earlier wear facet analyses (Crompton & Jenkins 1968, Mills 1971, Crompton 1974), were a byproduct of attrition of tooth-to-tooth contact. The tooth action related to these extensive facets represents only a part of total masticatory functions. Further, the extent of wear facets is dependent on ageing. It takes time for individuals of *Morganucodon* to develop extensive and fully matched wear facets (Crompton & Jenkins 1968, Butler & Clemens 2001), but other dental occlusal functions may start earlier than the formation of extensive wear facets.

Morganucodon and *Megazostrodon* differ in pattern of cusps opposition between the upper and lower molars. *Megazostrodon* is characterized by an embrasure occlusion with lower molar cusp a entering between two upper molars, and upper molar cusp A between two lower molars (Crompton & Jenkins 1968). For *Morganucodon*, a differing occlusal mode was described, with lower molar cusp a occluding between upper molar cusps B and A, and upper molar cusp A occluding between lower molar cusps a and c (one-to-one occlusion) (Crompton & Jenkins 1968, Crompton 1974). Jäger et al. (2019) corroborated the embrasure occlusion for *Megazostrodon* by their OFA analysis. However, in *Morganucodon* the occlusal relationships are

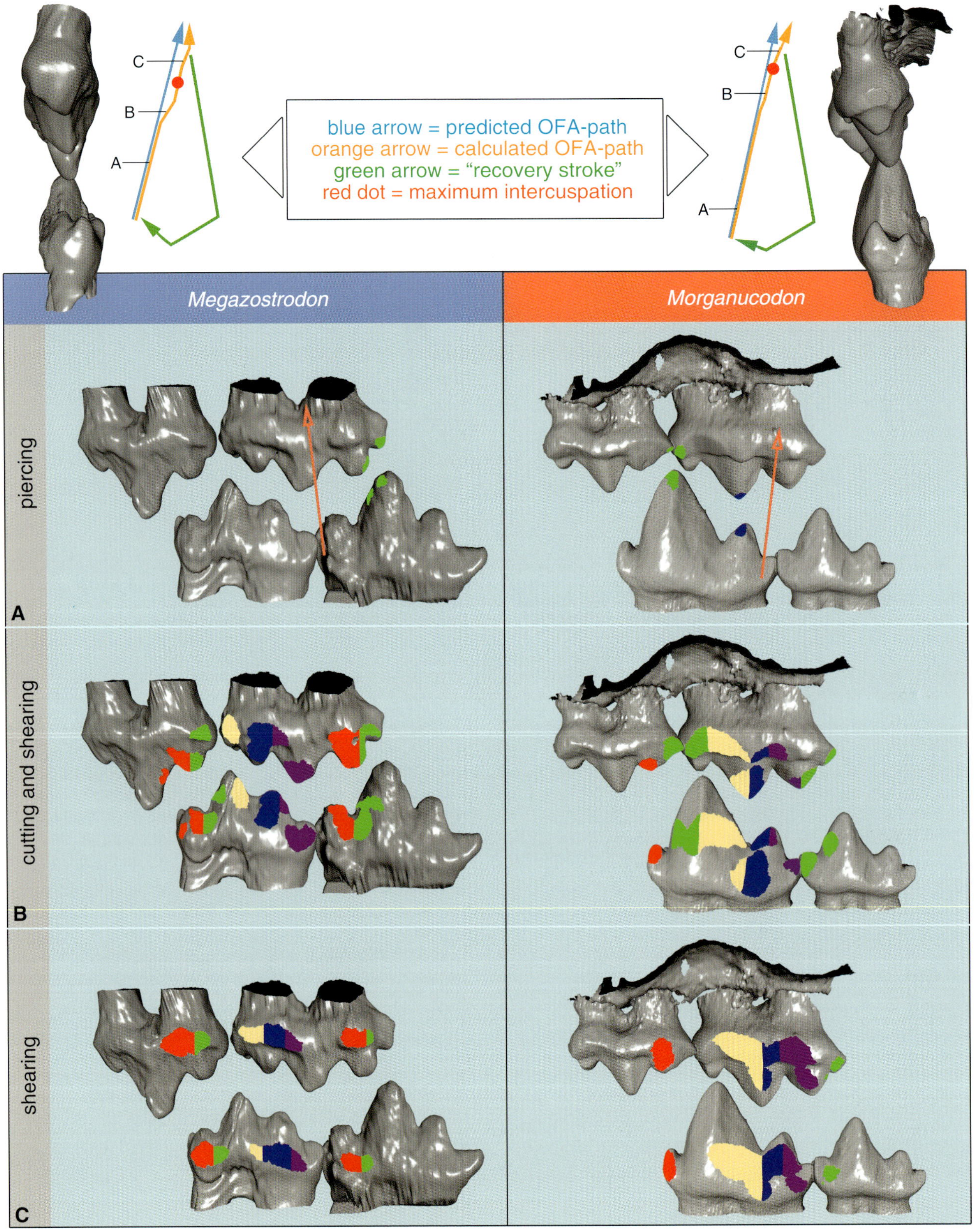

Fig. 10.3. OFA collision contacts of *Megazostrodon* and *Morganucodon* during the masticatory cycle. M1, M2, m2, m3 of *Megazostrodon* and P5, M1, m1, m2 of *Morganucodon*. **A,** initial contact with piercing of food items; **B,** all cusps are in contact, shear cutting occurs along the crests and shearing along the flanks; **C,** maximum intercuspation with pure shearing; lower molars mirrored. Not to scale, mesial to the left. Modified from Jäger et al. (2019).

more variable, by cusp differences among tooth positions and also by stages of wear facet development on the same teeth in opposition (Fig. 10.3). The occlusal pattern is less distinctive between *Morganucodon* and *Megazostrodon* than suggested by Mills (1971), who argued that these two taxa followed two different evolutionary pathways. According to Jäger at al. (2019), the occlusal differences between *Morganucodon* and *Megazostrodon* are not solely

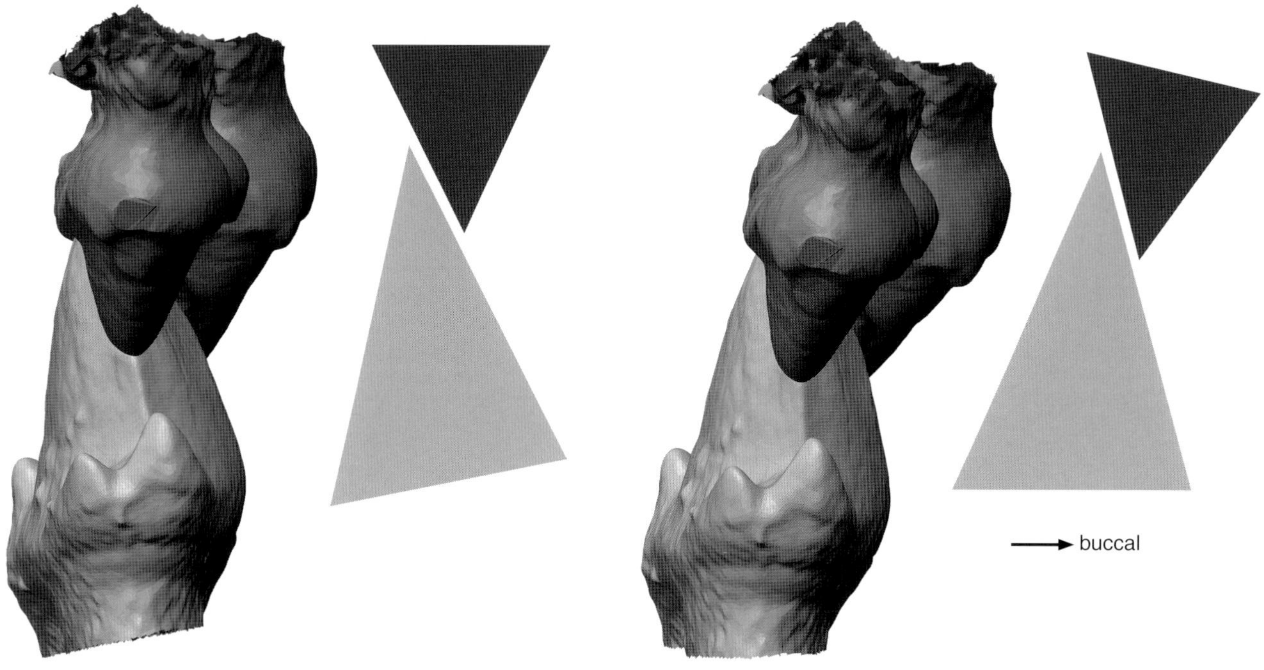

Fig. 10.4. Orientation of upper and lower molars of *Morganucodon* and implications for occlusion. Without inclination of upper molars (**A**) a considerable translation or roll of the mandible would be necessary. The observed inclination of 18° (**B**) enables the lower molars to pass orthally along their antagonists. Not to scale. Modified from Jäger et al. (2019).

caused by molar position as had been suggested earlier (Mills 1971, Butler & Sigogneau-Russell 2016), but also by variation of cusp size and cusp position. Because variations in cusp size and shape require only minor genetic changes (Jernvall 2000, Salazar-Ciudad & Jernvall 2010, Kavanagh et al. 2007, Harjunmaa et al. 2014), a difference of their patterns at this magnitude may not be sufficient to justify the separation of the *Morganucodon* and *Megazostrodon* lineages as suggested by Mills (1971). Other Morganucodonta overlap between the two occlusal modes, such as the megazostrodontids *Dinnetherium* (Crompton & Luo 1993) and *Brachyzostrodon* that have a "*Morganucodon*-like" occlusion, the latter differing by strong apical wear (Hahn et al. 1991, Debuysschere et al. 2015). The occlusion of the morganucodontid *Erythrotherium* closely resembles that of *Megazostrodon* (Crompton 1974, Jenkins et al. 1983, Kielan-Jaworowska et al. 2004), whereas *Bridetherium* has a unique wear pattern, although the molariforms are morphologically very close to those of *Morganucodon* (Clemens 2011). Bhullar et al. (2019) emphasized the importance of roll (rotation of the hemimandible around the long [horizontal] axis of the jaw) for the evolution of precise occlusion in mammals. They observed substantial roll of the hemimandible during the masticatory cycle of extant *Monodelphis* and suggested a "considerable roll" also for *Morganucodon* after the orientation of wear facets. However, the µCT observation of intact teeth in the maxilla demonstrates an inclination of the upper molars (Fig. 10.3), and the OFA analysis by Jäger et al. (2019) shows that the orientation angle of upper tooth wear facets is strongly influenced by this inclination. Therefore it may not be possible to infer a roll of the lower teeth based on the angle of wear facets on the upper molars (e.g., Crompton & Luo 1993). We interpret that it is not possible to differentiate between wear facets caused by transversal movements (Fig. 10.4) or by roll; the role of the latter during the masticatory cycle of *Morganucodon* therefore remains ambiguous.

Within Eutriconodonta, the molars of Triconodontidae are characterized by cusps of equal size (Fig. 10.5). In earlier studies that investigated occlusal relationships in *Trioracodon* it was assumed that the main cusp a of the lower molars occluded in between cusps B and A of the upper antagonists, whereas the upper main cusp A occluded between cups a and c of the lower antagonists (Mills 1971, Crompton 1974) – an occlusal pattern that corresponds to that of *Morganucodon*. The OFA analysis of the holotype specimen of *Priacodon fruitaensis* revealed that the main cusps of upper and lower molars occlude better, if fitting in between two antagonists, thus that its occlusion is different from that of *Morganucodon* (Jäger et al. 2020). Because the dentition of *Priacodon* is very similar to that of other stem Triconodontidae, it can be assumed that this is the general occlusal pattern for all Eutriconodonta. It was previously proposed that there is a considerable rolling of the mandible during the masticatory cycle, on the basis of more oblique orientation of wear facets on the upper molars than the lower molars of triconodontids (Crompton & Luo 1993). We offer a different interpretation – we note that there is an in situ inclination of the upper molars within the maxilla in eutriconodontans as in *Morganucodon* (Fig. 10.4) which is hitherto unnoticed. Taking this inclination of the upper molars into account, the OFA analysis shows a roll rate of the active hemimandible of ~10° during the power stroke of *Priacodon* (Jäger et al. 2020), which is less than previously suggested for eutriconodontans, but already more than the roll rate of the extant marsupial *Monodelphis* (Bhullar et al. 2019). We hypothesize that the roll in *Priacodon* and eutriconodontans was a passive process caused by the inclination of the upper molars (Jäger et al. 2020). The equal height of the molar cusps requires a

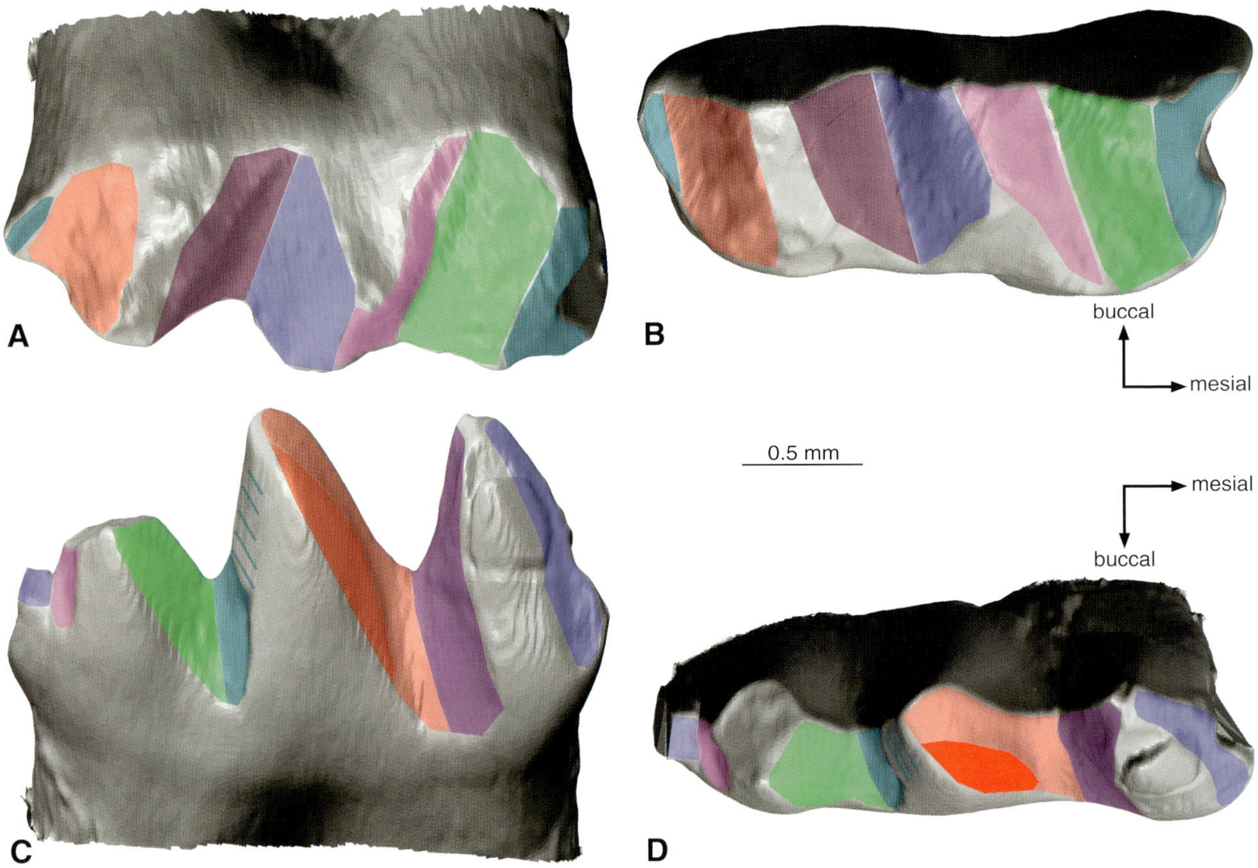

Fig. 10.5. *Priacodon* upper and lower molars with corresponding wear facets. Right M3/m3 of *P. fruitaensis*. **A,** lingual aspect and (**B**) occlusal aspect of M3. **C,** buccal aspect and (**D**) occlusal aspect of m3. M3 is mirrored. Modified from Jäger et al. (2020).

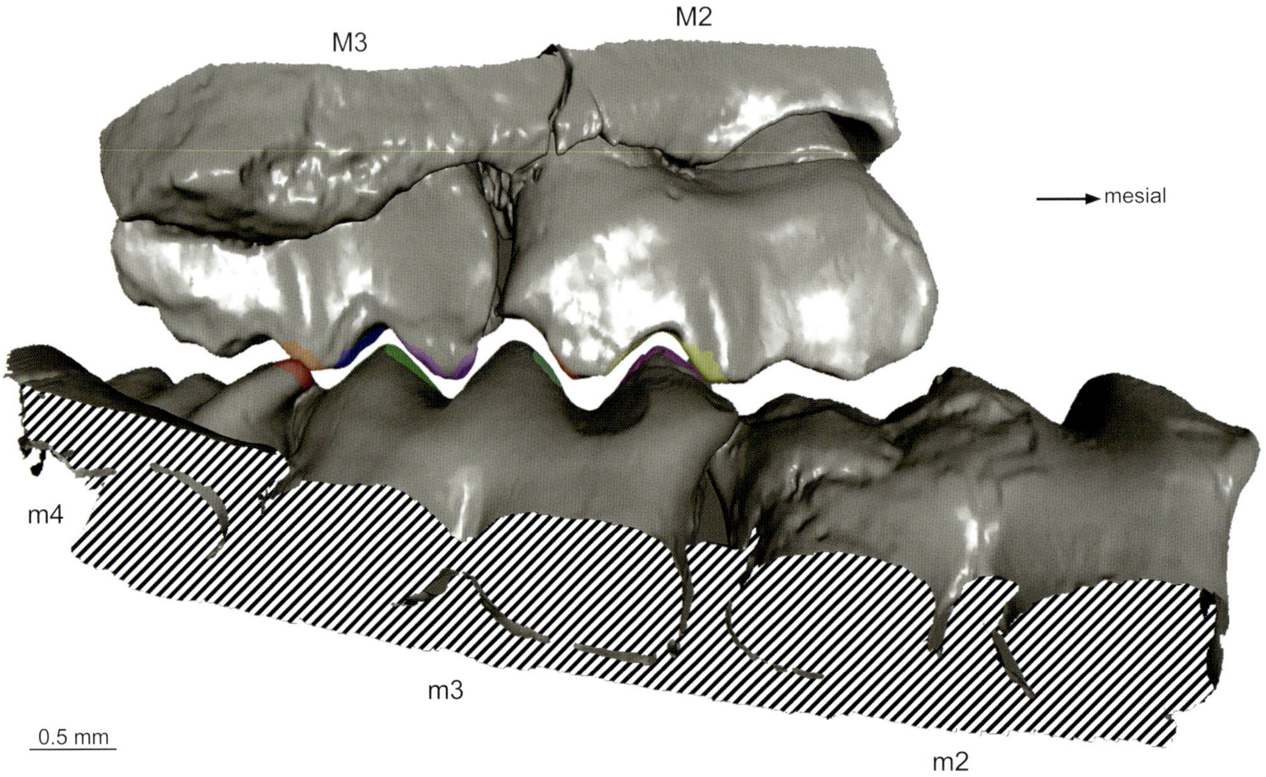

Fig. 10.6. Occlusal pattern of *Priacodon*. Right dentition of *P. fruitaensis* with zig-zag pattern, in the middle of the power stroke. Contact areas indicated by colors. Modified from Jäger et al. (2020).

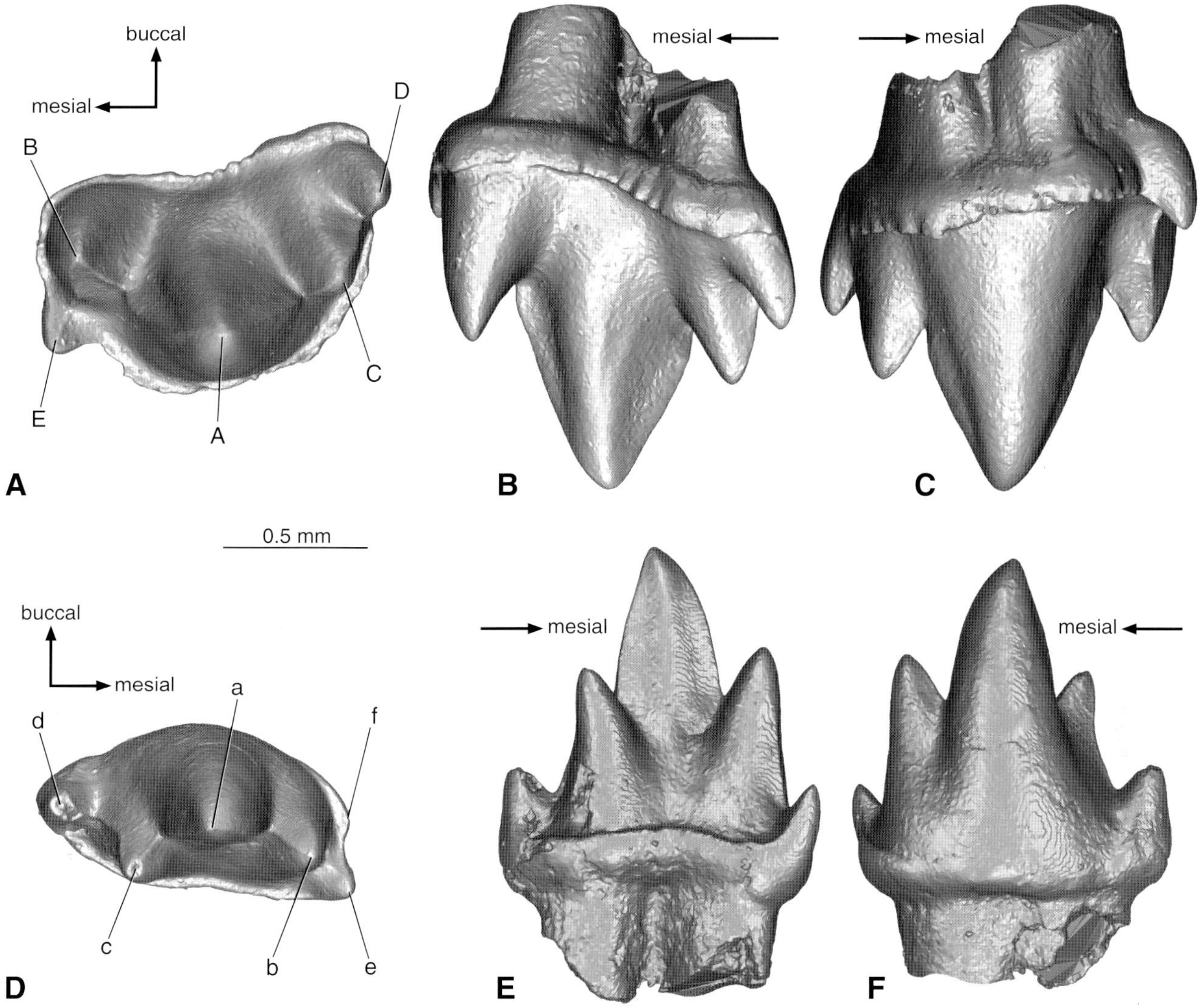

Fig. 10.7. Upper and lower molar of *Kuehneotherium*. Right upper M4 (mirrored) (A, B, C) and left lower m3 (D, E, F) of *K. praecursoris*. **A, D,** occlusal aspect; **B, F,** buccal aspect; **C, E,** lingual aspect. Modified from Plogschties (2020).

precise orientation of upper and lower teeth, because all cusps come into contact in short succession at the beginning of the power stroke. The function of the triconodontid molars can be compared to pinking shears. The molars form a mesio-distally oriented elongated cutting edge with linguo-buccal extensions (zig-zag pattern) (Fig. 10.6). This pattern of triconodontid molars combines the key characters of carnivorous extant mammals (mesio-distal cutting edges) and insectivorous mammals (prominent development of more linguo-buccally oriented cutting edges). From this we infer that triconodontid molars were capable of both carnivorous and insectivorous diets. The sharp cusps of the lower molars are self-sharpening when they occlude into the narrow valleys between the upper molar cusps. In comparison with the carnassials of carnivorans and creodontans, the molars of Triconodontidae are less specialized for cutting meat, and show instead a more generalized function for faunivorous diet (Jäger et al. 2020).

The general shape of the molars of Triconodontidae changed only little during evolution from the Late Jurassic to the Late Cretaceous. The number of molars was augmented, their crown height increased, and cusps D/d became larger which indicates a trend to maximize the cutting edges. The shape of the molars, however, is remarkably homogeneous and changed hardly through time. Apparently the homogenous *bauplan* and precise occlusion of the molars was subject to fitting constraints that prevented major changes such as the development of new cups (Jäger et al. 2020).

"Symmetrodontan" molars – reversed triangle pattern with increased shearing function

The symmetrodont molar pattern represents a more advanced evolutionary grade than the triconodont pattern, and is derived from the triconodont pattern. But Mesozoic mammals with symmetrodont molars do not designate a monophyletic clade. "Symmetrodontan" molars are characterized by angulation of the molar main cusps (A/a, B/b, C/c), and a very small cusp D/d, giving the teeth a symmetric appearance. Upper and lower molars are arranged as alternating reversed triangles, with cusp A in lingual position in the uppers and cusp a in buccal position in the lowers (Fig. 10.7). The "symmetrodontan" chewing cycle was characterized by embrasure occlusion, in which one upper molar occluded between two lowers (two-to-one occlusion, Fig. 10.8). Traditionally, two types

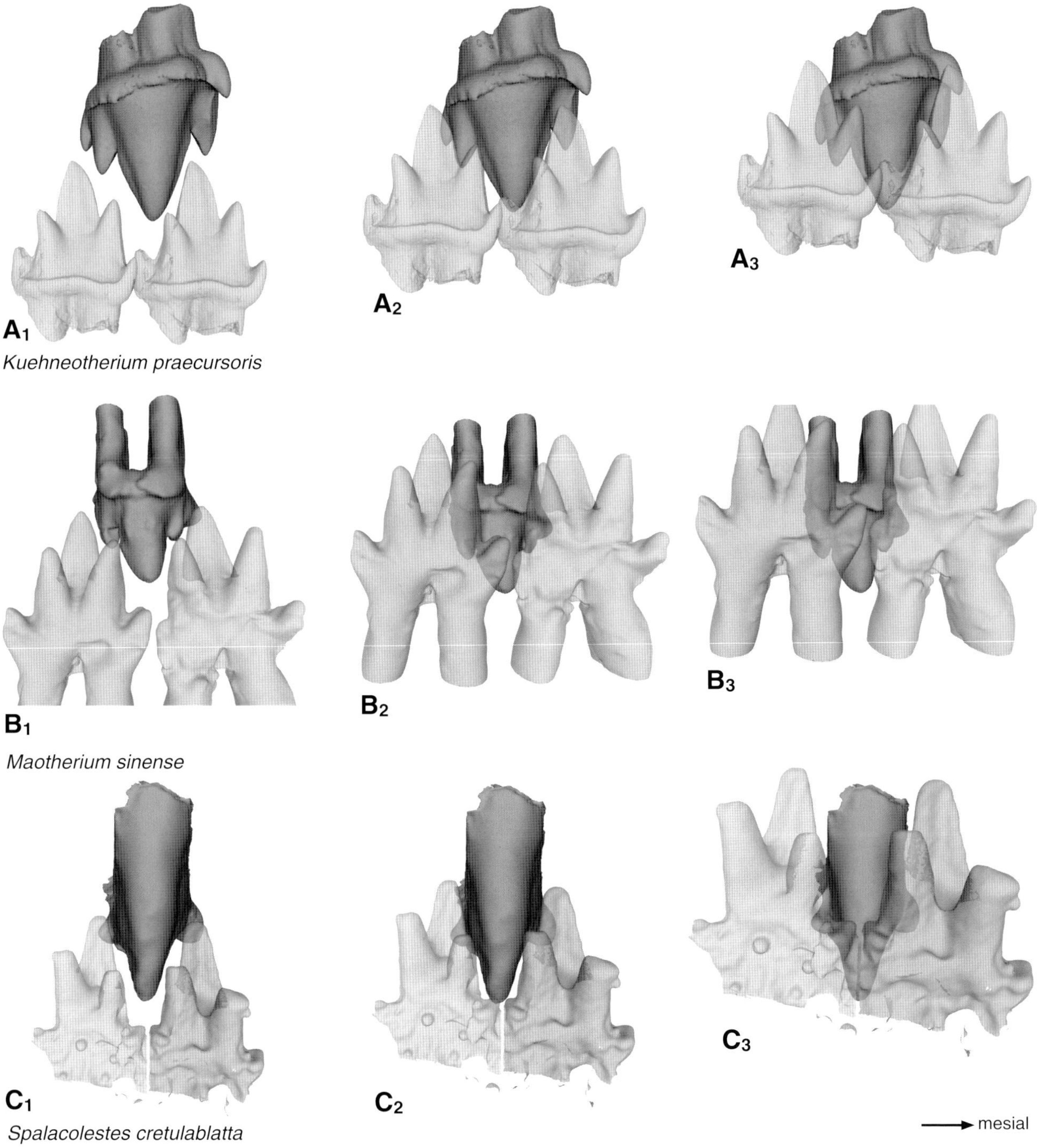

Fig. 10.8. Comparison of occlusal patterns of symmetrodont molars. **A,** *Kuehneotherium praecursoris*; **B,** *Maotherium sinense*; **C,** *Spalacolestes cretulablatta*. Stage 1: beginning of the power stroke; stage 2: closure of the intercuspid spaces; stage 3: end of power stoke, maximum intercuspation. Not to scale. Modified after Plogschties (2020).

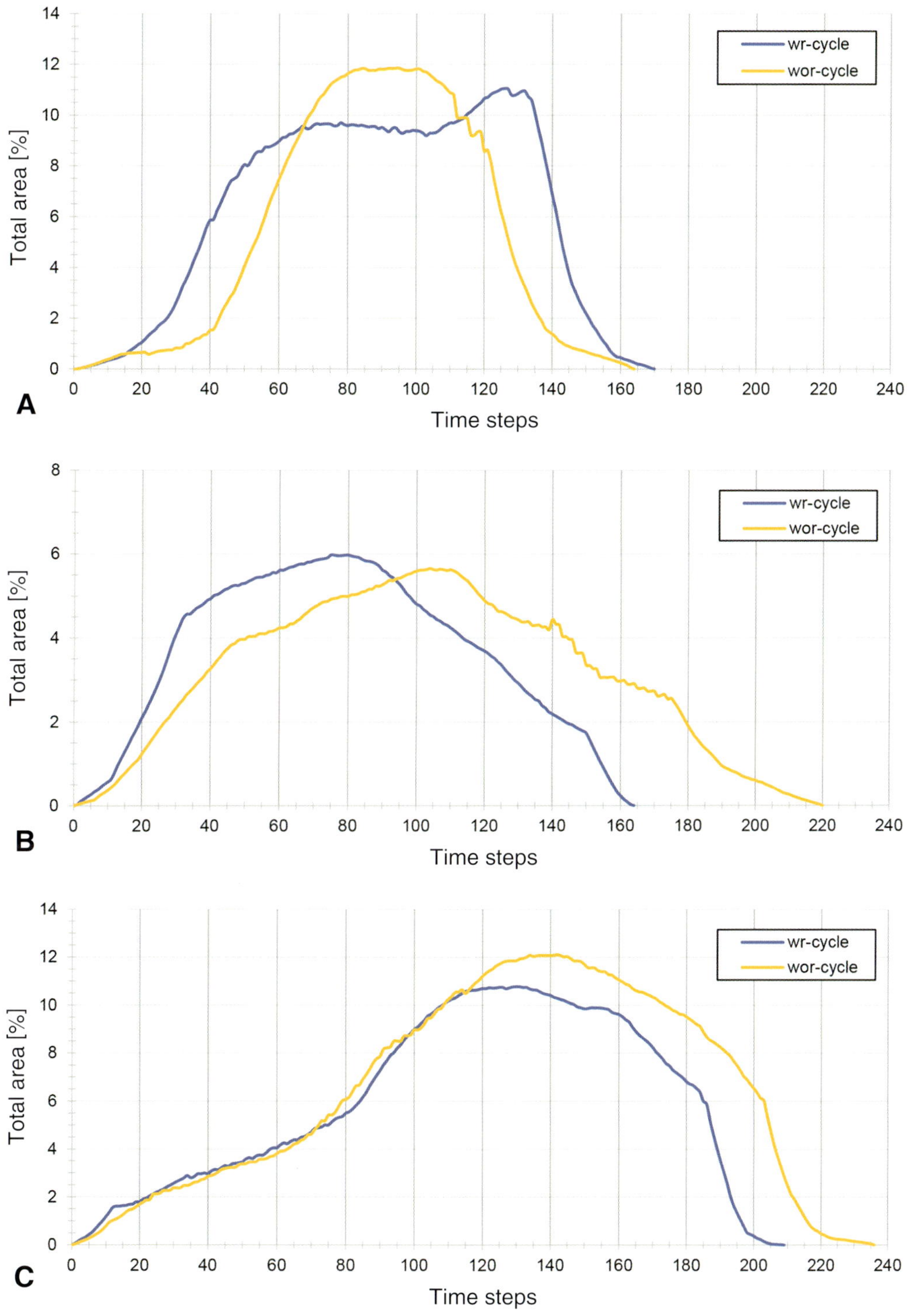

Fig. 10.9. Total collision areas during the masticatory cycle of "symmetrodontans". **A,** *Kuehneotherium praecursoris*; **B,** *Maotherium sinense*; **C,** *Spalacolestes cretulablatta* after the OFA analysis with roll (wr; blue line) and without roll (wor; yellow line) of the mandible. Modified from Plogschties (2020).

of symmetrodont molars have been distinguished, the obtuse angled molars of kuehneotheriids and tinodontids and the acute angled molars of spalacotheriids. In obtuse angled "symmetrodontans", the trigonid angle of lower molars is above 90° (Li & Luo 2006) (80° according to Averianov et al. 2013), whereas it is less in "acute angled symmetrodontans". The best-studied obtuse angled "symmetrodontan" is the early Jurassic *Kuehneotherium*, of which thousands of isolated teeth have been recovered from Welsh fissure fillings (Kermack et al. 1968, Gill 2004, Gill et al. 2014). Plogschties (2020) performed OFA analyses for *Kuehneotherium* with a roll of the hemimandible and without a roll; both chewing cycles are single-phased and differ only slightly (Fig. 10.9). Based on a finite element analysis of the mandible and microtexture analysis of the molars, Gill et al. (2014) concluded that *Kuehneotherium* fed on softer insects (such as lepidopterans) than *Morganucodon* which is also supported by its more slender jaw morphology. Conith et al. (2016) used simplified models of *Morganucodon* and *Kuehneotherium* molar tooth rows

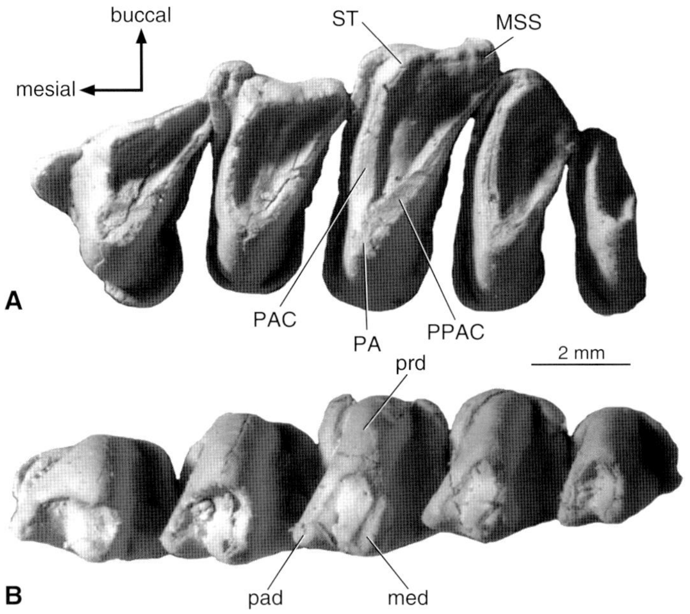

Fig. 10.10. Cheek teeth of the tenrecid *Setifer setosus*. **A,** left P4–M3 (ZMB MAM 44292); **B,** right p4–m3 (ZMB MAM 44293) in occlusal view. Abbreviations: med, metaconid; MSS, mestostyl; PA, paracone; PAC, paracrista; pad, paraconid; PPAC, postparacrista; prd, protoconid; ST, stylocone. Modified from Schwermann (2014).

for physical biting experiments into soft and hard gels as proxies for soft- and hard-shelled insect prey. They found the triangulated molars of *Kuehneotherium* to be more suited to processing soft foods and the blade-like molars of *Morganucodon* to be more suited to processing hard foods, because the blade-like *Morganucodon* cusps penetrated through the hard food items and propagated cracks more easily. More importantly, the triangulated, more complex *Kuehneotherium* molars were found to inflict more damage to the prey. Accordingly, Conith et al. (2016) concluded that the changes from the triconodont to the symmetrodont molar pattern were driven primarily by selection for maximizing damage, and secondarily for maximizing biomechanical efficiency for food property.

Somewhat intermediate between kuehneotheriids and spalacotheriids regarding molar cusp angulation are the Early Cretaceous zhangheotheriids. Plogschties & Martin (2017) and Plogschties (2020) studied the occlusion and masticatory cycles of the zhangheotheriid *Maotherium sinense* and the spalacotheriid *Spalacolestes cretulablatta* by OFA. The chewing cycles of both taxa are single-phased, and the movement of the mandible is mainly orthal with a medial shift that allowed for a deep intercuspation. The orientation of the striations on the wear facets is more variable in *M. sinense* than in *S. cretulablatta*, indicating a less precise guidance of the molars in *M. sinense*. According to the OFA-analysis with one upper and two lower molars, the maximum contact area during the chewing cycle is almost two times larger for *S. cretulablatta* than for *M. sinense* (Fig. 10.9). At the beginning of the chewing cycle, the prey was fixed, pierced, and puncture-crushed by the main cusps of the upper and lower molars. Subsequently, *M. sinense* mainly stretched and sheared the food particles between the blunt crests and consecutively between the molar flanks. In *S. cretulablatta* (and spalacotheriids in general), the newly developed sharp crests that worked like a cigar cutter with two concave blades, allowed for an increased shear-cutting function. The prey was first cut with the molar crests and then sheared between the molar flanks (prevallid/postvallum and postvallid/prevallum). The second shearing process of zhangheotheriids, in which the indentation of the lower molar main cusps (protoconid-paraconid and protoconid-metaconid) and the lingual cingulum of the upper molars were involved, was replaced by a more efficient prevallum/prevallid and postvallum/postvallid crushing and shearing complex, a more extensive tooth-to-tooth contact and a more accurate guidance of the teeth in spalacotheriids. The development of circumferential cingulids provided additional guidance and crushing function against the paracone (Plogschties 2020).

In spalacotheriids ("acute angled symmetrodontans"), the angles of the upper molar primary trigon and of the lower molar trigonid are much smaller than in the other "symmetrodontans". Spalacotheriids have a higher molar count (five or more) than the other "symmetrodontans", and their molars are more strongly shortened mesiodistally and widened linguo-buccally (zalambdodonty). The increase in number of molars and the acute angulation of the primary trigon and trigonid are interpreted as a strong insectivorous specialization ("hyperinsectivory"), as seen in extant insectivorous tenrecids (Afrosoricida).

A functional analogue for spalacotheriid molars are the molars of certain extant tenrecs such as *Setifer* or *Tenrec* which have been studied by Schwermann (2014). Tenrecs are placental mammals (afrotherians) with zalambdodont molars that have secondarily lost the protocone and other cusps on the uppers, and reduced the talonid to a single cusp at the lowers (Fig. 10.10); this way the molars (especially the uppers) resemble the pretribosphenic morphology of cladotherians such as dryolestidans (Asher & Sánchez-Villagra 2005) but also of trechnotherian spalacotheriids (the term pretribosphenic is here used for molars with acute angulation of main cusps and lacking a protocone). In upper molars of *Setifer* the trigon comprises the lingually placed paracone and the buccally situated stylocone and meta-

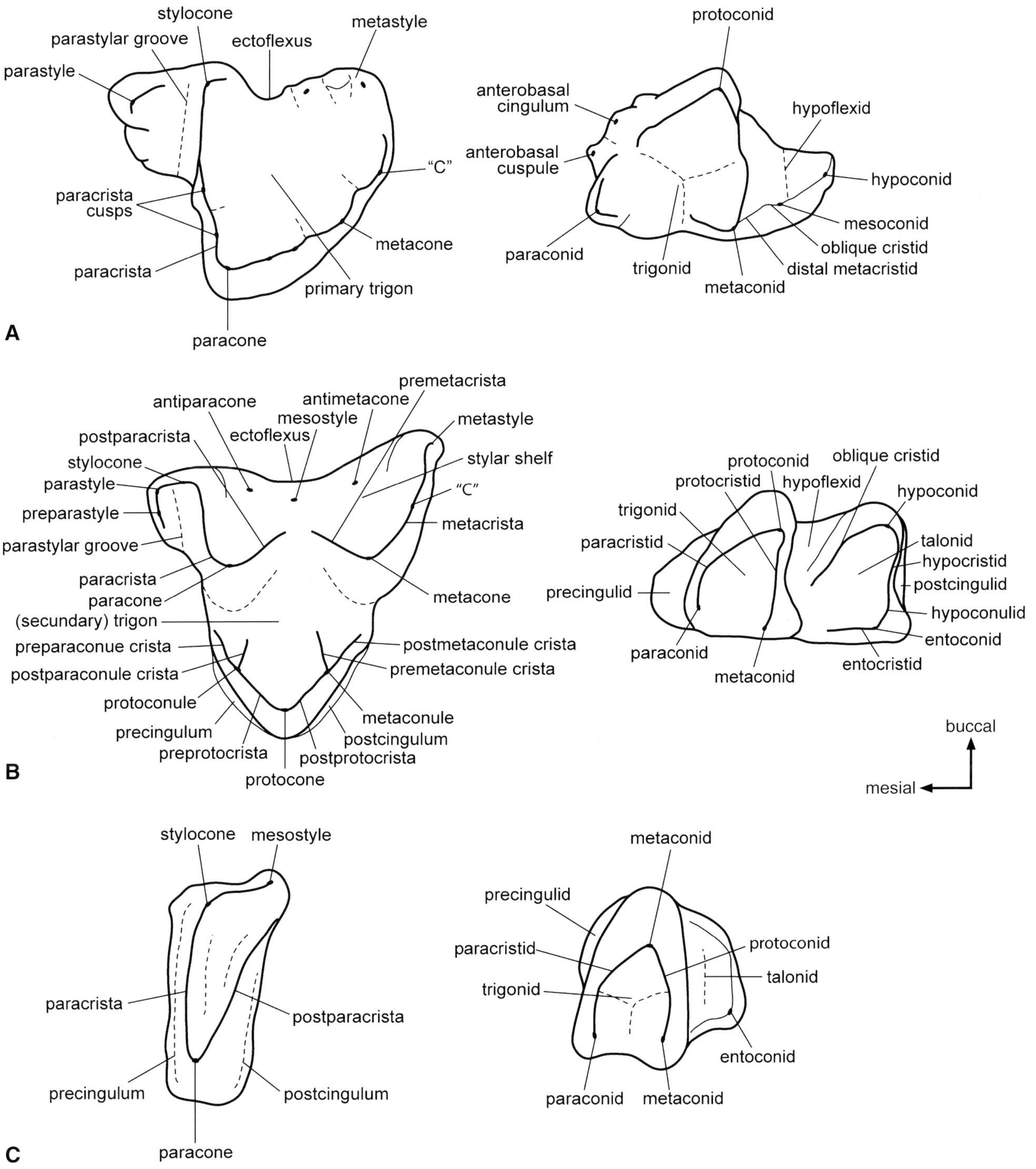

Fig. 10.11. Pretribosphenic and tribosphenic molar structure terminology. Upper (left) and lower (right) molars of **A,** pretribosphenic *Nanolestes*, **B,** of a generalized tribosphenid, **C,** of *Setifer*. Not to scale. Modified from Schwermann (2014).

style, with the mesial paracrista and distal postparacrista as connecting shear-cutting edges (Fig. 10.11). Protocone and metacone are fully reduced, and the tooth crown is shortened mesio-lingually and widened linguo-buccally, closely resembling the shape of spalacotheriid upper molars. The lower molars are likewise mesio-distally short and linguo-buccally wide and comprise the trigonid and a small talonid with a lingually placed single cusp (entoconid according to Thenius 1989). Schwermann (2014) analyzed the chewing stroke of *Setifer* by OFA and found a single-phased power stroke that ends when the paracone occludes with the talonid (centric occlusion) (Fig. 10.12). This is a general difference to pretribosphenic molars, which lack structures for a centric occlusion (*sensu* Schwermann 2014). The curved preparacrista and protocristid, respectively postparacrista and paracristid glide past each other and the space between is narrowed like in a cigar cutter while the food items are shear-cut. The second phase of the power stroke which is characteristic for most boreosphenidans is not present in *Setifer* and other tenrecs.

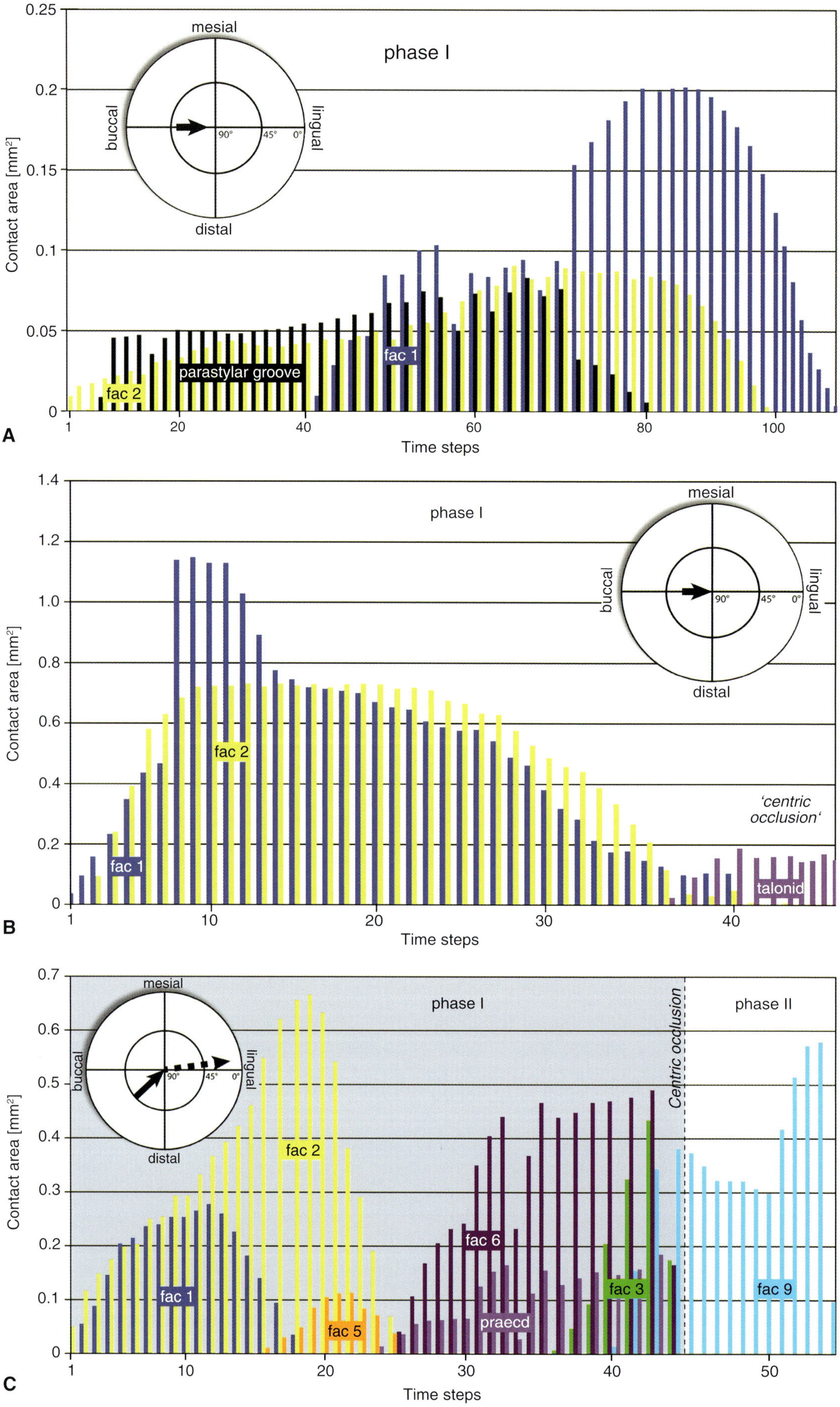

Fig. 10.12. Contact areas during the masticatory cycles of pretribosphenic and tribosphenic molars. **A,** *Nanolestes* (pretribosphenic); **B,** *Setifer* (tribosphenic [zalambdodont]); **C,** *Didelphis* (tribosphenic). Modified from Schwermann (2014). Abbreviations: fac, facet; praecd, precingulid.

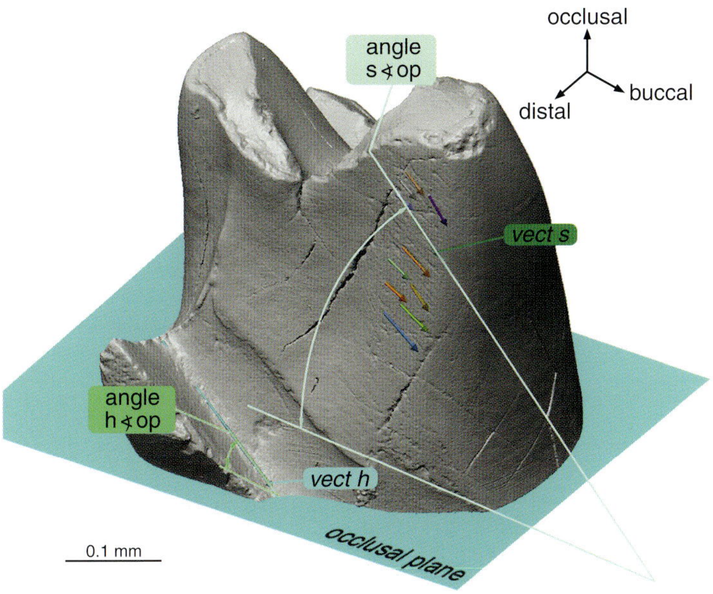

Fig. 10.13. Wear striation patterns on 3D surface model of lower molar of the dryolestid *Dryolestes leiriensis*. Wear in the hypoflexid groove and striations on facet 1 are caused by the paracrista of the upper antagonist. Molar main cusps with apical wear. The angle (∢) between the hypoflexid groove (h) and occlusal plane (op) differs slightly from the angle (∢) between the striations (s) and the occlusal plane (op). Directions of striations and inclination of the hypoflexid groove are traced by vectors (vect). Modified from Schultz & Martin (2011).

Pretribosphenic molars

The next functional step in mammalian molar evolution is represented by the pretribosphenic (*sensu lato* for acute angled molars without protocone and talonid basin) molars of dryolestidans (Cladotheria). The dryolestidan upper molar consists of the "primary trigon" which is not homologous to the trigon of tribosphenidans, because it lacks a protocone (Fig. 10.11A) which is a neomorphic cusp of tribosphenic molars. The "primary trigon" comprises paracone (cusp A), metacone (cusp C), and stylocone (cusp B) which are connected via the mesial paracrista and the distal metacrista. The parastyle (cusp E) is prominent and bulging, the metastyle (cusp D) generally less prominent, and on the buccal portion of the metacrista a cusp "C" of variable size is present which is a neomorph according to Davis (2011: 233). The lower molar trigonid consists of protoconid (cusp a), paraconid (cusp b), and metaconid (cusp c) and is homologous to that of boreosphenidans (Figs. 11A, 13). The mesial paracristid connects protoconid and metaconid, and the distal protocristid connects protoconid and metaconid (Martin 1999). At the mesial side, mesiolingual cuspule e and mesiobuccal cuspule f are variably present. Dryolestidans have a single-cusped talonid (cusp d, homologous either to hypoconulid or hypoconid, for discussion see Davis 2011) and a well-developed hypoflexid that is inclined in buccal direction. Like the dentition of "symmetrodontans", the dryolestidan dentition is characterized by an embrasure occlusion with the principal cusps of the upper and lower molars arranged in reversed triangles (Schultz & Martin 2011). The shearing function occurs on two almost vertically oriented shearing planes with sharp crests, on the mesial and distal sides of the triangles (Crompton 1971). This occlusal type has been described as embrasure shearing (Crompton et al. 1994). Dryolestidans developed new shearing surfaces which are not present in "symmetrodontans". These occur along the hypoflexid groove of the unicuspid talonid in the lower molars, and in the upper molars on the mesial side of the paracone (Schultz & Martin 2011, 2014). During mastication, the paracone travels along the hypoflexid groove where the main shearing occurs; due to the low inclination of the guiding groove, there is also a certain crushing component. The power stroke ends when the paracone reaches the buccal end of the hypoflexid and thus, without a point of centric occlusion. According to Schultz & Martin (2011), the food particles are first cut at the crests of facets 1 and 2 and then sheared and transported in cervical direction into the hypoflexid guiding groove. The inclination of the shearing surface and the change of transport direction crush the particles when they touch the upper part of the hypoflexid groove (Fig. 10.13). Schultz & Martin (2011, 2014) stated that only shearing and no grinding occurs when the paracone travels along the hypoflexid groove, because the chewing forces do not change their direction during the downward movement in the groove. The chewing stroke in dryolestidans is single-phased (only phase I present). The OFA analysis of *Dryolestes* revealed that the power stroke starts with a fast increase of contact area between upper and lower molars reaching maximum size when the teeth occlude deepest into the interdental spaces of the lowers, ending soon after maximal intercuspation and without centric occlusion (Fig. 10.14). In the state of maximal intercuspation the steep shearing surfaces of primary trigon and trigonid are in full contact. After maximal intercuspation, the recovery stroke begins with lowering the lower jaw and all contact is lost at once (Schultz & Martin 2011, 2014).

In comparison, in the pretribosphenic stem zatherian *Nanolestes* with elongated single-cusped talonid (Martin

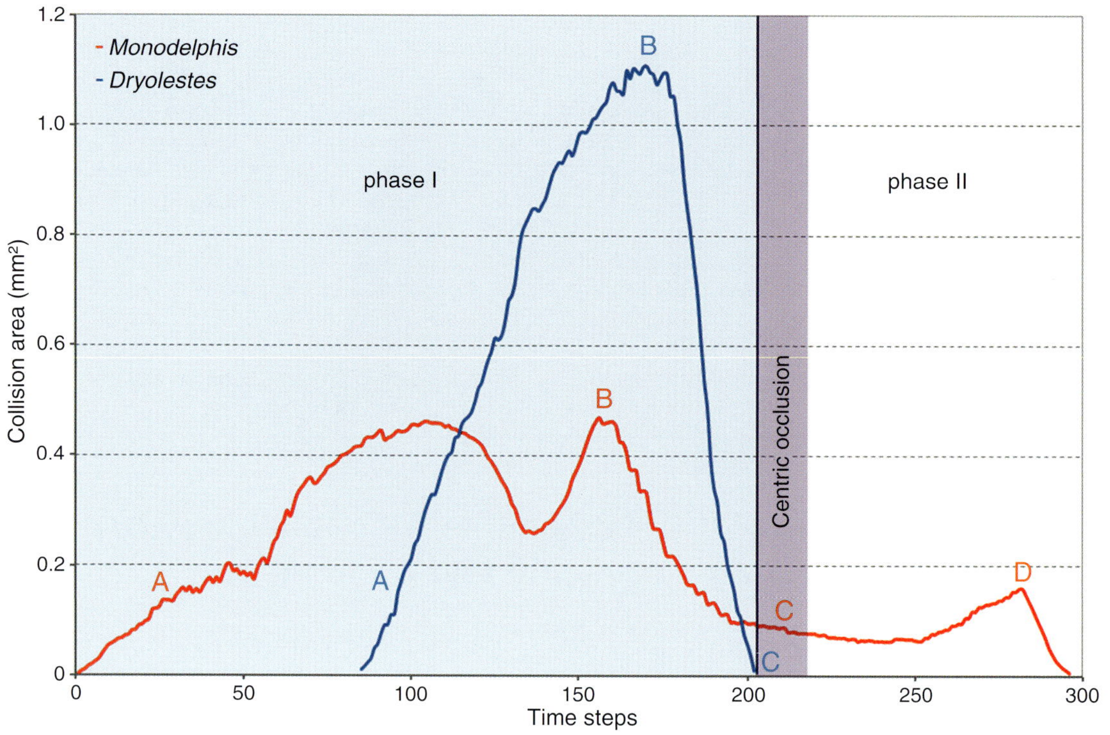

Fig. 10.14. Comparison of collisional molar contacts of Jurassic *Dryolestes* and extant *Monodelphis* after OFA detection. Pretribosphenic *Dryolestes* with single-phased power stroke ending with maximum intercuspation. Power stroke of tribosphenic *Monodelphis* is two-phased with phase II occurring after centric occlusion when the protocone moves out of the talonid basin. Abbreviations: A, entrance phase; B, maximum intercuspation; C, centric occlusion; D, phase II occlusion. Modified from Schultz & Martin (2014).

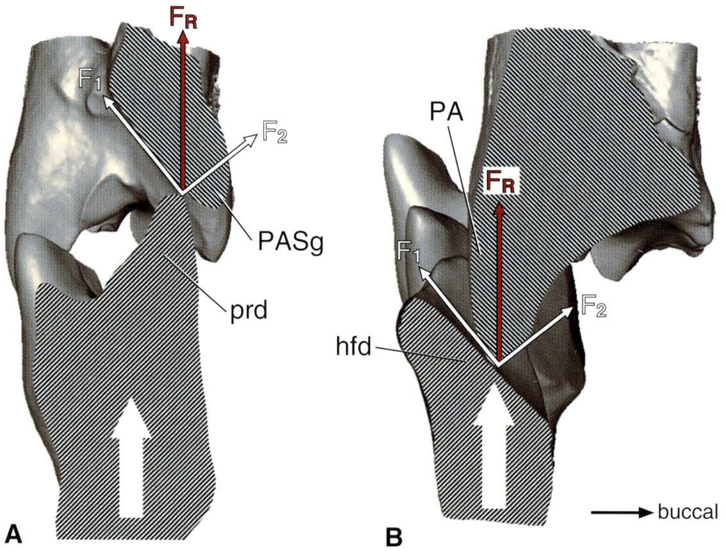

Fig. 10.15. Contact points and inferred bite forces during mastication of *Nanolestes*. Bite forces are identical in the (**A**) first and (**B**) second part of the occlusal movement. In the first part the protoconid and parastylar groove are in contact, in the second part the hypoflexid groove slides across the paracone. Abbreviations: F1, deflection force; F2, pressing force; FR, resulting bite force; hfd, hypoflexid; prd, protoconid; PA, paracone; PASg, parastylar groove. Not to scale. Modified from Schwermann (2014).

2002) (Fig. 10.11A), the parastylar groove of the upper molars is the guiding structure for the protoconid during the first part of phase I (Schwermann 2014, Schwermann et al. 2014) (Fig. 10.12A). In the second part of phase I the guiding of the chewing motion is performed by the hypoflexid groove in which the paracone travels along, just like in dryolestidans. This second part of phase I shows a considerable increase in contact area in the OFA simulation (Fig. 10.12A) which is reflected in extensive wear in the hypoflexid groove (Fig. 10.15B). The orthal bite force is differentiated into partial forces, which press the teeth vertically against each other and deflect the lower jaw. Due to the steep orientation of the parastylar and hypoflexid grooves, the deflection force is higher than the pressing force, resulting in a more powerful deflection of the lower jaw than occlusal biting force between the antagonistic molars (Fig. 10.15). Like in the other pretribosphenidans *sensu lato*, in *Nanolestes* phase I ends without a definitive point of centric occlusion, and a phase II is not present (Fig. 10.12A).

Due to their thin enamel, the dryolestidan molars exhibit a characteristic wear pattern with extensive exposure of dentin along the lower molar shearing edges, paracristid and protocristid (Schultz & Martin 2011, 2014). The striae on the facets are more steeply inclined near the protoconid tip and the protocristid edge (about 45°) than the more cervical striae (about 35°) (Fig. 10.13). The lower dentition is raised at about 45° at the beginning of the dryolestidan chewing cycle, and the first occlusal contact occurs between the v-shaped protocristid and crescent antagonistic paracrista (Schultz & Martin 2014). Protocristid and paracrista function like the blades of a cigar cutter when they pass each other and perform a shear-cutting function. According to Schultz & Martin (2014), the term "shear-cutting" is more appropriate than "cutting" for the function performed along the edges, because cutting is either defined by two blades that meet along their entire length or by a single blade that is pressed vertically against a horizontal surface (Lucas 1979, Sibbing 1991). The contact between paracristid and metacrista occurs slightly later.

Transition from pretribosphenic to tribosphenic pattern – towards crushing and grinding

During the transition from the pretribosphenic to the tribosphenic condition a fundamental morphological and functional transformation occurred (Crompton 1971, Davis 2011, Schultz & Martin 2014, Grossnickle 2020). In the upper molars, the neomorphic protocone developed in a lingual position, and the talonid basin formed at the distal side of the lower molars, surrounded by the three talonid cusps hypoconulid, hypoconid, and entoconid (Fig. 10.1). The hypoflexid groove lost its function as sole shearing surface and became part of the buccal rim of the talonid basin. The power stroke (and the contact phase) were prolonged by the addition of a second phase that occurred after occlusion of the protocone into the talonid basin (centric occlusion); Kay & Hiiemäe (1974) suggested that primitive tribosphenic taxa only had one phase, and that phase II convergently evolved in multiple therian lineages. Phase I is an upward, anterior, and medial movement of the lower molars, terminating in centric occlusion (here equivalent to maximum intercuspation) followed by phase II with an anterior, medial, and downward movement, followed by the recovery stroke (Hiiemäe & Crompton 1971, Hiiemäe 1976, 1978, Schultz & Martin 2014) (Fig. 10.14). Functionally, the development of a second phase to the chewing cycle added crushing and grinding function to the shear cutting function of phase I, enabling boreosphenidans to exploit a much wider range of food. An OFA study of the molars of extant *Monodelphis* documented phase II (Fig. 10.16) by detecting collisional contacts on the buccal wall within the talonid basin which occurred after a directional change in the chewing path, which is in accordance with the striations found on the surface (Schultz & Martin 2014; see also Hielscher et al. 2015 for bats). The additional collision contacts of the protocone and the talonid basin during phase II provide an increased potential for food reduction. Although it had been argued that the occlusal forces are much lower during phase II (Hiiemäe 1984, Hylander et al. 1987) compared to phase I (Wall et al. 2006), the observed small contact areas of phase II indicate a more versatile way of food comminution (Schultz & Martin 2014) which is dominated by food-to-tooth contacts and abrasive wear (Schwermann 2014). Although phase II added only a moderate increase of contact area to the chewing cycle in the tribosphenic pattern as exemplified by *Monodelphis* (Fig. 10.16), it was a crucial precondition for the modifications of transverse jaw movement in more derived boreosphenidans. According to Spoutil et al. (2010) the (functionally) basic pattern of tribosphenic molars is retained almost unmodified in many insectivorous mammals such as afrosoricidans, macroscelideans, and lipotyphlans. A number of studies demonstrated that the crushing function is important for the comminution of hard insect exoskeletons (Fraenkel & Rudall 1940; Freeman 1979, 1981; Moore & Sanson 1995; Evans & Sanson 1998, 2005a, b). The OFA analysis of the pretribosphenic dryolestidan molars shows mainly piercing and cutting function, with lesser ability to breakdown insect exoskeletons than found for tribosphenic molars (Schultz & Martin 2014). The tribosphenic molars with crushing function squeezed out the soft tissue of the insect prey in addition to sectioning the insect bodies into pieces, and thus allowed a more efficient extraction of the nutritious substances resulting in a higher energy gain in the digestive system. Schwermann (2014) conducted comparative feeding experiments with the didelphid *Monodelphis* and the scandentian *Tupaia* with molars resembling the tribosphenic groundplan on one side, and the tenrecid *Setifer* with zalambdodont molars (Fig. 10.10) resembling the dryolestidan dentition (reduced protocone and talonid basin) on the other. By analyzing feces Schwermann (2014) found some evidence that the taxa with tribosphenic molars broke the chitinous exoskeleton of mealworms into smaller pieces than *Setifer*. If corroborated, this would support the hypothesis that tribosphenic molars perform a better food breakup in comparison to zalambdodont or pretribosphenic molars. Throughout most of the Cretaceous, tribosphenids remained remarkably conservative in molar morphology and stayed mostly insectivorous (Kielan-Jaworowska et al. 2004, Martin 2018). In the latest Cretaceous, several lineages developed carnivorous dental adaptations such as some large stagodontids (body sizes up to 2000 g, Gordon 2003) and *Nanocuris* (Wilson & Riedel 2010) among metatherians, and the eutherian *Altacreodus* (Lillegraven 1969). The metatherian *Didelphodon* has large and bulbous third upper and lower premolars that indicate a durophagous specialization (Clemens 1968, Wilson et al.

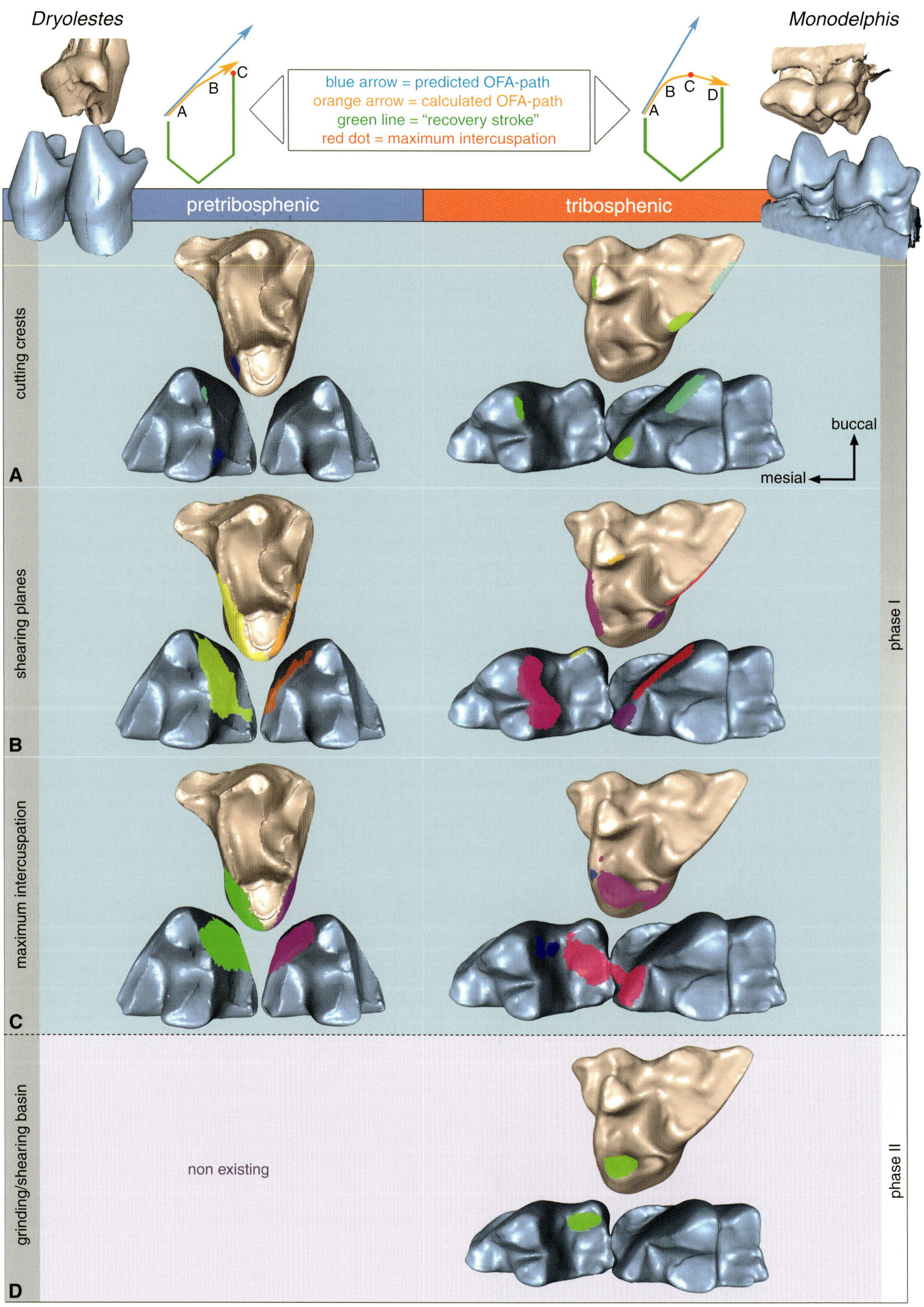

Fig. 10.16. Collision areas during the power stroke of pretribosphenic (*Dryolestes*) and tribosphenic (*Monodelphis*) molars as detected by OFA analysis. Not to scale. From Schultz & Martin 2014.

2016, Cohen 2017). Others evolved adaptations towards omnivory/herbivory, such as the metatherian *Glasbius* (Wilson 2013) and the eutherian taeniodont *Schowalteria* (Fox & Naylor 2003), indicating parallel radiations and ecomorphological diversifications in metatherians and eutherians in the Late Cretaceous. This also probably occurred in parallel with ecological diversifications of other mammal groups (Grossnickle et al. 2019), like the meridiolestidan *Reigitherium* (Harper et al. 2018), cimolodontan multituberculates (Wilson et al. 2012; Grossnickle & Polly 2013), and gondwanatherians (Krause et al. 2014). Therians were likely diversifying ecologically prior to the K-Pg boundary (Grossnickle & Newham 2016), and their dietary flexibility was fostered by the phase II occlusion which allowed an increase of the crushing and grinding function.

"Pseudotribosphenic" molars – convergent evolution of crushing and grinding structures

Molars with protocone-like and talonid-like structures ("pseudotribosphenic" molars) developed several times independently in mammalian evolutionary history (see Martin 2018 and references therein). The most prominent example are the Australosphenida of the Southern Hemisphere. Discovery of the first australosphenidan teeth (Rich et al. 1997, 1999; Flynn et al. 1999) led to the hypothesis of a dual origin of tribosphenic molars in the Northern and Southern Hemispheres (Luo et al. 2001). Australosphenidans such as *Asfaltomylos* and *Hesnoferus* from the Middle to Late Jurassic Cañadon Asfalto Formation of Argentina possess a fully basined and distally situated talonid or "pseudotalonid" structure in the lower molars (Rauhut et al. 2002, Martin & Rauhut 2005, Rougier et al. 2007). However, talonid wear facets occur only on the top of the talonid rim (apical wear) and not within the basin as in tribosphenic molars of Northern Hemisphere boreosphenidans (Fig. 10.17). Upper molars of australosphenidans have not been published yet, but an undescribed isolated upper molar of *Hesnoferus* possesses a linguo-buccally compressed and hook-like protocone or "pseudoprotocone" (G. W. Rougier, personal communication). Martin & Rauhut (2005) hypothesized that the tip of the protocone of *Asfaltomylos* occluded on the buccal side of the talonid and not into the talonid basin. If so, the crushing function occurred between the lingual flank of the protocone and the apices of the talonid cusps, respectively the talonid rim. According to Martin & Rauhut (2005), the same apical wear pattern is evident on the mandibular fragment of the australosphenidan *Ambondro* from the Middle Jurassic of Madagascar (Flynn et al. 1999).

Well-developed basined talonids are also present in the Australian australosphenidans *Ausktribosphenos* and *Bishops* (Rich et al. 1999, 2001a). Like in the South American australosphenidans and Malagasy *Ambondro*, the wear pattern of *Ausktribosphenos* is restricted to the

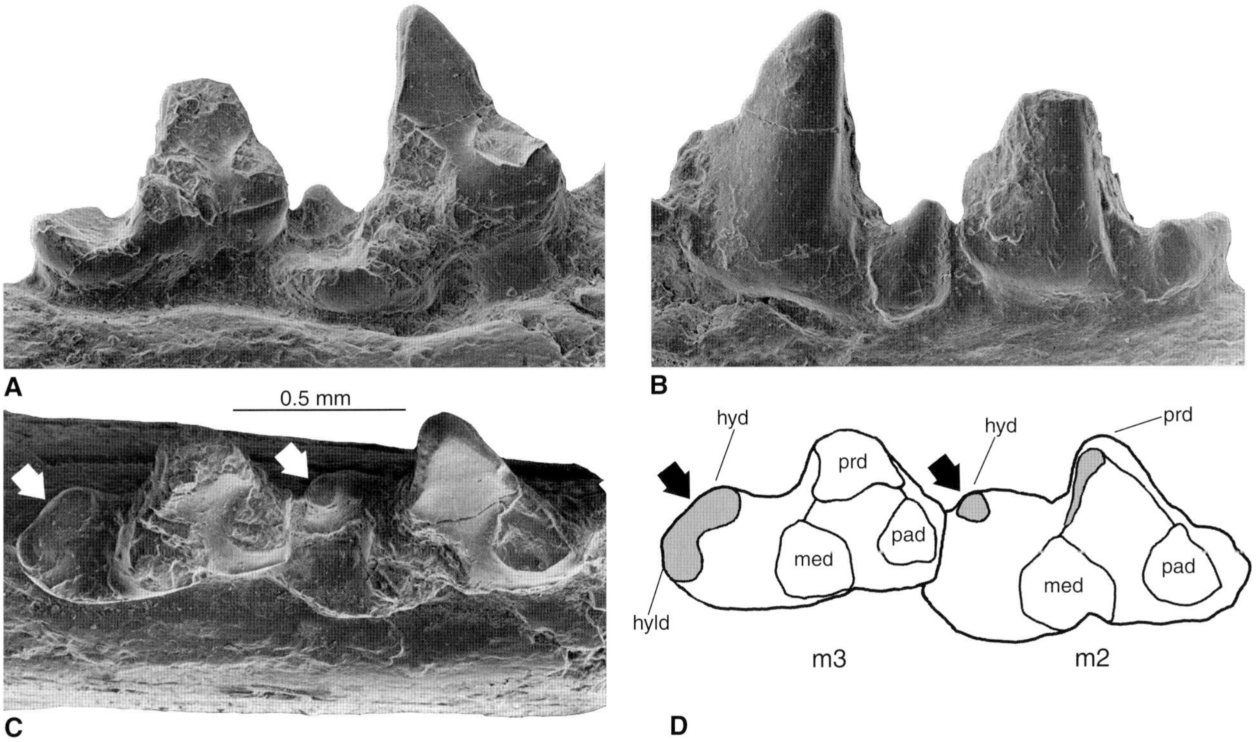

Fig. 10.17. Lower molars of *Asfaltomylos patagonicus*. SEM micrographs (A–C) and schematic drawing (D) of lower m2 and m3 with apical wear (white arrows) on the talonid rim in **A,** lingual, **B,** buccal, **C, D,** occlusal views. Abbreviations: hyd, hypoconid; hyld, hypoconulid; med, metaconid; pad, paraconid; prd, protoconid. Modified from Martin & Rauhut (2005).

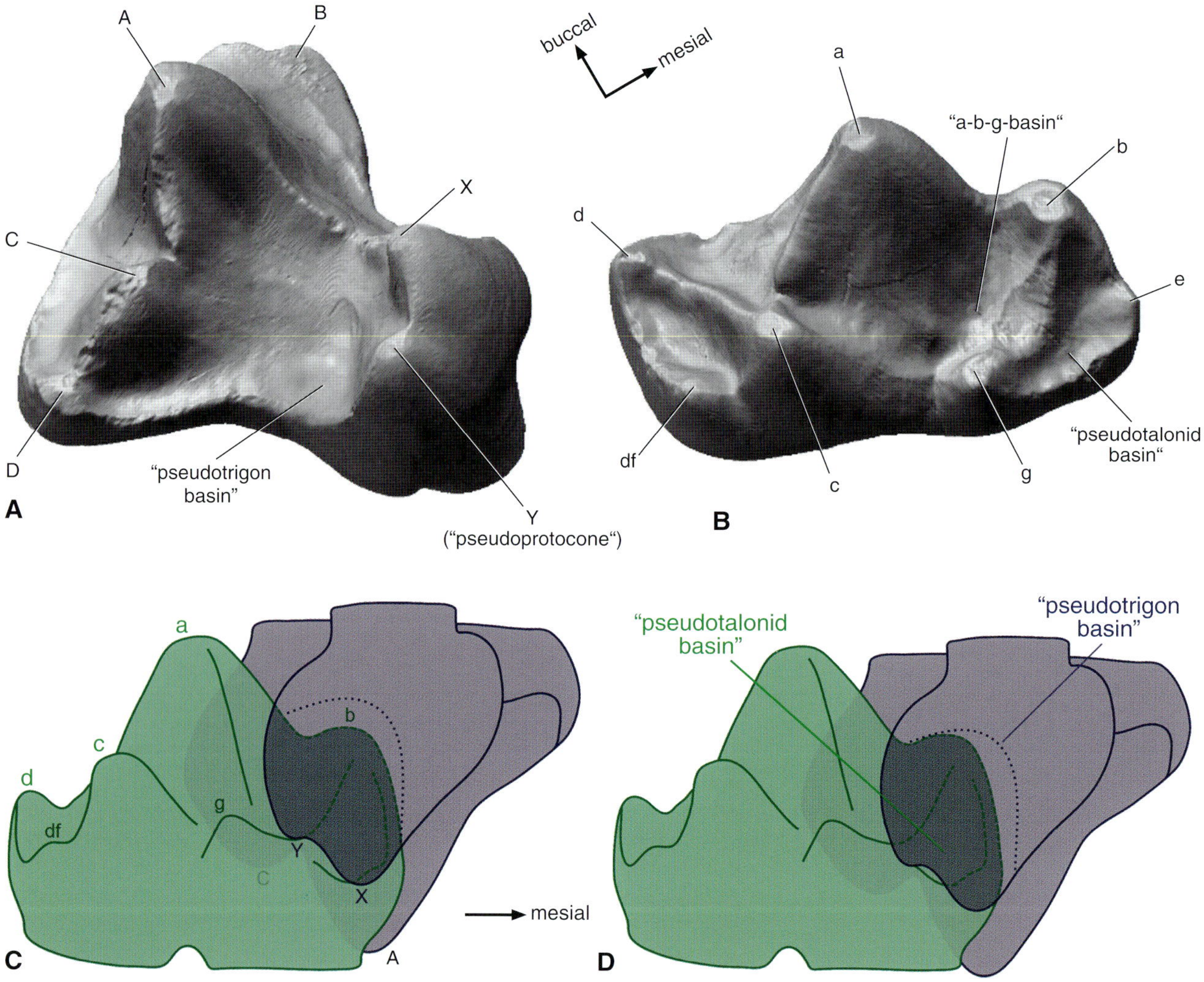

Fig. 10.18. Terminology of docodontan (*Haldanodon*) molar morphology and occlusion of cusp Y into the "pseudotalonid basin". **A,** left upper molariform DP4; **B,** left lower molar; **C, D,** left upper molariform and left lower molar in centric occlusion with designation of cusps (**C**) and basins (**D**). Not to scale. Modified from Brinkkötter (2018).

talonid rim and no wear facets occur within the talonid basin (Martin & Rauhut 2005). Upper dentitions are not known for *Ausktribosphenos* and *Bishops*, but the lower molar wear pattern of *Ausktribosphenos* suggests a similar occlusal pattern like in *Ambondro* and the South American taxa. The Early Cretaceous monotreme *Steropodon* from Australia (Archer et al. 1985) is represented by a mandible fragment with three molars. The molars have a dilambdodont pattern, formed by trigonid and talonid. Like in the australosphenidans, the talonid of *Steropodon* exhibits wear only on the apices. A similar molar pattern occurs in the Early Cretaceous (Aptian) monotreme *Teinolophos* from Victoria (Rich et al. 1999, 2001b, 2016). Upper and lower dentitions are known for the late Early Cretaceous (Albian) Australian monotreme *Kollikodon*, the assumed sister taxon of Monotremata (Flannery et al. 1995, Pian et al. 2016). *Kollikodon* has polybunodont molars with strikingly thick enamel adapted to a durophagous diet with a crushing function of the molars.

Shuotherium and *Pseudotribos* from the Northern Hemisphere appear in cladistic analyses as members of Australosphenida (e.g., Luo et al. 2007). Both are characterized by talonid basins that sit at the mesial end of the lower molars (Chow & Rich 1982, Luo et al. 2007). Wang et al. (1998) reported an isolated upper molar of *Shuotherium* with a "pseudoprotocone" and wear facets 5' and 6' (analogous to Crompton's 1971 scheme) at the mesiolingual and distolingual flanks of the "pseudoprotocone". More recently, Luo et al. (2007) reported a partial skeleton of the shuotheriid *Pseudotribos* from the Middle Jurassic of China with mesially placed "pseudotalonids". The "pseudoprotocones" are preserved in occlusion with the "pseudotalonids" and demonstrate a pestle-to-mortar crushing function which differs from that of the Southern Hemisphere Australosphenida and is similar to that of Boreosphenida.

Docodonta form a major clade of stem mammals (mammaliaforms) that evolved molars with complex shearing, crushing and grinding function (Luo & Martin 2007) which have been studied in comparison to tribosphenic molars (Gingerich 1973, Butler 1988). The origin of the docodontan molar is still under dispute, and there is no consensus on the homologization of cusps (Luo & Martin 2007). The upper molars of docodontans are linguo-buccally widened with a mesiolingual cusp X and a mesiodistal cusp Y, of which the distolingual cusp Y has been homologized with the protocone and mesiolingual cusp X with the hypocone in the original description of *Simpsonodon* (Kermack et al.

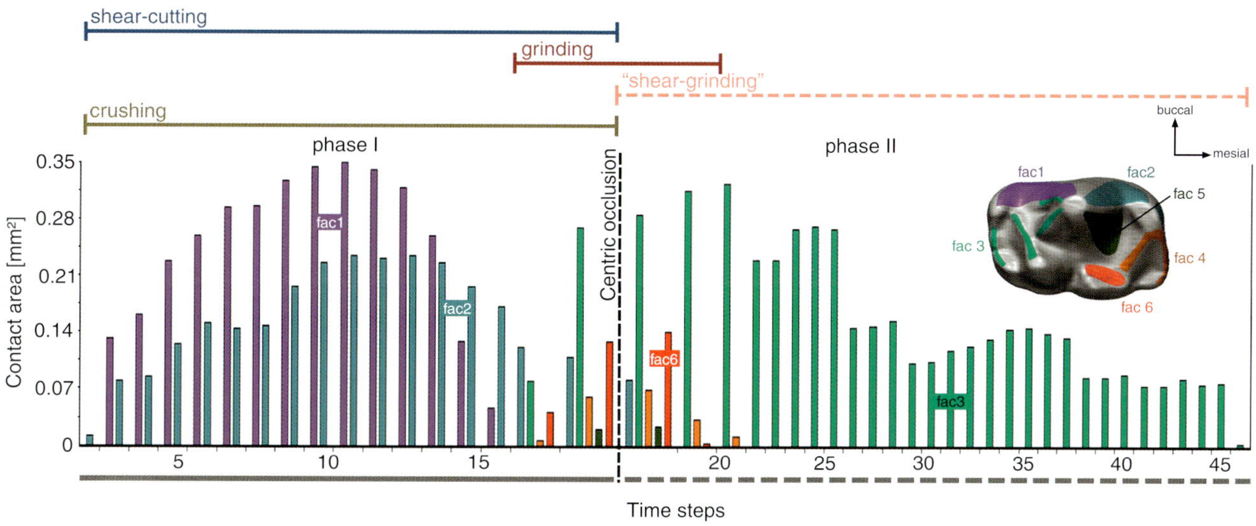

Fig. 10.19. Contact areas during the two-phased power stroke of *Haldanodon* as detected by OFA analysis. Modified after Brinkkötter (2018).

1987: fig. 45) (Fig. 10.18). The lower docodontan molars have a mesial basin demarcated by cusps a-b-g that was named "pseudotalonid basin" by Kermack et al. (1987: 34). This term originally had been coined for the lower molars of *Shuotherium* (Chow and Rich 1982), and Kermack et al. (1987) were the first to apply it to docodontan molars (*Simpsonodon*). They designated distolingual cusp Y as "pseudoprotocone" by occlusal and functional reasons, because it bites into the "pseudotalonid basin" formed by cusps a-b-g in *Simpsonodon*. However, most subsequent authors (see Brinkkötter 2018 for discussion and references) considered larger mesiodistal cusp X to be the "pseudoprotocone" because it is in a similar mesiolingual position as the protocone of tribosphenic molars. Because cusp X does not occlude into the "pseudotalonid basin" formed by cusps a-b-g, it does not appear functionally equivalent to the protocone of tribosphenic molars. The different naming of the mesiolingual and mesiodistal cusps and varying morphology of molars in different docodontan taxa (e. g., presence or absence of crest b-g on lower molars, see Fig. 10.18B) caused confusion and major discussions on the identity of the docodontan "pseudoprotocone" and "pseudotalonid basin" (Luo & Martin 2007, Wang & Li 2016, Brinkkötter 2018). According to Brinkkötter's (2018) occlusal analysis of the Late Jurassic docodontan *Haldanodon*, smaller cusp Y is the cusp that actually occludes into the basin formed by cusps a-b-g ("pseudotalonid basin") of the lower molars; therefore cusp Y is considered the "pseudoprotocone" here. An OFA analysis and virtual reconstruction of the mastication cycle of *Haldanodon* (Brinkkötter et al. 2014, 2017, Brinkkötter 2018) basically show two phases in the power stroke (Fig. 10.19). In phase I the lower molars move upwards from buccal to lingual into centric occlusion and occlude in between the upper molars (Fig. 10.20). Phase II most probably is a downward movement in palinal (backward) direction, but a separate upward movement in proal (forward) direction cannot be excluded with certainty. During phase I the lingual flanks of the upper molar main cusps and the buccal flanks of the lower molar main cusps perform a shear-cutting function, and cusp b of the lower molar occludes into the "pseudotrigon basin" of the upper antagonistic molar performing a crushing function. Phase I ends when the lingual cusp Y of the upper molar slides into the "pseudotalonid basin" and grinding is performed. When the mesial border of the talonid basin is obliterated by wear, cusp Y glides over the basin rim, so that lower molar cusp b contacts the "pseudotrigon basin" performing additional grinding function. Most interestingly, according to Brinkkötter (2018) in *Haldanodon* the majority of crushing and grinding occurs within the large "pseudotrigon basin" of the upper molars and not in the "pseudotalonid basin" of the lowers. During the subsequent palinal downward movement of phase II, the lower molars perform a shear-grinding function when the distal crests move along the mesial flank of cusp X of the upper molars. Brinkkötter (2018) assumed a similar chewing stroke also for other docodontans, and she concluded that crushing and grinding were much more important in docodontan chewing than expected before.

Schultz et al. (2017) re-studied the mandible and dentition of *Docodon victor* and performed an OFA analysis of the chewing cycle which suggests that the trajectory of the lower molars has a continuous palinal (disto-lingual) component starting from the early contact during phase I and proceeding throughout phase II after centric occlusion. This slight difference from *Haldanodon*, in which phase I of the power stroke is described as an upward movement strictly from buccal to lingual with a palinal component only at the very end (Brinkkötter 2018), may be related to minor occlusal structural differences.

Differences in docodontan molar morphology (Luo & Martin 2007) suggest a possible diversification of dietary preferences apart from the insectivory that has been suggested for *Haldanodon* and *Docodon*. Pfretzschner et al. (2005) conducted an analogous chewing experiment based on epoxy casts of upper and lower molars and suggested a mainly omnivorous diet for the Late Jurassic docodontan *Dsungarodon* from China. The same dietary adaptation is assumed for *Simpsonodon*, which shows an extensive grinding area at the distal end of the lower molars. Although these docodontan molars performed in a similar efficient way as tribosphenic molars, their thin enamel made them

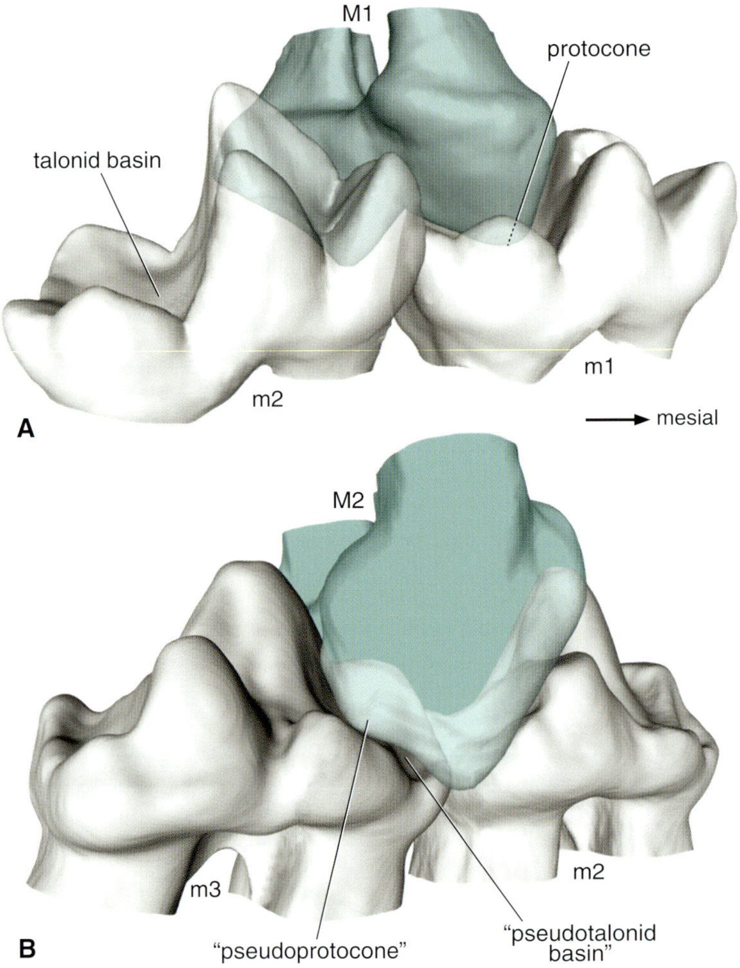

Fig. 10.20. Comparison of upper and lower molars of *Didelphis* and of *Haldanodon* in centric occlusion. **A,** M1, m1 and m2 of *Didelphis* and **B,** M2, m2 and m3 of *Haldanodon*. Not to scale. From Brinkkötter (2018).

much more prone to abrasion and therewith loss of efficiency; tribosphenic molars likely provided a longer-lasting functionality during lifetime (Brinkkötter et al. 2017, Brinkkötter 2018). The large Middle Jurassic docodontan *Itatodon* from Siberia (interpreted as a shuotheriid by Wang & Li 2016) has linguo-buccally compressed lower molars with well-developed cutting edges (Lopatin & Averianov 2005) that suggest a faunivorous or even carnivorous adaptation. The first and second lower molars (uppers are not preserved) of the semi-aquatic *Castorocauda lutrasimilis* from the Middle Jurassic of Liaoning Province (China) have linguo-buccally compressed tooth crowns, each with five cusps in straight alignment, three of which are slightly curved backwards. They are convergently similar to molars of Eocene whales or extant seals, and are interpreted as an adaptation for catching and holding slippery prey like fish and other aquatic animals (Ji et al. 2006).

Reduced peg-like molariforms without enamel

A peculiar, highly reduced dentition has been reported for the Late Jurassic *Fruitafossor*, a mammaliaform of uncertain systematic position from the Morrison Formation (Luo & Wible 2005). *Fruitafossor* has tubular and single-rooted molariforms with open-ended roots that completely lack enamel. The open root suggests that *Fruitafossor* probably had continuous tooth growth during life. These teeth are unique among Mesozoic mammaliaforms and convergently resemble the teeth of extant aardvarks and armadillos. Luo & Wible (2005) suggested that *Fruitafossor* possibly fed on colonial insects similar to modern insectivorous xenarthrans. It is important to note that according to the parsimony analysis by Luo & Wible (2005), *Fruitafossor* is not closely related to both eutherians or placental xenarthrans.

Multicusped teeth as early adaptation to omnivory-herbivory

The dentitions of Haramiyida and Multituberculata differ from those of other Mesozoic stem mammals and crown mammals by their multicusped premolars and molars with a large number of cusps arranged in longitudinal rows (Fig. 10.21). The unique similarity of the molariforms of both groups led to the concept of Allotheria that unites haramiyidans and multituberculates in one clade (Hahn et al. 1989). Later it had been put forward that Haramiyida are stem mammals not closely related to Multituberculata which belong to crown Mammalia (Zhou et al. 2013, Luo et al. 2015, Puttick et al. 2017), but some authors keep up a multituberculate-haramiyidan clade (Meng et al. 2014, Bi et al. 2014, Krause et al. 2014). Recently, King & Beck (2019) proposed polyphyly of haramiyidans, with *Haramiyavia* and *Thomasia* being stem mammals (Luo et al. 2015) and euharamiyidans being crown mammals in a clade of their own separate from multituberculates which corroborates polyphyly of allotherians.

Haramiyidan molars show a central basin with a lingual and buccal bordering row of cups (in upper anterior molars a lingual offset with an additional cusp row can be present) (Zhou et al. 2013). The cusps are designated as buccal row A (upper)/a (lower) and lingual row B (upper)/b (lower), and are numbered consecutively from mesial to distal. The upper lingual row occludes into the basin of the lower molars. The jaw movements in haramiyidans were orthal (vertical) with a palinal (posterior) component as evident from striation analysis of the enamel surface (Butler 2000, Meng et al. 2014, Mao & Meng 2019) and from the kinematic animation study of tooth occlusion (Luo et al. 2015). Haramiyidans of the Late Triassic and Early Jurassic have brachyodont tooth crowns, but molars of Middle-Late Jurassic euharamiyidans show multiple hypsodont roots which are proximally fused, called dentin or root hypsodonty (Kermack et al. 1998, Zhou et al. 2013, Luo et al. 2017). Some Asian euharamiyidans such as the arboreal and gliding *Vilevolodon* have a unique dual mortar-pestle occlusal pattern of the molars. The large anteriorly positioned cusp a1 of the lower molars occludes into the anteriorly positioned basin of the upper antagonists. At the same time the large posteriorly positioned cusp A1 of the upper molars occludes into the posteriorly positioned basin of the lower antagonists (Luo et al. 2017), as first recognized by Kermack et al. (1998). This molar morphology and occlusal pattern indicates crushing and grinding function and suggests a mainly herbivorous diet, with feeding on soft plant tissues and possible granivory (Luo et al. 2017).

Among multituberculates, Paulchoffatiidae have molars with cusps of differing heights and a basin-shaped m2 with only one cusp (Hahn & Hahn 2004, Yuan et al. 2013) which represents the most plesiomorphic tooth pattern among Multituberculata (Lazzari et al. 2010). Lazzari et al. (2010) studied the mastication of Paulchoffatiidae, and they reconstructed two different masticatory cycles, a puncture-crushing cycle and a grinding cycle that consists of two different phases. The first phase of the grinding cycle is oblique and can be seen as a precursor of the slicing-crushing cycle in ptilodontoid Cimolodonta (Krause 1982), whereas the second phase is fully palinal and homologous to the ptilodontoid grinding cycle. Paulchoffatiids exhibit fewer attrition facets than Cimolodonta and paulchoffatiid occlusion apparently was less precise than that of other multituberculates. Lazzari et al. (2010) noticed similarities to the occlusal pattern of Haramiyida (Butler & McIntyre 1994) but pointed out that these are symplesiomorphic and do not imply a close relationship.

Gingerich (1977) and Krause (1982) analyzed the enamel microwear pattern, dentin wear pattern, and the cusp morphology of derived Cenozoic ptilodontoid Cimolodonta and reconstructed two chewing cycles: one that involves the ultimate premolars and a second one that involves the molars. During the first (slicing and crushing) cycle, the large arcuate and laterally compressed blade-like p4 cuts vertically deep into the food items that are held up against the upper premolar (P4). Subsequently, the sliced food items pass down on both sides of the p4 and cause vertical subparallel striations in the valleys between the almost vertical enamel ridges. The second (grinding) cycle is a palinal (backwards) movement of the mandible with the molars in occlusion (Gingerich 1977). This causes flat attrition facets with longitudinal striations inside the deep valleys between the cusp rows of upper and lower molars (Krause 1982). Multituberculates have enlarged lower incisors that are separated from the cheek teeth by a wide diastema. The incisors of derived multituberculates such as ptilodontoids and taeniolabidids are rootless and evergrowing and have the enamel restricted to the buccal side of the tooth. Due to the similarities of multituberculate and rodent dental form and function (Krause 1986), multituberculates have been dubbed "rodents of the Mesozoic" with similar dietary adaptations. However, in rodents with longitudinal chewing motion such as muroids, the power stroke is oriented proal (forward) and not palinal (backward) as in multituberculates.

Hypsodont molariforms with flat occlusal surface

The enigmatic Late Cretaceous-Paleocene Gondwanatheria from the southern landmasses evolved the first truly hypsodont molars in mammalian evolutionary history (Fig. 10.22). Monophyly of Gondwanatheria is generally accepted, but their phylogenetic relationship is under discussion. Cladistic analyses by Gurovich & Beck (2009) and Krause et al. (2014) support multituberculate affinities of Gondwanatheria, but the presence of four molariform positions in the mandible of the Paleocene gondwanatherian *Sudamerica* (Pascual et al. 1999) contradicts this assignment to multituberculates which have only two molars and no molariform lower premolars. A recent analysis using 84 cynodont taxa and 530 morphological characters recognized Gondwanatheria as sister taxon of Multituberculata

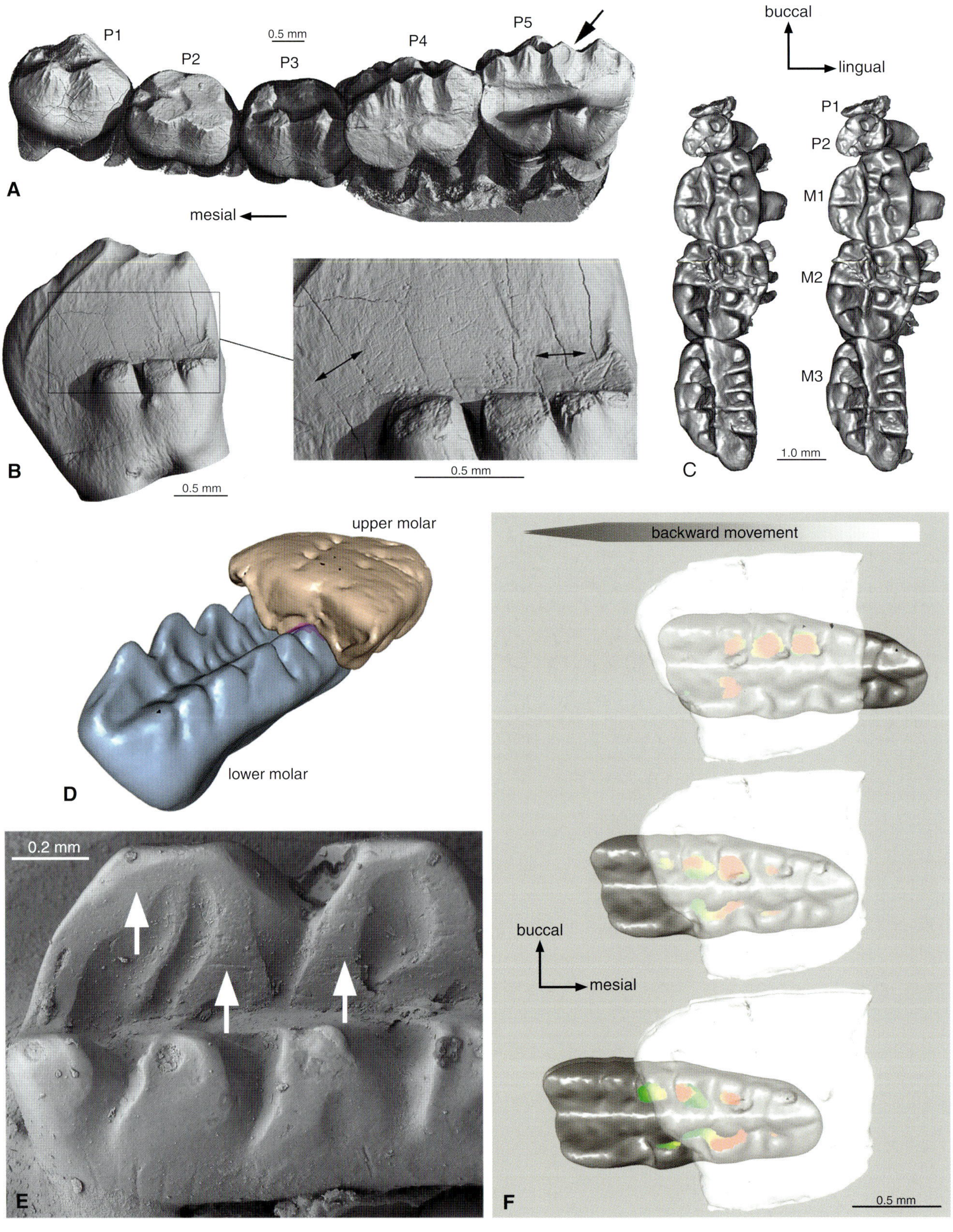

Fig. 10.21. Multicusped molars of multituberculata and Haramiyida. **A,** upper left tooth row of *Pseudobolodon oreas* with worn cusp showing the trailing edge (black arrow). From Lazzari et al. (2010), mirrored. **B,** buccal view of a lower p4 of a paulchoffatiid multituberculate with striations (close up) on the surface reflecting two directions of relative jaw movements (black double arrows). **C,** stereo-pair of the upper right tooth row of the haramiyidan *Megaconus*, modified from Zhou et al. (2013). **D–F,** occlusion pattern of multituberculates as represented by *Neoplagiaulax*. **D,** two rows of cusps of the lower molar (blue) interlock in the three rows of cusps of the upper molars (pink). **E,** close up of wear facets along the cusp rows of a lower molar. White arrows indicate striations. **F,** OFA simulation of the backward chewing movement typical for multituberculates, colored patches indicate occlusal contacts along the cusp rows.

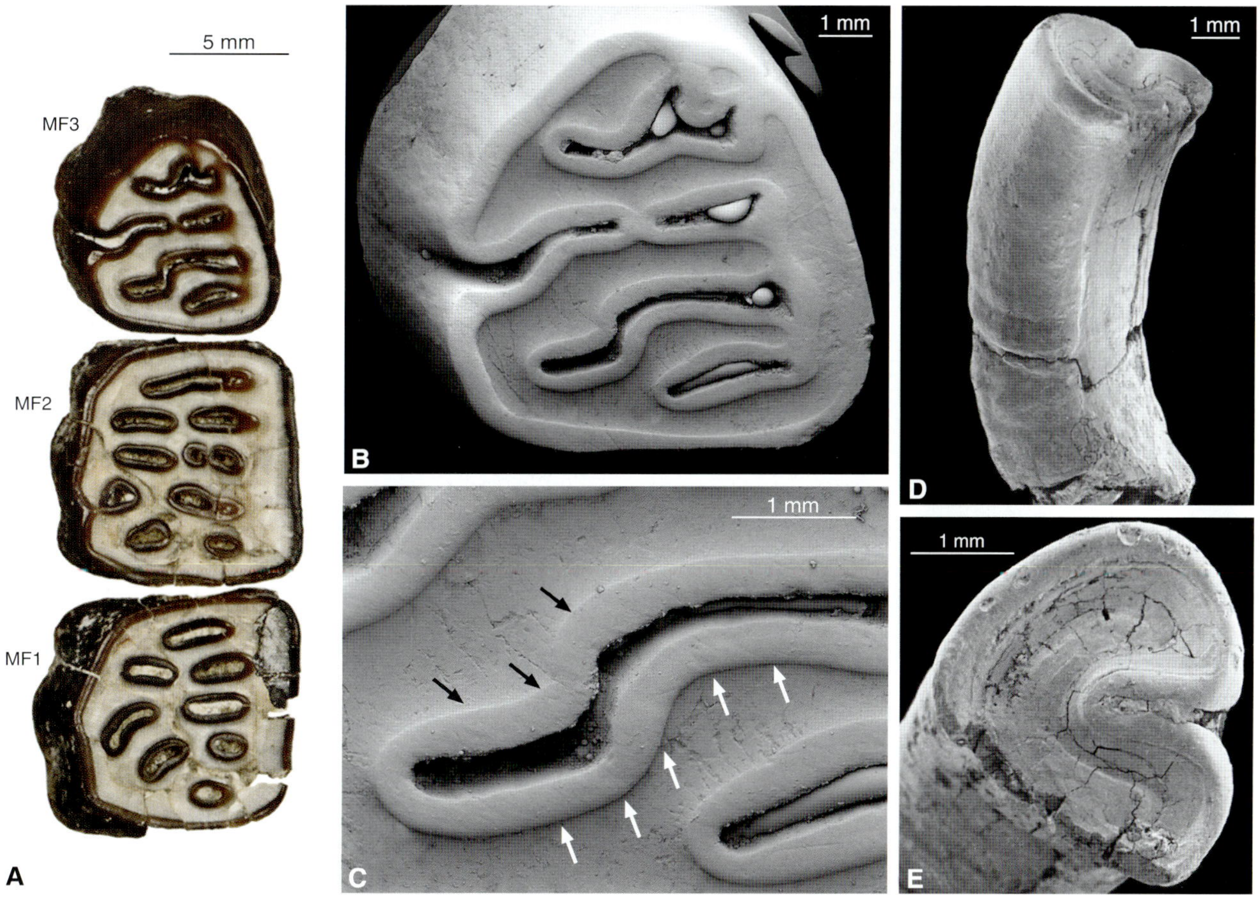

Fig. 10.22. Hypsodont gondwanatherian molars. **A,** upper left tooth row of *Vintana sertichi* showing its unique flat occlusal surface with several enamel islets, modified from Krause et al. (2014). **B, C,** close ups of the last upper molariform (MF3) of *Vintana sertichi*, illustrating the uneven excavation of the enamel bands by wear, modified after Schultz et al. (2014). Black arrows indicate the leading edge, white arrows indicate the trailing edge. **D, E,** high crowned molariform of *Lavanify* sp. with long root (**D**) and flat occlusal surface (**E**), from Koenigswald & Krause (2014).

(Krause et al. 2020). Before the discovery of the skull of the sudamericid *Vintana sertichi* from the Late Cretaceous of Madagascar (Krause et al. 2014), Gondwanatheria were known only by fragmentary mandibles (*Sudamerica*) and isolated teeth. The skull of *Vintana* has three hypsodont molariforms preserved in both maxillae. The molariforms are characterized by enamel infoldings and enamel islets on the flat occlusal surfaces. Schultz et al. (2014) analyzed the occlusal relationships and reconstructed the chewing stroke of *Vintana*. The palinal (posteriorly directed) power stroke had a significant buccally directed component during the unilateral chewing. Parallel wear striations and flat occlusal surfaces without cusps suggest that the power stroke was almost horizontal, without a separation in different masticatory phases. The buccal component of the power stroke differentiates *Vintana* from other mammals and mammaliaforms with palinal chewing movements such as multituberculates (Krause 1982, Gambaryan & Kielan-Jaworowska 1995, Lazzari et al. 2010), haramiyidans (Butler 2000, Butler & Hooker 2005, Zhou et al. 2013), and other gondwanatherians (Krause et al. 1992, Krause and Bonaparte 1993, Koenigswald et al. 1999). According to the hypsodont tooth morphology and striation analysis, Schultz et al. (2014) suggested a herbivorous diet for *Vintana* with a wide range of herbaceous material. Convergently to multituberculates and rodents, gondwanatherians evolved ever-growing incisors that are separated by wide diastemata from the hypsodont cheek teeth. Open-rooted upper and lower incisors and wide diastemata are also seen in the recently described gondwanatherian *Adalatherium* from the Maastrichtium of Madgascar (Krause et al. 2020). The presence of brachyodont postcanines with four cusps connected by prominent crests in *Adalatherium*, a dental pattern otherwise unknown from Mesozoic mammaliaforms, suggests that Gondwanatheria were dentally more diverse than hitherto believed.

Conclusion

Mesozoic mammals show a remarkable variety of dental morphologies and functional adaptations. Plesiomorphically, mammaliaforms had a single-phased (phase I) chewing stroke with a mainly piercing and cutting function. A second phase (phase II) that added a crushing and/or grinding function was evolved independently in several groups, including meta- and eutherians, "pseudotribosphenic" lineages, and multituberculates. The more complex chewing cycles with enhanced crushing and grinding in these lineages likely permitted greater dietary diversity and helped catalyze their ecological diversification.

Almost all morphological and functional types of teeth that are seen in Cenozoic therians were present in the Mesozoic, including peg-like dentin teeth, ever-growing incisors, and hypsodont molars. Not realized in Mesozoic mammals are dentitions with premolars and molars forming extensive crushing or grinding surfaces suited for processing large quantities of plant materials, as found in large herbivores such as perissodactyls, artiodactyls, and proboscideans. Mesozoic mammals were small, the majority had a body mass between 10 and 250 g, with the Early Cretaceous carnivorous *Repenomamus* being the largest known Mesozoic mammal weighting about 10 kg (see Martin 2018 and references therein). Large herbivores and subsequently large carnivorous mammals appeared only some million years after the extinction of the non-avian dinosaurs at the K/Pg-boundary. Among the earliest mega-herbivores is the Middle Eocene (Bridgerian, about 48 Myr) rhinoceros-sized *Uintatherium* with an estimated body mass of 1500 to 2000 kg (Turnbull 2002). The majority of Mesozoic mammaliaforms and mammals with non-tribosphenic mammalian tooth types became extinct before the K/Pg-boundary. A prominent exception are the multituberculates with multicusped cheek teeth that survived until the Late Eocene. The vast diversity of Cenozoic and extant therian molars is derived from the generalized tribosphenic molar with *sensu lato* insectivorous adaptation that remained almost unchanged from its first appearance in the Late Jurassic (*Juramaia*) until the K/Pg boundary when the hypothetical ancestor of crown placentals existed (O'Leary et al. 2013). Initial step towards a more complex molar pattern was the addition of a fourth main cusp, the hypocone, at the upper molars as seen, among others, in Paleocene stem macroscelideans such as *Prolouisina*, *Adapisorex*, and *Walbeckodon*. The quadritubercular pattern provided increased crushing area for more efficient processing of a wider array of food sources (Hunter & Jernvall 1995, Engler & Martin 2018). This modification marks the beginning of the success story of the tribosphenic molar in the Cenozoic which has its roots deep in the Mesozoic.

Acknowledgments

We would like to thank the reviewers David Grossnickle (Seattle) and Zhe-Xi Luo (Chicago) for their constructive criticism and helpful suggestions that greatly improved the manuscript. Robert Asher (Cambridge), Richard Cifelli (Norman), Pam Gill (Bristol), and the Yizhou Fossil & Geology Park are thanked for providing specimens. Georg Oleschinski (Bonn) assisted at the SEM. Financial support was provided by Deutsche Forschungsgemeinschaft (DFG) grants MA 1643/12-1, 12-2, MA 1643/15-1, MA 1643/16-1, MA 1643/17-1, MA 1643/20-1, and MA 1643/21-1 within and associated with the DFG Research Unit 771.

References

* indicates publications that originated from the DFG Research Unit 771.

Archer, M., Flannery, T. F., Ritchie, A. & Molnar, R. E. (1985): First Mesozoic mammal from Australia – an Early Cretaceous monotreme. Nature 318: 363–366.

Asher, R. J. & Sánchez-Villagra, M. R. (2005): Locking yourself out: diversity among dentally zalambdodont therian mammals. Journal of Mammalian Evolution 12: 265–282.

Averianov, A. O., Martin, T. & Lopatin, A. V. (2013): A new phylogeny for basal Trechnotheria and Cladotheria and affinities of South American endemic Late Cretaceous mammals. Naturwissenschaften 100: 311–326.

Bhullar, B.-A. S., Manafzadeh, A. R., Miyamae, J. A., Hoffman, E. A., Brainerd, E. L., Musinsky, C. & Crompton, A. W. (2019): Rolling of the jaw is essential for mammalian chewing and tribosphenic molar function. Nature 566: 528–532. https://doi.org/10.1038/s41586-019-0940-x

Bi, S., Wang, Y., Guan, J. & Meng, J. (2014): Three new Jurassic euharamiyidan species reinforce early divergence of mammals. Nature 514: 679–584.

*Brinkkötter, J. J. (2018): Molar dentition of the docodontan *Haldanodon* (Mammaliaformes) as functional analog to tribosphenic teeth. Doctoral dissertation, Universität Bonn, i–vi +1–140. http://hss.ulb.uni-bonn.de/2019/5487/5487.htm

*Brinkkötter, J., Schwermann, A. & Martin, T. (2014): Comparison of crushing function in pseudotribosphenic molars of *Haldanodon exspectatus* (Mammaliaformes, Docodonta) with tribosphenic molars of *Didelphis marsupialis* (Mammalia, Marsupialia). Society of Vertebrate Paleontology 74th Annual Meeting, Berlin. Meeting Program and Abstracts: 97–98.

*Brinkkötter, J., Schwermann, A. & Martin, T. (2017): Molar function of Jurassic pseudotribosphenic docodonts (Mammaliaformes) and tribosphenic *Monodelphis* (Didelphidae) is similar in crushing and grinding efficiency. Society of Vertebrate Paleontology 77th Annual Meeting, Calgary. Meeting Program and Abstracts: 87.

Butler, P. M. (1988): Docodont molars as tribosphenic analogues (Mammalia, Jurassic). In: Russell, D. E., Santoro, J.-P., and Sigogneau-Russell, D. (eds.). Teeth Revisited: Proceedings of the VIIth International Symposium on Dental Morphology. Mémoirs du Muséum national d'Histoire naturelle Paris C 53: 329–340.

Butler, P. M. (2000): Review of the early allotherian mammals. Acta Palaeontologica Polonica 45: 317–342.

Butler, P. M. & Clemens, W. A. (2001): Dental morphology of the Jurassic holotherian mammal *Amphitherium*, with a discussion of the evolution of mammalian post-canine dental formulae. Palaeontology 44: 1–20

Butler, P. M. & Hooker, J. J. (2005): New teeth of allotherian mammals from the English Bathonian, including the earliest multituberculates. Acta Palaeontologica Polonica 50: 185–207.
Butler, P. M. & McIntyre, G. T. (1994): Review of the British Haramiyidae (Mammalia, Allotheria), their molar occlusion and relationships. Philosophical Transactions of the Royal Society of London B 345: 433–458.
Butler, P. M. & Sigogneau-Russell, D. (2016): Diversity of triconodonts in the Middle Jurassic of Great Britain. Palaeontologia Polonica 67: 35–65.
Chow, M. & Rich, T. H. (1982): *Shuotherium dongi*, n. gen. and sp., a therian with pseudo-tribosphenic molars from the Jurassic of Sichuan, China. Australian Mammalogy 5: 127–142.
Clemens, W. A. (1968): A mandible of *Didelphodon vorax* (Marsupialia, Mammalia). Los Angeles County Museum Contributions in Science 133: 1–11.
Clemens, W. A. (2011): New morganucodontans from an Early Jurassic fissure filling in Wales (United Kingdom). Palaeontology 54: 1139–1156.
Cohen, J. E. (2017): Earliest divergence of stagodontid (Mammalia: Marsupialiaformes) feeding strategies from the Late Cretaceous (Turonian) of North America. Journal of Mammalian Evolution 25: 165-177 (2018). https://doi.org.10.1007/s10914-017-9382-0.
Conith, A. J., M. J. Imburgia, A. J. Crosby & Dumont, R. R. (2016): The functional significance of morphological changes in the dentitions of early mammals. Journal of the Royal Society Interface 13: 20160713.
Crompton, A. W. (1971): The origin of the tribosphenic molar. In: Kermack, D. M. & Kermack, K. A. (eds.). Early Mammals. Zoological Journal of the Linnean Society 50, supplement 1: 65–87.
Crompton, A. W. (1974): The dentition and relationships of the southern African Triassic mammals, *Erythrotherium parringtoni* and *Megazostrodon rudnerae*. Bulletin of the British Museum (Natural History), Geology 24: 397–437.
Crompton, A. W. & Jenkins, F. A. Jr. (1968): Molar occlusion in Late Triassic mammals. Biological Reviews 43: 427–458.
Crompton, A. W. & Luo, Z.-X. (1993): Relationships of the Liassic mammals *Sinoconodon, Morganucodon*, and *Dinnetherium*. In: Szalay, F. S., Novacek, M. J. & McKenna, M. C. (eds.). Mammal Phylogeny: Mesozoic Differentiation, Multituberculates, Monotremes, Early Therians, and Marsupials. Springer, New York: 30–44.
Crompton, A. W. & Parker, P. (1978): Evolution of the mammalian masticatory apparatus. American Scientist 66: 192–201.
Crompton, A. W., Wood, C. B. & Stern, D. N. (1994): Differential wear of enamel: a mechanism for maintaining sharp cutting edges. Advances in Comparative and Environmental Physiology 18: 321–346.
Davis, B. M. (2011): Evolution of the tribosphenic molar pattern in early mammals, with comments on the "dual-origin" hypothesis. Journal of Mammalian Evolution 18: 227–244.
Debuysschere, M., Gheerbrant, E. & Allain, R. (2015): Earliest known European mammals: a review of the Morganucodonta from Saint-Nicolas-de-Port (Upper Triassic, France). Journal of Systematic Palaeontology 13: 825–855.
*Engler, T. & Martin, T. (2018): Differentiation in dental morphology and function of Paleocene small mammals after the Cretaceous-Paleogene mass extinction. Society of Vertebrate Paleontology 78th Annual Meeting, Albuquerque. Meeting Program and Abstracts: 122.
Evans, A. R. & Sanson, G. D. (1998): The effect of tooth shape on the breakdown of insects. Journal of Zoology 246: 391–400.
Evans, A. R. & Sanson, G. D. (2005a): Biomechanical properties of insects in relation to insectivory: cuticle thickness as an indicator of insect 'hardness' and 'intractability'. Australian Journal of Zoology 53: 9–19.
Evans, A. R. & Sanson, G. D. (2005b): Correspondence between tooth shape and dietary biomechanical properties in insectivorous microchiropterans. Evolutionary Ecology Research 7: 453–478.
Evans, A. R. & Sanson, G. D. (2006): Spatial and functional modeling of carnivore and insectivore molariform teeth. Journal of Morphology 267: 649–662.
Flannery, T. F., Archer, M., Rich, T. H. & Jones, R. (1995): A new family of monotremes from the Cretaceous of Australia. Nature 377: 418–420.
Flynn, J. J., Parrish, J. M., Rakotosamimanana, B., Simpson, W. F. & Wyss, A. E. (1999): A Middle Jurassic mammal from Madagascar. Nature 401: 57–60.
Fox, R. C. & Naylor, B. G. (2003): A Late Cretaceous taeniodont (Eutheria, Mammalia) from Alberta, Canada. Neues Jahrbuch für Geologie und Paläontologie, Abhandlungen 229: 393–420.
Fraenkel, G. & Rudall, K. M. (1940): A study of the physical and chemical properties of the insect cuticle. Proceedings of the Royal Society London B 129: 1–35.
Freeman, P. W. (1979): Specialized insectivory: beetle-eating and moth-eating molossid bats. Journal of Mammalogy 60: 467–479.
Freeman, P. W. (1981): Correspondence of food habits and morphology in insectivorous bats. Journal of Mammalogy 62: 164–166.
Gambaryan, P. P. & Kielan-Jaworowska, Z. (1995): Masticatory musculature of Asian taeniolabidoid multituberculate mammals. Acta Palaeontologica Polonica 40: 45–108.
Gill, P. G. (2004): *Kuehneotherium* from the Mesozoic fissure fillings of South Wales. PhD dissertation, University of Bristol. http://research-information.bristol.ac.uk
Gill, P. G., Purnell, M. A., Crumpton, N., Brown, K. R., Gostling, N. J., Stampanoni, M. & Rayfield E. J. (2014): Dietary specializations and diversity in feeding ecology of the earliest stem mammals. Nature 512: 303–305.
Gingerich, P. D. (1973): Molar occlusion and function in the Jurassic mammal *Docodon*. Journal of Mammalogy 254: 1008–1013.
Gingerich, P. D. (1977): Patterns of evolution in the mammalian fossil record. In: Hallam, A. (ed.). Patterns of Evolution. Elsevier, Amsterdam: 469–500.
Gordon, C. L. (2003): A first look at estimating body size in dentally conservative marsupials. Journal of Mammalian Evolution 10: 1–21.
Grossnickle, D. M. (2020): Jaw roll and jaw yaw in early mammals. Nature 582: E6–E8. https://doi.org/10.1038/s41586-020-2365-y
Grossnickle, D. M. & Newham, E. (2016): Therian mammals experience an ecomorphological radiation during the Late Cretaceous and selective extinction at the K-Pg boundary. Proceedings of the Royal Society B 283: 20160256. http://dx.doi.org/10.1098/rspb.2016.0256
Grossnickle, D. M. & Polly, P. D. (2013): Mammal disparity decreases during the Cretaceous angiosperm radiation. Proceedings of the Royal Society B 280: 20132110. https://dx.doi.org/10.1098/rspb.2013.2110
Grossnickle, D. M., Smith, S. M. & Wilson, G. P. (2019): Untangling the multiple ecological radiations of early mammals. Trends in Ecology & Evolution 34. https://doi.org/10.1016/j.tree.2019.05.008
Gurovich, Y. & Beck, R. M. D. (2009): The phylogenetic affinities of the enigmatic mammalian clade Gondwanatheria. Journal of Mammalian Evolution 16: 25–49.
Hahn, G. & Hahn, R. (2004): The dentition of Plagiaulacida (Multituberculata, Late Jurassic to Early Cretaceous). Geologica et Palaeontologica 38: 119–156.
Hahn, G., Sigogneau-Russel, D. & Godefroit, P. (1991): New data on *Brachyzostrodon* (Mammalia; Upper Triassic). Geologica et Palaeontologica 25: 237–249.
Hahn G., Sigogneau-Russell D. & Wouters G. (1989): New data on Theroteinidae – their relations with Paulchoffatiidae and Haramiyidae. Geologica et Palaeontologica 23: 205–215
Harjunmaa, E., Seidel, K., Häkkinen, T., Renvoise, E., Corfe, I. J., Kallonen, A., Evans, A. R., Zhang, Z. Q., Mikkola, M. L., Salazar-Ciudad, I., Klein, O. D. & Jernvall, J. (2014): Replaying evolutionary transitions from the dental fossil record. Nature 512: 44–48.
Harper, T., Parras, A. & Rougier, G. W. (2018): *Reigitherium* (Meridiolestida, Mesungulatoidea) an enigmatic Late Cretaceous mammal from Patagonia, Argentina: morphology, affinities, and dental evolution. Journal of Mammalian Evolution 26: 447–478 (2019). https://doi.org/10.1007/s10914-018-9437-x

*Hielscher, R. C., Schultz, J. A. & Martin, T. (2015): Wear pattern of the molar dentition of an extant and an Oligocene bat assemblage with implications on functionality. Palaeobiodiversity and Palaeoenvironments 95: 597–611. https://doi 10.1007/s12549-015-0186-z

Hiiemäe, K. M. (1976): Masticatory movements in primitive mammals. In: Anderson, D. J. & Matthews, B. (eds.). Mastication. Proceedings of a Symposium on the Clinical and Physiological Aspects of Mastication held at the Medical School, University of Bristol on 14–16 April 1975. Wright, Bristol: 105–118.

Hiiemäe, K. M. (1978): Mammalian mastication: a review of the activity of the jaw muscles and the movements they produce in chewing. In: Butler, P. M. & Joysey, K. A. (eds.). Development, Function, and Evolution of Teeth. Academic Press, London: 359–398.

Hiiemäe, K. M. (1984): Functional aspects of primate jaw morphology. In: Chivers D. J. & Hladik, C. H. (eds.). Food acquisition and processing in nonhuman primates. Academic Press, London: 257–281.

Hiiemäe, K. M. & Crompton, A. W. (1971): A cinefluorographic study of feeding in the American opossum, *Didelphis marsupialis*. In: Dahlberg, A. A. (ed.). Dental Morphology and Evolution. University of Chicago Press, Chicago: 299–334.

Hiiemäe, K. & Kay, R. F. (1972): Trends in the evolution of primate mastication. Nature 240: 486–487.

Hiiemäe, K. M. & Kay, R. F. (1973): Evolutionary trends in the dynamics of primate mastication. In: Montagna, W. & Zingeser, M. R. (eds,). Symposia of the International Congress of Primatology, vol. 3. Karger, Basel: 28–64.

Hu, Y.-M., Meng, J., Wang, Y.-Q. & Li, C.-K. (2005): Large Mesozoic mammals fed on young dinosaurs. Nature 433: 149–152.

Hunter, J. P. & Jernvall, J. (1995): The hypocone as a key innovation in mammalian evolution. Proceedings of the National Academy of Sciences of the United States of America 92: 10718–10722.

Hylander, W. L., Johnson, K. R. & Crompton, A. W. (1987): Loading patterns and jaw movements during mastication in *Macaca fascicularis*: a bonestrain, electromyographic, and cineradiographic analysis. American Journal of Physical Anthropology 72: 287–314.

*Jäger, K. R. K., Cifelli, R. L. & Martin, T. (2020): Molar occlusion in the Jurassic mammal *Priacodon*. Scientific Reports. https://doi.org/10.1038/s41598-020-79159-4.

*Jäger, K. R. K., Gill, P. G., Corfe, I. & Martin, T. (2019): Occlusion and dental function of *Morganucodon* and *Megazostrodon*. Journal of Vertebrate Paleontology e1635135. https://doi 10-1080/02724634.2019.1635135.

Jenkins, F. A., Crompton A. W. & Downs W. R. (1983): Mesozoic mammals from Arizona: new evidence on mammalian evolution. Science 222: 1233–1235.

Jenkins, F. A. Jr. & Schaff, C. R. (1988): The Early Cretaceous mammal *Gobiconodon* (Mammalia, Triconodonta) from the Cloverly Formation in Montana. Journal of Vertebrate Paleontology 8: 1–24.

Jernvall, J. (2000): Linking development with generation of novelty in mammalian teeth. Proceedings of the National Academy of Sciences of the United States of America 97: 2641–2645.

Ji, Q., Luo, Z.-X., Yuan, C.-X.& Tabrum, A. R. (2006): A swimming mammaliaform from the Middle Jurassic and ecomorphological diversification of early mammals. Science 311: 1123–1127.

Kavanagh, K. D., Evans, A. R. & Jernvall, J. (2007): Predicting evolutionary patterns of mammalian teeth from development. Nature 449: 427–432.

Kay, R. F. & Hiiemäe, K. M. (1974). Jaw movement and tooth use in Recent and fossil primates. American Journal of Physical Anthropology 40: 227–256.

Kermack, D. M., Kermack, K. A. & Mussett, F. (1968): The Welsh pantothere *Kuehneotherium praecursoris*. Zoological Journal of the Linnean Society 47: 407–423.

Kermack, K. A., Kermack, D. M., Lees, P. M. & Mills, J. R. E. (1998): New multituberculate-like teeth from the Middle Jurassic of England. Acta Palaeontologica Polonica 43: 581–606.

Kermack, K. A., Lee, A. J., Lees, P. M. & Mussett, F. (1987): A new docodont from the Forest Marble. Zoological Journal of the Linnean Society 89: 1–39.

Kielan-Jaworowska, Z., Cifelli, R. L. & Luo, Z.-X. (2004): Mammals from the Age of Dinosaurs. Origins, Evolution, and Structure. Columbia University Press, New York.

King, B. & Beck, R. M. D. (2019): Bayesian tip-dated phylogenetics: Topological effects, stratigraphic fit and the early evolution of mammals. (preprint) https://www.biorxiv.org/content/10.1101/533885v1.abstract).

Koenigswald, W. v. & Krause, D. W. (2014): Enamel microstructure of *Vintana sertichi* (Mammalia, Gondwanatheria) from the Late Cretaceous of Madagascar. Journal of Vertebrate Paleontology Memoir 14. Journal of Vertebrate Paleontology 34 (6, Supplement): 166–181.

Koenigswald, W. v., Goin, F. J. & Pascual, R. (1999): Hypsodonty and enamel microstructure in the Paleocene gondwanatherian mammal *Sudamerica ameghinoi*. Acta Palaeontologica Polonica 44: 263–300.

Krause, D. W. (1982): Jaw movement, dental function, and diet in the Paleocene multituberculate *Ptilodus*. Paleobiology 8: 265–281.

Krause, D. W. (1986): Competitive exclusion and taxonomic displacement in the fossil record: the case of rodents and multituberculates in North America. Contributions to Geology, University of Wyoming, Special Paper 3: 95–117.

Krause, D. W. & Bonaparte, J. F. (1993): Superfamily Gondwanatherioidea: a previously unrecognized radiation of multituberculate mammals in South America. Proceedings of the National Academy of Sciences of the United States of America 90: 9379–9383.

Krause, D. W., Hoffmann, S., Wible, J. R., Kirk, E. C., Schultz, J. A., Koenigswald, W. v., Groenke, J. R., Rossie, J. B., O'Connor, P. M.,Seiffert, E. R., Dumont, E. R., Holloway, W. L., Rogers, R. R., Rahantarisoa, L. J., Kemp, A. D. & Andriamialison, H. (2014): First cranial remains of a gondwantherian mammal reveal remarkable mosaicism. Nature 515: 512–517.

Krause, D. W., Hoffmann, S., Hu, Y.-M., Wible, J. R., Rougier, G. W., Kirk, E. K., Groenke, J. R., Rogers, R. R., Rossie, J. B., Schultz, J. A., Evans, E. R., Koenigswald, W. v. & Rahantarisoa, L. J. (2020): Skeleton of a Cretaceous mammal from Madagascar reflects long-term insularity. Nature 581: 421–427.

Krause, D. W., Kielan-Jaworowska, Z. & Bonaparte, J. F. (1992): *Ferugliotherium* Bonaparte, the first known multituberculate from South America. Journal of Vertebrate Paleontology 12: 351–376.

*Kullmer, O., Benazzi, S., Fiorenza, L., Schulz, D., Bacso, S. & Winzen, O. (2009): Technical Note: Occlusal Fingerprint Analysis (OFA): Quantification of tooth wear pattern. American Journal of Physical Anthropology 139: 600–605.

*Kullmer, O., Menz, U. & Fiorenza, L. (2020): Occlusal Fingerprint Analysis (OFA) reveals dental occlusal behavior in primate molars. In: Martin, T. & Koenigswald, W. v. (eds.). Mammalian Teeth – Form and Function. Pfeil, Munich: 25–43.

*Lazzari, V., Schultz, J. A., Tafforeau, P. & Martin, T. (2010): Occlusal pattern in paulchoffatiid multituberculates and the evolution of cusp morphology in mammaliamorphs with rodent-like dentitions. Journal of Mammalian Evolution 17: 177–192.

Li, G. & Luo, Z.-X. (2006): A Cretaceous symmetrodont therian with some monotreme-like postcranial features. Nature 439: 195–200.

Lillegraven, J. A. (1969): Latest Cretaceous mammals of upper part of Edmonton Formation of Alberta, Canada, and review of marsupial-placental dichotomy in mammalian evolution. University of Kansas Paleontological Contributions 50: 1–122.

Lopatin, A. V. & Averianov, A. O. (2005): A new docodont (Docodonta, Mammalia) from the Middle Jurassic of Siberia. Doklady Biological Sciences 405: 434–436.

Lucas, P. W. (1979): The dental-dietary adaptations of mammals. Neues Jahrbuch für Geologie und Paläontologie, Monatshefte 8: 486–512.

Luo, Z.-X. (2007): Transformation and diversification in early mammal evolution. Nature 450: 1011–1019.

Luo, Z.-X. & Martin, T. (2007): Analysis of molar structure and phylogeny of docodont genera. Carnegie Museum of Natural History Bulletin 39: 27–47

Luo, Z.-X. & Wible, J. R. (2005): A new Late Jurassic digging mammal and early mammalian diversification. Science 308: 103–107.

Luo, Z.-X., Cifelli, R. L. & Kielan-Jaworowska, Z. (2001): Dual origin of tribosphenic mammals. Nature 409: 53–57.

Luo, Z.-X., Gatesy, S. M., Jenkins, F. A., Amaral, A. A. & Shubin, N. H. (2015): Mandibular and dental characteristics of Late Triassic mammaliaform *Haramiyavia* and their ramifications for basal mammal evolution. Proceedings of National Academy of Sciences of the Unites States of America 112 (51): E7101–E7109. https://doi 10.1073/pnas.1519387112.

Luo, Z.-X., Ji, Q. & Yuan, C.-X. (2007): Convergent dental evolution in pseudotribosphenic and tribosphenic mammals. Nature 450: 93–97.

Luo, Z.-X., Kielan-Jaworowska, Z. & Cifelli, R. L. (2002): In quest for a phylogeny of Mesozoic mammals. Acta Palaeontologica Polonica 47: 1–78.

Luo, Z.-X., Meng, Q.-J., Grossnickle, D. M., Liu, D., Neander, A. I., Zhang, Y.-G. & Ji, Q. (2017): New evidence for mammaliaform ear evolution and feeding adaptation in a Jurassic ecosystem. Nature 548: 326–329.

Luo, Z.-X., Yuan, C.-X., Meng, Q.-J. & Ji, Q. (2011): A Jurassic eutherian mammal and divergence of marsupials and placentals. Nature 476: 442–445.

Mao, F. & Meng, J. (2019): Tooth microwear and occlusal modes of euharamiyidans from the Jurassic Janliao Biota reveal mosaic tooth evolution in Mesozoic allotherian mammals. Palaeontology 62: 639–660. https://doi 10.1111/pala.12421

Martin, T. (1999): Dryolestidae (Dryolestoidea, Mammalia) aus dem Oberen Jura von Portugal. Abhandlungen der senckenbergischen naturforschenden Gesellschaft 550: 1–119.

Martin, T. (2002): New stem-line representatives of Zatheria (Mammalia) from the Late Jurassic of Portugal. Journal of Vertebrate Paleontology 22: 332–348.

*Martin, T. (2018): Mesozoic mammals – early mammalian diversity and ecomorphological adaptations. In: Zachos, F. E. and Asher, R. J. (eds.). Handbook of Zoology, Mammalia. Mammalian Evolution, Diversity and Systematics. De Gruyter, Berlin: 199–299.

Martin, T. & Rauhut, O. W. M. (2005): Mandible and dentition of *Asfaltomylos patagonicus* (Australosphenida, Mammalia) and the evolution of the tribosphenic teeth. Journal of Vertebrate Paleontology 25: 414–425.

Mellett, J. S. (1985): Autocclusal mechanisms in the carnivore dentition. Australian Mammalogy 8: 233–238.

Meng, J., Bi, S., Wang, Y., Zheng, X. & Wang, X. (2014): Dental and mandibular morphologies of *Arboroharamiya* (Haramiyida, Mammalia): a comparison with other haramiyidans and *Megaconus* and implications for mammalian evolution. PLoS One 9 (12): e113847.

Mills, J. R. E. (1971): The dentition of *Morganucodon*. In: Kermack, D. M. & Kermack, K. A. (eds.). Early Mammals. Zoological Journal of the Linnean Society 50, supplement 1: 29–63.

Moore, S. J. & Sanson, G. D. (1995): A comparison of the molar efficiency of two insect-eating mammals. Journal of Zoology 235: 175–192.

O'Leary, M. A., Bloch, J. I., Flynn, J. J., Gaudin, T. J., Giallombardo, A., Giannini, N. P., Goldberg, S. L., Kraatz, B. P., Luo, Z.-X., Meng, J., Ni, X., Novacek, M. J., Perini, F. A., Randall, Z. S., Rougier, G. W., Sargis, E. J., Silcox, M. T., Simmons, N. B., Spaulding, M., Velazco, P. M., Weksler, M., Wible, J. R. & Cirranello, A. L. (2013): The placental mammal ancestor and the post-K-Pg radiation of placentals. Science 339: 662–667.

Pascual, R., Goin, F. J., Krause, D. W., Ortiz-Jaureguizar, E. & Carlini, A. A. (1999): The first gnathic remains of *Sudamerica:* implications for gondwanathere relationships. Journal of Vertebrate Paleontology 19: 373–382.

Pfretzschner, H. U., Martin, T., Maisch, M., Matzke, A. & Sun, G. (2005): A new docodont from the Late Jurassic of the Junggar Basin in Northwest China. Acta Palaeontologica Polonica 50: 799–808.

Pian, R., Archer, M., Hand, S. J., Beck, R. M. D. & Cody, A. (2016): The upper dentition and relationships of the enigmatic Australian Cretaceous mammal *Kollikodon ritchiei*. Memoirs of Museum Victoria 74: 97–105.

*Plogschties, T. (2020): Functional analysis of molars in "symmetrodontan" mammals. Doctoral dissertation, Universität Bonn, 133 pp. http://hss.ulb.uni-bonn.de/2020/5772/5772.pdf

*Plogschties, T. & Martin, T. (2017): The masticatory cycle in acute angled symmetrodonts. Society of Vertebrate Paleontology 77th Annual Meeting, Calgary. Meeting Program and Abstracts: 178.

Puttick, M. N., O'Reilly, J. E., Tanner, A. R., Fleming, J. F., Clark, J., Holloway, L., Lozano-Fernandez, J., Parry, L. A., Tarver, J. E., Pisani, D. & Donoghue, P. C. J. (2017): Uncertain tree: discriminating among competing approaches to the phylogenetic analysis of phenotype data. Proceedings of the Royal Society B 284: 20162290.

Rauhut, O. W. M., Martin, T., Ortiz-Jaureguizar, E. O. & Puerta, P. (2002): A Jurassic Jurassic mammal from South America. Nature 416: 165–168.

Rich, T. H., Flannery, T. F., Trusler, P., Constantine, A., Kool, L., van Klaveren, N. & Vickers-Rich, P. (2001a): An advanced ausktribosphenid from the Early Cretaceous of Australia. Records of the Queen Victoria Museum 110: 1–9.

Rich, T. H., Hopson, J. A., Gill, P. A., Trusler, P., Rogers-Davidson, S., Morton, S., Cifelli, R. L., Pickering, D., Kool, L., Siu, K., Burgmann, F. A., Senden, T., Evans, A. R., Wagstaff, B. E., Seegets-Villiers, D., Corfe, I. J., Flannery, T. F., Walker, K., Musser, A. M., Archer, M., Pian, R. & Vickers-Rich, P. (2016): The mandible and dentition of the Early Cretaceous monotreme *Teinolophos trusleri*. Alcheringa 40: 475–501.

Rich, T. H., Vickers-Rich, P., Constantine, A., Flannery, T. F. & Kool, L. (1997): A tribosphenic mammal from the Mesozoic of Australia. Science 278: 1438–1442.

Rich, T. H., Vickers-Rich, P., Constantine, A., Flannery, T. F., Kool, L. & van Klaveren, N. (1999): Early Cretaceous mammals from Flat Rocks, Victoria, Australia. Records of the Queen Victoria Museum 106: 1–35.

Rich, T. H., Vickers-Rich, P., Trusler, P., Flannery, T. F., Cifelli, R. L., Constantine, A., Cool, L. & van Klaveren, N. (2001b): Monotreme nature of the Australian Early Cretaceous mammal *Teinolophos trusleri*. Acta Palaeontologica Polonica 46: 113–118.

Rougier, G. W., Martinelli, A. G., Forasiepi, A. M. & Novacek, M. J. (2007): New Jurassic mammals from Patagonia, Argentina: A reappraisal of australosphenidan morphology and interrelationships. American Museum Novitates 3566: 1–54.

Salazar-Ciudad, I. & Jernvall, J. (2010): A computational model of teeth and the developmental origins of morphological variation. Nature 464: 583–586.

*Schultz J. A. & Martin T. (2011): Wear pattern and functional morphology of dryolestoid molars (Cladotheria, Mammalia). Paläontologische Zeitschrift 85: 269–285.

*Schultz, J. A. & Martin, T. (2014): Function of pretribosphenic and tribosphenic mammalian molars inferred from 3D animation. Naturwissenschaften 101: 771–781.

*Schultz, J. A., Bhullar, B.-A. S. & Luo Z.-X. (2017): Re-examination of the Jurassic mammaliaform *Docodon victor* by computed tomography and occlusal functional analysis. Journal of Mammalian Evolution 26: 9–38 (2019). https://doi.org/10.1007/s10914-017-9418-5

*Schultz, J. A., Koenigswald, W. v. & Dumont, E. R. (2014): Dental function and diet of *Vintana sertichi* (Mammalia, Gondwanatheria) from the Late Cretaceous of Madagascar. Journal of Vertebrate Paleontology Memoir 14. Journal of Vertebrate Paleontology 34 (6, Supplement): 182–202.

*Schwermann, A. H. (2014): Über die Funktionsweise prätribosphenischer und tribosphenischer Gebisse. Doctoral dissertation, Universität Bonn, Bonn. http://hss.ulb.uni-bonn.de/2015/4027/4027.htm

*Schwermann, A., Schultz, J., Kullmer, O. & Martin, T. (2014): Molar form and function of the Late Jurassic stem-zatherian *Nanolestes* (Mammalia). Society of Vertebrate Paleontology 74th Annual Meeting, Berlin. Meeting Program and Abstracts: 226.

Sibbing, F. A. (1991): Food Capture and Oral Processing. In Winfield, I. J. and J. S. Nelson (eds.): Cyprinid Fishes: Systematics, Biology and Exploitation. Chapman & Hall, London: 377–412.

Spoutil, F., Vlcek, V. & Horacek, I. (2010): Enamel microarchitecture of a tribosphenic molar. Journal of Morphology 271: 1204–1218.

Thenius, E. (1989): Zähne und Gebiß der Säugetiere. Handbuch der Zoologie, Band VIII Mammalia. De Gryuter, Berlin.

Turnbull, W. D. (2002): The mammalian faunas of the Washakie Formation, Eocene Age, of southern Wyoming. Part IV. The Uintatheres. Fieldiana 47: 1–189.

Wall, C. E., Vinyard, C. J., Johnson, K. R., Williams, S. H. & Hylander, W. L. (2006): Phase II jaw movements and masseter muscle activity during chewing in *Papio anubis*. American Journal of Physical Anthropology 129: 215–224.

Wang, Y.-Q., Clemens, W. A., Hu, Y.-M. & Li, C.-K. (1998): A probable pseudo-tribosphenic upper molar from the late Jurassic of China and the early radiation of the Holotheria. Journal of Vertebrate Paleontology 18: 777–787.

Wang, Y.-Q. & Li, C.-K. (2016): Reconsideration of the systematic position of the Middle Jurassic mammaliaforms *Itatodon* and *Paritatodon*. Palaeontologia Polonica 67: 249–256.

Wilson, G. P. (2013): Mammals across the K/Pg boundary in northeastern Montana, USA: dental morphology and body-size patterns reveal extinction selectivity and immigrant-fueled ecospace filling. Paleobiology 39: 429–469.

Wilson, G. P. & Riedel, J. A. (2010): New specimen reveals deltatheroidan affinities of the North American Late Cretaceous mammal *Nanocuris*. Journal of Vertebrate Paleontology 30: 872–884.

Wilson, G. P., Ekdale, E. G., Hoganson, J. W., Calede, J. J. & and Van der Linden, A. (2016): A large carnivorous mammal from the Late Cretaceous and the North American origin of marsupials. Nature Communications 7: 13734. https://doi: 10.1038/ncomms13734.

Wilson, G. P., Evans, A. R., Corfe, I. J., Smits, P. D. Fortelius, M. & Jernvall, J. (2012): Adaptive radiation of multituberculate mammals before the extinction of dinosaurs. Nature 483: 457–460. https:// doi:10.1038/nature10880.

Yuan, C.-X., Ji, Q., Meng, Q.-J., Tabrum, A. R. & Luo, Z.-X. (2013): Earliest evolution of multituberculate mammals revealed by a new Jurassic fossil. *Science* 341: 779–783.

*Zhou, C.-F., Wu, S., Martin, T. & Luo, Z.-X. (2013): A Jurassic mammaliaform and the earliest mammalian evolutionary adaptations. Nature 500: 163–167.

Evolutionary trends in the mastication patterns in some perissodactyls, cetartiodactyls, and proboscideans

Julia A. Schultz, Sandra Engels, Leonie C. Schwermann, and Wighart v. Koenigswald

Introduction

Both tooth shape and mode of mastication gained an enormous variability during the evolution of large herbivorous mammals, which allowed them to exploit diverse ecological niches. The variability in mastication includes specific adaptations to diet and is linked to the evolutionary history of a group. In addition, teeth are composed of the hardest tissue of the vertebrate body and thus are the most abundant vertebrate remains in the fossil record. Because of that and the resistance of the enamel to the wear process, teeth are preferred objects in paleontology.

Teeth are worn and partly abraded by the continuous chewing processes and need to serve the entire lifespan of an animal. Wear creates facets typically appearing on the crown of molars forming light reflecting polished surfaces (Mills 1955, Butler 1973). These wear facets occur as matching pairs on upper and lower teeth; several numbering systems and nomenclatures were erected for descriptive and comparative purposes (see summary in Schultz et al. 2018, 2020, this volume). Attritional facets are generated during mastication by tooth-tooth contact between antagonistic molars, abrasional facets occur during tooth-food interaction without a direct contact to the antagonistic tooth structure (Maier & Schneck 1981, Academy of Prosthodontics 2005, Grippo et al. 2004). Both types of facets generally occur on the same occlusal surface. Erupting teeth usually do not show the fully functional shape right from the beginning. Many teeth need at least some initial use to develop precise functioning surfaces. Despite the continuous abrasion of the surface over time, such functional surfaces need to function with the same precision for a relatively long period, at least for most of the animal's life.

In this chapter, we concentrate on the functional surfaces that last for a long period and which characterize various large herbivorous mammals. In the following we distinguish between functional surfaces being composed of primary and secondary surfaces. The two conditions were introduced by Fortelius (1985): (1) teeth with *primary functional crown shape* are shaped and functional as they erupt with full enamel cover, and (2) teeth with *secondary functional crown shape* are formed and maintained by wear. Follow-

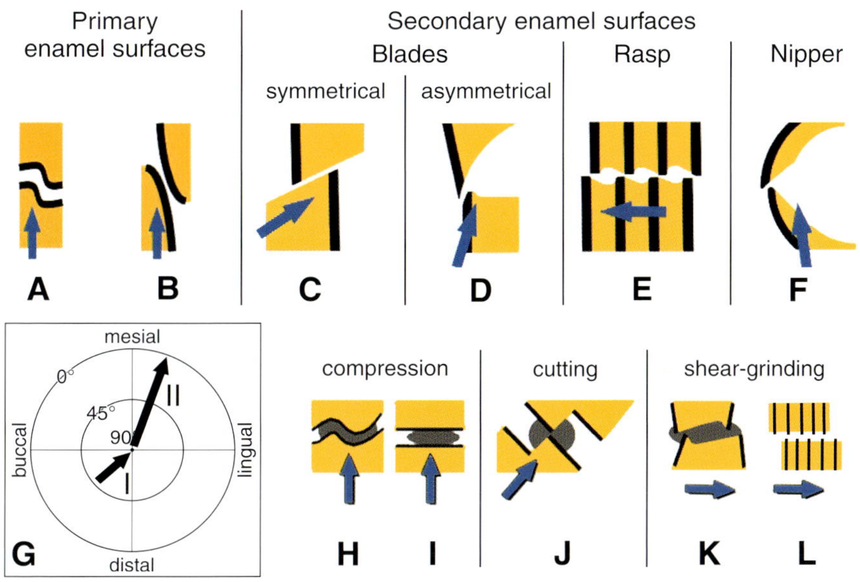

Fig. 11.1. Difference between primary and secondary functional enamel surfaces. **A, B,** all teeth that retain a primary functional surface mostly compress the food between the enamel coated (black layer) basins of the upper and lower teeth, or shear food along steep walls. **C–F,** secondary functional surfaces are characterized by enamel crests (black lines) and dentin basins (yellow), functional equivalent tools are blades (**C, D**), rasps (**E**) or nippers (**F**). **G,** the mastication compass is used to illustrate the direction of the lower jaw movements during the power stroke. The center of the circle depicts the maximal intercuspation between phase I and phase II. The arrows indicate the direction, the length of each arrow indicates the inclination during the jaw motion. In case phase II is descending the arrow is illustrated in gray. **H–L,** summary of tooth functions distinguished in this chapter: compression between primary occlusal surfaces (**H, I**), cutting along the enamel edges of a secondary occlusal surface (**J**), and shear-grinding along rasp-like secondary occlusal surfaces (**K, L**).

T. Martin & W. v. Koenigswald (eds.): Mammalian Teeth – Form and Function. Pp. 215–230, 9 figs.
 – ISBN 978-3-89937-266-3
DOI: http://doi.org/10.23788/mammteeth.11 Published 22 December 2020

ing this definition, the closed enamel layer is the dominant functioning part in primary surfaces, which is also preserved for most of the individual's lifespan. Its morphology functions either as shearing surface or compressing chamber (Fig. 11.1). The enamel layer of primary functional surfaces is always impacted on the natural surface (outside), until heavy wear leads to the breakdown of the closed enamel layer. Secondary functional surfaces are characterized by widely exposed dentin in combination with enamel bands surrounding the dentin fields, enclosing either the whole occlusal surface or forming infolds and enamel islets (see Ruf et al. 2020, this volume). The difference in hardness of enamel and dentin creates protruding enamel crests during the wear process, which have functional significance for food breakdown. Once the initially closed enamel layer is opened, the cross section of the enamel becomes a functional surface. The protruding enamel crests can form sharp edges with blade-like functions or rounded crests with rasp-like functions (Fig. 11.1). Such protruding enamel bands of secondary surfaces can also be impacted on the natural inner side, when they form solid vertical walls and the dentin in front is scoured. Simple tooth forms with a primary functional surface can be compared to an anvil or a mortar (Lucas 1979), more complex enamel patterns of secondary functional surfaces function like a rasp. Symmetrical enamel blades are similar to scissors, whereas asymmetrical enamel blades are reminiscent of a paper cutter. Rodent incisors form nipper-facets along their incisors (secondary functional surfaces) functioning like pliers (Koenigswald 2016).

This chapter focuses on the evolutionary changes of the cheek tooth morphology of some mammal groups and the change in their function observed from reconstructed mastication processes. That includes the evolutionary transition from bunodont teeth with primary functional surfaces, to specialized teeth with blades or rasps formed by secondary functional surfaces.

The process of mastication is essential to herbivorous mammals, because plant material generally is nutrient-poor and has a high fiber content (Batzli & Hume 1994, Sanson 2006). Tough cellulose cell wall material is processed poorly in mammalian guts and needs to be broken down by tools first to make the nutritive cell content available and to increase the surface area for digestive enzymes and bacteria (Crompton & Hiiemäe 1969, Lucas 1979, Wright et al. 2008). The evolutionary development of the cheek teeth of many large herbivores is dominated by opening the primary functional surface to form enamel bands and crests utilized to sufficiently break down plant-matter. In the mammalian dentition, premolars and molars are mainly involved in the mastication process (Crompton & Hiiemäe 1969, Lucas 1979). Due to natural selection, the complex crown morphology of cheek teeth in herbivorous mammals correlates with the type of diet, because the mechanical properties of the food strongly influence the shape of tooth crown elements after eruption (Maier 1980, Lucas 2004).

Most tools work only well when moved in a specific direction. Therefore most differentiated teeth follow a very distinct pathway during mastication. Of the different phases in the complex chewing cycle (Hiiemäe 1978, Hiiemäe & Kay 1972) only the power stroke is important for the function of teeth. It directly can be observed by watching the chewing animal or more accurately using cineradiography, or indirectly from muscle activity in living animals (Crompton et al. 2008, 2010), but in fossil mammals the movement needs to be reconstructed from the morphology of the occlusal surface, the wear facets, and striations. Describing the movements as vertical, horizontal, proal, palinal, or lateral is not sufficient, because most movements do not follow these traditionally used directions but are oblique in three dimensional space. The movements can be reconstructed with digital models of antagonistic teeth in the Occlusal Fingerprint Analyser software (OFA) (Kullmer et al. 2009, 2020, this volume) developed in the DFG Research Unit 771, and the directions of the movements can be visualized in two dimensions using the mastication compass (Koenigswald et al. 2012).

Institutional abbreviations

BSPG	Bayerische Staatssammlung für Paläontologie und Geologie, München
HLMD	Hessisches Landesmuseum Darmstadt
IGPB	Institute of Geosciences, Section Paleontology, Rheinische Friedrich-Wilhelms-Universität Bonn
TAMU	University of Texas at Austin, Texas Memorial Museum
UM	University of Michigan, Paleontological Museum

Perissodactyla

The earliest perissodactyls used the primary occlusal surface in combination with a two-phased power stroke (Janis 1990, Engels 2011, Mihlbachler et al. 2011). In the different groups of perissodactyls discussed here (Fig. 11.2), exposed dentin and the cross-section of enamel (i.e., the secondary functional surfaces) are increasingly utilized, and enamel bands are the dominant cutting devices for the mastication process. Despite the great diversity, the ectoloph remained the main element with shearing function in various groups, often combined with a shear-grinding function in the second phase of the mastication cycle. However, the ectoloph in modern horses is integrated in the almost flat occlusal surface of the hypsodont teeth, and the alignment of the different chewing directions and inclinations of the two chewing phases to one horizontal movement of the power stroke led to the evolution of large rasp-like structures extended over the entire tooth surface. In rhinocerotoids the occurrence of modified enamel (i.e., vertical Hunter-Schreger bands) correlates with a higher degree of compression of the cusps and extreme lophodonty in this group (Rensberger & Koenigswald 1980). The absence of such modified enamel structure in the equoids, even in the ectoloph, correlates with the lesser importance of the ectoloph in equoids relative to brontotherians, chalicotherians,and rhinocerotoids.

Equoidea

Many studies investigated the evolution of equoid hypsodonty in particular, and the mammalian hypsodont dentition in general, which have been shown to be driven by the increasing grazing habit (e.g., MacFadden & Hulbert 1988, Fortelius et al. 2003, Strömberg 2006, Koenigswald 2011). The Equoidea comprise the two families Equidae and Palaeotheriidae. Both groups show a distinct evolutionary transformation of their molar morphology (Stirton 1941, Rensberger et al. 1984, MacFadden 1988, 2005, Joomun et al. 2008, Engels 2011, Evans & Janis 2014, Semprebon et al. 2016, Engels & Schultz 2018). The earliest forms have multi-cusped teeth that mainly compress the plant materials. Shearing and cutting along an enhanced ectoloph evolved in the Palaeotheriidae, whereas Equidae developed hypsodont teeth with a rasp like occlusal surface.

Early equoids (e.g., *Hyracotherium*, *Eurohippus*, *Propalaeotherium*) are mostly folivorous and frugivorous and show a mastication cycle consisting of two phases. In those forms the upper molars show larger contact surfaces and a cutting edge on the buccal side. These shearing surfaces are more strongly developed compared to more generalized functional precursory forms like the 'condylarth' *Phenacodus*. By developing a "mesostyle-like" structure the guidance of the power stroke is stronger in early equoids than in their bunodont ancestors with a simpler rounded cusp pattern (Fig. 11.3). The enhancement of the buccal cutting function leads to a rather effective reduction of leaves. In early equoids the buccal cusps of the upper premolars and molars show distinctly larger facet groups in combination with well-developed enamel crests compared to the 'condylarth' *Phenacodus*.

Eocene to Miocene Equidae (e.g., *Mesohippus* and *Anchitherium*) and derived Palaeotheriidae (e.g., *Palaeotherium* and *Plagiolophus*) develop an elevated ectoloph with a higher occlusal relief and a distinct cutting function compared to the early forms, in combination with a stronger guidance of the lower molars sliding along the mesostyle. However, each family developed different adaptations to the vegetation of their changing environment. During evolution, the members of the more derived Equidae show a tendency to develop a one phase power stroke with emphasis of the cutting function on the buccal ectoloph (Fig. 11.4). In addition, enhancement of the shearing function was expressed by a double-blade on the buccal side of the upper molars and steep inclined shearing facets. The development of this functional complex enhancing cutting and shearing function exhibits similarities to the extant hypsodont members of the family with a flat occlusal rasp.

Derived Palaeotheriidae show different adaptations and maintain the two-phased power stroke. The occlusal surface of the molars clearly shows a separation of a buccal cutting/shearing function and a lingual crushing function constituted by a well-developed buccal facet group for processing leaves. Due to the more versatile teeth with distinct large shearing surfaces on the buccal side and a second phase in the power stroke, they are also able to process harder food items (e.g., twigs, hard fruits). The reduction of leaves appears not as efficient as in Paleogene-Neogene Equidae such as the Eocene-Oligocene *Mesohippus* and Miocene *Anchitherium*. Those equids are in general specialized leaf eaters, but are able to also comminute grasses. In comparison, modern hypsodont horses (genus *Equus*) with a flat occlusal surface and protruding enamel crests are adapted to a diet predominantly consisting of grass.

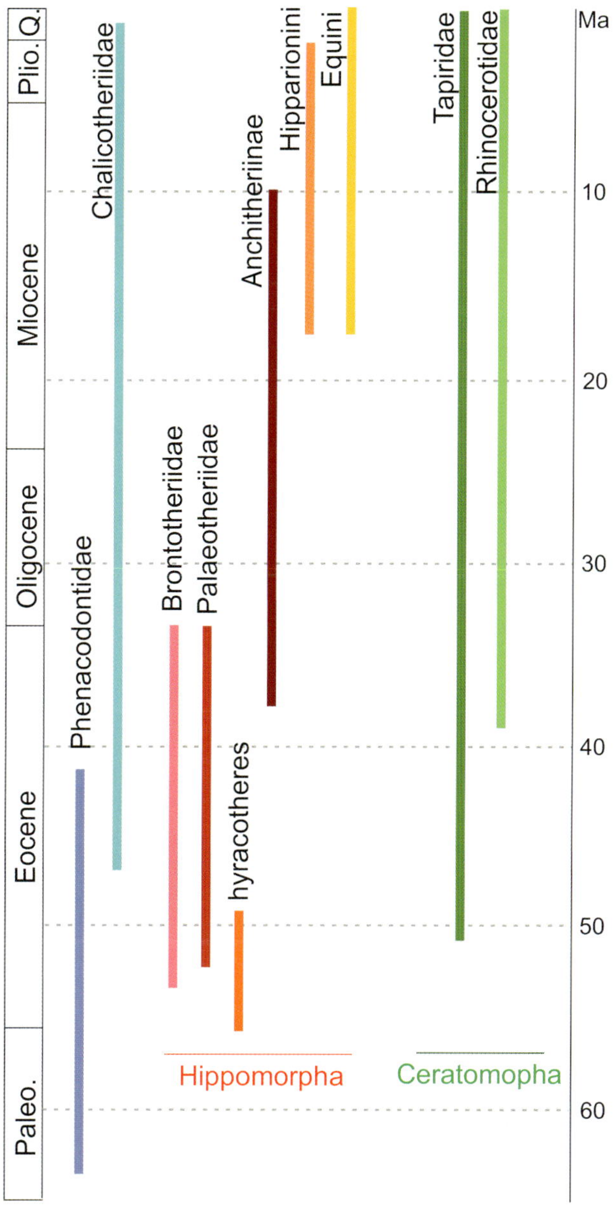

Fig. 11.2. Biostratigraphic ranges of perissodactyl groups discussed in this chapter. This diagram is based on data from MacFadden (2005), Strömberg (2006), Damuth & Janis (2011) and Mihlbachler et al. (2011). Abbreviations: Ma, mega-annums (million years); Paleo., Paleocene; Plio., Pliocene; Q., Quaternary.

Equinae (i.e., Hipparionini and Equini) evolved hypsodont molars (MacFadden 2005, Strömberg 2006) in a heterochronic process of the tooth formation (Koenigswald 2011, Madden 2014). The direction of the power stroke of phase I and phase II becomes similarly adjusted during evolution. Thus, the lower molars cross the uppers from buccal to lingual horizontally in one stroke with no distinct separation of the two phases (e.g., centric occlusion) (Fig. 11.4). The occlusal surfaces function as rasps because softer materials cementum and dentin fill the gaps between the protruding harder enamel tooth elements. Such teeth

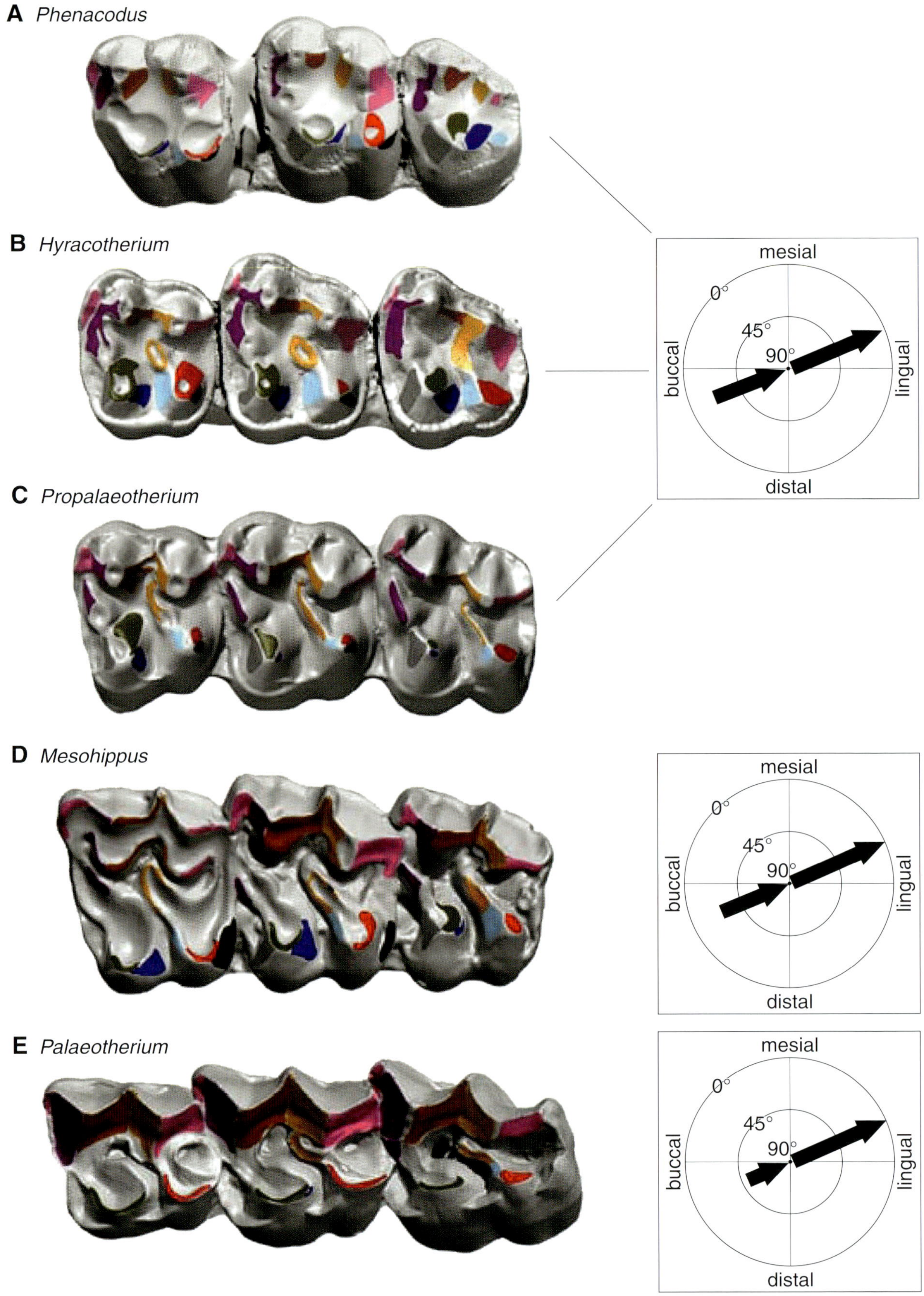

Fig. 11.3. Molar series of *Phenacodus* and early Equoidea illustrating the enhancement of the ectoloph during the early equoid evolution. Molar facets colored after Butler (1951). **A,** *Phenacodus primaevus* (AMNH 15285); **B,** *Eohippus pernix* (USNM 522769); **C,** *Propalaeotherium hassiacum* (HLMD Me54a); **D,** *Mesohippus bairdii* (BSPG 1959 XXIII 169); **E,** *Palaeotherium euzetense* (NHMB StH 2046). The mastication compasses show the stepwise steeper inclination of phase I reflecting the elevation of the ectoloph. Phase II is almost horizontal in all taxa. Not to scale. Virtual 3D surface models. Modified from Engels (2011) and combined with Koenigswald et al. (2013).

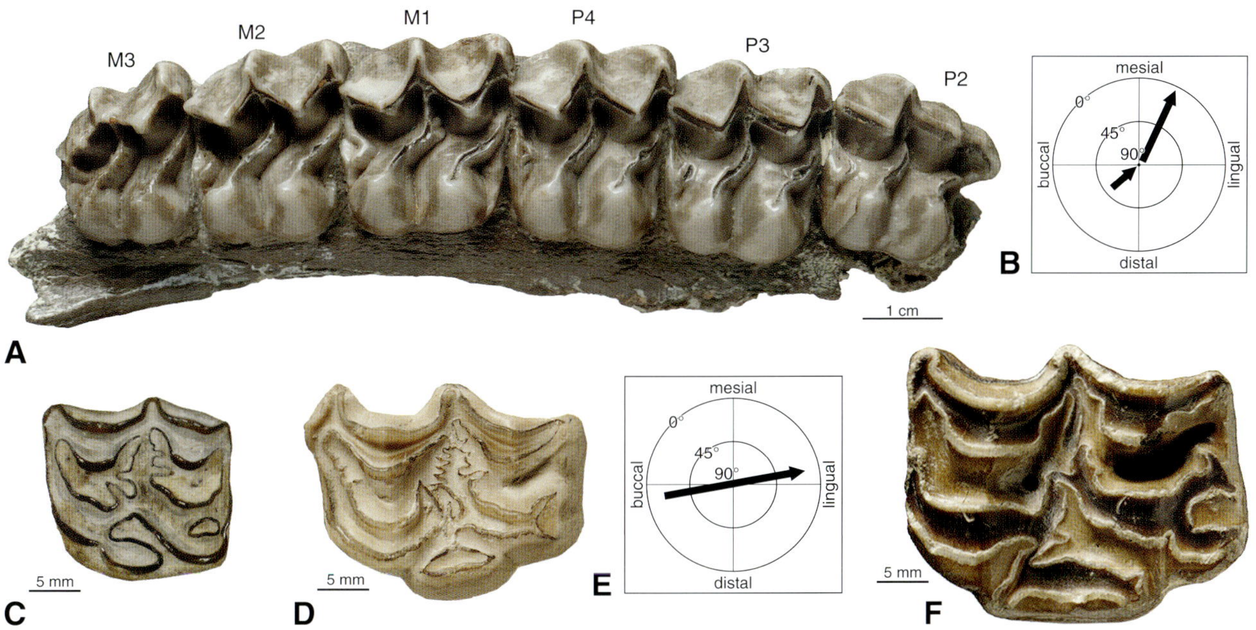

Fig. 11.4. Examples of the changing tooth morphology and mastication pattern in advanced Equoidea. **A,** *Anchitherium aurelianense steinheimense*, right maxilla with P2–M3, Miocene, Steinheim am Albuch, Germany (BSPG 1897 XIII4a); the premolars are molariform and the steep ectoloph is cutting during phase I. Abrasion along the crosslophs mark the beginning shear-grinding during phase II. **B,** mastication compass for *Anchitherium* illustrating the mesial shift of phase II. **C,** *Calippus (Merychippus) francisi*, left M1, Miocene, Noble Farm, New Mexico (TAMU 3016). **D,** *Hipparion* sp., right M3, Miocene, Samos (IGPB M7812). **E,** mastication compass for derived Equoidea with a flat occlusal surface. The power stroke is strictly directed bucco-lingually, phase I and phase II are continuous. **F,** *Equus caballus*, subfossil, right upper molar, Rhineland, Germany (IGPB M1850, mirrored); the two phases of the power stroke merged into a strict transverse movement, as indicated in the mastication compass (E). Photos: C, K. Bader; A, D, E, G. Oleschinski.

are regarded as highly specialized for grasses as diet (MacFadden 1988). The hypsodonty forms a reservoir of tooth material that compensates the high abrasion due to phytoliths and external abrasives such as grit or soil.

Tapiroidea and Rhinocerotoidea

In the Eocene the molar morphology of early rhinocerotoids and tapiroids is very similar. The upper molars show the typical pattern with a strong and elevated buccal ectoloph and two linguo-buccally oriented crosslophs, the antero- and posteroloph (Rose 2006). The lower molars show two transverse lophids linked on the buccal side. Nevertheless, the mastication in these two groups develops in different directions as indicated by the facets on the lophs and lophids in the secondary occlusal surfaces. It has been shown that the heterogeneous occlusal surfaces may facilitate the comminution of heterogeneous browse, whereas uniform, broad grinding surfaces may enhance the comminution of physically more homogeneous grass (Taylor et al. 2013). In order to highlight the differences, the most primitive genera are not considered and the more derived ones of each group are discussed here.

The tapiroid mastication pattern was investigated in the Eocene *Lophiodon lautricense* from France, the Pliocene *Tapirus lundelius* from Florida, and the extant *Tapirus terrestris* from Brazil. *Lophiodon* is a rather large tapiroid (body size of a horse or a small rhino) with typical bilophodont lower molars and well-marked steeply inclined facets on the distal sides of the lophids (Koenigswald 2014). In the upper cheek teeth the facets are located on the mesial sides of the crosslophs. In the functional facets the dentin is exposed and the shearing crests are formed by the sharpened cross-sections of the enamel. The posterior sides of the lophids cut like blades against the anterior sides of the crosslophs. Striations indicate a jaw movement in a mesio-lingual direction with a steep inclination. A facet on the buccal side of the paralophid fits to this direction. The crosslophs of the upper molars correspondingly have steep facets on their anterior sides (Fig. 11.5C). The ectoloph is not continuous, but it directs the movement of the lower jaw in mesiolingual direction, especially along the pronounced parastyle. Attrition sharpens the crests of the various facets indicating that the food items were primarily cut by symmetrical blades during phase I of the power stroke, whereas no facets represent a second phase (Fig. 11.5D).

In the molars of the upper Pliocene *Tapirus lundelius* a very similar set of facets was observed and a similar mastication pattern can be assumed. The transverse lophs and lophids are functioning like blades. The guiding function of the parastyle, however, is somewhat minor and the buccal facets on the paralophid are minimal.

In the extant *Tapirus terrestris* the cutting function of the symmetrical blades is the same, but the lateral aspect seems to be even more reduced than in *T. lundelius*. No lateral movement during the chewing process could be detected in an X-ray observation of a living tapir (personal comm. Marcus Clauss). Therefore, the tapiroid mastication pattern concentrates on the shearing function of the transverse symmetrical blades.

Rhinocerotoidea share the basic pattern of the molars with tapirs (and several other perissodactyls) but during evolution the function developed in a different direction.

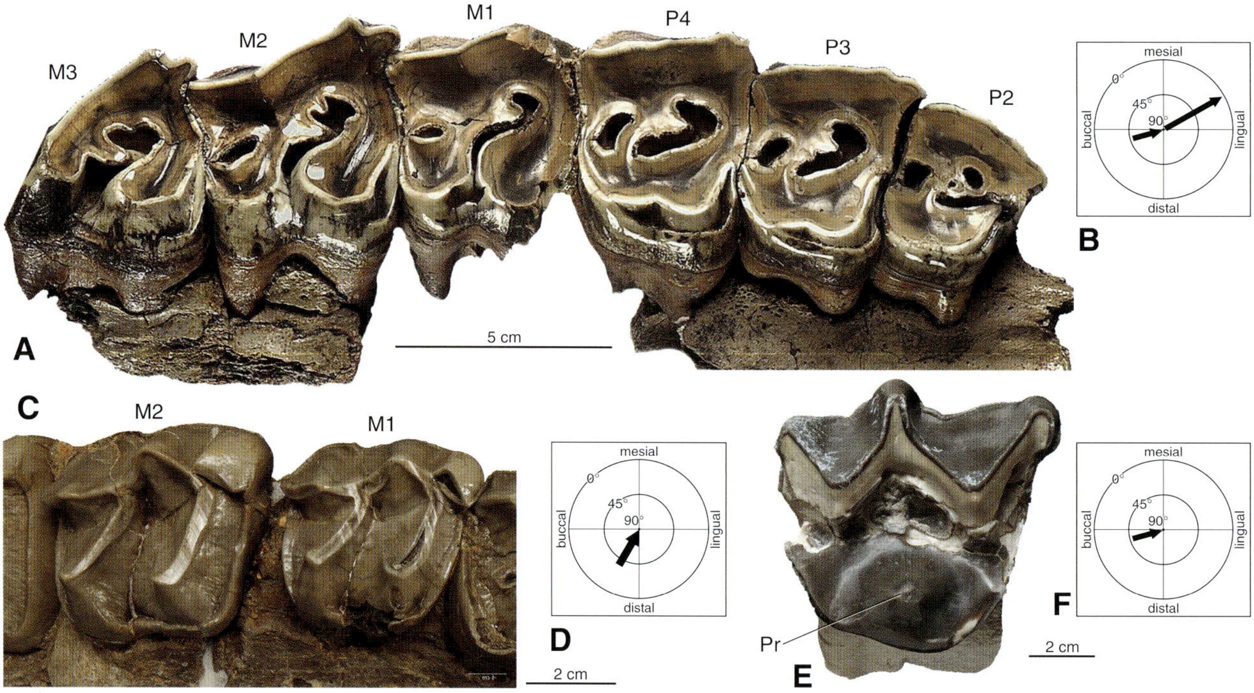

Fig. 11.5. In many perissodactyls the ectoloph has a shearing function. **A,** *Stephanorhinus etruscus* (Rhinocerotidae), right upper tooth row with P2–M3, Pleistocene, Brüggen, Germany (IGPB M4863); the ectolophs form the typical ectoloph-guillotine for phase I. During phase II the lower teeth shear across the crosslophs. The inclination between phase I and II is very distinct (**B**). **C,** *Lophiodon lautricense* (Tapiroidea), M1 and M2, Eocene, Robiac, France (HLMD-Ro 2, inverted); the ectoloph is of minor significance, but the crosslophs show blade facets of phase I of the power stroke. After cutting the food is compressed, but a phase II is lacking (**E**). **D,** *Megacerops* sp. (Brontotheriidae), right M2, Orellan, Badlands, South Dakota (HLMD WT735); the W-shaped ectoloph forms a cutting edge for phase I that ends at the massive protocone (Pr) functioning as a bumper. Thus, its mastication compass (**F**) shows no phase II. Photos: G. Oleschinski.

In the Late Eocene *Trigonias* from the White River Group the upper molars show a prominent ectoloph and two crosslophs. The dentin is exposed in an early stage of wear, but differently from tapirs. The buccal enamel of the ectoloph forms a continuous steep cutting edge supported by the dentin. In the two transverse lophs the dentin is exposed almost horizontally. The cutting ectoloph occurs in both premolars and molars (Fig. 11.5A). The antagonists for this device are shaped differently. Only a small facet is seen on the buccal sides of the lower molars and premolars. Thus, the ectoloph forms an asymmetrical blade with an abutment as antagonist, similar to a guillotine or paper cutter (Koenigswald 2016).

The cutting during phase I of the power stroke is followed by shear-grinding during phase II. The crosslophs in the upper molars as well as the lophids in the lower molars with the exposed dentin areas are well suited for shear-grinding with a low inclination. The direction of this grinding is oriented mesio-lingually (Fig. 11.5B).

The mastication process combining cutting and shear-grinding is very stable within tapiroids and rhinocerotoids, and occurs in the various genera of the Oligocene to the Pleistocene. This function is also found in the Eocene rhinocerotoid *Hyrachyus* and most of the later rhinoceroses, including the few extant species that are browsing: *Dicerorhinus sumatrensis*, *Rhinoceros sondaicus*, and the African *Diceros bicornis* (Hullot et al. 2019). Grazing and browsing extant rhino species are distinguished by tooth shape and microwear, and also by individual head posture (Schellhorn 2018), an additional indicator to identify grazing and browsing rhinos in the fossil record.

The extant white rhino *Ceratotherium simum* differs in its mastication pattern as does the Pleistocene *Coelodonta antiquitatis*. Their dentition is specialized for grazing and shear-grinding. The ectoloph is less prominently developed, allowing a shear-grinding movement across the entire occlusal surface which is rather flattened, and more complex transverse lophs increase the length of enamel crests in the rasp. Both genera show a trend towards hypsodonty to compensate for the increased wear. The Pleistocene *Elasmotherium* from Asia progressed in this trend for more intensive grinding even further (Kosintsev et al. 2018). The molars became euhypsodont and the crenulation of the enamel ridges in both crosslophs (and also the lower molar lophids) raised the effectiveness of the rasp.

Brontotheriidae and Chalicotheriidae

The upper molars of Brontotheriidae show a dominant ectoloph with a distinct W-shape and strongly folded prominent para- and metastyles, which resemble vertical pillars. The protocone forms an isolated large cone on the lingual side and the hypocone is somewhat smaller and often connects to the ectoloph via a minor crossloph (Fig. 11.5E, F).

The function of these teeth is mainly cutting along the W-shaped ectoloph (Fortelius 1985, Fahlke et al. 2013). The sharp edge is formed by enamel from the buccal side. It is so strongly inclined that the antagonist meets the enamel almost perpendicularly. The antagonist is formed by the continuous crest of the lower molar that forms a zig-zag line corresponding to the W-shaped ectoloph. In the lower molars slight facets occur on the buccal side that form an abutment.

The asymmetrical blade in the upper molars and the matching abutment in the lower molars perform a guillotine-like (or paper-cutter-like) motion as in other perissodactyls with an enhanced ectoloph. Because the direction of movement is almost vertical, it functions more like a punching process. The cutting along the ectoloph blade represents phase I that is followed by compression of the cut food parts against the rounded protocone and hypocone. With increasing wear the ectoloph facet shifts gradually to the buccal side and the space between the protocone and hypocone increases; however, both the cutting and compression functions are stable even in very old individuals.

Chalicotheres, although not closely related to brontotheres, possess a similar ectoloph morphology, but have lingual lophs instead of rounded cusps. A combination of punching, with a subsequent compression and shear along the lophs occurs in their dentition.

Cetartiodactyla

The complex mastication pattern of early terrestrial artiodactyls is characterized by the enhancement of a fourth cusp (i.e., the metaconule), from which the classical selenodont tooth pattern derived (Geisler et al. 2007). As in perissodactyls, the alignment of direction and inclination of the two chewing phases of the power stroke allowed a continuous jaw movement in a transverse direction, with shearing across a secondary occlusal surface. A trend in modern groups of cetartiodactyls (e.g., camelids, bovids, cervids, antilocaprids) exists to increase the efficiency of

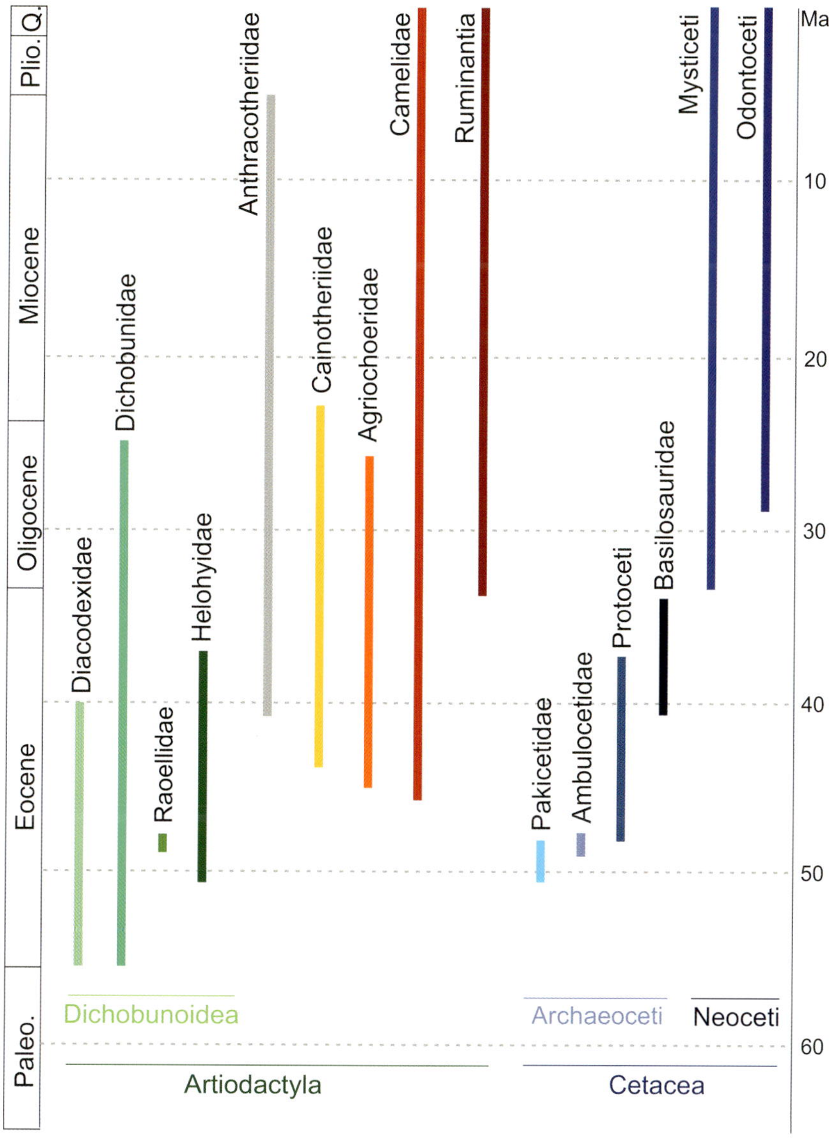

Fig. 11.6. Biostratigraphic ranges of cetartiodactyl groups discussed in this chapter. This diagram is based on data from Janis et al. (1998), Prothero & Foss (2007) and Houssaye et al. (2015). Abbreviations: Ma, mega-annums (million years); Paleo., Paleocene; Plio., Pliocene; Q., Quaternary.

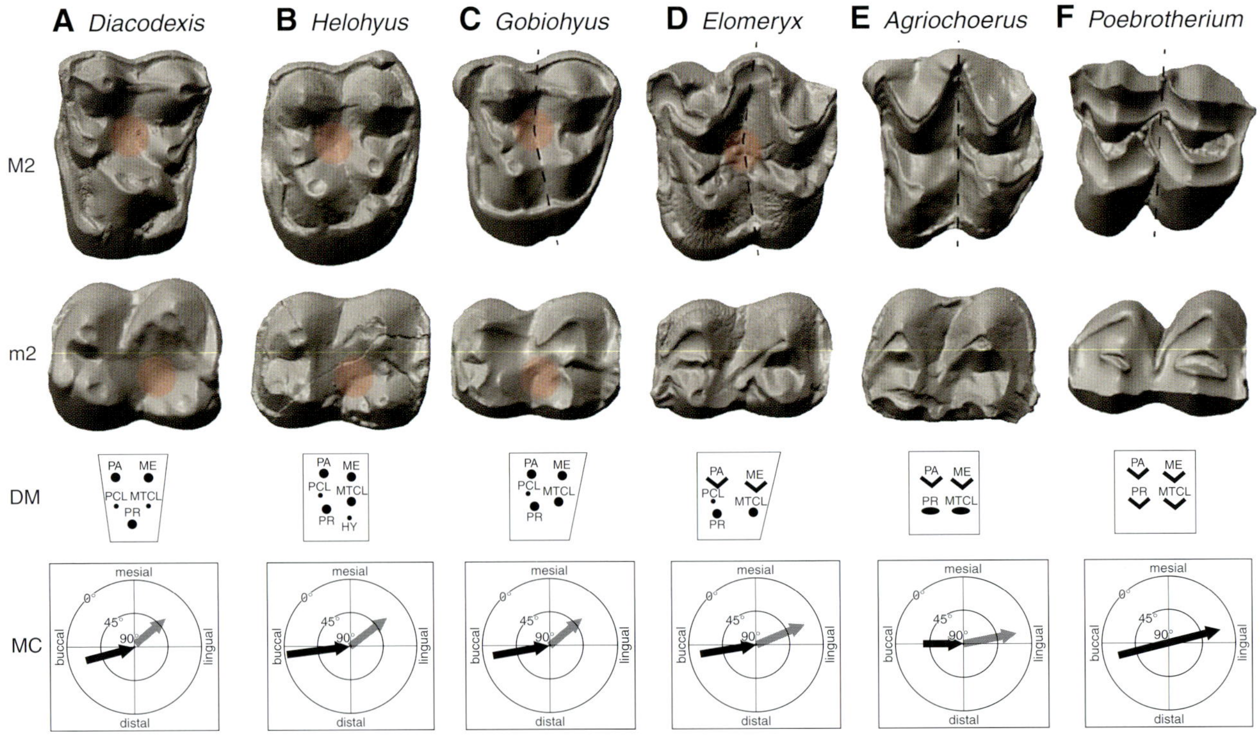

Fig. 11.7. Modification of trigon and trigonid basin (red dot) of the upper and lower molars during artiodactyl evolution to form the selenodont tooth type. In the derived types D to E the former basins form a trench (black dashed lines). **A,** *Diacodexis* sp. (M2, NMNH 5611; m2, CM 62046). **B,** *Helohyus plicodon* (M2, m2; AMNH 12147). **C,** *Gobiohyus orientalis* (M2, AMNH 26279; m2, AMNH 16283). **D,** *Elomeryx borbonicus* (M2, m2; NMB Mar 550). **E,** *Agriochoerus* sp. (M2, m2; NMNH 215024). **F,** *Poëbrotherium* sp. (M2, m2; AMNH 63821). In the first row upper second molars are illustrated, lower second molars follow in the second row below. Third row depicts simplified dental morphotypes (DM) of each taxon's upper molar. Last row, mastication compass (MC) for each taxon illustrating the shift from the ancestral two-phased power stroke to the one-phased power stroke of derived taxa. Not to scale. Virtual 3D surface models based on CT scans. Abbreviations: HY, hypocone; ME, metacone; MTCL, metaconulus; PA, paracone; PCL, paraconulus; PR, protocone.

the teeth processing fibrous diets for longer periods by the evolving hypsodont teeth. An exception are cetaceans. Marine cetaceans evolved from terrestrial artiodactyls with a two-phased mastication pattern and a primary functional occlusal surface (Gingerich et al. 2001, Thewissen & Bajpai 2001). During the adaptation to a piscivorous diet the tooth morphology was reduced to teeth with a single cusp that lost even their enamel cover in various species. The mastication pattern was simplified to a mainly orthal movement. Baleen whales, adapted to feed on very small crustaceans (krill), reduced their dentition completely.

Early terrestrial Cetartiodactyla

During the Paleogene radiation of cetartiodactyls numerous tooth shapes evolved. The selenodont dental pattern evolved convergently within several early terrestrial cetartiodactyl lineages (Fig. 11.6) (Janis & Fortelius 1988, Schwermann, 2016). The early cetartiodactyl *Diacodexis* exhibits a rather basic cheek dentition more similar to the tribosphenic tooth pattern of early therians with a typical two-phased power stroke. The two phases differ in inclination and direction but have equal significance for the power stroke: cutting (during phase I) and crushing (during phase II) the structural and reproductive plant parts of the dietary spectrum. A two-phased power stroke as reconstructed for *Diacodexis* is considered the ancestral pattern for North American and Eurasian terrestrial cetartiodactyls.

From this basal condition different evolutionary steps can be observed in other Eocene and later artiodactyls (Fig. 11.7), for example the enlargement of the occlusal surface by adding cusps or developing new functions by the conversion from bunodont to more selenodont cusps (Lihoreau & Ducrocq 2007). While the primary functional surface is of great significance with its complex morphology in the basal condition, the secondary surface is crucial for the functionality of selenodont cusps.

The dentitions of both *Dichobune* and *Helohyus* represent a stage of evolution, where the functional surface of the upper molars is enlarged by an additional cusp. The upper molars evolved a small hypocone and thus differ from the basic tribosphenic type (Fig. 11.7A,B). However, the power stroke is still similar to that of *Diacodexis*. In general the cusps are more rounded, and the crushing function is increased due to the presence of a hypocone.

The Asian *Gobiohyus* has an enlarged and lingually shifted metaconule (Fig. 11.7C). The buccal cusps of the upper molars show the beginning of a developing ectoloph and the enamel edge is extended by a large parastyle. The cusps are more pointed than the rounded cusps of *Dichobune* and *Helohyus* and wear produces doubled enamel edges on the distal and mesial crests of paracone and metacone. Despite the fact that the chewing motion is similar to that of *Diacodexis*, *Dichobune*, and *Helohyus*, structural plant elements probably constituted the greater part of the dietary spectrum. The teeth of *Gobiohyus* are the first in the fossil record of artiodactyls that indicate the beginning significance of the secondary occlusal surface.

This proceeding development can be observed in *Elomeryx*. The molars are intermediate between the bunodont and selenodont pattern. *Elomeryx* has selenodont buccal cusps on the upper molars on which wear produces doubled enamel edges (Fig. 11.7D). The lingual cusps in the upper molars are bunodont, and the trigon basin is partly reduced.

The similarities in the power strokes of *Elomeryx*, *Dichobune*, *Helohyus*, and *Gobiohyus* probably evolved convergently in their separate lineages (Schwermann 2016). The structural changes of the tooth surface with an improved shearing function indicate a greater portion of structural plant parts in the dietary spectrum. The reduction of the basins suggests a decreased importance of the crushing function.

Caenomeryx differs in its chewing function from the two-phased power stroke of the other taxa because of the parallel arrangement of cusps that allow a simple transverse movement. The molars of *Caenomeryx* function with the secondary surface. The cusps are selenodont and the basins are completely reduced. There is no distinct centric occlusion, but inclination changes after maximum intercuspation. A second phase is characterized by tooth-tooth-contact and compression following maximum intercuspation (Schwermann 2016). Overall, the dentition suggests a diet composed mainly of structural plant elements. In contrast, the small body size (around 2 kg, Tsubamoto 2014) indicates that fruits, seeds, sprouts and possibly insects were also part of the dietary spectrum.

The North American agriochoerid *Agriochoerus* and camelid *Poëbrotherium* both have (buno-)selenodont dentitions with four-cusped molars and a uniform power stroke resulting from combining chewing phase I and II into one simple movement (Fig. 11.7E, F). There is no functional differentiation. In upper and lower molars the enamel crests are double-bladed which improves the shear-cutting function. Structural plant elements like leaves and grass were main parts of the diet of both genera.

The described taxa except *Caenomeryx* occur in the Eocene and thus represent an early evolutionary chapter of the very diverse group of artiodactyls. In modern artiodactyls the selenodont tooth morphology is dominant, and in many taxa within the Camelidae and Bovidae selenodonty is paired with hypsodonty.

The functional comparison of the basal bunodont teeth and the early selenodont teeth (and stages in between) shows that the masticatory movement changes from the basal two-phased condition to a unified, single chewing motion, which is performed in a stereotypic manner due to the complex selenodont cusp and crest morphology. As a result, the wear facets develop homogeneously and the ability to break down harder plant parts with each power stroke increases. The crushing function, which is performed by the protocone/talonid and hypoconid/trigon basin in tribosphenic and early artiodactyl bunodont teeth is reduced and shifted to the crests of the selenodont morphology. The combination of enamel bands and scoured dentin basins shears, crushes, and simultaneously cuts the food items, but in a different way than the mortar and pestle function of tribosphenic or bunodont artiodactyl molars (Schwermann 2016). The cutting and shearing function becomes increasingly important (Butler 1972). With the complete change to the selenodont tooth type the functionality of the molars thus shifts from a primary occlusal surface of the bunodont dentition to a secondary occlusal surface of the selenodont dentition, using the exposed enamel ridges, enamel islets, and worn dentin basins in a rasp facet. The functional changes of the teeth go along with a shift in the dietary spectrum: While in early artiodactyls (e.g., *Diacodexis*) with blunt cusps reproductive plant elements (i.e., nuts, fruits, seeds, flowers) were the preferred diet, a shift to structural plant parts (i.e., leafs, nodes, internodes) as preferred diet occurred in artiodactyls with selenodont dentitions (Lucas & Luke 1984, Lucas 2004).

Aquatic Cetartiodactyla

The greatest evolutionary modification in tooth morphology and mastication occurred during the evolution of whales. Whales evolved from terrestrial Cetartiodactyla (Gingerich et al. 2001), possibly Raoellidae (Thewissen & Bajpai 2001), that had four-cusped molars like other early herbivorous terrestrial cetartiodactyls such as *Diacodexis*.

Eocene terrestrial Cetartiodactyla gave rise to the marine and carnivorous Odontoceti with single-cusped teeth, as well as to Mysticeti which reduced the teeth and use baleen for filter feeding (Fig. 11.6).

Modifications that happened during the evolution from terrestrial raoellids or raoellid-like cetartiodactyls to fully aquatic cetaceans included various changes, such as the simplification of the tooth morphology leading to single-cusped teeth that finally lost their enamel, modification in the dentition leading to homodonty, polydonty, monophyodonty, and a functional shift from mastication to filter feeding. The fossil record provides information on several intermediate stages.

Raoellidae from the Eocene of India (Fig. 11.8A) have molars with four distinct cusps that enable mastication (Koenigswald et al. 2012). The main function during phase I of the power stroke is compression of the plant material, and there is a slight mesial movement during phase II (Thewissen et al. 2011). The wear is limited and thus the primary surface persists. According to an analysis of stable isotopes the raoellid *Indohyus* was proven to be herbivorous (Thewissen et al. 2011).

Pakicetidae, Ambulocetidae and Protocetidae were semiaquatic with fully functional hind limbs and – most probably – webbed feet. In the dentition, the premolars are enlarged and have a triangular shape, if viewed from the side, with shearing facets. Their size increases from front to back. The lowers intercalate with the upper ones, which is useful for catching fish. The molars are smaller than the premolars and the uppers have one cusp on the lingual side (i.e., the protocone) and no talonid basin in the lowers, reducing their function in mastication. Pakicetids have numerous adaptations for swimming and became piscivorous (Gingerich 1983, Thewissen et al. 2009). But they were not fully aquatic yet, at least for giving birth they came on land, which is also assumed for the protocetid *Maiacetus inuus* (Gingerich et al. 2009).

During the Middle Eocene, the next evolutionary level was reached by the fully aquatic Basilosauridae with reduced hind legs. They still had an almost complete dental formula lacking only M3, and replaced the deciduous premolars (Uhen 2004). The incisors and canines are pointed for grasping and catching slippery pray. The

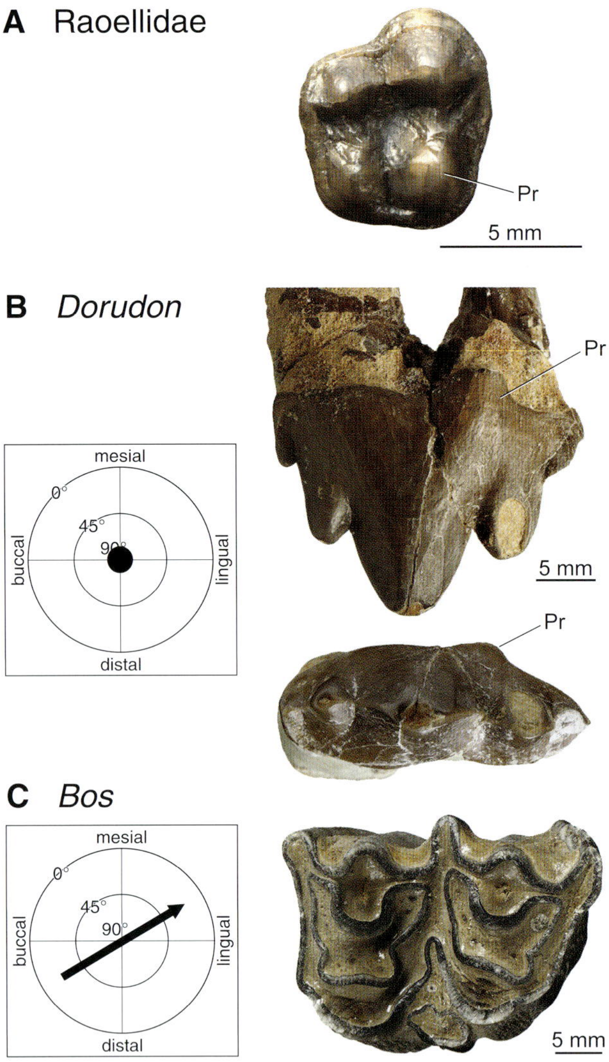

Fig. 11.8. Illustration of the drastic modification of tooth morphology and mastication pattern during the evolution of Cetartiodactyla, and in particular cetaceans. **A,** Raoellids (Middle Eocene, India, IGPB-KOE 2025) have square-shaped simple four-cusped molars with a primary functional surface. **B,** in the early whale *Dorudon atrox* (M2, Eocene, Wadi Hitan, Egypt, UM 101215) the premolars and molars are adapted to catching fish and molar morphology is reduced. The protocone (Pr) is a small swelling on the lingual side (lower picture occlusal view). The jaw movement is mainly orthal as shown in the mastication compass (central black circle). Terrestrial artiodactyls such as (**C**) *Bos primigenius* (upper molar, Pleistocene, Groß-Rohrheim, Germany, IGPB M 7824) evolved hypsodont teeth with functional flat secondary functional surfaces and one continuous power stroke. Photos: G. Oleschinski.

premolars are enlarged, laterally compressed, and have denticulate shearing crests. In the molars the lingual cusps (i.e., protocone and hypocone) were further reduced and the molars became premolariform. In upper molars the protocone is just a small bulge (Fig. 11.8B). The molars are distinctly smaller than the premolars and functionally only phase I of the chewing cycle is retained. In the Upper Eocene deposits from Egypt all teeth of *Basilosaurus* and *Dorudon* are fully covered with enamel (Fig. 11.8B). Larger prey was undoubtedly sectioned by the large denticulate premolars.

Coronodon havensteini, from the Oligocene of South Carolina (USA), which resembles basilosaurids in the dentition, was described as the first mysticete, although it does not have baleen (Geisler et al. 2017). The denticulate premolars of the lower jaw are of similar size and overlap with slots between them, through which water can pass. This indicates, together with other characters, that with slightly closed jaws water was pressed through and the small prey was kept in the mouth cavity like in a creel. The incisors show some tooth wear, probably resulting from sediment particles within the extruded water. Tooth function thus had totally changed because the process of chewing or catching and holding prey was no longer necessary. Modern Mysticeti reduced their teeth completely and replaced them with baleen, a novel integumentary structure (Peredo et al. 2018). They use ram feeding or suction feeding to get their food into the giant mouth (Marx et al. 2016).

From the basilosaurids another lineage led to the toothed whales. The dolphins are characterized by a large number of single-cusped teeth. In *Stenella attenuata* 44 teeth are present per quadrant. Homodonty replaced the traditional diversity seen in the mammalian dentition. The gene expression of BMP4 is regarded as a trigger for the single cusped teeth in the rear of the tooth row (Armfield et al. 2013). Thus, polydonty has replaced the tooth formula that in most lineages of terrestrial mammals is stable or shows reduction more often than augmentation. The pointed teeth are often covered by thin enamel. In addition, tooth replacement was suppressed, making toothed whales monophyodont. These teeth have lost their chewing ability. The numerous teeth are suitable for catching and holding fish and other prey until they are swallowed.

In the large toothed whales, such as *Physeter microcephalus* or *Orcinus orca* the number of teeth is not exaggerated but each tooth is much larger. Their function for catching and holding the prey is recognizable from their intensive wear. The initial enamel cap is mostly worn off and the dentin tooth grows continuously. In the extant *Monodon* the continuously growing incisor is protruding from the mouth forwardly and used for various functions. A similar tooth, but turned backwards, was evolved by the Late Miocene *Odobenocetops* from Peru (Muizon 1993).

Proboscidea

Although proboscideans are fairly similar in their overall appearance and have many characteristic dental features in common (e.g., tusks, horizontal tooth replacement, and dominance of the last molars) the various lineages developed very different molar morphologies and mastication patterns. Certainly the various tooth morphologies were adapted to different diets, but the enormous amount of plant material required daily by the large body size may prohibit the specialization into a very narrow food source.

The earliest proboscideans come from Africa but their fossil record is rather incomplete. *Eritherium* from the Paleocene of Morocco (Gheerbrant et al. 2012) has low-crowned molars. Lingual and buccal cusps oppose each other forming the prerequisite for transverse lophs that evolved in later taxa. In *Phosphatherium* (Gheerbrant et al. 2005) the molar cusps form transverse ridges that regularly have facets, in uppers on the anterior side and in lowers on the distal side. The transverse crests between the cusps are used for cutting. Thus the lower jaw is lifted with a slight anterior movement.

In the Early Miocene three proboscidean families expanded to Eurasia: Deinotheriidae, Mammutidae, and Gomphotheriidae. Their fossil record is more complete and the mastication process is better understood (Koenigswald 2016).

Deinotheriidae

Deinotheres retain their bilophodont molar structure throughout the Miocene without adding additional lophs at the posterior end of their molars (Harris 1975). Functionally, the deinothere pattern of mastication is a completion of the mastication pattern of the Phosphatheriidae. The lophs intercuspate deeply and create a broad facet in the lower molars on the distal side, and in the uppers on the anterior side. The sharp transverse crests are formed by the enamel of the leading side of each loph, where the food items are cut during phase I of the power stroke (Fig. 11.9A). Thus, deinothere molars form typical symmetrical blades. Phase I continues with progressing intercuspation. The cut food items are compressed between the enamel-covered sides of the following loph. No facets indicate a second phase of the power stroke in lateral direction. When lophs are worn down the remaining enamel ridges function as a rasp. Since the deinothere dentition has no horizontal shift, the entire battery of premolars and molars functions as a rasp in very old individuals.

Mammutidae

Mammutids are very conservative in their molar morphology and thus in their mastication pattern (Fig. 11.9B). The molars have up to three transverse lophs that intercuspate during mastication, but differently from those in deinotheres. Lophs are abraded by wear, but no development of facets on any specific side occurs. Therefore, during mastication the lophs meet the valley of the antagonist more or less vertically during phase I of the power stroke. Thus, the ingested plant material is strongly compressed. A slight lateral movement during phase II is indicated by striations and by the different abrasion in the pretrite and posttrite (the labial and lingual half-lophids) sides of the molars (Vacek 1877). During evolution only the M3/m3 are slightly elongated by an additional loph. Due to the horizontal tooth replacement, worn out premolars or molars are discarded, and in old individuals only the second and third molar are in function.

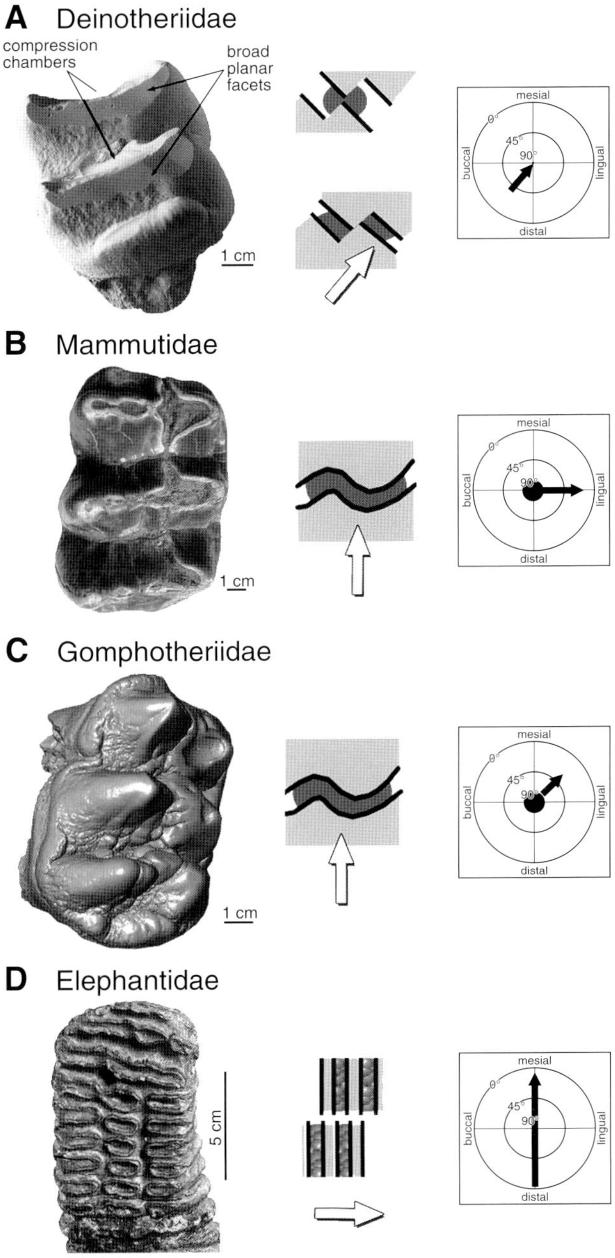

Fig. 11.9. Within proboscideans different tooth morphologies and mastication patterns evolved. **A,** in deinotheres during phase I, the food is cut along the lophs and consequently compressed between the primary functional surfaces. The power stroke ends with phase I, phase II does not exist. **B,** Mammutidae simply compress food without initial cutting during an orthal movement of the lower jaw. A slight lingual movement exists during phase II. **C,** gomphotheres compress food between the primary functional surfaces like mammutids by moving the lower jaw mesio-lingually. **D,** elephantids shear-grind the food along the typical rasp facets (secondary functional surface) performing a proal movement of the lower jaw. Modified from Koenigswald (2016).

Gomphotheriidae

Gomphothere molars have a rather bulbous structure. The main cusps form transverse lophs including additional conules (Tobien 1978). The valleys are filled with cusps that hinder a transverse movement (Fig. 11.9C). A thick enamel layer covers the cusps making the occlusal surface more solid, and extending the use of the primary occlusal surface for mastication. The mastication pattern of Gomphotheriidae is not obvious because the wear facets are mostly rounded and do not indicate a distinct direction of the jaw movement. Manipulation of isolated antagonistic molars indicates a complex chewing motion composed of mesial and mesio-lingual movements. Virtual 3D models of the occlusal surfaces allow a much more detailed analysis in the OFA (Kullmer et al. 2009, 2020, this volume). According to the OFA analysis, the teeth occlude during phase I more or less in an orthal movement. This is followed by a movement in mesio-lingual direction with low inclination during phase II. In each chewing cycle, only one side of the tooth row is in occlusion. The main function of the molars is to reduce the size of food items by compression and squeezing. Due to the movement of the lower tooth row in mesio-lingual direction, the contact is more intense on the pretrite side than on the posttrite side (Koenigswald 2016). When the teeth are worn the thick enamel ridges function like a rasp. Due to the horizontal tooth replacement, the second and third molars are functionally very important and the M3/m3 are elongated by additional lophs.

Elephantidae

Elephantidae evolved during the Late Miocene. The molars of derived Elephantidae are characterized by hypsodonty and have an almost flat occlusal surface. The transverse ridges are formed by the enamel band enclosing the exposed dentin of the mesio-distally compressed lophs (Fig. 11.9D). The molar construction is characteristic for multicusped hypsodonty, because the individual lophs are elongated and fuse only at their very base (Koenigswald 2011). The space between the lamellae is filled with cementum. The primary surface is functionally unimportant, because the secondary one forms as soon as the lamellae are integrated into the chewing process and thus the occlusal surface. The sequence of lamellae forms an efficient rasp with ridges oriented almost perpendicularly to the jaw movement in mesial direction (Maglio 1972, Fortelius 1985).

The cheek teeth are replaced horizontally. The teeth are forming at the posterior end of the jaw and push forward. Because of the great length of the teeth, the anterior part may be worn down while lamellae are being added at the posterior end. The functional area in a jaw included either one full molar or two halves of adjacent molars. The functional area increases continuously with age corresponding to the increasing body size (Anders et al. 2011).

The occlusal surface functions like a rasp for breaking down the tough plant food. The power stroke is horizontal and mainly in a mesial direction perpendicular to the crests. Phases I and II cannot be distinguished.

Conclusion

The evolution of tooth function and adaptation of chewing movements is driven by different selective processes in the different groups of large-bodied herbivores, and it is difficult to find general rules in such diverse groups. However, the groups discussed above all have a brachydont four-cusped tooth morphology as ancestral pattern in common. Generally, three trends are apparent: (1) a simple ancestral state, using the primary tooth surface (closed enamel layer) and a two phased chewing cycle with considerable crushing function, inherited from a most probably omnivorous origin (MacFadden 2000); (2) modification of the primary surface by restructuring the cusp arrangement of the cheek teeth integrating secondary tooth surfaces (enamel layer partly opened to form enamel crests and dentin basins) in combination with a simplified one-phased chewing cycle; (3) hypsodonty and euhypsodonty (tooth crown higher than wide) integrating cementum, to add a third level of tooth tissue of different hardness and in order to fill the space between tall tooth lophs for stability.

Morphology-related aspects

In order to access the fibrous herbivorous diet the groups discussed above followed different evolutionary pathways, involving the modification of cusp numbers, arrangement of cusps, crown height, direction and length of the chewing path (among others). Hypsodonty (high crowned teeth) has evolved in several large herbivores as an adaptation to the abrasive plant food, and the proportion of appearance of hypsodont teeth during mammal evolution increased over time (Janis & Fortelius 1988, Jernvall & Fortelius 2002, Damuth & Janis 2011, Samuels & Hopkins 2017). Looking at the extant species of grazers, the majority shows selenodont molars (utilizing the secondary functional surface) in combination with a transverse movement of the lower jaw, reflecting the advantage of selenodonty in connection with plant-based diet. Being able to access more nutrients benefits larger body sizes, and this in turn allows for increased tooth size (Janis & Fortelius 1988). A higher complexity level of the secondary occlusal surface is a relevant evolutionary process that was realized several times during evolution. While grazing species tend to have more elaborated and complex folded enamel ridges, browsers have less complex enamel ridges (Gailer & Kaiser 2014, Winkler & Kaiser 2015).

One main step in evolution was to achieve a larger occlusal surface, realized by the enlargement of the hypocone in early forms of Perissodactyla or the metaconule in most Cetartiodactyla (Hunter & Jernvall 1995). In early forms a primarily polished enamel surface (i.e., facets) works against an opposing enamel surface during chewing. The secondary functional surface is formed by opening the initially closed primary functional surface during the chewing process early in ontogeny. This wear process exposes enamel cross-sections across the occlusal surface forming continuous enamel bands and dentin basins, which act as shearing surfaces and compression chambers. The less hard dentin is excavated during the lifetime of an animal and the enamel bands protrude and in some forms rasp-like structures are formed (e.g., molars of the proboscidean *Elephas*, the equid *Equus*, and some modern bovids). The structural difference between enamel and dentin is therefore an advantage for specialized occlusal surfaces, and tough plant food is sheared along enamel crests realized in most of the large herbivore groups discussed in this chapter. The proboscidean rasp with multiple tranverse enamel crests is an extreme example for this utilization: however, in several perissodactyls only parts of the occlusal surface of a tooth form horizontal enamel bands (e.g., protoloph and metaloph in the upper molars of rhinoceroses) while the other parts are enhanced steep shearing surfaces (e.g., the ectoloph), and the combination of both is key to plant food reduction. The occurrence of a pronounced ectoloph in equoids, tapiroids and rhinocerotoids reflects the fact that phylogeny limits the possibilities to modify the occlusal surface to adapt to a changing environment.

An exception of the evolutionary scenario from simple to complex enamel patterns is seen in the aquatic Cetartiodactlya. While also having dental origins in a simple cusped bunodont tooth pattern and a two-phased chewing cycle (e.g., Raoellidae), aquatic Cetartiodactyla developed an even more simple morphology of the teeth as an adaptation to the aquatic life style, where catching and swallowing prey supersedes mastication. In accordance with the process of adapting to a new habitat with a different environment, a change in feeding strategy occurred during the transition of terrestrial raoellids or raoellid-like cetartiodactyls to fully aquatic cetartiodactyls in the Middle Eocene. Toothed whales are active hunters, feeding on fish and octopus. Their derived archaeocete ancestors (e.g., Late Eocene *Dorudon*) retained only one phase of the chewing cycle and their cheek teeth are uniformly modified into cusp rows fully covered with enamel. Baleen whales modified their dentition into a filter feeding apparatus consisting of thin keratinized lamellae (baleen).

Diet related aspects

The rise of grasslands in the later Cenozoic triggered a number of dental adaptations in herbivorous mammals as described for the perissodactyls and cetartiodactyls discussed above. The groups of mammals discussed here are mainly herbivorous and can be broadly classified into two groups: browsers preferring leaves and grazers with a diet mostly consisting of grass. To differentiate between these two categories is important, because the mechanical properties, as well as the architecture and physical structure of grass and browse determine how the animals forage and masticate. In addition, the intake and procession rate of food is important for the maximization of energy production over shorter time scales (i.e., minutes, hours) which in turn is linked to different foraging strategies (Stephens & Krebs 1986). It is well known that grazers tend to have more robust mandibles and larger masseters compared to browsing species, because they have to comminute the relatively tough, highly fibrous plant food more thoroughly (Janis 1995, Fletcher et al. 2010). In general, plant food has low nutritive value until the cell contents of plant parts are released. Mastication is therefore an important process to break down the material to make nutritives available. Chewing reduces the size of the particles, which in turn crucially reduces the digestive time in the intestines. It has been shown that Perissodactyla have stiffer mandibles than ruminant species (Zhou et al. 2019), which is linked to more heavy chewing of food in Perissodactyla before ingestion. This demands higher bite forces and greater stress resistance in perissodactyl mandibles, while ruminants shift comminution to a later state (rumination) with less mechanical effort. In fact, the smaller the particles are the larger is the surface area for the digestive enzymes to be effective. Digestion time is a significant constraint (Searle & Shipley 2008). Ruminants utilize regurgitation (i.e., re-exposure to saliva, chemical breakdown) and reprocessing (i.e., re-masticate) of food as an essential part of the digestion process, to regulate the rate of reduction of the particle size and hence the timing of the passage of the particles through the digestive system (Clauss et al. 2009, Fritz et al. 2009). Zhou et al. (2019) suggested that ruminants chew more "sloppily" regardless of the type of food, while hindgut fermenters chew more precisely. However, in modern ruminants feeding mainly on grasses, selenodonty (typical secondary occlusal surface) dominates with sharp enamel blades.

The large herbivore groups discussed in this chapter evolved their dentition from a simple cusped brachydont tooth pattern by modifying cusp arrangements and facet patterns of the primary functional surface and, developing complex shear and rasp structures of the secondary functional surfaces over time. Those groups increased the ability to break down the fibrous plant food. In addition, large herbivores like equoids and a large number of terrestrial cetartiodactyls adapted to the spreading of open range grasslands by developing hypsodont (high-crowned) teeth (MacFadden 2005). Hypsodont molars can tolerate intensive abrasion and their occlusal surfaces function constantly over a long period of time (Janis & Fortelius 1988, Damuth & Janis 2011, Koenigswald 2011). Two main factors that drive intensive wear and are known to trigger hypsodonty during evolution are phytoliths and grit (Williams & Kay 2001, Kaiser et al. 2013, Damuth & Janis 2014). Generally, grasses contain more phytoliths (hard opaline silicates) than browse (Hodson et al. 2005, Sanson et al. 2007). Because of a lower feeding height compared with browsing species, grazers tend to ingest more soil abrasives with plant matter, additionally increasing tooth wear (Janis & Fortelius 1988, Damuth & Janis 2011). It has been shown in rodents that continuously growing teeth require persistent stem cell presence at the base of the tooth (cervical loop) promoting tooth growth when the crown material is lost to wear by attrition (Jernvall & Thesleff 2012, Renvoisé & Michon 2014).

All ruminants with a diet mainly composed of grasses have high-crowned selenodont teeth (Williams & Kay 2001). During evolution, integration of the secondary functional surface (enamel crests and dentin basins) into the molar morphology for shearing tough fibrous plants in combination with a high molar crown to compensate for extensive tooth wear was crucial for herbivorous mammals to adapt to the changing environment with expanding grasslands. In correlation with the morphological change, a shift from a two-phased power stroke to a simple one-phased movement occurred several times in different groups independently. In all groups the tendency to increase the size of occlusal elements and functional surfaces is notable.

Acknowledgments

We thank Peter Göddertz, Georg Oleschinski, Olaf Dülfer, and Beate Mühlens for technical and organizational support (Bonn, Germany). We benefitted from the generosity of curators and colleagues around the world for providing material and discussion for this study: Elmar P. J. Heizmann, Reinhard Ziegler (Stuttgart, Germany), Gertrud Rößner (München, Germany), Ernest L. Lundelius (Austin, Texas, USA), Philipp D. Gingerich (Ann Arbor, Michigan, USA), Kenneth D. Rose, François Gould (Baltimore, Maryland, USA), David Bohaska (Washington D.C., USA), Marisa Blume, Eric Milsom (Darmstadt, Germany), Elvira Brahm, Jörg Habersetzer, Michael Ackermann (Frankfurt, Germany), Loïc Costeur, Martin Schneider (Basel, Switzerland), Sally Shelton (Rapid City, South Dakota, USA), Judy Galkin (New York, USA), Rico Schellhorn (Bonn, Germany), Amy Henrici (Pittsburgh, USA). We are grateful to Philipp D. Gingerich for his advice and insights on the evolution of whales part of this chapter. We thank referees Christine M. Janis and Jessica M. Theodor for their input and advice that greatly improved the manuscript of the chapter.

References

* indicates publications that originated from the DFG Research Unit 771.

Academy of Prosthodontics. (2005): Glossary-of-prosthodontic-terms (8th edition). Journal of Prosthetic Dentistry 94: 10–92.

*Anders, U., Koenigswald, W. v., Ruf, I. & Smith, B. H. (2011): Generalized individual dental age stages (IDAS) for fossil and extant placental mammals. Paläontologische Zeitschrift 85: 321–341.

Armfield, B. A., Zheng, Z., Bajpai, S., Vinyard, C. & Thewissen, J. G. M. (2013): Development and evolution of the unique cetacean dentition. PeerJ 1:e24. https://doi.10.7717/peerj.24

Batzli, G. & Hume, I. (1994): Foraging and digestion in herbivores. In: Chivers, D. & Langer, P. (eds.). The Digestive System in Mammals: Food, Form and Function. Cambridge University Press, Cambridge: 313–314.

Butler, P. M. (1972): Some functional aspects of molar evolution. Evolution 26: 474–483.

Butler, P. M. (1973): Molar wear facets of early tertiary North American primates. In: Zingeser, M. R. (ed.). Craniofacial Biology of Primates. Symposia of the Fourth International Congress of Primatology, Portland, Oregon, August 15–18, 1972, vol. 3. Karger, Basel: 1–27.

Clauss, M., Nunn, C., Fritz, J., & Hummel, J. (2009): Evidence for a tradeoff between retention time and chewing efficiency in large mammalian herbivores. Comparative Biochemistry and Physiology Part A: Molecular & Integrative Physiology, 154: 376–382.

Crompton, A. W. & Hiiemäe, K. (1969): How mammalian teeth work. Discovery 5: 23–34.

Crompton, A. W., Barnet, J., Lieberman, D. E., Owerkowicz, T., Skinner J. & Baudinette R. V. (2008): Control of jaw movements in two species of macropodines (*Macropus eugenii* and *Macropus rufus*). Comparative Biochemistry and Physiology A 150: 109–123.

Crompton, A. W., Owerkowicz, T. & Skinner, J. (2010): Masticatory motor pattern in the koala (*Phascolarctos cinereus*): a comparison of jaw movements in marsupial and placental herbivores. Journal of Experimental Zoology 313A: 564–578.

Damuth, J., & Janis, C. M. (2011): On the relationship between hypsodonty and feeding ecology in ungulate mammals, and its utility in palaeoecology. Biological Reviews 86: 733–758.

Damuth, J. & Janis, C. M. (2014): A comparison of observed molar wear rates in extant herbivorous mammals. Annales Zoologici Fennici 51: 188–200.

*Engels, S. (2011): Funktionelle und morphologische Transformationen der Molaren bei frühen Hippomorpha im Hinblick auf den Mastikationsprozess. Doctoral dissertation, Rheinische Friedrich-Wilhelms-Universität Bonn. 200 pp. http://hss.ulb.uni-bonn.de/2011/2722/2722.htm

*Engels, S. & Schultz, J. A. (2019): Evolution of the power stroke in early Equoidea (Perissodactyla, Mammalia). Palaeobiodiversity and Palaeoenvironments 99: 271–291.

Evans, A. R. & Janis, C. M. (2014): The evolution of high dental complexity in the horse lineage. Annales Zoologici Fennici 51: 73–79.

Fahlke, J. M., Coombs, M. C. & Semprebon, G. M. (2013): *Anisodon* sp. (Mammalia, Perissodactyla, Chalicotheriidae) from the Turolian of Dorn-Dürkheim 1 (Rheinhessen, Germany): morphology, phylogeny, and palaeoecology of the latest chalicothere in Central Europe. Palaeobiodiversity and Palaeoenvironments 93: 151–170.

Fletcher, T. M., Janis, C. M., & Rayfield, E. J. (2010): Finite element analysis of ungulate jaws: can mode of digestive physiology be determined. Palaeontologica Electronica 13, 3; 21A: 15 p. http://palaeo-electronica.org/2010_3/234/index.html

Fortelius, M. (1985): Ungulate cheek teeth: developmental, functional and evolutionary interrelations. Acta Zoologica Fennica 180: 4–76.

Fortelius, M., Eronen, J., Liu, L., Pushkina, D., Tesakov, A., Vislobokova, I. & Zhang, Z. (2003): Continental-scale hypsodonty patterns, climatic paleobiogeography and dispersal of Eurasian Neogene large mammal herbivores. In: Reumer, J. W. F. & Wessels, W. (eds.). Distribution and Migration of Tertiary Mammals in Eurasia – A Volume in Honor of Hans de Bruijn. Deinsea 10: 1–12.

*Fritz, J., Hummel, J., Kienzle, E., Arnold, C., Nunn, C., & Clauss, M. (2009): Comparative chewing efficiency in mammalian herbivores. Oikos 118: 1623–1632.

*Gailer, J. P. & Kaiser, T. M. (2014): Common solutions to resolve different dietary challenges in the ruminant dentition: The functionality of bovid postcanine teeth as a masticatory unit. Journal of Morphology 275: 328–341.

Geisler, H., Boessenecker R. W., Brown M. & Beatty, B. L. (2017): The origin of filter feeding in whales. Current Biology 27: 1–7

Geisler, J. H., Theodor, J. M., Uhen, M. D. & Foss, S. E. (2007): Phylogenetic relationships of cetaceans to terrestrial artiodactyls. In: Prothero, D. R. & Foss, S. E. (eds.). The Evolution of Artiodactyls. Johns Hopkins University Press, Baltimore: 19–31.

Gheerbrant, E., Bouya, B. & Amaghzaz, M. (2012): Dental and cranial anatomy of *Eritherium azzouzorum* from the Paleocene of Morocco, earliest known proboscidean mammal. Palaeontographica A 297: 151–183.

Gheerbrant, E., Sudre, J., Tassy, P., Amaghzaz, M., Bouya, B. & Iarochene, M. (2005): Nouvelles données sur *Phosphatherium escuilliei* (Mammalia, Proboscidea) de l'Eocène inférieur du Maroc, apports à la phylogénie des Proboscidea et des ongulés lophodontes. Geodiversitas 27: 239–333.

Gingerich, P. D. (1983): Evidence for evolution from the vertebrate fossil record. Journal of Geological Education 31: 140–144.

Gingerich, P. D., Ul-Haq, M., Koenigswald, W. v., Sanders, W. J. Smith, B. H. & Zalmout, I. S. (2009): New protocetid whale from the middle Eocene of Pakistan: Birth on land, precocial development, and sexual dimorphism. PloS One 4:e4366: 1–20.

Gingerich P. D., Ul-Haq, M., Zalamout, I. S., Hussain Khan I. & Malkani M. S. (2001): Origin of whales from early artiodactyls: Hands and feet of Eocene Protocetidae from Pakistan. Science 293: 2239–2242.

Grippo, J. O., Simring, M. & Schreiner, S. (2004): Attrition, abrasion, corrosion and abfraction revisited. A new perspective on tooth surface lesions. Journal of the American Dental Association 135: 1109–1118.

Harris, J. M. (1975): Evolution of feeding mechanisms in the family Deinotheriidae (Mammalia: Proboscidea). Zoological Journal of the Linnean Society 56: 331–362.

Hiiemäe, K. M. (1978): Mammalian mastication: a review of the activity of the jaw muscles and the movements they produce in chewing. In: Butler, P. M. & Joysey, A. (eds.). Development, Function and Evolution of Teeth. Academic Press, London: 359-398

Hiiemäe, K. & Kay, R. F. (1972): Trends in the evolution of primate mastication. Nature 240: 486-487.

Hodson, M. J., White, P. J., Mead, A. & Broadley, M. R. (2005): Phylogenetic variation in the silicon composition of plants. Annali di Botanica 96: 1027-1046.

Houssaye, A., Tafforeau, P., De Muizon, C., & Gingerich, P. D. (2015): Transition of Eocene whales from land to sea: evidence from bone microstructure. PLoS One 10(2): e0118409.

Hullot, M., Antoine, P. O., Ballatore, M. & Merceron, G. (2019): Dental microwear textures and dietary preferences of extant rhinoceroses (Perissodactyla, Mammalia). Mammal Research 64: 397-409.

Hunter, J. P. & Jernvall, J. (1995): The hypocone as a key innovation in mammalian evolution. Proceedings of the National Academy of Sciences 92: 10718-10722.

Janis, C. M. (1995): Correlations between craniodental morphology and feeding behavior in ungulates: reciprocal illumination between living and fossil taxa. In: Thomason, J. J. (ed.). Functional Morphology in Vertebrate Paleontology. Cambridge University Press, Cambridge: 76-98.

Janis, C. M. (1990): The correlation between diet and dental wear in herbivorous mammals, and its relationship to the determination of diets of extinct species. In: Boucot, E. J. (ed.). Evolutionary Paleobiology of Behaviour and Coevolution. Elsevier, Amsterdam: 241-260.

Janis, C. M. & Fortelius, M. (1988): On the means whereby mammals achieve increased functional durability of their dentitions, with special reference to limiting factors. Biological Reviews 63: 197-230.

Janis, C. M., Scott, K. M., Jacobs, L. L., Gunnell, G. F., & Uhen, M. D. (1998): Evolution of Tertiary Mammals of North America. Volume 1: Terrestrial Carnivores, Ungulates, and Ungulatelike Mammals. Cambridge University Press, Cambridge.

Jernvall, J. & Fortelius, M. (2002): Common mammals drive the evolutionary increase of hypsodonty in the Neogene. Nature 417: 538-540.

Jernvall, J. & Thesleff, I. (2012): Tooth shape formation and tooth renewal: Evolving with the same signals. Development 139: 3487-3497.

Joomun, S. C., Hooker, J. J. & Collinson, M. E. (2008): Dental wear variation and implications for diet: an example from Eocene perissodactyls (Mammalia). Palaeogeography, Palaeoclimatology, Palaeoecology 263: 92-06.

Kaiser, T. M., Müller, D. W. H., Fortelius, M., Schulz, E., Codron, D. & Clauss, M. (2013): Hypsodonty and tooth facet development in relation to diet and habitat in herbivorous ungulates: Implications for understanding tooth wear. Mammal Review 43: 34-46.

Koenigswald, W. v. (2011): Diversity of hypsodont teeth in mammalian dentitions – Construction and classification. Palaeontographica A 294: 63-94.

Koenigswald, W. v. (2014): Mastication and wear *Lophiodon* (Perissodactyla, Mammalia) compared to lophodont dentitions in some other mammals. Annales Zoologici Fennici 51: 162-176.

Koenigswald, W. v. (2016): The diversity of the mastication patters in the Neogene and Quaternary proboscideans. Palaeontographica A 307: 1-41.

*Koenigswald, W. v. (2020): Construction and wear of mammalian teeth in terms of heterochrony. In: Martin, T. & Koenigswald, W. v. (eds.). Mammalian Teeth – Form and Function. Pfeil, Munich: 171-186.

*Koenigswald, W. v., Anders, U., Engels, S., Schultz, J. A. & Kullmer, O. (2012): Jaw movement in fossil mammals: analysis, description and visualization. Paläontologische Zeitschrift 87: 141-159.

Kosintsev, P., Mitchell, K. J., Deviese, T., van der Plicht, J., Kuitems, M., Petrova, E., Tikhonov, A., Higham, T., Comeskey, D., Turney, C., Cooper, A., van Kolfschoten, T., Stuart, A. J. & Lister, A. M. (2019): Evolution and extinction of the giant rhinoceros *Elasmotherium sibiricum* sheds light on late Quaternary megafaunal extinctions. Nature Ecology & Evolution 3: 31-38.

*Kullmer, O., Bennazzi, S., Fiorenza, L., Schulz, D., Basco, S. & Winzen, O. (2009): Technical Note: Occlusal Fingerprint Analysis (OFA): Quantification of tooth wear pattern. American Journal of Physical Anthropology 139: 600-605.

*Kullmer, O., Menz, U. & Fiorenza, L. (2020): Occlusal Fingerprint Analysis (OFA) reveals dental occlusal behavior in primate molars. In: Martin, T. & Koenigswald, W. v. (eds.). Mammalian Teeth – Form and Function. Pfeil, Munich: 25-43.

Lihoreau, F. & Ducrocq, S. (2007): Family Anthracotheriidae. In: Prothero, D. R. & S. E. Foss (eds.). The Evolution of Artiodactyla. Johns Hopkins University Press, Baltimore: 89-105.

Lucas, P. W. (1979): The dental-dietary adaptions of mammals. Neues Jahrbuch für Geologie Paläontologie Monatshefte 8: 486-512.

Lucas, P. W. (2004): DFunctional Morphology – How Teeth Work. Cambridge University Press, Cambridge.

Lucas, P. W. & Luke, D. A. (1984): Chewing it over: Basic principles of food breakdown. In: Chivers, D. J., Wood, B. A. & Bilsborough, A. (eds.). Food Acquisition and Processing in Primates. Plenum Press, New York: 283-302.

MacFadden, B. J. (1988): Fossil horses from "*Eohippus*" (*Hyracotherium*) to *Equus*, 2: rates of dental evolution revisited. Biological Journal of the Linnean Society 35: 37-48.

MacFadden, B. J. (2000): Cenozoic mammalian herbivores from the Americas: reconstructing ancient diets and terrestrial communities. Annual Review of Ecology and Systematics 31: 33-59.

MacFadden, B. J. (2005): Fossil horses-evidence for evolution. Science 307: 1728-1730.

MacFadden, B. J. & Hulbert, R. C. (1988): Explosive speciation at the base of the adaptive radiation of Miocene grazing horses. Nature 336: 466-468.

Madden, R. H. (2014): Hypsodonty in Mammals. Cambridge University Press, Cambridge.

Maier, W. (1980): Konstruktionsmorphologische Untersuchungen am Gebiß der rezenten Prosimiae (Primates). Abhandlungen der senckenbergischen naturforschenden Gesellschaft 538: 1-158.

Maier, W. & Schneck, G. (1982): Functional morphology of hominoid dentitions. Journal of Human Evolution 11: 693-696.

Maglio, V. J. (1972): Evolution of mastication in the Elephantidae. Evolution 26: 638-658.

Marx, F. G, Hocking, D. P., Park, T., Ziegler, T., Evans, A. R. & Fitzgerald, E. M. G. (2016): Suction feeding preceded filtering in baleen whale evolution. Memoirs of Museum Victoria. 75: 71-82.

Mihlbachler, M. C., Rivals, F., Solounias, N. & Semprebon, G. (2011): Dietary change and evolution of horses in North America. Science 331: 1178-1181.

Mills, J. R. E. (1955): Ideal dental occlusion in the primates. Dental Practitioner 6: 47-61.

Muizon, C. de (1993): Walrus-like feeding adaptation in a new cetacean from the Pliocene of Peru. Nature 365: 745-748.

Peredo, C. M., Pyenson, N. D., Marshall, C. D. & Uhen, M. D. (2018): Tooth loss precedes the origin of baleen in whales. Current Biology 28: 3992-4000.

Prothero, D. R., & Foss, S. E. (2007). The evolution of artiodactyls. Johns Hopkins University Press, Baltimore.

Rensberger, J. M. & Koenigswald, W. v. (1980): Functional and phylogenetic interpretation of enamel microstructure in rhinoceroses. Paleobiology 6: 477-495.

Rensberger, J. M., Forsten, A. & Fortelius, M. (1984): Functional evolution of the cheek tooth pattern and chewing direction in Tertiary horses. Paleobiology 10: 439-452.

Renvoisé, E. & Michon, F. (2014): An Evo-Devo perspective on evergrowing teeth in mammals and dental stem cell maintenance. Frontiers in Physiology 5: 324.

Rose, K. D. (2006): The Beginning of the Age of Mammals. Johns Hopkins University Press, Baltimore.

*Ruf, I., Schubert, A. M. & Koenigswald, W. v. (2020): Case studies on functional aspects and constraints in early and late tooth ontogeny. In: Martin, T. & Koenigswald, W. v. (eds.). Mammalian Teeth – Form and Function. Pfeil, Munich: 102-124.

Samuels, J. X. & Hopkins, S. S. B. (2017): The impacts of Cenozoic climate and habitat changes on small mammal diversity of North America. Global and Planetary Change 149: 36-52.

Sanson, G. D. (2006): The biomechanics of browsing and grazing. American Journal of Botany 93: 1531–1545.

Sanson, G. D., Kerr, S. A. & Gross, K. A. (2007): Do silica phytoliths really wear mammalian teeth? Journal of Archaeological Science 3: 526–531.

Schellhorn, R. (2018): A potential link between lateral semicircular canal orientation, head posture, and dietary habits in extant rhinos (Perissodactyla, Rhinocerotidae). Journal of Morphology 279: 50–61.

*Schultz, J. A., Anders, U., Braune, C., Brinkkötter, J. J., Calandra, I., Engels, S., Findeisen, E., Gailer, J. P., Hummel, J., Jäger, K. R. K., Kaiser, T. M., Kalthoff, D. C., Koenigswald, W. v., Kullmer, O., Landwehr, C. Mau, M., Menz, U., Ruf, I., Schubert, A. M., Schulz-Kornas, E., Schwermann, A. H., Schwermann, L. C., Skiba, M., Steuer, P., Südekum, K.-H., Winkler D. E. & Martin, T. (2020): A new wear facet terminology for mammalian dentitions. In: Martin, T. & Koenigswald, W. v. (eds.). Mammalian Teeth – Form and Function. Pfeil, Munich: 11–24.

*Schultz, J. A., Menz, U., Winkler, D. E. Schulz-Kornas, E., Engels, S., Kalthoff, D. C., Koenigswald, W. v., Ruf, I., Kaiser, T., Kullmer, O., Südekum, K.-H. & Martin, T. (2018): Modular wear facet nomenclature for mammalian post-canine dentitions. Historical Biology 30: 30–41.

*Schwermann, L. C. (2016): Evolutionsstadien der Kaufunktion früher Artiodactyla. Doctoral dissertation, Rheinische Friedrich-Wilhelms-Universität Bonn. 276 pp. http://hss.ulb.uni-bonn.de/2016/4453/4453.htm

Searle, K. R. & Shipley, L. A. (2008): The comparative feeding behavior of large browsing and grazing herbivores. In: Gordon, I. J. & Prins, H. H. T. (eds.). The Ecology of Browsing and Grazing. Ecological Studies 195. Springer, Heidelberg: 117–148.

Semprebon, G. M., Rivals, F., Solounias, N. & Hulbert, R. C. (2016): Paleodietary reconstruction of fossil horses from the Eocene through Pleistocene of North America. Palaeogeography, Palaeoclimatology, Palaeoecology 442: 110–127.

Stephens, D. W. & Krebs, J. R. (1986): Foraging Theory. Princeton University Press, Princeton.

Stirton, R. A. (1941): Development of characters in horse teeth and the dental nomenclature. Journal of Mammalogy 22: 434–446.

Strömberg, C. A. E. (2006): Evolution of hypsodonty in equids: testing a hypothesis of adaptation. Paleobiology 32: 236–258.

*Taylor, L. A., Kaiser, T. M., Schwitzer, C., Müller, D. W. H., Codron, D., Clauss, M. & Schulz, E. (2013): Detecting inter-cusp and inter-tooth wear patterns in rhinocerotids. PLoS One 8: e80921.

Thewissen, J. G. M. & Bajpai, S. (2001): Whale origins as a poster child for macroevolution. BioScience 51:1037–1049.

Thewissen, J. M. G., Cooper, L. N., George, J. C. & Bajpai, S. (2009): From land to water: the origin of whales, dolphins, and porpoises. Evolution: Education and Outreach 2: 272–288.

Thewissen, J. M. G., Sensor, J. D., Clementz, M. T. & Bajpai, S. (2011): Evolution of dental wear and diet during the origin of whales. Paleobiology 37: 655–669.

Tobien, H. (1978): On the evolution of mastodonts (Proboscidea, Mammalia). Part 2: The bunodont tetralophodont groups. Geologisches Jahrbuch Hessen 106: 159–208.

Tsubamoto, T. (2014): Estimating body mass from the astragalus in mammals. Acta Palaeontologica Polonica 59: 259–265.

Uhen, M. D. (2004): Form, function, and anatomy of *Dorudon atrox* (Mammalia, Cetacea): An archaeocete from the Middle to Late Eocene of Egypt. Papers in Paleontology 34: 1–222.

Vacek, M. (1877): Ueber österreichische Mastodonten und ihre Beziehungen zu den Mastodonten Europas. Abhandlungen der Geologischen Reichsanstalt 7: 1–47.

*Winkler, D. E. & Kaiser, T. M. (2015): Structural morphology of molars in large mammalian herbivores: Enamel content varies between tooth positions. PLoS One 10: e0135716.

Williams, S. H. & Kay, R. F. (2001): A comparative test of adaptive explanations for hypsodonty in ungulates and rodents. Journal of Mammalian Evolution 8: 207–229.

Wright, B. W., Ulibarri, L., O'Brien, J., Sadler, B., Prodhan, R., Covert, H. H., & Nadler, T. (2008): It's tough out there: variation in the toughness of ingested leaves and feeding behavior among four Colobinae in Vietnam. International Journal of Primatology 29: 1455–1466.

*Zhou, Z., Winkler, D. E., Fortuny, J., Kaiser, T. M. & Marcé-Nogué, J. (2019): Why ruminating ungulates chew sloppily: Biomechanics discern a phylogenetic pattern. PLoS One 14: e0214510.

Short review of dental microstructure and dental microwear in xenarthran teeth

Daniela C. Kalthoff

Introduction

Dentin is one of three tooth tissues occurring in mammals, the other two being the hypermineralized tooth enamel and tooth cementum. In respect to volume, dentin usually makes up the largest part of the tooth. Interestingly, a number of mammal taxa have reduced the resistant layer of tooth enamel during their evolutionary history, thus focusing on dentin as the main tooth building tissue. Enamelless teeth can be found in various extant and extinct taxa of odontocete whales (Odontoceti), aardvarks (Tubulidentata), armadillos, glyptodonts and pampatheres (Cingulata), sloths (Folivora), walruses (Odobenidae), sea cows (Dugongidae), and elephants (Proboscidea).

This review will give an overview of recent results on histology as well as dietary adaptations using the distinctly built dentin teeth in armadillos and sloths and their fossil relatives (in the following referred to with the informal term "xenarthran(s)"). Toothed xenarthrans show an astonishing diversity regarding tooth shapes (e.g., complex to simplified; Fig. 12.1) and number of teeth (severely reduced to abundant supernumerary teeth).

Afrotheria, Laurasiatheria, Euarchontoglires, and Xenarthra are the four major clades within placental mammals (Murphy et al. 2001, O'Leary et al. 2013). Showing unique characters in e.g., skeletal morphology and life history traits, both extinct and extant xenarthrans are well investigated and even have experienced increased scientific attention in the last two decades (e.g., Fariña et al. 2003, 2013, Vizcaíno & Loughry 2008, Bargo & Nyakatura 2018). On the other hand, research on dental histology in this group is rather meagre. Historically, studies on xenarthran dentin date back to the first half of the 19th century (Retzius 1837, Owen 1840-1845, Owen 1842). Only 80 years later, some more papers were published (Schmidt 1924, Arsuffi 1938, Schmidt & Keil 1958, Keil & Venema 1963), all focusing on one or few taxa and not revealing dentin microstructure in necessary detail, mainly because all these studies relied on light microscopy. Ferigolo (1985) was the first investigating a larger number of both fossil and extant species and also the first one applying scanning electron microscopy (SEM). Also using SEM technique, the most comprehensive and detailed histological study to date is from Kalthoff (2011), comprising teeth from about 50 xenarthran taxa dating from late Eocene to Holocene.

Dental microwear analysis is the study of microscopic scars on the chewing surface of cheek teeth. Last consumed food items (and potentially also exogenous grit; Hoffman et al. 2015) leave these scars, mainly different kinds of pits and scratches, during mastication and different food categories (leaves, grass, seeds, fruits, meat, insects, etc.) produce different kinds of scars. Extant species with known diets are used to calibrate microwear scars. From the beginning of dental microwear analysis, the highly mineralized enamel had been the target tissue, and this technique has been employed to various mammal clades (e.g., Walker 1976, Walker et al. 1978, Rensberger 1978, Teaford 1991, Semprebon et al. 2004), but has also been used to recognize dietary adaptations in other vertebrates such as reptiles and fish (e.g., Purnell et al. 2012, Winkler et al. 2019). The much softer dentin was thought not suitable to reliably record microwear features; however, this assumption was challenged and proven incorrect. Beginning with Oliveira (2001), several studies successfully applying dental microwear on dentin have been published, leading to new and exciting insights in xenarthran paleobiology (e.g., Green 2009a, b, Green & Resar 2012, Haupt et al. 2013, Resar et al. 2013, Green & Kalthoff 2015, Kalthoff & Green 2018). Thus, the analysis of dentin has been proven feasible and very interesting as independent proxy for feeding ecology in taxa with enamelless teeth.

In the following, established knowledge and recent results on dentin microstructure analysis as well as dentin microwear analysis will be reviewed. If not mentioned differently, they are based on Kalthoff (2011), Green & Kalthoff (2015), and Kalthoff & Green (2018). Appendices 1 and 2 provide an overview about specimen details and their most conspicuous characters in respect to dentin microstructure and/or dental microwear.

Preparation techniques

In the fossil record, dentin microstructures – similar as those of enamel – are usually preserved without alterations. As a general phenomenon of taphonomic processes, a tooth can be impregnated with different minerals (e.g., Mn, Fe) and thus change color, but the internal structures usually are not affected. This means that dentin of extinct and extant taxa can be analyzed with the same techniques. However, if alterations occur, they are obvious and easy to recognize, regardless whether they occurred at an early (Kalthoff et al. 2011) or later post-mortem stage.

T. Martin & W. v. Koenigswald (eds.): Mammalian Teeth – Form and Function. Pp. 231-241, 3 figs.

DOI: http://doi.org/10.23788/mammteeth.12 Published 22 December 2020

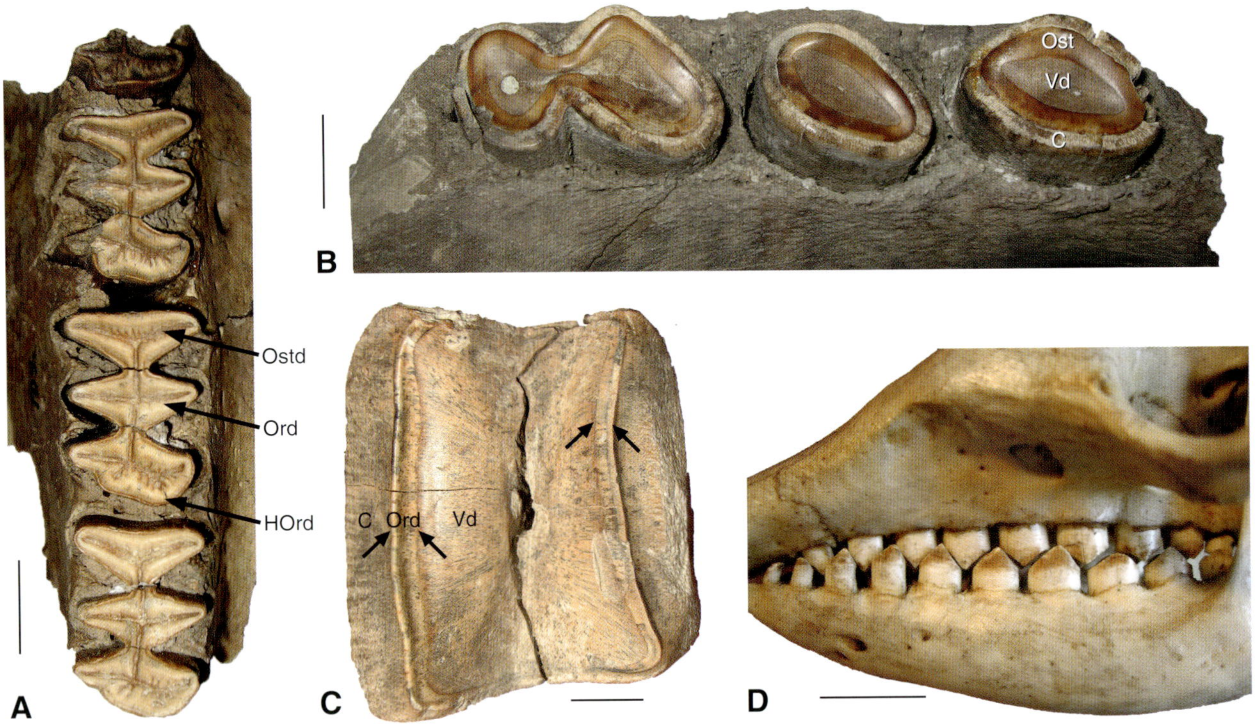

Fig. 12.1. Tooth morphology in extant and extinct xenarthrans. **A,** Left maxilla fragment of *Glyptodon clavipes* (Glyptodontidae) in occlusal view with three complete, "Christmas tree-shaped" cheek teeth (NHMD-ZMK 70/1885). The slightly darker, ramifying, and elevated material in the tooth center is osteodentin (Ostd), outward follows orthodentin (Ord), then a thin, slightly darker and elevated layer of hard orthodentin (HOrd). Rostral is up. **B,** Fragment of right mandible of *Lestodon armatus* (Mylodontidae) in oblique view from lateral with three complete molariform teeth (NHMD-ZMK 52/1888). The center layer is vasodentin (Vd), the hazelnut-brown layer orthodentin (Ord), the outer, gray-brown later is tooth cementum (C). Rostral is to the right. **C,** *Megatherium americanum* (Megatheriidae), undetermined cheek tooth (NHMD-ZMK 56/1885) in occlusal view. The thin layer between the arrows is orthodentin (Ord), forming sharp cutting edges. Outward is a thick layer of softer tooth cementum (C), the inner part is made of softer vasodentin (Vd). **D,** Left lateral view of dentition of the living six-banded armadillo *Euphractus sexcinctus* (NHMD L62). The peg-like teeth show anterior- and posterior-facing facets and interlocking of maxillary and mandibular teeth. NHMD, Natural History Museum of Denmark, Copenhagen, Denmark; NHMD-ZMK, Zoologisk Museum Kvartærzoologi (Quaternary Zoology collection), Natural History Museum of Denmark, Copenhagen. Scale bars equal 1 cm.

Dentin needs "special attention" when it comes to preparation in order to expose the full set of microstructural details. Dentin microstructures can be analyzed with transmitted light microscopy using thin sections or by scanning electron microscopy (SEM) using pseudo-3D surfaces produced through etching. In contrast to enamel, dentin is a friable tissue and easily fractured under preparation stresses. Therefore, tooth samples should be completely embedded and thus stabilized in translucent epoxy resin; for thin sectioning, it is even advised to use the resin for repeated vacuum impregnation during the different preparation steps. For SEM preparation, it has proven most advantageous to etch the trimmed surfaces with Mutvei's solution (Schöne et al. 2005). Mutvei's solution comprises (1) acetic acid for gentle and slow etching, (2) glutardialdehyde as fixation agent, and (3) alcian blue for staining. A detailed description of preparation procedures can be found in Kalthoff (2011).

Different types of dentinal tissues in xenarthran cheek teeth

In mammalian teeth, dentin (Fig. 12.2) occurs in three different varieties, namely orthodentin, vasodentin, and osteodentin. Orthodentin is the most common variety (often simply referred to as "dentin") and occurs in all mammalian teeth (regardless whether enameled or enamelless), whereas vasodentin and osteodentin are restricted to xenarthrans. Compared to the almost completely mineralized tooth enamel, the moderately porous orthodentin is much softer and elastic making it an ideal counter bearing tissue in enameled teeth. Orthodentin contains about 20 % by weight of collagenous proteins (Carlson 1990) and is not vascularized. Its microstructure is characterized by subparallel, slightly S-curved dentinal tubules raising radially in direction of the tooth crown in a rather homogeneous dentin matrix. Each dentinal tubule accommodates a thin cytoplasmic process, the so-called odontoblastic process or Tomes' process, and dentinal fluid. From the outer wall of the odontoblastic process protrude even thinner extensions, which normally are short and occasionally bifurcate (Fig. 12.2E). The odontoblastic process and its extensions are commonly preserved as calcified tubes (Fig. 12.2A, E).

Orthodentin can further be split into two subtypes of different biomechanical properties: firstly, the organic matter-rich intertubular dentin, and secondly, the almost inorganic peritubular dentin (Ten Cate 1998) (Fig. 12.2C, D). Intertubular dentin makes up the dentin matrix being a network of collagenous fibers with embedded mineral crystals. In contrast, peritubular dentin is a collagen-free and highly mineralized dentinal tissue (Habelitz et al. 2007), and in this respect resembles tooth enamel. However, in contrast to the dense enamel, peritubular dentin is rather porous. Peritubular dentin is located at the inner periphery of the individual tubules (Fig. 12.2C) and assumed to be precipitated during dentinogenesis (Ten Cate 1998).

Orthodentin occurs in all xenarthrans and is the hardest dental tissue in sloths (Fig. 12.2B). Xenarthran orthodentin is microstructurally different from that of other mammals. Firstly, dentinal tubules are arranged in close proximity throughout the entire tooth height, herewith differing from e.g., human teeth in which tubule density is wider at the enamel-dentin junction (Mjör & Nordahl 1996). Secondly, individual dentinal tubules are on average wider as in other mammals, measuring up to 7 µm in, e.g., the extant three-toed sloth *Bradypus*. Both SEM and synchrotron radiation X-ray tomographic microscopy results suggest that large parts of the lumen in the wider dentinal tubules are occupied by the highly mineralized peritubular dentin (Fig. 12.2D), thus suggesting that xenarthran orthodentin is biomechanically more resistant than that of other mammals. Thirdly, in xenarthran orthodentin and here especially in sloths, the small extensions protruding from the odontoblastic process are very long, regularly reaching and piercing through the tubule wall and thereby entering the neighboring tubule. In addition, extensions show frequent branching and even are connected to each other and, as a consequence, form a dense but irregular meshwork (Fig. 12.2E). In other mammals, side-extensions are either absent (e.g., diprotodont marsupials) or fewer and shorter without forming networks (many placental mammals).

The reason behind the differing orthodentin architecture in xenarthran teeth is far from being understood and needs further investigation. As mentioned above, one possible explanation lies in the improvement of the biomechanical properties of the tooth, which evolved to counterbalance the lack of a resistant enamel layer. Alternatively, the biochemical regulating transport system of the tooth is enhanced, in which peritubular dentin plays an important role (Gotliv & Veis 2007, 2008). Both explanations can be true, of course.

A conspicuous type of orthodentin exclusively occurs in extinct cingulates (Glyptodontidae, Pampatheriidae, *Proeutatus*, *Eutatus*). It is termed 'hard orthodentin' for it forms an outer rim surrounding the tooth and thus being more wear resistant than the neighbouring 'regular orthodentin' (Figs. 1A, 2F, G). Microstructurally, the 'hard orthodentin' is undistinguishable from the regular type but µCT images reveal that the 'hard orthodentin' is optically more dense, which can be translated to being higher mineralized (Fig. 12.2G). The fact that the hard orthodentin type only occurs in fossil cingulate teeth makes it difficult to reveal the reasons for its increased hardness. Analysis of mineral composition and hardness tests by indentation seem appropriate techniques to approach this question; however, taphonomic processes have to be taken into account when interpreting the results. Still, the increased hardness is very likely to be primary because the prominent outer rim is a consistent feature.

The second dentinal tissue, true osteodentin, occurs within Mammalia exclusively in extinct cingulates (Glyptodontidae, Pampatheriidae, *Proeutatus*, *Eutatus*). This dentin variety consists of densely packed primary osteons and is permeated with blood vessels and nerves, both of which are often arranged in rows. True osteodentin is always located in the center of the tooth, followed toward the outer tooth surface by orthodentin and hard orthodentin (Fig. 12.2F, G). In glyptodontids and pampatheres, the area occupied by osteodentin is elongated and shows branching toward the convexities (Fig. 12.1A). The osteodentin area always forms the highs in the tooth relief, suggesting it being hard and resistant to abrasion. The teeth of the Miocene horned armadillo *Peltephilus*, the Pliocene armadillo *Macroeuphractus*, and, unexpectedly, also the extant three-banded armadillo *Tolypeutes*, the hairy armadillo *Chaetophractus*, and the pichi *Zaedyus*, show a swelled, central cloud-shaped region encircling a tissue, harder than the surrounding orthodentin. Although being devoid of vascularization and identifiable primary osteons, the microstructural appearance reminds of osteodentin, which is why this central tissue is interpreted as 'degenerated osteodentin'.

The third dentin variety is vasodentin (Fig. 12.2B). As the term suggests, vasodentin is vascularized and shows relatively wide vascular canals (20–25 µm on average) and an overall porous appearance. Vasodentin lacks dentinal tubules and is relatively soft, compared to orthodentin and osteodentin. It occurs as central tissue in the teeth of extant and extinct sloths (Fig. 12.1B, C) and some living armadillos such as *Dasypus*. Vasodentin is absent in fossil cingulates and it has never been proven to be present in any other mammal.

The so-called "globular dentin" of the walrus *Odobenus rosmarus* (Kastelein 2009) is currently under study by the author.

Form and function of xenarthran cheek teeth

Xenarthran teeth are unusual in several aspects. Above, it has already been mentioned that they generally are devoid of enamel; despite the great systematic diversity of fossil xenarthrans, only two Eocene armadillo species show remnants of enamel (Simpson 1932; Ciancio et al. 2014), likewise do the milk teeth of the extant genus *Dasypus* (Martin 1916). The reasons for xenarthran teeth becoming enamelless are still not solved, however, the fossil record in armadillos shows that this happened synchronous with teeth becoming rootless and continuously growing. To date, no fossil sloth teeth with an outer enamel coating have been discovered.

Genetic studies suggest that reduction and loss of enamel occurred independently in cingulates and folivores (Meredith et al. 2009). This result makes sense in the light of their different strategies of dentin tissue organization: in sloths and their fossil relatives (Folivora), we see a layered composition of a central vasodentin core, which is surrounded by a thick orthodentin layer, followed by an outer collar of cementum (Figs. 1B, 2B). In this make-up, orthodentin is the most wear-resistant (=hard) layer forming the highest rims of the tooth, while vasodentin and cementum are comparably less wear-resistant (=soft). By contrast, fossil cingulates combine an inner layer of osteodentin (or 'degenerated osteodentin' in some taxa, see above), surrounded by orthodentin as middle layer and followed by an outer layer of 'hardened orthodentin' (Figs. 1A, 2F, G). In this make-up, both osteodentin types and the 'hardened orthodentin' are the most wear resistant, rim-forming tissues (=very hard) while the regular orthodentin is comparably less wear-resistant (=less hard). Living armadillos may show divergent tissue combinations, i.e. they lack true osteodentin, never show 'hardened orthodentin', might have an outer collar of cellular (*Dasypus*) or acellular cementum (*Priodontes*, *Cabassous*, *Zaedyus*), or have an inner vasodentin core (*Dasypus*).

The combination of dentin varieties (and cementum) of differing biomechanical properties, mainly correlated with hardness, allows xenarthrans to develop a proper tooth relief depicting food choice and mastication movements (Fig. 12.1). Regarding function, xenarthrans with crushing-grinding dentitions show a forward directed jaw movement and gently sloped tooth relief (e.g., *Glyptodon*: Fig. 12.1A; *Lestodon*: Fig. 12.1B), contrasting strongly with xenarthrans with shear-cutting dentitions (e.g., *Euphractus*: Fig. 12.1D) showing a simpler vertical (open-close) jaw movement but pronounced tooth relief with cusps, valleys, and cutting edges. The latter occur as self-sharpening cutting blades, e.g., in the very large-bodied fossil ground sloths *Eremotherium* and *Megatherium* (Fig. 12.1C). The self-sharpening effect is achieved by considerable thinning of the orthodentin and orientation of the cutting blade perpendicular to the jaw movement. In addition, cheek teeth of *Megatherium* and *Eremotherium* show an especially dense meshwork of the odontoblastic process extensions.

Another peculiarity can be seen in species of the giant ground sloths *Scelidotherium*, *Lestodon*, *Glossotherium*, *Paramylodon*, and *Mylodon*. Microstructurally, the outer orthodentin portion shows fusion of five to fifteen individual dentinal tubules to cone-shaped bundles, each bundle raising slightly above the chewing surface. Raised parts are radially orientated and perpendicular to the outer margin of the tooth, thus having a washboard-like appearance. Although a functional interpretation does not suggest itself, the structure might have a positive effect on food processing. Morphologically similar structures are known from enamel in rhinoceros cheek teeth, here caused by a special microstructural arrangement of the enamel prisms (vertical Hunter-Schreger bands; Koenigswald 1997).

Dental microwear

Dental microwear depicts the relationship of the scaring of the chewing surface and the animal's last food items. Different techniques have been successfully employed to explore dietary adaptations in xenarthrans: scanning electron microscopy, stereomicroscopy, and confocal microscopy, each of them being different from the others in regard to magnification, area and parameters analyzed, and illumination (see references in Kalthoff & Green 2018), but all of them using orthodentin as target tissue. Green (2009a) used stereomicroscopy to categorize living armadillos and sloths in four dietary classes, namely folivore, folivore-frugivore, insectivore, and carnivore-omnivore, and evaluated in a follow-up study (Green 2009b) inter-tooth variations of microwear signals depending on tooth position and facet analyzed. Since then and with these prerequisites at hand, various fossil ground sloths have been analyzed in respect to their feeding ecology.

Two of the stratigraphically earliest definite sloths, *Octodontotherium grande* and *Orophodon hapaloides*, are known from the Late Oligocene of Patagonia, and

Fig. 12.2. Dentin microstructures in xenarthran cheek teeth. **A,** Glyptodontidae indet. (KOE 2879), longitudinal section, scanning electron micrograph. The subparallel oriented, calcified odontoblastic processes are densely packed and show short but strong side extensions. **B,** *Hapalops* sp. (KOE 3012), longitudinal section, scanning electron micrograph. In sloths, cheek teeth show an outer collar of tooth cementum (C), followed by a layer of orthodentin (Ord), and a layer of vasodentin (Vd) in the tooths' center. A crack in the vasodentin is filled by epoxy resin, appearing dark grey in the image. **C,** *Nothrotheriops texanus* (KOE 3053), transverse section, scanning electron micrograph. Detail of the orthodentin layer showing dentinal tubules with odontoblastic processes (OP); the orthodentin matrix consists of intertubular dentin (ITD), dentinal tubules are lined with the highly mineralized peritubular dentin (PTD). **D,** *Bradypus tridactylus* (KOE 3070), unprepared cheek tooth, three dimensional synchrotron radiation X-ray tomographic micrograph. Light grey color represent the densely spaced, peritubular dentin (PTD) filled dentinal tubules with the odontoblastic process (OP, black spot). Darker grey color surrounding the tubules represent intertubular dentin (ITD). **E,** Megatheriidae indet. (KOE 3003), longitudinal section, scanning electron micrograph. The calcified tubes of the odontoblastic processes show very long extensions, the latter forming a dense network. **F,** *Proeutatus* sp. (KOE 3425), transverse section, scanning electron micrograph. In a number of fossil cingulates, cheek teeth show a central, cloud-shaped core of osteodentin (Ostd), followed by a thick layer of orthodentin (Ord), and an outer rim of hardened orthodentin (HOrd). The orthodentin layer displays well preserved incremental lines. **G,** *Glyptodon clavipes* (KOE 3060), transverse section, µ-CT image. The tooth fragment shows the same built-up as described in (F), the hardened orthodentin (HOrd) is light-grey colored and thus optically more dense. KOE, Enamel collection, established by Wighart von Koenigswald, housed in the Institute of Geosciences, Section Paleontology, Rheinische Friedrich-Wilhelms-Universität Bonn. ▷

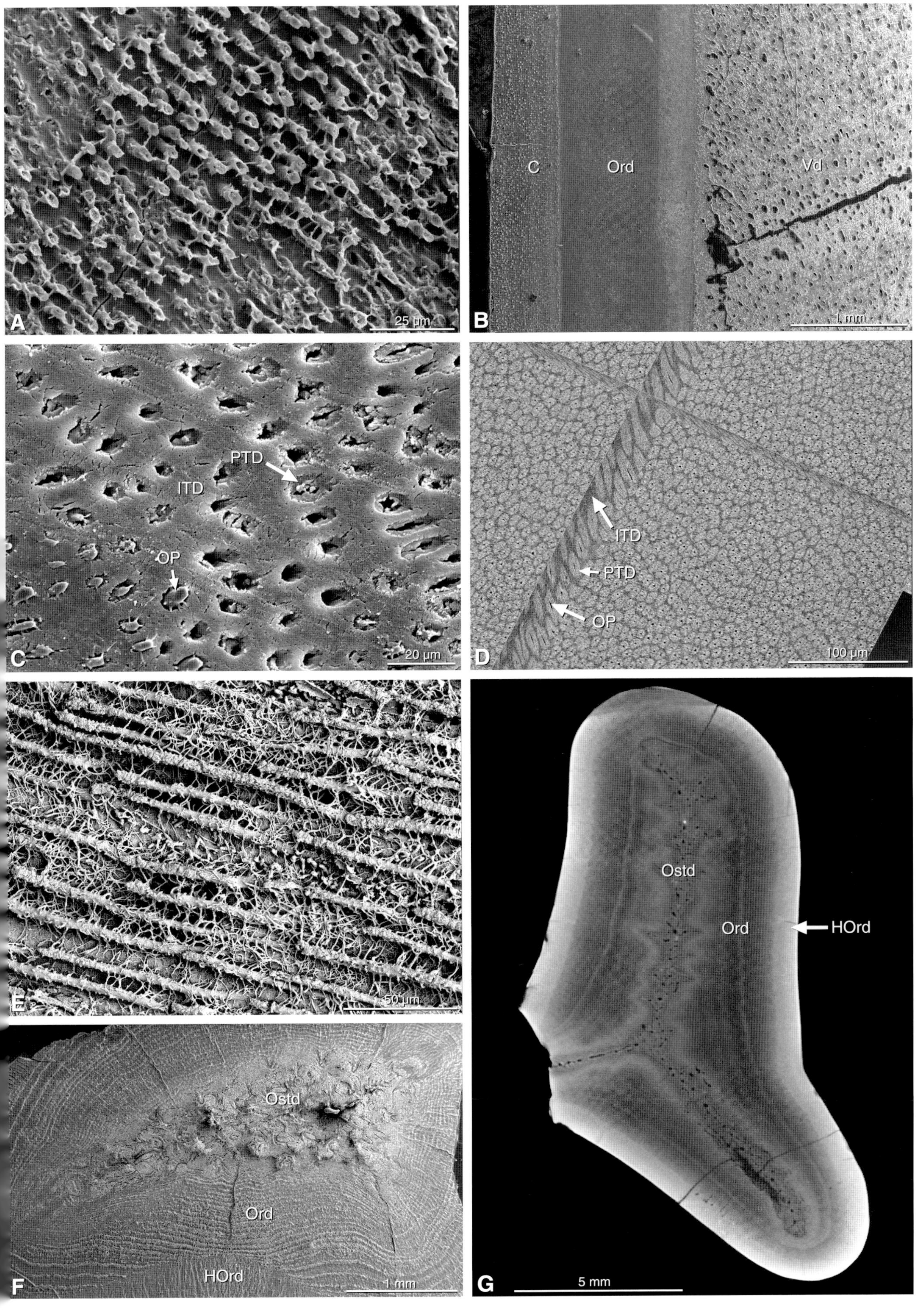
C
Ord
Vd
PTD
ITD
OP
ITD
PTD
OP
Ostd
Ord
HOrd
Ostd
Ord
HOrd
25 µm
1 mm
20 µm
100 µm
50 µm
1 mm
5 mm
A
B
C
D
E
F
G

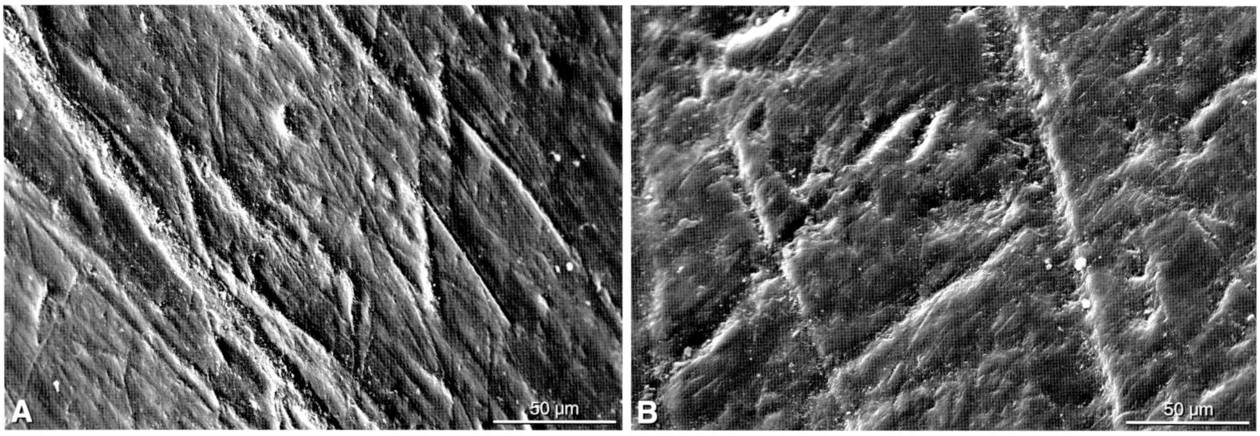

Fig. 12.3. Orthodentin surface of the stratigraphically earliest definite sloths showing representative microscopic scars. **A,** *Orophodon hapaloides* (MNHN.F.DES 267), right lower cheek tooth, scanning electron micrograph. **B,** *Octodontotherium grande* (MNHN.F.DES 238), right upper cheek tooth, scanning electron micrograph. Both wear surfaces display fine (i.e. narrow), coarse (i.e. wide), and hypercoarse (i.e. very wide scratches). Small (i.e. shallow) and large (i.e. deep) pits are scattered over the surfaces but not so easy to discern. MNHN: Muséum national d'Histoire naturelle, Paris.

systematically belong to the Mylodontoidea. Both species show microwear signals (Fig. 12.3), which are very different from all other analyzed sloths to date, both fossil and living. The results in Kalthoff & Green (2018) suggest a generalized herbivorous diet consisting of softer (foliage, fruit pulp) to harder (fruit, seeds, twigs) food items that were taken both from higher vegetation levels as well as from ground level. On the base of their wide muzzle, *Octodontotherium* and *Orophodon* are interpreted as so-called bulk feeders, meaning that the morphology and/or dimensions of their feeding apparatus prevents selective foraging; this is supported by their unspecialized, varied menu. In the early days of their evolution, sloths clearly were herbivorous but still unspecific in their food choice distinguishing them from both stratigraphically younger sloths and the living species, all of which are much more specialized in respect to their diet.

A number of Pleistocene ground sloths have been analyzed and compared to the living two- and three-toed sloths, i.e. the frugivore-folivore *Choloepus* and the folivore *Bradypus* (Green 2009a, Green & Resar 2012, Resar et al. 2013, Green & Kalthoff 2015). The Antillean sloth *Acratocnus odontrigonus* and the North American *Megalonyx wheatleyi* have pit-dominated microwear signals with *A. odontrigonus* falling in the dietary ecospace together with *Choloepus*, while *M. wheatleyi* falls between the latter and the all folivore *Bradypus*. A browsing behavior was also identified in the North American Shasta ground sloth *Nothrotheriops shastensis*, being in its food choice most similar to extant *Bradypus*. In contrast, the North American mylodontid *Thinobadistes segnis* and the giant ground sloth *Megatherium americanum* are characterized by scratch-dominated microwear signals, suggesting a more abrasive herbivorous diet comprising mature leafs, woody plants, and fruit. In both species, high scratching might be related to contamination of food items with grit, which is in accordance with habitat reconstructions suggesting a dryer, more open landscape, at least for *Megatherium* (Bargo 2001). These results fit well with the idea that large megatheriid sloths ecologically represented the "giraffe niche" (McDonald 2005) feeding in the higher levels of the canopy, in which dust particle accumulation might occur (Ungar et al. 2005).

The above results clearly show a dietary niche partitioning in Pleistocene ground sloths. However, only a fraction of their diversity has been analyzed to date, not to mention fossil cingulates, i.e. fossil Dasypodidae, Glyptodontidae, Pampatheriidae), for which microwear studies are currently unavailable. Much work remains to be done, and we have only opened a small window into the feeding ecology of fossil xenarthrans.

Summary

The monophyletic Xenarthra show a large number of unusual morphological synapomorphies distinguishing them from all other placental clades. Within toothed xenarthrans (sloths, armadillos, and their fossil relatives), their ever-growing, enamelless teeth with a layered composition of dentin varieties are unique. Each of these dentin varieties is identified by a specific microstructural make-up, representing different biomechanical properties from weaker (vasodentin) to strong (orthodentin) to very strong (hard orthodentin and osteodentin) capability of resistance against abrasion. These specific microstructural make-ups lead to diverse tooth reliefs, ranging from dentitions with lobed, rather shallow teeth having a predominant grinding function (e.g., Mylodontidae) to dentitions with bilophodont teeth having a predominant shear-cutting function (Megatheriidae). This result implies that the loss of the hard outer enamel layer has never been a drawback in the evolution and diversification of xenarthrans, probably due to hypselodonty and also relaxed selection on the dentition. Recent studies on the feeding ecology of extant and extinct ground sloths revealed a variety of dietary adaptations along a line of specialized leaf browsing to more generalized browsing comprising also food items with a higher intrinsic toughness. They allow exciting insights into the biology of this clade and independent evidence for testing existing assumption on life history traits of fossil xenarthrans.

Acknowledgments

I deeply thank Wighart von Koenigswald and Thomas Martin (both Institute of Geosciences, Section Paleontology, Rheinische Friedrich-Wilhelms-Universität Bonn) as speakers of the DFG Research Unit FOR 771, for excellent guiding of our group through six years of funding. I am also grateful to all members of FOR 771 for inspiring discussions and fruitful collaborations. I thank all curators and collection managers of a large number of South and North American as well as European museums for their generosity to make material available for studies on microstructure and microwear by DCK. Thanks also go to Jeremy Green (Kent State University at Tuscarawas) for an excellent collaboration on two microwear projects. I am deeply indebted to Georg Oleschinski and Olaf Dülfer (both Institute of Geosciences, Section Paleontology, Rheinische Friedrich-Wilhelms-Universität Bonn) for professional help with photography and specimen preparation. I would like to express my sincere thanks to the two reviewers Helder Gomes-Rodrigues (Muséum national d'Histoire naturelle, Paris) and Ryan Haupt (Smithonian Institution, Washington) for their insightful comments.

This research was funded by grants KA 1556/4-1 and 5-1 of the Deutsche Forschungsgemeinschaft (DFG, German Research Foundation) Research Unit 771 "Function and performance enhancement in the mammalian dentition – phylogenetic and ontogenetic impact on the masticatory apparatus".

References

* indicates publications that originated from the DFG Research Unit 771.

Arsuffi, E. (1938): Beiträge zur Kenntnis des Vasodentins. Zeitschrift für Anatomie und Entwicklungsgeschichte 108: 749–760.

Bargo, M. S. (2001): The ground sloth *Megatherium americanum*: skull shape, bite forces and diet. Acta Palaeontologica Polonica 46: 173–192.

Bargo, M. S. & Nyakatura, J. A. (eds.) (2018): Special Issue: Morphology and Evolution of Xenarthra. Journal of Mammalian Evolution 25: 445–588.

Carlson, S. J. (1990): Vertebrate dental structures. In: Carter, J. G. (ed.). Skeletal Biomineralisation: Patterns, Processes and Evolutionary Trends. Vol. 1. Van Nostrand Reinhold, New York: 531–556.

Ciancio, M. R., Vieytes, E. C. & Carlini, A. A. (2014): When xenarthrans had enamel: insights on the evolution of their hypsodonty and paleontological support for independent evolution in armadillos. Naturwissenschaften 101: 715–725.

Fariña, R. A, Vizcaíno, S. F. & Iuliis, G. de (2013): Megafauna. Giant Beasts of Pleistocene South America. Indiana University Press, Bloomington.

Fariña, R. A, Vizcaíno, S. F. & Storch, G. (eds.) (2003): Morphological Studies in Fossil and Extant Xenarthra (Mammalia). Senckenbergiana Biologica 83: 1–101.

Ferigolo, J. (1985): Evolutionary trends in the histological pattern in the teeth of Edentata (Xenarthra). Archives of Oral Biology 30: 71–82.

Gotliv, B. A. & Veis, A. (2007): Peritubular dentin, a vertebrate apatitic mineralized tissue without collagen: Role of a phospholipid-proteolipid complex. Calcified Tissue International 81: 191–205.

Gotliv, B. A. & Veis, A. (2008): The composition of bovine peritubular dentin: Matching TOF-SIMS, scanning electron microscopy and biochemical component distributions. Cells Tissues and Organs 189: 12–19.

Green, J. L. (2009a): Dental microwear in the orthodentine of the Xenarthra (Mammalia) and its use in reconstructing the paleodiet of extinct taxa: the case study of *Nothrotheriops shastensis* (Xenarthra, Tardigrada, Nothrotheriidae). Zoological Journal of the Linnean Society 156: 201–222.

Green, J. L. (2009b): Intertooth variation of orthodentine microwear in armadillos (Cingulata) and tree sloths (Pilosa). Journal of Mammalogy 90: 768–778.

*Green, J. L. & Kalthoff, D. C. (2015): Xenarthran tooth architecture and dietary adaptations from analyses of dental microstructure and microwear, with new data for the giant sloth *Megatherium americanum* (Megatheriidae). Journal of Mammalogy 96: 645–657.

Green, J. L. & Resar, N. A. (2012): The link between dental microwear and feeding ecology in tree sloths and armadillos. Biological Journal of the Linnean Society 107: 277–294.

Habelitz, S., Rodriguez, B. J., Marshall, S. J., Marshall Jr, G. W., Kalinin, S. V. & Gruverman, A. (2007): Peritubular dentin lacks piezoelectricity. Journal of Dental Research 86: 908–911.

Haupt, R. J., DeSantis, L. R. G., Green, J. L. & Ungar, P. S. (2013): Dental microwear texture as a proxy for diet in xenarthrans. Journal of Mammalogy 94: 856–866.

Hoffman, J. M., Fraser, D. & Clementz, M. T. (2015): Controlled feeding trials with ungulates: a new application of in vivo dental molding to assess the abrasive factors of microwear. Journal of Experimental Biology 218: 1538–1547.

*Kalthoff, D. C. (2011): Microstructure of dental hard tissues in fossil and Recent xenarthrans (Mammalia: Folivora and Cingulata). Journal of Morphology 272: 641–661.

*Kalthoff, D. C. & Green, J. L. (2018): Feeding ecology in Oligocene mylodontidoid sloths (Mammalia, Xenarthra) as revealed by orthodentine microwear analysis. Journal of Mammalian Evolution 25: 551–564.

*Kalthoff, D. C., Rose, K. D. & Koenigswald, W. v. (2011): Dental microstructure in *Palaeanodon* and *Tubulodon* (Palaeanodonta) and bioerosional tunneling as a widespread phenomenon in fossil mammal teeth. Journal of Vertebrate Paleontology 31: 1303–1313.

Kastelein, R. A. (2009): Walrus: *Odobenus rosmarus*. In: Perrin, W. F., Würsig, B. & Thewissen, J. G. M. (eds.). Encyclopedia of Marine Mammals (Second Edition). Academic Press in Elsevier, Burlington, San Diego, London: 1212–1217.

Keil, A. & Venema, B. (1963): Struktur und Mikrohärteuntersuchungen an Zähnen von Gürteltieren. Zoologische Beiträge 9: 173–195.

Koenigswald, W. v. (1997): Evolutionary trends in the differentiation of mammalian enamel ultrastructure. In: Koenigswald, W. v. & Sander, P. M. (eds.). Tooth Enamel Microstructure. Balkema, Rotterdam: 203–235.

Martin, B. E. (1916): Tooth replacement in *Dasypus novemcinctus*. Journal of Morphology 27: 647–682.

McDonald, H. G. (2005): Paleoecology of extinct xenarthrans and the Great American Biotic Interchange. Bulletin of the Florida Museum of Natural History 45: 313–333.

Meredith, R. W., Gatesy, J., Murphy, W. J., Ryder, O. A. & Springer, M. S. (2009): Molecular decay of the tooth gene enamelin (ENAM) mirrors the loss of enamel in the fossil record of placental mammals. PLoS Genetics 5: 1–12.

Mjör, I. A. & Nordahl, I. (1996): The density and branching of dentinal tubules in human teeth. Archive of Oral Biology 41: 401–412.

Murphy, W. J., Eizirik, E., Johnson, W. E., Zhang, Y. P., Ryder, O. A. & O'Brien, S. J. (2001): Molecular phylogenies and the origin of placental mammals. Nature 409: 614–618.

O'Leary, M. A., Bloch, J. I., Flynn, J. J., Gaudin, T. J., Giallombardo, A., Giannini, N. P., Goldberg, S. L., Kraatz, B. P., Luo, Z.-X., Meng, J., Ni, X., Novacek, M. J., Perini, F. A., Randall, Z. S., Rougier, G. W., Sargis, E. J., Silcox, M. T., Simmons, N. B., Spaulding, M., Velazco, P. M., Weksler, M., Wible, J. R. & Cirranello, A. L. (2013): The placental mammal ancestor and the post-K-Pg radiation of placentals. Science 339: 662–667.

Oliveira, E. V. (2001): Micro-desgaste dentario em alguns Dasypodidae (Mammalia, Xenarthra) [Dental microwear in some Dasypodidae]. Acta Biologica Leopoldensia 23: 83–91.

Owen, R. (1840–1845): Odontography; or, a Treatise on the Comparative Anatomy of the Teeth; their Physiological Relations, Mode of Development, and Microscopic Structure, in Vertebrate Animals. Hippolyte Bailliere, London.

Owen, R. (1842): Description of the Skeleton of an Extinct Gigantic Sloth, *Mylodon robustus*, Owen, with Observations on the Osteology, Natural Affinities, and Probable Habits of the Megatheroid Quadruped in General. Taylor, London.

Purnell, M., Seehausen, O. & Galis, F. (2012): Quantitative three-dimensional microtextural analyses of tooth wear as a tool for dietary discrimination in fishes. Journal of the Royal Society Interface 9: 2225–2233.

Rensberger, J. M. (1978): Scanning electron microscopy of wear and occlusal events in some small herbivores. In: Butler, P. M. & Joysey, K. A. (eds.). Development, Function, and Evolution of Teeth. Academic Press, New York: 415–438

Resar, N. A., Green, J. L. & McAfee, R. K. (2013): Reconstructing paleodiet in ground sloths (Mammalia, Xenarthra) using dental microwear analysis. Kirtlandia 58: 61–72.

Retzius, A. (1837): Bemerkungen über den inneren Bau der Zähne mit besonderer Rücksicht auf den in Zahnknochen vorkommenden Röhrenbau. Archiv für pathologische Anatomie und Physiologie und für klinische Medizin 1837: 486–566.

Schmidt, W. J. (1924): Über das Dentin von *Bradypus tridactylus*. Anatomischer Anzeiger 58: 97–107.

Schmidt, W. J. & Keil, A. (1958): Die gesunden und die erkrankten Zahngewebe des Menschen und der Wirbeltiere im Polarisationsmikroskop. Hanser, München.

Schöne, B. R., Dunca, E., Fiebig, J. & Pfeiffer, M. (2005): Mutvei's solution: An ideal agent for resolving microgrowth structures of biogenic carbonates. Palaeogeography, Palaeoclimatology, Palaeoecology 228: 149–166.

Semprebon, G. M., Godfrey, L. R., Solounias, N., Sutherland, M. R. & Jungers, W. L. (2004): Can low-magnification stereomicroscopy reveal diet? Journal of Human Evolution 47: 115–144.

Simpson, G. G. (1932): Enamel on the teeth of an Eocene edentate. American Museum Novitates 567: 1–4.

Teaford, M. F. (1991): Dental microwear: what can it tell us about diet and dental function. In: Kelley, M. A. & Larsen, C. S. (eds.). Advances in Dental Anthropology. Wiley-Liss, New York: 341–356.

Ten Cate, A. R. (1998): Oral Histology, 5th ed. Mosby, St. Louis.

Ungar, P. S., Teaford, M. F., Glander, K. E. & Pastor, R. F. (1995): Dust accumulation in the canopy: a potential cause of dental microwear in primates. American Journal of Physical Anthropology 97: 93–99.

Vizcaíno, S. F. & Loughry, W. L. (eds.) (2008): The Biology of the Xenarthra. University Press of Florida, Gainesville.

Walker, P. L. (1976): Wear striations on the incisors of cercopithecoid monkeys as an index of diet and habitat preference. American Journal of Physical Anthropology 45: 299–308.

Walker, A., Hoeck, H. N. & Perez, L. (1978): Microwear of mammalian teeth as an indicator of diet. Science 201: 908–910.

Winkler, D. E., Schulz-Kornas, E., Kaiser, T. M. & Tütken, T. (2019): Dental microwear texture reflects dietary tendencies in extant Lepidosauria despite their limited use of oral food processing. Proceedings of the Royal Society B 286. https://doi.10.1098/rspb.2019.0544

Appendix 1. Specimen details of the extinct and extant sloths mentioned in the text. Dietary adaptations follow Green (2009a), Resar et al. (2013), Green and Kalthoff (2015), and Kalthoff and Green (2018). Country codes refer to ISO 3166-1 alpha-2. Abbreviations: LMA, Land Mammal Age; m, lower molariform; M, upper molariform. For Acronyms of collections, see Appendix 2.

Taxon	Stratigraphical age/LMA
Mylodontidae	
Scelidotherium sp.	Pleistocene ("Pampeano")
Scelidotherium	Early Pleistocene (Ensenadense)
Mylodontidae indet.	No data
Mylodontidae indet.	Late Miocene-Pliocene (Mesopotamian)
Mylodon sp.	No data
Thinobadistes segnis	Miocene
Glossotherium robustus	Late Miocene/Early Pliocen ('Mesopotamiense')
Paramylodon harlani	Late Pleistocene (Rancholabrean)
Paramylodon harlani	Late Pleistocene (Rancholabrean)
Lestodon armatus	Early Pleistocene (Pampas inferior)
Octodontotherium sp.	?Late Oligocene (Deseadar
Octodontotherium grande	Late Oligocene (Deseadan)
Orophodon hapaloides	Late Oligocene (Deseadan)
Megatheriidae	
Eremotherium eoimigrans	Early Pleistocene
Eremotherium laurillardi	Late Pleistocene
Megatheriidae indet., small (*Promegatherium* vel *Pliomegatherium* vel *Pyramiodontherium*)	Late Miocene-Pliocene
Megatheriidae indet.	Pleistocene
Megatheriidae indet.	No data
Megatherium americanum	Pleistocene
Nothrotheriidae	
Nothrotheriops shastensis	Late Pleistocene (Rancholabrean)
Nothrotheriops texanus	Early Pleistocene (Irvingtonian)
Megatheroidea indet.	
Hapalops sp.	Late Early Miocene (Santacrucian)
Megalonychidae	
Choloepus sp.	Recent
Acratocnus odontrigonus	Pleistocene
Megalonyx wheatleyi	Pleistocene
Bradypodidae	
Bradypus torquatus	Recent
Bradypus tridactylus	Recent

Sample provenance	Sample no.	Original coll. no.	Tooth position	Special dentin character	Diet based on dentin microwear
No data	KOE 3577	MLP 50-VIII-1-1	Probably M2	Cone-shaped bundles of dentinal tubules	
Entre Pto. de Olivos y Pta. nchore, prov. Buenos Aires, AR	KOE 3433	MLP 54-VI-19-8	Molariform	Cone-shaped bundles of dentinal tubules	
No data	KOE 3004	AMNH 11282	Molariform	Cone-shaped bundles of dentinal tubules	
Paraná, prov. Entre Rios, AR	KOE 3432	MLP 41-XII-13-2080	Molariform	Cone-shaped bundles of dentinal tubules	
No data	KOE 3000	AMNH 132682	Molariform	Cone-shaped bundles of dentinal tubules	
ʌixon's Bone Bed, Florida, US	–	AMNH FAM (2 teeth); FMNH (3 teeth)	molariforms (5) (2 MF, 3 mf)		Plants with low to moderate intrinsic toughness; possible influence of grit/dust
No data	KOE 3541	AMNH 11273	Molariform	Cone-shaped bundles of dentinal tubules	
ancho La Brea, California, US	KOE 3569	LACMHC 130 757	Left M2	Cone-shaped bundles of dentinal tubules	
ancho La Brea, California, US	KOE 3573	LACMHC 130 761	Right m2	Cone-shaped bundles of dentinal tubules	
Rio Arrecifes, AR	–	NHMD-ZMK 52/1888 (former ZMUC 154)	Right m1-3	Cone-shaped bundles of dentinal tubules	
Ez. Co. Alto. zona del Rio eseado, Prov. Santa Cruz, AR	KOE 3434	MLP 52-XI-3-15, diverse numbers	Molariform	Small amount of central vasodentin	Generalized herbivorous diet; bulk feeding
La Flecha, Patagonia, AR	–	MNHN.F.DES	molariforms (9) (5 MF, 3 mf, 1 MF/mf)	Small amount of central vasodentin	Generalized herbivorous diet; bulk feeding
La Flecha, Patagonia, AR	–	MNHN.F.DES diverse numbers	molariforms (5) (2 MF, 3 mf)	Small amount of central vasodentin	Generalized herbivorous diet; bulk feeding
L033: Haile 16A, Alachua Co., Florida, US	KOE 3044	UF 46361	Molariform	Strong thinning of orthodentin layer	
LV028: Waccassassa River, Levy Co., Florida, US	KOE 3045	UF 16416	Molariform	Strong thinning of orthodentin layer	
Paraná, Prov. Entre Rios, AR	KOE 3431	MLP 325	Molariform	Strong thinning of orthodentin layer	
Quebrada de Agua, ?AR	KOE 3003	AMNH 123683	Molariform	Strong thinning of orthodentin layer	
No data	KOE 3578	YPM 25011 PU	Molariform	Strong thinning of orthodentin layer	
"Plata-Landene" territory of today's Argentina, Paraguay, Uruguay, and parts of Bolivia)	–	NHMD-ZMK 56/1885 (former ZMUC 19)	Molariform	Strong thinning of orthodentin layer	Plants with low to moderate intrinsic toughness; possible influence of grit/dust
Potter Creek Cave, California, US	–	UCMP 8141, 8336, 8704, 8715	molariforms (4) (all M2/M3)	Thinning of orthodentin layer	Herbivorous diet
HI007: Leisey Shell pit 1A, ʜillsborough Co., Florida, US	KOE 3053	UF 87136	Molariform	Thinning of orthodentin layer	
Estancia La Costa, layer ?3, Prov. Santa Cruz, AR	KOE 3012	IGPB uncat.	Molariform		
No data	KOE 3590	SMF 35.784	Molariform		Frugivore - folivore
Puerto Rico, PR	–	AMNH	molariforms (3)		Frugivore - folivore
Smith Pit, Florida, US	–	AMNH	molariforms (6)		Folivore
No data	KOE 3018	ZSM 1903/9534	Molariform		Folivore
"Vivarium", supposedly from Ménagerie du Jardin des Plantes, Paris, FR	KOE 3070	MNHN CG 1954-268	Molariform		Folivore

Appendix 2. Specimen details of the extinct and extant cingulates mentioned in the text. Dietary categories follow Green (2009a). Country codes refer to ISO 3166-1 alpha-2.
Abbreviations: LMA, Land Mammal Age; M, upper molariform. Acronyms to museum collections: AMNH, American Museum of Natural History, New York, USA; FMNH, Field Museum of Natural History, Chicago, USA; IGPB, Institute of Geosciences, Section Paleontology, Rheinische Friedrich-Wilhelms-Universität Bonn, Bonn, Germany; KOE, Enamel collection, established by Wighart von Koenigswald, Institute of Geosciences, Section Paleontology, Rheinische Friedrich-Wilhelms-Universität Bonn, Germany; LACMHC, Los Angeles County Museum Hancock Collection (Page Museum), Los Angeles, USA; MACN, Museo Argentino de Ciencias Naturales "Bernardino Rivadavia", Buenos Aires, Argentina; MLP, Museo de La Plata, La Plata, Argentina; MNHN, Muséum national d'Histoire naturelle, Paris, France; NRM, Swedish Museum of Natural History, Stockholm, Sweden; NHMD, Natural History Museum of Denmark, Copenhagen, Denmark; NHMD-ZMK, Natural History Museum of Denmark: Quaternary Zoology collection, Copenhagen, Denmark; SMF, Senckenberg Forschungsinstitut und Naturmuseum Frankfurt, Germany; UCMP, University of California Museum of Paleontology, Berkeley, USA; UF, Florida Museum of Natural History, University of Florida, Gainesville, USA; YPM, Peabody Museum of Natural History, Yale University, New Haven, USA; ZMB, Museum für Naturkunde (Zoologisches Museum), Berlin, Germany; ZSM, Zoologische Staatssammlung München, Germany. Uncat.: uncataloged.

Taxon	Stratigraphical age/LMA	Sample provenance	Sample n
Cingulata: Dasypodinae			
Dasypus hybridus	Recent	No data	KOE 301
Dasypus novemcinctus	Recent	No data	KOE 344
Dasypus novemcinctus	Recent	Golf Course near Friesenhahn Ranch, Texas, US	KOE 353
Euphractinae			
Chaetophractus villosus	Recent	No data	KOE 304
Chaetophractus villosus	Recent	No data	KOE 344
Chlamyphorus truncatus	Recent	No data	KOE 302
Euphractus sexcinctus	Recent	No data	KOE 359
Euphractus sexcinctus	Recent	Lagoa Santa, Minas Gerais, BR	–
Eutatus seguini	Pleistocene ("Pampeano")	No data	KOE 342
Macroeuphractus sp.	Pliocene	No data	KOE 358
Proeutatus sp.	Late Early Miocene (Santacrucian)	Monte León, Santa Cruz, AR	KOE 342
Zaedyus pichiy	Recent	Puerto Jenkins, Santa Cruz, AR	KOE 302
Tolypeutinae			
Cabassous unicinctus	Recent	Ipitingua, Rio Acara, BR	KOE 301
Priodontes maximus	Recent	El Cerro, Chiguitos, BO	KOE 301
Tolypeutes matacus	Recent	La Urbana, Tapikiolé, lower Rio Pilcomayo, PY	KOE 302
Peltephilidae			
Peltephilus ferox	Late Early Miocene (Santacrucian)	No data	KOE 342
Pampatheriidae			
Pampatherium sp.	No data	No data	KOE 300
Pampatherium sp.	Pleistocene ("Pampeano")	Algarrobo, Ptdo C. Casares, Buenos Aires province, AR	KOE 342
Pampatherium (= *Holmesina*) *floridanus*	Early Pleistocene (Irvingtonian)	HI007: Leisey Shell pit 1A, Hillsborough County, Florida, US	KOE 304
Vassallia minuta	Pliocene ("Araucanense")	Catamarca province, AR	KOE 342
Glyptodontidae			
Propalaehoplophorinae indet.	No data	Santa Cruz province, AR	KOE 343
Propalaehoplophorus australis	Late Early Miocene (Santacrucian)	Santa Cruz province, AR	KOE 353
Glyptodontidae indet.	Early Pleistocene (Ensenadan)	Buenos Aires province, AR	KOE 287
Glyptodontidae indet.	No data	No data	KOE 343
Glyptodon sp.	No data	No data	KOE 300
Glyptodon sp.	No data	South America	KOE 353
Glyptodon reticulatus	Pleistocene	Tarija, BO	KOE 302
Glyptodon clavipes	Pleistocene	Arroyo del medio, Pampa intermedio, AR	KOE 306
Hoplophorus ornatus	Pleistocene	Pergamino, Province Buenos Aires, AR	KOE 306

Original coll. no.	Tooth position	Special dentin character	Dietary catagory based on dentin microwear
MLP uncat.	Molariform	Milk tooth generaton with enamelled teeth; ?central vasodentin	
–	Molariform	Milk tooth generaton with enamelled teeth; central vasodentin	Insectivore
IGPB M-6357	Molariform	Milk tooth generaton with enamelled teeth; central vasodentin	Insectivore
MLP uncat.	Molariform	'Degenerated osteodentin'	
MLP uncat.	Molariform	'Degenerated osteodentin'	
ZSM AM/1125	Right M7 and M8		Insectivore
SMF 908	Molariform		Carnivore - omnivore
NHMD L62 (former ZMUC L62)	Skull and mandibles		Carnivore - omnivore
MLP 16-146	Molariform	Hard orthodentin; osteodentin	
FMNH P 14493	Molariform	'Degenerated osteodentin'	
MLP 67-X-31-1	Molariform	Hard orthodentin; osteodentin	
ZMB 48626	Molariform	'Degenerated osteodentin'	Carnivore - omnivore
ZSM 1910/252	Molariform		Insectivore
ZSM 1926/359	Right M3		
ZSM 1925/595	Molariform	'Degenerated osteodentin'	Insectivore
MLP 69-VIII-13-3	Molariform	'Degenerated osteodentin'	
–	Molariform	Hard orthodentin; osteodentin	
MLP 70-III-10-1	Molariform	Hard orthodentin; osteodentin	
UF 65890	Molariform	Hard orthodentin; osteodentin	
MLP 29-X-10-71	Molariform	Hard orthodentin; osteodentin	
MACN A-11229	Upper molariform	Hard orthodentin; osteodentin	
AMNH 132681	Molariform	Hard orthodentin; osteodentin	
–	Molariform	Hard orthodentin; osteodentin	
MLP s/n	Molariform	Hard orthodentin; osteodentin	
–	Molariform	Hard orthodentin; osteodentin	
AMNH 11301	Molariform	Hard orthodentin; osteodentin	
NRM-PZ M2519	Molariform	Hard orthodentin; osteodentin	
NHMD-ZMK 70/1885 (former ZMUC 109)	Molariform	Hard orthodentin; osteodentin	
NHMD-ZMK 77/1888 (former ZMUC 170)	Molariform	Hard orthodentin; osteodentin	

T. Martin & W. v. Koenigswald: Mammalian Teeth – Form and Function. – München (Pfeil) 2020 – ISBN 978-3-89937-266-3

Index

A

abrasional contacts 11
abrasional facets 215
abrasives 48, 90
Aceratherium 120
Acratocnus odontrigonus 236, 239
Adalatherium 209
Adapisorex 210
adult 117, 119
Aegyptopithecus 12, 35
Aepyceros melampus 127, 130 ff., 137, 140, 153
African bushmen 36
African elephant 91, 93 f.
Afrosoricida 16
Afrotheria 112
afrotherians 17
Agriochoerus 222 f.
Agriotherium africanum 70
Alcelaphus buselaphus 127, 130, 132, 137 f., 150
allotherians 207
Altacreodus 201
altricial 113, 115 f., 121
Ambondro 187, 203 f.
Ambulocetidae 223
ameloblasts 103
amelogenesis 104
amphilestids 187
Anchitherium 180 f., 217, 219
Antidorcas marsupialis 127, 130 ff., 137, 139, 151
Antilocapra americana 127, 130 ff., 137 f., 149
Antilope cervicapra 127, 130, 132, 137, 139, 152
apical wear 203
Aplodontia 182
Aplodontidae 110
Apodemus flavicollis 16, 18
aquatic Cetartiodactyla 223
Archaeomys 106
Arsinoitherium 180
Artiodactyla 20, 104, 112, 120
Arvicola 14, 106
arvicolids 38
Asfaltomylos 187, 203
Asian elephant 93
Astrapotheria 180
asymmetrical blade facets 180
attritional contacts 11
attritional facets 215
Ausktribosphenos 203 f.
Australian aborigines 36
Australopithecus africanus 38
Australosphenida 203

B

baboon 81
Babyrousa 176
Bactrian camel 92
balanced occlusion 38
baleen 224
Basilosauridae 223
Basilosaurus 224
beam theory 58
Bighorn basin 34
bilophodont 34, 38
bilophodont scissors 180
biomechanics 56
Bishops 203
Bison bonasus 20, 22
Bison priscus 182
bite forces 126, 129, 200
black rhino 93
blade facets 180
blue wildebeest 92
body mass 92
bone morphogenetic proteins 104
Bos 106, 174
Bos primigenius 224
Bos taurus 175
Bos taurus primigenius 92
Bos taurus taurinus 95
Bovidae 38, 119, 223
brachydont 102
Brachyzostrodon 191
Bradypus 233 f., 236, 239
Bridetherium 191
Brontotheriidae 220
browsers 93, 227
bulk feeders 236
bunodont 38

C

Cabassous 73 ff., 234, 240
Caenomeryx 119 f., 223
Cainotheriidae 120
Cainotherium 119 f.
Calippus (Merychippus) francisi 219
camel 81
Camelidae 223
Camelus bactrianus 92, 127, 130, 132, 137, 140, 154
Canis lupus 20 f., 55, 69, 71
Cantius 34
Capra aegagrus hircus 92
Capra ibex 127, 130, 132, 137, 141, 155
Capreolus capreolus 118
Carbonic anhydrases 83
carnassials 180
Carnivora 20
Carodnia 180
Castor 106 f., 110, 112, 182
Castor canadensis 16, 20, 107, 176
Castor fiber 107, 109
Castorocauda 187, 206
cattle 81
Cavia 114 ff., 117
cementum 232, 234
centric occlusion 26, 201
Ceratotherium simum 92 f., 106, 127, 130, 132, 134, 137, 141, 156, 220
cercopithecines 35
Cercopithecoidea 34 f.
cervical loop 104
Cervidae 38, 119, 134
Cetartiodactyla 125, 221
Chaetophractus 59, 60, 70 ff., 73 ff., 233, 240
Chalicotheriidae 180, 220
chewing cycle 26
chimpanzee 83
Chinchilla 106, 115 f.
Chlamyphorus truncatus 73 ff., 240
Chlorocebus aethiops 31
Choloepus 236, 239
Cimolodonta 207
Coelodonta antiquitatis 105 f., 182, 220
collision areas 195
collision detection 38
colobines 35
Colobus guereza 31
comminution 49
comminution resistance 88
Common tree shrew 16
Connochaetes taurinus 55, 92, 127, 130 ff., 137, 142, 157

Cormohipparion 125
Coronodon havensteini 224
Crocuta crocuta 69, 71
crushing-grinding dentitions 234
Cryptomys 182
Ctenodactylidae 110
Ctenohystrica 106
Cuniculus 106 ff.
cusp “C” 199
Cuvieronius 176

D

Damaliscus pygargus 127, 130, 132, 137, 143, 159
Daouitherium 180
Dasyprocta 106 ff., 112
Dasypus 59, 60, 70 f., 73 ff., 233 f., 240
Dasyurus maculatus 70
Daubentonia 175, 183
deciduous 113, 116 f., 119
degenerated osteodentin 233
Deinohyus 173 f.
Deinotheriidae 225
Deinotherium 180
dental adaptation 34
dental arch 37
dental articulator 38
dental cusp nomenclature 12
dental eruption 115
dental microwear 234
dental microwear analysis 231
dental wear analyses 50
dental wear patterns 25
dentin 184, 231
dentinal tubules 232 ff.
dentin hypsodonty 176
dentin microstructures 231
dentinogenesis 104, 233
dentin teeth 176
dentin tracts 175
dentistry 39
Desmostylus hesperus 174
Diacodexis 222 f.
Dicerorhinus 180 f.
Dicerorhinus sumatrensis 220
Diceros bicornis 93, 127, 130 ff., 137, 142, 158, 220
Diceros diceros 117
Dichobune 222, 223
Dicrostonyx torquatus 176
Didelphidae 179
Didelphis 14, 16, 45, 198, 206
Didelphodon 201
Didelphodus 12
dietary adaptations 110
Dinnetherium 191
Dinocrocuta gigantea 69, 71
Dinomys 182
Diprotodon 175, 180
displacements 64
Docodon 205
Docodonta 204
dolphins 176
domestic goat 92
domestic horse 92
domestic sheep 92
Dorcatherium 120
Dorudon 224 f.
Dryolestes 199 f., 202
dryopithecine 35
Dsungarodon 205

E

Elaphurus davidianus 127, 130, 132, 134, 137, 143, 160
Elasmotherium 106, 182, 220
elasticity 57
Elephantidae 112, 173, 226
elephantids 38
Elephas 173, 227
Elomeryx 222 f.
embrasure occlusion 189, 194, 199
embrasure shearing 199
enamel 103, 106, 117
enamel-band hypsodonty 175
enamel cover 180
enamel-dentin junction 103
enamel islets 104, 106 ff., 121, 182
enamelless teeth 231
enamel ridge alignment 108, 110 ff.
enamel ridges 108, 171
enamel surface 185
energy budget 95
entoconid 201
Eohippus pernix 218
Equidae 38, 104, 217
equilibrium 175
Equoidea 217
Equus 106, 174, 217, 227
Equus caballus 95, 105 f., 175, 219
Equus ferus caballus 92
Equus ferus przewalskii 92
Equus grevyi 92
Equus quagga 125, 127, 130 ff., 137, 144, 161
Eremotherium 177, 184, 234, 239
Erethizon dorsatum 107 f.
Eritherium 225
eruption rate 175 f.
Erythrotherium 191
euharamiyidans 207
euhypsodont 102, 104, 106, 172, 174, 177
euhypsodont teeth 174
euhypsodonty 121
Euphractus 73 ff., 232, 234, 240
euprimates 34
Eurohippus 217
European bison 20
European mole 20
Eutatus 233, 240
eutherian 117
Eutriconodonta 187
evolutionary success 180
external hypsodonty 172 f.

F

facet 12
facet description 12
fecal particle size 91
feeding experiments 47, 201
fermentation 89
fetuses 120
Ficus 88
finite element analysis (FEA) 39, 61
flexids 104
flexus 104
Folivora 234
folivorous 35
forest buffalo 92
fossettes 104
fossettids 104
fracture patterns 88
free-body diagram 56
frugivorous 34
Fruitafossor 206
functionality 108, 112, 121
functional structural density 108, 110
functional symmetry 108

T. Martin & W. v. Koenigswald: Mammalian Teeth – Form and Function. – München (Pfeil) 2020 – ISBN 978-3-89937-266-3

G

Galago 33
gelada baboon 81
gemsbok 92
giant ground sloth 236
Giraffa camelopardalis 95
giraffe 93
Glasbius 203
Glis 183
globular dentin 233
Glossotherium 234, 239
Glyptodon 176, 184, 232, 234, 240
Glyptodontidae 184, 232 f.
GMM 29
goat 81
Gobiconodon 188
Gobiohyus 222 f.
Gomphotheriidae 225
Gondwanatheria 207
Gorilla 35
grazer 93
grazers 227
great apes 35
Greater hedgehog tenrec 16
Grevy's zebra 92
Groeberia 183
guillotine cutters 180

H

hairy armadillo 233
Haldanodon 15 f., 204 ff.
Halophila ovalis 88
hamadryas baboon 81
Hapalops 234, 239
haplorrhines 33
Haramiyavia 207
Haramiyida 207
hardened orthodentin 234
hard orthodentin 232, 236
Helohyus 222 f.
Hemitragus jemlahicus 127, 130, 132, 137, 144, 162
herbivores 91
herbivorous 102
Hesnoferus 203
heterochronic 104, 121
heterochronies 102
heterochrony 11, 171 f., 177, 186
Heteromyidae 110
Hipparion 219
Hippopotamus 175 f.
Hippotherium 120
Hippotragus niger 92
Holmesina floridanus 184, 240
hominids 35
hominin 36
Hominoidea 35
hominoid wear facet pattern 36
Homo 36
Homodonty 224
Homo neanderthalensis 30, 36
Homo sapiens 28, 36, 37, 39, 83, 113, 179
Hoplophorus ornatus 240
horizontal tooth displacement 172
horned armadillo 233
hunter-gatherer 28, 36
Hunter-Schreger bands 234
Hydrochoerus 173 f., 182
hyperinsectivory 196
hypocone 33
hypoconid 201
hypoconulid 201
hypoflexid groove 199, 201
hypselodont 172
hypsodont 104, 106, 177
hypsodont molars 207, 209
hypsodonty 102, 121, 177, 226
Hyrachyus 220
Hyracotherium 217
Hystrix 107 ff, 112

I

IDAS 117, 119, 121
Iguana 88
incisors 116 f., 175
inclination angle 37
inclination direction 37
incremental lines 234
individual dental age stages (IDAS) 117 f., 121
Indohyus 223
infant 119
infantile 117
insectivorous 33
internal hypsodonty 172, 184
intertubular dentin 233 f.
intervals' method 76
Inuit 36
Itatodon 206

J

Juramaia 187, 210
juvenile 117, 119

K

Kobus ellipsiprymnus 92, 127, 130, 132, 137, 145, 163
Kollikodon 204
Kolponomos 69
kuehneotheriids 195
Kuehneotherium 70, 193 ff.

L

lactation 94
Lagomorpha 113, 115
lagomorphs 117, 175
Lagostomus 106
Lama glama 127, 130, 132, 137, 145, 164
Lama huanachus 127, 130, 132, 146, 165
late adult 119
lateral infolds 104, 182
lateroprotrusion 29
lateroretrusion 29
laterotrusion 29
Lavanify 209
leading and trailing edges 106
Lemuridae 33
Lemur mongoz 37
Lepus brachyurus brachyurus 117
Lepus europaeus 115, 117, 174 f., 183
Lestodon 232, 234, 239
lever mechanics 56
life history 117
Lion 20
Litocranius walleri 127, 130, 132, 134, 137, 146, 166
Litopterna 180
longevity 120
lophids 106
Lophiodon 180 f., 219 f.
lophodont 38
lophs 106
Loris tardigradus 33, 37
Loxodonta africana 92

M

Macaca fascicularis 55
Macaca maurus 31
Macaca rhesus 72

Macroeuphractus 233, 240
Macropodidae 112, 180
macrowear 26
Maiacetus inuus 223
Mammuthus 173, 176, 182
Mammuthus primigenius 106, 117, 174, 177
Mammutidae 225
Maotherium sinense 194 ff.
Marsupialia 16, 183
marsupials 172, 180
Martes 58
mastication compass 33, 37
masticatory cycle 195, 198
maximum intercuspation 26, 200 f.
mechanical advantage 56
medioprotrusion 29
mediotrusion 29
Megacerops 180 f., 220
Megaconus 208
Megalonyx wheatleyi 236, 239
Megatheriidae 232, 234
Megatherium 176, 184, 232, 234, 236, 239
Megazostrodon 188 ff.
mesh-weighted average mean (MWAM) 74
mesh-weighted median (MWM) 74
Mesocricetus 113, 115 f.
Mesohippus 217 f.
mesowear 47
metacone 35
Micromys minutus 113, 115
Microtus gregalis 19 f., 183
microwear 25, 45
Mimomys 106 f., 175
mineralization 104, 108, 110
Miotragocerus 120
modular system 13
Modular Wear Facet Nomenclature 11, 13 ff., 17 ff., 21 ff.
Monodelphis 191, 200 ff.
*Monodo*n 117, 176, 224
Morganucodon 70, 178, 188 ff., 195 f.
Morganucodonta 187
Morganucodon watsoni 14 f., 179
morphogenesis 103
Moschus 173
multicusped hypsodonty 173
multicusped molars 208
Multituberculata 207
Mus musculus 115
Mutvei's solution 232
Mylagaulus 107 ff., 182
Mylodon 234, 239
mylodontid 236
Mylodontidae 232
Mylodontoidea 236
Myocastor coypus 106 ff., 116
Myoprocta pratti 115
Mysticeti 223

N

Naemorhedus goral 127, 130, 132, 137, 147, 167
Nanocuris 201
Nanolestes 198 ff.
narrow-headed vole 20
Nasalis larvatus 31
Neanderthals 28, 36, 39
neonate 117
Neoplagiaulax 208
Nimbacinus dicksoni 70
nipper facets 183
non-mammalian Mammaliaformes 14
North American beaver 246
Nothrotheriops 234, 236, 239

O

obtuse angled "symmetrodontans" 195
occlusal 3D compass 29
occlusal compass 26
occlusal fingerprint 27
Occlusal Fingerprint Analysis (OFA) 28, 32
occlusal surface 48, 107, 112
occlusal topometry 47
Octodon degus 114 ff.
Octodontotherium 234, 236, 239
Odobenocetops 224
Odobenus 117, 176 f., 233
odontoblastic process 232, 234
odontoblasts 103
Odontoceti 223
OFA 28, 29
OFA 2.0 33
OFA software 33, 38
OFA workflow 29
Old World monkeys 35
Ondatra zibethicus 105
one-to-one occlusion 189
ontogenetic 125, 134
ontogenetic phase I 173
ontogenetic phases 171, 177 f.
ontogeny 102, 106, 112, 171 f., 177 f., 185 f.
oral food processing 47
Orcinus orca 224
Orophodon 234, 236, 239
orthodentin 184, 232, 234, 236
Oryctolagus cuniculus 115, 117
Oryx gazella 92
osteodentin 184, 232 f., 236
Otomys 106, 173, 183
Ovis orientalis aries 92
Ozotoceros bezoarticus 127, 130 ff., 134, 137, 147, 168

P

Pakicetidae 223
Palaeolagus 107
Palaeotheriidae 217
Palaeotherium 180, 217 f.
palinal 207
Pampatheriidae 233
Pampatherium 240
Pan 35, 83
Panthera leo 14, 20
Papio 31, 83 f., 173
para-masticatory activity 37
Paramylodon 234, 239
Patagonia 183
Paulchoffatiidae 207
Peltephilus 233, 240
perikymata 117
perinatal 113
perinatal tooth eruption 116
periodontal ligament 175
Perissodactyla 120, 125, 216
peritubular dentin 233 f.
Peromyscus maniculatus 115 f.
pestle-to-mortar 204
Petaurus 173
Petromus typicus 115
Phacochoerus 92, 173 f., 182
Phascolonus 175
phase I 12, 26, 187, 201, 210
phase II 12, 26, 187, 200 f., 210
Phenacodus 217 f.
Phoberomys 173
Phosphatherium 225
Phyllostomus hastatus 179
Physeter 176, 184, 224
physiological cross-sectional area 62

phytoliths 47, 102, 227
pichi 233
Plagiolophus 217
plane elasticity 61
Plesiadapis 12
Pliomegatherium 239
Poëbrotherium 222 f.
Poisson coefficient 57
Polar bear 20
polydonty 224
Pongo 35
postnatal 117
power stroke 25, 38, 182
precocial 113 ff., 121
prenatal 116, 119
Presbytis cristatus 31
pretribosphenic 196, 199, 201
pretribosphenic molar 197, 202
Priacodon 191 f.
primary functional surfaces 216
primary osteons 233
primary relief 25
primary tooth elements 11
primary trigon 199
primate 38
primate molars 26
primates 82, 112
principal stress 65
Priodontes 73 ff., 234, 240
Proboscidea 38, 224
proboscideans 172
Procervulus 175
Proconsulidae 35
Procyon lotor 117
Proeutatus 233 f., 240
Prolouisina 210
Promegatherium 239
Propalaeohoplophorus australis 240
Propalaeotherium 217 f.
prosimians 34 f., 38
Protocetidae 223
protocone 35, 201
protrusion 29
Przewalski horse 92
Pseudobolodon oreas 208
pseudoprotocone 203, 205 f.
pseudotalonid basin 204 ff.
Pseudotribos 204
pseudotribosphenic 203
pseudotrigon basin 205
Pyramiodontherium 239
Pyrotherium 180 f., 183

Q

quantitative wear analysis 44
quasi-ideal meshes (QIM) 74

R

Rangifer tarandus 127, 130, 132, 134, 137, 148, 170, 182
Raoellidae 223
rasp facets 180, 182
Redunca fulvorufula 127, 130, 132, 137, 148, 169
Reigitherium 203
Repenomamus 188, 210
retrusion 29
Retzius lines 117
Rhinoceros sondaicus 220
Rhinocerotidae 104
Rhinocerotoidea 219
Rodentia 16, 38, 104, 113, 117
rodents 106, 172, 175, 207
roll 191, 195
rootless 172
Rudapithecus 38
Ruminantia 38

S

sable antelope 92
salivary proteins 81
Sarcophilus harrisii 70
scale-sensitive fractal analysis 45
Scandentia 16
Scelidotherium 234, 239
Schowalteria 203
secodont 38
secondary functional surfaces 216
secondary occlusal surface 185
secondary relief 25
selenodont 38, 125
self-sharpening effect 234
senile 117, 119
Setifer 16 f., 196 f., 198, 201
Shasta ground sloth 236
shear-cutting 201
shear-cutting dentitions 234
shearing edges 112
Shetland pony 92
Shuotherium 204 f.
Siamogale melilutra 64, 69
sidewall hypsodonty 174
Simpsonodon 204 f.
single-cusped talonid 199
single-phased power stroke 200
Sirenia 16
sirenians 172
Sivapithecus 35
six-banded armadillo 232
Smilodon 69 f.
Soricidae 179
Soricomorpha 20
Spalacolestes cretulablatta 194 ff.
spalacotheriids 196
specialized wear facets 171, 178, 180
Stenella attenuata 224
Steneofiber 107 ff.
Stephanorhinus etruscus 220
Steropodon 204
strain 57
Strepsiceros 58
strepsirrhines 33
Strepsirrhini 34
stress 57
Sudamerica 107 f., 182, 207, 209
surface profilometry 45
surface texture 45
Sus scrofa 118
symmetrical blade facets 180
symmetrodontan 194
symmetrodont molars 194
Synceros caffer nanus 92

T

Taeniodonta 177, 183
Taeniolabis 175
talonid basin 201
Talpa europaea 20 f.
Talpidae 179
Tapiroidea 219
tapirs 180
Tapirus 95, 219
Tarsiidae 33
Teinolophos 204
temporomandibular joint 135
tenrecs 196
tensor 64
terrestrial Cetartiodactyla 222

Theropithecus gelada 84
Thinobadistes segnis 236, 239
Thomasia 207
Thomomys 182
three-banded armadillo 233
Thryonomys swinderianus 107f., 114f.
Thylacinus 69f.
Thylacoleo 180
Thylacosmilus 70, 175
Tillodus 183
tinodontids 195
Tolypeutes matacus 73ff., 233, 240
Tomes' process 232
tooth eruption 112f.
tooth replacement 112
tooth volume loss 49
tooth wear 117
tribology 46
tribosphenic 187f., 201
tribosphenic molars 178, 197
Trichechus manatus 16f.
triconodont 187
Triconodontidae 187, 193
trigon 199
Trigonias 220
Trioracodon 191
Tupaia 16, 18, 201
two-to-one occlusion 194

U

Uintatherium 210
ungulate 125
unicusped hypsodont 173
upper third molars (M3) 37
Ursus maritimus 20, 22
Ursus spelaeus 178f.

V

Vancouver Islanders 36
vasodentin 184, 232ff., 236
Vassallia minuta 240
vector 64
Victoriapithecus 35
Vilevolodon 207
Vintana 107, 182, 187, 209
Virginia opossum 16
virtual 3D models 38
Vombatus 175, 183
von Ebner lines 117
von Mises stress 65

W

Walbeckodon 210
waterbuck 92
wear 175
wear facets 11, 26, 215
wear stages 29, 107, 178
West Indian manatee 16
whales 176
white rhino 92f.
Wolf 20

X

Xenarthra 184
xenarthrans 176
Xenosmilus hudsonae 181
X-ray 110

Y

Yellow-necked fieldmouse 16
Young's modulus 57
Y-pattern fissure pattern 35

Z

Zaedyus 73ff., 233f., 240
zalambdodont 196
zalambdodonty 196
zhangheotheriids 196
zig-zag pattern 192f.

T. Martin & W. v. Koenigswald: Mammalian Teeth – Form and Function. – München (Pfeil) 2020 – ISBN 978-3-89937-266-3